of the Elements

	IB	IIB	IIIA	IVA	VA	VIA	VIIA	VIIIA
								2 He 4.00
			5 B 10.8	6 C 12.0	7 N 14.0	8 O 16.0	9 F 19.0	10 Ne 20.2
			13 Al 27.0	14 Si 28.1	15 P 31.0	16 S 32.1	17 Cl 35.5	18 Ar 39.9
28 Ni 58.7	29 Cu 63.5	30 Zn 65.4	31 Ga 69.7	32 Ge 72.6	33 As 74.9	34 Se 79.0	35 Br 79.9	36 Kr 83.8
46 Pd 106.4	47 Ag 107.9	48 Cd 112.4	49 In 114.8	50 Sn 118.7	51 Sb 121.8	52 Te 127.6	53 I 126.9	54 Xe 131.3
78 Pt 195.1	79 Au 197.0	80 Hg 200.6	81 Tl 204.4	82 Pb 207.2	83 Bi 209.0	84 Po (210)	85 At (210)	86 Rn (222)

63 Eu 152.0	64 Gd 157.2	65 Tb 158.9	66 Dy 162.5	67 Ho 164.9	68 Er 167.3	69 Tm 168.9	70 Yb 173.0	71 Lu 175.0
95 Am (243)	96 Cm (247)	97 Bk (247)	98 Cf (249)	99 Es (254)	100 Fm (253)	101 Md (256)	102 No (256)	103 Lr (257)

Understanding Chemistry

Understanding Chemistry: A Preparatory Course

J. Dudley Herron
Purdue University

Random House New York

First Edition
98765432

Library of Congress Cataloging in Publication Data

Herron, James Dudley, 1936–
Understanding chemistry.

Includes bibliographies and index.
1. Chemistry. I. Title.
QD31.2.H49 540 80-21249

ISBN: 0–394–32087–5

Manufactured in the United States of America. Composed by Monotype Composition Co., Baltimore, Md. Printed and bound by R. R. Donnelley & Sons, Crawfordsville, Ind.

Text Design: Arthur Ritter

Text Art: Russell Peterson

To
PETER PRESCOTT,
the person responsible for me writing this book,
DEBORAH CONNOR,
the person responsible for the book being what it is,
and my wife,
JOYCE,
the person who kept me sane during the process.

Preface

When I began teaching a course for underprepared students in 1973, I had no idea that I would write a book for the course. Rather, I wanted to learn more about difficulties students have with chemistry. I did, and this book is an attempt to overcome them.

There are many reasons why students fail chemistry. Apart from lack of motivation, poor study habits, difficulty adjusting to life away from home, and diversions resulting from newly-found love—problems that plague college students in every subject—chemistry presents students with special problems.

Chemistry relies on reasoning that normally develops in early adolescence, but does not develop in some individuals until later in life. Without experiences to encourage the development of this reasoning, it may not develop at all. Some students haven't had those experiences.

We compare things quantitatively in two ways: we say that one stick is 3 cm longer than another and we say that a third stick is twice the length of the first. We may recognize an equality between two sticks that differ in length by 3 cm and two people who differ in length by 3 cm. Similarly, we may recognize an equality between two sticks, one of which is twice the length of the other, and two people, one of whom is twice as tall as the other. Understanding the first kind of equality develops early; however, the proportional relationship implied by the latter equality is not understood until early adolescence or later. Results from the National Assessment of Educational Progress suggest that it is understood by no more than half of the adults in this country. And yet proportional reasoning is involved in scores of important concepts in chemistry. Therefore, leading students to understand proportions is a necessary condition for their understanding of chemistry.

From Chapter 1, where students use proportional relationships to convert from one measurement to another, until Chapter 12, where students deal with direct and inverse proportions in a formal sense, a concerted effort has been made to make proportions sensible.

A second problem in chemistry is that so much of our reasoning is done using concepts that have no concrete reality. Until one can visualize atoms and molecules, it is doubtful that processes like melting, boiling, dissolving, or reacting can mean what they mean to a chemist. Nor can reaction kinetics, dynamic equilibrium, or thermodynamics be understood to explain chemical processes until one can think about unseen, microscopic particles.

Understanding begins with direct experience. Only gradually do we learn to invent abstract models to explain and organize our experience. Until early adolescence, all of our reasoning is in terms of direct experience. We have no difficulty with "if . . . then" statements such as, "If you drop the glass, then it will break," because such statements represent direct extrapolation from past experience. However, statements such as, "An ideal gas would have zero volume at zero temperature," are not easily understood because they depend on "if . . . then" relationships that are hypothesized and that contradict experience. Even after we learn to think in hypothetical terms, we have difficulty understanding relationships that are new and unfamiliar.

If students are to understand chemistry, they must begin with ideas that can be developed from direct experience, and move gradually to those ideas that rely on imagination. Consequently, an understanding of atoms is developed gradually, and the more abstract and elusive notions are not introduced until the end of this book. This has required an adjustment in the standard treatment of some topics, most notably a resurrection of the use of combining numbers when writing formulas is introduced. Many chemists may view such a move as regressive. If you are among them, I hope you will take time to read the arguments for this treatment presented in the instructor's manual. In case you are not convinced by those arguments, you are also told how to sequence chapters for a more traditional approach.

Students differ in aptitude for chemistry. We all know that, but we do not all view aptitude in the same way. Traditionally, aptitude has been viewed as a capacity for learning; those with high aptitude can learn a lot, those with low aptitude only a little. An alternative view of aptitude considers the time required for a person to learn a given amount of material; those with high aptitude learn in a short period of time, those with low aptitude take longer. In an average classroom students with the lowest aptitude require up to five times as much time and effort as students with the highest aptitude, in order for both groups to reach the same level of understanding.

Most texts written for preparatory courses appear to ignore the fact that students in those courses are likely to require more time studying an idea in order to understand it. Typically, texts written for a preparatory course present all of the ideas found in an ordinary introductory text but deal with them superficially. Students with low aptitude in chemistry are almost forced to memorize words, formulas, and equations without understanding what they mean. When students get to subsequent courses where they are asked to apply concepts and principles, they are doomed to failure.

This book takes a different approach. Fewer ideas are introduced, but those ideas are treated extensively so that the student can understand the concepts and principles at a level that allows application in later courses. Even so, students and instructors must recognize that some students will need far more study and practice than others. To succeed there must be a mutual faith that students can understand chemistry, and a realization that the time and energy required for understanding may be extensive.

In addition to the major pedagogical considerations just mentioned, you will find other characteristics of this book that will help students learn.

- Examples are presented in a programmed text format that encourages students to think through the example rather than reading it passively.
- Questions and problems are inserted throughout the text to encourage students to practice new skills immediately after they are introduced. Solutions are given at the end of each chapter so that students can check their results.
- Hundreds of additional questions are given at the end of the chapters for those who need additional practice. More challenging problems are identified as such.
- Objectives are given at the beginning of each chapter to help students identify important skills and ideas to be learned. The instructor can add to or delete from these lists according to the requirements of a particular course.
- Word lists are provided to help students identify key concepts.
- Each chapter ends with a summary that provides a review of important concepts and principles.
- Each chapter contains an article from a periodical, which can be used to show applications of ideas that students are learning, to provide a historical perspective, to introduce unresolved social issues related to science, or to simply inform students about periodical literature that can be used to continue their chemical education.
- The text is accompanied by a study guide and a laboratory manual, which were developed along with the text and are compatible with it.
- There is an instructor's manual that provides more than answers to homework. Sample tests with item analysis data, suggestions for demonstrations, suggestions for sequencing, and discussions of research on teaching are included. Many of the comments are keyed to specific sections in the text, and symbols such as    and  found in the margin of the text indicate when specific suggestions are found in the manual.

I am grateful to have shared the wisdom of those who reviewed draft manuscripts of *Understanding Chemistry*. Richard Lungstrom of American River College, George Goth of the College of San Mateo, Norman Rose of Portland State University, Tamar Susskind of Oakland Community College, Lucy Pryde of Southwestern College, Jeff Davis of the University of South Florida, and Lynn James of the University of Northern Colorado all made thoughtful suggestions. Those suggestions were incorporated to the extent that it was possible to do so in a book of limited length and consistent philosophy.

Betsy Kean at the University of Wisconsin/Madison and Jane Copes of Hamline University deserve special thanks. Their task of writing a study guide and laboratory manual while the text was being reviewed and rewritten was a difficult one. Not only did they produce outstanding ancillary materials, but they made significant contributions to the text as well.

To Keith, Alan, and Tom who took photographs and worked problems; to the secretaries who typed various drafts of the manuscript; to the editors and editorial assistants who struggled through differences in philosophy and style; I offer my sincere thanks and humble apologies.

A final word of appreciation must go to my wife, Joyce. She didn't type, collate, check problems, or otherwise participate in the writing of this book. What she did was endure long hours without a husband, tolerate absent stares when he was present, and keep our house a home while the book was being written. Hers was the most important task of all.

JDH
West Lafayette, IN
August 1980

A Word to Students

It is important that students and teachers have an understanding from the start. It saves headaches later on. With this in mind, I want to tell you what this book is designed to do and how you should use it.

This book is a *preparation* for general chemistry. Students get into a course like this for several reasons. Most are in the course because chemistry is required in their major and they haven't had high school chemistry. Some had chemistry long ago and fear they have forgotten (or just want to take things slower in the beginning). For all of these students, *the purpose of this course is to prepare you for success in an introductory college chemistry course.*

What is it that will make you successful? Do you need to memorize facts and solve problems or can you learn to beat the system? A little bit of each, I expect.

Students have trouble with chemistry for several reasons. One problem is mathematics. It is not so much a problem of basic math skills, although this may be a problem. It is usually difficulty with *mathematical reasoning* that is the obstacle. If you are to succeed in chemistry, you need some preparation in mathematical reasoning, and one of the major goals of this book is to provide it.

Another difficulty that students have with chemistry is with reading. Now I know that you think this doesn't apply to you, but you may be surprised. Reading a chemistry book isn't like reading a novel. It takes different skills and you need to develop them if you haven't done so already. This book will help you understand sentences like "$Pb + PbO_2 + 2\ H_2SO_4 \rightarrow 2\ PbSO_4 + 2\ H_2O$" and "$PV = nRT$" as well as the ones you have been reading.

Some students *can* do math and *can* read, but don't. They have poor study habits and fail because they are unwilling to believe that it is necessary to change those habits. They got by in high school and they expect to repeat the performance now. Perhaps they are right. Some students *do* get through chemistry with poor study habits. Most don't.

If your study habits need improvement, ask your instructor where you can get good advice. Then follow the advice when you get it. The following are suggestions that will help you study this book.

a) Schedule your study time. You have a class schedule and most of you feel obligated to meet your classes. You need to be just as disciplined about your study. Look at your class schedule and write in times that you will study chemistry.

Be realistic. Some people have exceptional ability in music, some appear to be natural athletes, and others have great aptitude in science. Each of us is able to do a great deal in all of these areas, but those of us with low aptitude in a subject need *three to five times* as much time to learn as those with a high aptitude in the subject. The *average* college student should spend two hours studying for every hour spent in class. *Schedule this much study time now.* After a few weeks, add more study hours to your schedule if they are needed.

b) One of the best times to study lecture material is immediately after class. If you can't study because of another class, make a habit of reviewing what was covered as you walk to your next class.

c) Read assigned text material *before* it is treated in lecture. As you read, write down questions in the margin of the book or underline confusing points. When you attend class, listen for clarification of ideas that you did not understand from the reading.

d) Work practice problems *immediately* after you are taught how to do them. If you wait a day to practice new skills, you will forget details and work problems incorrectly, or waste time going back to learn the details again. Immediate practice reinforces ideas and helps you to remember them.

e) Don't study a lot of chemistry at one sitting. For most students, one hour at a time is enough. Space your practice. One hour a day, five days a week, will be much more effective than five hours on the weekend.

f) Schedule for variety. If you *must* study several hours at a time, separate study on subjects like math and chemistry from work on English or social studies. Start with the things that you find *least* interesting and take an occasional break for physical activity.

g) Cram sessions before a major exam are not very effective. If you keep up with your work every day, a short review just before the exam is all you need. Take the night off and relax. Get a good night's sleep. You will be alert for the exam and you will make a higher score.

h) You can't relax before an exam if you know you haven't learned the material as you go along. As you read a textbook, make a habit of reading each heading and guessing what the section will be about. At the end of each section or two, restate the major ideas in your own words. (These major ideas, stated in your own words, are excellent for your last-minute review.) At the end of each chapter, review the objectives given in the beginning. Be sure you can do all of them.

Good study habits are necessary but not sufficient for success. Here are other suggestions that will help you.

Don't be bashful. Students don't like to ask dumb questions. Consequently they sit through lecture without understanding what is being said. They don't want others to know how little they know.

Keep in mind that all of us are born ignorant. People who are afraid to ask questions are likely to stay ignorant. Can you afford to protect your ego at the expense of never understanding?

Memorizing versus Thinking. Memorizing is an important part of any education. Certain information is used over and over and must be remembered. However, you should not place too much emphasis on memorization. Some students believe that all they need to do is memorize what they are told and be ready to recall that information when test time rolls around. This isn't true. You must be able to *use* information, to apply facts and skills to situations that are different from those given in class.

Students think they understand when they do not. Follow these suggestions to see if *you* understand.

a) State what you have read or heard in your own words. If you can't, you probably don't understand.

b) Think of an example, illustration, or application of the idea you have just learned.

c) Make up your own test questions about the material.

d) Explain the idea to someone who doesn't know it.

Overlearn basic skills. Practice. I can illustrate what I mean by telling you about a student in my class. The student did poorly on examinations because he made many math errors. When I pointed this out, he replied that he understood the math and that he had always done well in math class. I gave the student a diagnostic math test and he made several errors. When I pointed this out, he looked over the problems and said, "Oh, I see where I made the error here . . . I really know how to do that . . . I just made a silly mistake."

He really did understand the math. The problem was that he could only work the problems *when he had plenty of time and had nothing else to think about.* On a test he did not have all the time he wanted and had *many* other things to think about. As he tried to remember how to calculate molecular weight, what was meant by mole, the symbol of copper, and how to determine combining numbers, he had too many things on his mind to really think about the math that he was doing. As a result, he made mistakes.

Skills that are used repeatedly—basic math skills, units of metric measurement, the mole concept, atomic and molecular weights, and formula writing, for example—must be practiced to the point where you can almost do them in your sleep. If you do not, you will spend so much mental energy on these fundamental ideas that you will not be able to attend to *new* ideas that require the applications of these basic skills.

Be objective. Many students do poorly because they are unrealistic. Some students have an unrealisticly high opinion of their ability and background. When they get poor scores on a quiz or don't understand what they read, they attribute their lack of success to poorly written test questions or a poorly written text. This *may* be true, but if others in the class are understanding and doing well on the quiz, it probably means that you are missing something the professor considers important. Find out what it is, and make the effort to overcome the problem.

Other students have an unrealistically low opinion of their ability and background. When they get poor scores on a quiz or don't understand what they read, they decide that they are stupid and can't understand chemistry. Usually the problem is that the professor is using terms, symbols, and reasoning that is unfamiliar to the student. When this happens, you *can* learn, but it may require more time and more effort. The key to success is having faith in your ability, finding out what information you are missing, and not giving up the first time you have difficulty.

I honestly believe that every student who is reading this book *can* understand chemistry, even though the time required for some will be great. If *you* share that faith and have the time to devote to study, this book will provide the background you need to succeed in college chemistry.

Good luck!

Contents

1
Numbers Large and Small 3

Objectives 3
1.1 How we use numbers 4
1.2 How big is a million? 6
1.3 "Counting" a million grains of rice 7
1.4 Why are units important? 8
1.5 Why use metric (SI) units? 10
1.6 How to solve problems with factor-label 10
1.7 Ratios as unit-factors 14
1.8 Metric (SI) units of mass 16
1.9 Converting from one unit to another 17
1.10 How small is a millionth? 19
1.11 Metric (SI) units of length 20
1.12 One-millionth of a page 22
1.13 Summary 24
Questions and Problems 24
Answers to Questions in Chapter 1 26
Reading 26
How to Solve Dosage Problems in One Easy Lesson 27

2
Numbers from Measurement 31

Objectives 31
2.1 Uncertainty in numbers 32
2.2 Accuracy and precision 33
2.3 Some reasons for uncertainty 35
2.4 Significant digits 38
2.5 Calculations with uncertain measures 40
2.6 Rounding fives 42
2.7 Infinite significant digits 43
2.8 When are zeros significant? 43
2.9 More units of measurement: volume 45
Using containers to measure volume 47
2.10 Temperature: an intensive property 49
2.11 Heat: a form of energy 50
2.12 Density: another intensive property 53

2.13 Using relationships 54
2.14 Nonproportional relationships 57
2.15 Still larger and smaller numbers 59
2.16 Using scientific notation 62
2.17 Reducing uncertainty 65
2.18 Summary 67
Questions and Problems 68
Answers to Questions in Chapter 2 70
Reading 72
Early Research on the Freezing Point of Mercury 72

3
Questions About Matter 74

Objectives 74
3.1 What chemistry is about 75
3.2 Properties 76
3.3 Heat of fusion and heat of vaporization 77
3.4 What do we mean by purity? 80
3.5 Checking on purity 82
3.6 Destructive distillation of wood 84
3.7 Chemical and physical changes 86
3.8 Is charcoal pure? 87
3.9 Elements and compounds 88
3.10 Mixtures and compounds 91
3.11 Summary 91
Questions and Problems 93
Answers to Questions in Chapter 3 94
Reading 96
The Wonder of Water 96

4
Atoms and Molecules 99

Objectives 99
4.1 Macroscopic observations and microscopic models 100
4.2 Atoms 100
4.3 The nature of elements 101
4.4 How temperature affects atoms 102
4.5 Atoms and molecules 103
4.6 Other molecules: compounds 104
4.7 Chemical and physical changes 106
4.8 Mixtures of atoms and molecules of atoms 109
4.9 Microscopic changes during melting and distillation 111

4.10 Chemical changes of elements 114
4.11 Chemical changes of compounds 115
4.12 Summary 116
Questions and Problems 118
Answers to Questions in Chapter 4 119
Reading 121
Ein Neues Metall 121

5

Introduction to Chemical Language 123

Objectives 123
5.1 Chemical symbols 124
5.2 The periodic table 127
Atomic radius 127
Metals and nonmetals 130
5.3 Chemical formulas 130
5.4 Multiple proportions 132
5.5 Things to come 134
5.6 Nomenclature 135
5.7 Naming compounds formed from nonmetals 136
5.8 Combining power 138
5.9 Rules for assigning combining numbers 140
5.10 Predicting formulas from combining numbers 141
5.11 Naming compounds of metals 143
5.12 Older names 144
5.13 Summary 144
Questions and Problems 145
Answer to Questions in Chapter 5 147
Reading 148
Periodic Table with Emphasis 148

6

Atomic Mass and Moles 150

Objectives 150
6.1 Atomic mass 152
6.2 Relative mass 153
6.3 Determining relative mass 154
6.4 How many is a mole? 157
6.5 Mole mass 159
6.6 Molecular mass 160
6.7 Mole mass of molecules 162
A mnemonic for mole relationships 163
6.8 A short history of atomic mass and moles 164

6.9 Some confusing points 167
Atoms or molecules 167
Does the compound form molecules? 167
Mole and molecule 168
6.10 Empirical formula 169
6.11 When is a number whole? 174
6.12 Empirical formula vs. molecular formula 175
6.13 Percent composition 176
6.14 Summary 178
Questions and Problems 179
Answers to Questions in Chapter 6 181
Reading 182
New Avogadro's Number 182

7

The Electrical Nature of Matter 184

Objectives 184
7.1 Electrical Charge 185
7.2 Electrons and atoms 187
7.3 The Rutherford experiment 188
7.4 The difference between atoms: atomic number 190
7.5 Giving atoms a charge: ions 190
7.6 Static and current electricity 193
Conductors and insulators 193
7.7 Conductivity by ions 193
7.8 Polyatomic ions 198
7.9 Names and formulas of compounds containing polyatomic ions 200
Formulas 200
Molecular mass 203
7.10 Summary 203
Questions and Problems 204
Answers to Questions in Chapter 7 205
Reading 206
The Inside Story on Those No-Fill, No-Fuss Batteries 206

8

Chemical Equations 213

Objectives 213
8.1 Chemical equations and conservation of mass 214
8.2 How chemical equations are written 215
8.3 How chemical equations are read 218
8.4 Chemical equations and chemical reactions 221
8.5 Things equations don't say 223

8.6 Everyday chemical changes 224
Rusting 224
Silver tarnish 226
Protecting aluminum 227
Burning 229
Burning in water 230
Metabolism 231
Baking 231
Carbonated drinks 232
Cave formation 233
Car batteries 234
8.7 Predicting products of reactions 235
Combination reactions 235
Decomposition 235
8.8 Summary 236
Questions and Problems 237
Answers to Questions in Chapter 8 238
Reading 239
Corrosion 240

9

What Do We Get and How Much: Predicting Products and Stoichiometry 244

Objectives 244
9.1 Calculations involving combination reactions 246
9.2 Burning magnesium 251
9.3 Stoichiometry 253
9.4 Predicting decomposition products 254
9.5 Stoichiometry involving decomposition reactions 255
9.6 Replacement reactions 255
9.7 Limiting reagents 257
9.8 Activity series 260
9.9 Summary 261
Questions and Problems 262
Answers to Questions in Chapter 9 265
Reading 268
Julia B. Hall and Aluminum 268

10

Reactions in Solution 270

Objectives 270
10.1 Solutions and suspensions 271
Types of solutions 273

10.2 Why solutions are important 274
10.3 Characteristics of solutions 275
10.4 Concentration of solutions 277
Molarity 278
Making a solution of known concentration 282
10.5 Solubility and temperature 282
10.6 Solubility and chemical reactions 286
Solubility rules 289
10.7 Ionic and net ionic equations 290
Other ionic reactions 291
Limitations of net ionic equations 293
10.8 Summary 293
Questions and Problems 294
Answers to Questions in Chapter 10 296
Reading 296
Chemistry in Oral Health 297

11

Classifying Compounds as Acids or Bases 300

Objectives 300
11.1 Why we classify 302
11.2 Acids 304
11.3 Properties of acids 305
11.4 Defining an acid 306
11.5 Bases 307
11.6 Properties of bases 310
Bases taste bitter 310
Bases neutralize acids 310
Bases affect indicators 310
OH^-, the most common base 310
11.7 Strong and weak 311
"Strong" and "weak" reactions 311
11.8 Reversible reactions and equilibrium 314
11.9 Strong and weak: a summary 318
11.10 Indicators 319
11.11 Hydrogen ion concentration 320
11.12 A scale for large changes: pH 321
11.13 Titration 326
Indicators and titration 327
11.14 Summary 332
Questions and Problems 334
Answers to Questions in Chapter 11 335
Reading 337
Phosphate Process Treats Acid Mine Drainage 337

12
The Behavior of Gases 340

Objectives 340
12.1 Reading mathematical sentences 342
Direct proportionality 342
Inverse proportionality 347
12.2 Introduction to gases 348
12.3 Pressure 349
Atmospheric pressure 350
Units for pressure 353
12.4 The spring in air: Boyle's law 354
12.5 The microscopic explanation of pressure 359
12.6 Temperature effects: Charles' law 361
12.7 The microscopic explanation of temperature 366
12.8 Combined gas laws 367
Remembering equations vs. regenerating them 370
12.9 Molar volume and Dalton's law of partial pressure 373
12.10 Vapor pressure 376
12.11 Finding molecular weight 379
12.12 Summary 381
Questions and Problems 382
Answers to Questions in Chapter 12 384
Reading 387
Physical versus Chemical Change 387

13
The Structure of Atoms 390

Objectives 390
13.1 A review of atomic structure 391
13.2 Mass number 392
Isotopes 392
13.3 Electron arrangement 393
Why we care about electron arrangement 394
13.4 Evidence for electron arrangement: line spectra 394
13.5 Energy levels: the bookcase analogy 397
13.6 Writing electron configurations 404
Valence electrons 406
13.7 Periodicity and electron arrangement 407
Group IA (s^1): the alkali metals 407
Group IIA (s^2): beryllium, magnesium, calcium, strontium, and barium 407
Groups IIIA (s^2p^1), IVA (s^2p^2), VA (s^2p^3), and VIA (s^2p^4) 407
Group VIIA (s^2p^5): the halogens 408
Group O (s^2p^6): the noble gases 408
13.8 Valence electrons, electron arrangement, and the periodic table 409

13.9 Electron configuration of transition metal ions 411
13.10 Shorthand notation for electron configurations 412
13.11 Lewis structures 413
13.12 Summary 414
Questions and Problems 415
Answers to Questions in Chapter 13 416
Reading 417
Nothing Is Very Important 417

14
Chemical Bonding 419

Objectives 419
14.1 What we need to know about bonding 420
How many atoms are connected? 420
Which atoms are connected? 420
What is the shape of the molecule? 421
How strong is the bond? 421
14.2 What causes a bond? 422
14.3 Orbitals 423
14.4 Lewis structures and the octet rule 424
Rules for writing Lewis structures 430
14.5 Shapes of molecules 434
14.6 Covalent and polar covalent bonds 438
Polar bonds and polar molecules 440
14.7 Electronegativity 441
Electronegativity and bond polarity 441
14.8 Ionic bonds 441
Ionic bonds and ionic compounds 442
14.9 Oxidation number 443
Rules for assigning oxidation numbers 446
14.10 Summary 446
Questions and Problems 449
Answers to Questions in Chapter 14 450
Reading 452
Adhesive Bonding 452

An Epilogue 459

Appendices

A. Basic Math Review 461
B. Decimal Numbers, Multiplication by Ten, and Rules of Signed Numbers 468
C. Exponential Notation 477
D. Concentration Terms 486
E. Rules for Good Graphing 489
F. Solving Word Problems 497

Understanding Chemistry

Numbers, Large and Small

Each chapter of this text begins with an overview of the chapter, a few hints for study, a list of new words, and a list of objectives. The objectives tell you what you should learn from the chapter. Read them before you study the chapter and then review them at the end to be sure that you have acquired the knowledge and skills the objectives describe.

Students often read an objective and think that they can do what it calls for when, in fact, they cannot. It is a good idea to check. To help you do this, I have included supplementary questions and problems at the end of the chapter. Working those problems is a good way to see whether you have mastered the objectives.

OBJECTIVES

Chapter 1 is about numbers. You must learn to get numbers from measurement, to change numbers from one form of notation to another, and to use numbers to calculate other numbers. We have tried to introduce these skills in an interesting and challenging way. As you learn the skills that you need, you will also get a better idea of how large numbers can be—and how small.

Metric Measurement

Our goal is that you become so familiar with the metric system of measurement that you are as comfortable using meters and grams as using yards and pounds. You should be, if you can do the following:

1. Name the basic metric units for length and mass, and give the correct abbreviation for each unit.
2. Explain the meaning of commonly used metric prefixes such as kilo-, centi-, and milli-.

3. Convert any measurement from one metric unit to another metric unit. For example, convert 20.2 g to kilograms. (See Problem 1-24.)
4. Convert any metric measurement to its equivalent in an English unit when you are given a conversion factor. (See Problem 1-25.)

Using Factor-Label

We will introduce you to a problem-solving system that we expect you to use throughout this course. You must learn to do the following:

5. Given the equivalence between two units (for example, 2.2 lb = 1 kg), state two unit-factors derived from that equivalence.
6. Use unit-factors to convert from one unit of measurement to another. (See Problem 1-25.)
7. Use units as a guide to correct answers when converting from one unit to another. (See Problem 1-25.)

Words You Should Know

unit
SI
milli-
centi-
kilo-
mega-
micro-
metric system
unit-factor
factor-label
ratio
gram
conversion factor
meter
area
square root

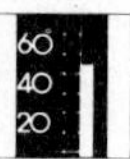

1.1 HOW WE USE NUMBERS

When we describe things around us, we often use numbers as an important part of our descriptions. We say that the average height of women is a little over 5 feet or that the average weight of men is about 160 pounds. It is 120 miles from my house to Chicago. The living space in an average house is 2,000 square feet. The national debt is rapidly approaching $1 trillion.

You understand these numbers—with the possible exception of the last—because you have seen a length of 5 feet, tried to lift a weight of 100 pounds, traveled distances of 100 miles, and live in a house. You have experienced numbers of this size many times in your life. The national debt is a different matter. You may have seen several thousand dollars at once, but you have probably never seen a million. A billion or a trillion is something else altogether. You may wonder whether you have ever seen a trillion of *anything.*

If we have some difficulty imagining a million, a billion, or a trillion, we certainly have difficulty with many of the numbers we encounter in science. The distance from the earth to the nearest star (the sun) is about 93 million miles. How far *is* that? The distance to other stars is measured in light years. A light year is the distance that light travels in 1 year (about 6,000,000,000,000

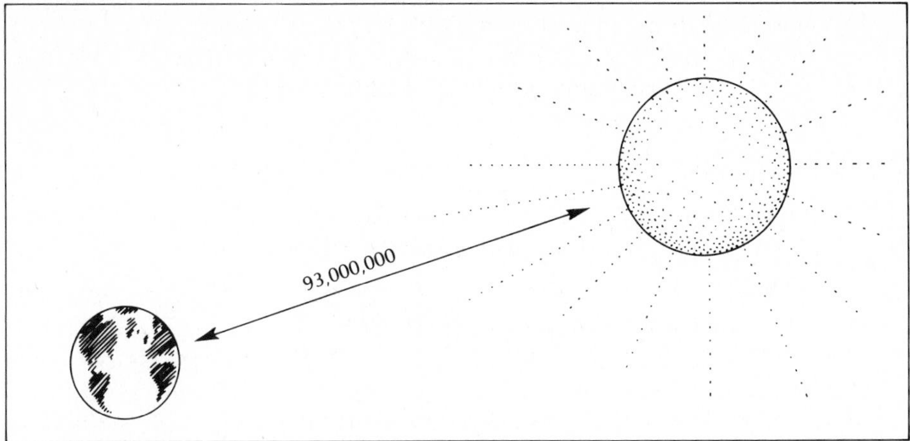

Figure 1.1
The distance from the earth to the sun is 93,000,000 miles. Do you have any idea how far that is?

miles). How far is *that?* One of the most important numbers in chemistry is the number of hydrogen atoms contained in a gram (a gram is a unit of mass in the metric system) of hydrogen: 602,000,000,000,000,000,000,000. You can see that this number is much larger than 1 million (1,000,000). Just how large *is* that number?

We have mentioned only large numbers so far, but there is a problem with small numbers, too. You are aware that the red blood cells in your body keep you alive by distributing oxygen throughout the body and returning carbon dioxide to the lungs. How big is one blood cell? About 0.0001 inch. But how big *is* that, really? Can you see it with a magnifying glass? An ordinary microscope? An electron microscope?

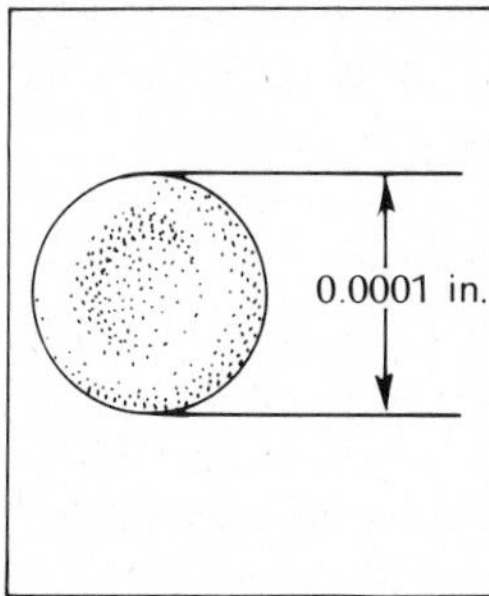

Figure 1.2
A red blood cell (depicted magnified at left) is about 0.0001 in. in diameter. Just how big is that?

Take another example of a small number. A single atom is about 0.000000001 inch in diameter. Do you have any idea how small that is?

Our point is this: Many of the things that we want to describe in science are either very large or very small. To describe them, we use big numbers with a lot of zeros at the end and small numbers with a lot of zeros at the beginning.

Such numbers will mean little to you until you develop a basic "feel" for their size. The material that follows is intended to help you do that. In the process, you will learn to make measurements and do arithmetic with large and small numbers as well.

1.2 HOW BIG IS A MILLION?

All of you have a good idea of what a hundred is, and most of you can visualize a thousand, but few know the size of a million. You are about to find out. Your first assignment is to identify a million of something. If you can, bring a million of something to class. If a million of what you count is too large or too expensive to bring, be prepared to explain how you counted it. For example, you may decide to "count" a million grains of sand or salt or sugar or rice. If this turns out to be a spoonful or a cup, you can bring it to class. If it turns out to be a ton, the cost might be prohibitive. In that case you can simply tell about it: "There are a million grains of sugar in a five pound bag," or "There are a million bricks in the chemistry building."

I suppose it goes without saying that you aren't expected to actually *count* to a million. That would be a very dull job, and you would probably lose count before you finished. You will need to find some other way. There are several techniques, and you are free to use any that you find.

To get you started, I will tell you how I "counted" a million grains of rice. First, I counted 300 grains. Next, I weighed the 300 grains and found the average weight of one grain of rice. Finally, I multiplied the weight of one grain by 1,000,000 to get the weight of 1 million grains. What did I get? Come on, now! That would spoil all of the fun. Find out for yourself if you really want to know.

Here are a few ideas that other students have tried:

One student found the number of acoustical ceiling tiles that contain 1 million holes.

Another student found the number of these books that contain 1 million pages.

Still another student found the number of pages in this book that contain 1 million letters.

A student who liked popcorn found the weight of 1 million kernels.

One ambitious student programmed a computer to print out 1 million A's. He then brought them to me so that I wouldn't run out at grading time!

Another student found the dimensions of a wire screen that contains 1 million holes.

BEFORE YOU CONTINUE, COUNT A MILLION OF SOMETHING.

1.3 "COUNTING" A MILLION GRAINS OF RICE

As I indicated earlier, I found the amount of rice that contains 1 million grains by weighing 300 grains of rice, calculating the weight of 1 grain, and then multiplying by 1,000,000 to get the weight of 1 million grains of rice. Let's work it out in steps.

Step 1: I placed 300 grains of rice on a balance and weighed them. I found that 300 grains weighed 5.28 grams so I wrote down the following:

$$300 \text{ grains of rice} = 5.28 \text{ grams of rice}$$

As long as everybody knows that we are talking about rice, the last statement can be shortened to read

$$300 \text{ grains} = 5.28 \text{ grams}$$

Step 2: Once I knew how much 300 grains of rice weigh, I found the weight of 1 grain as follows:

$$\frac{\overset{1}{\cancel{300 \text{ grains}}}}{\underset{1}{\cancel{300 \text{ grains}}}} = \frac{5.28 \text{ grams}}{300 \text{ grains}}$$

$$300\overline{)5.28}\quad \to 0.0176$$

$$1 \text{ grain} = 0.0176 \text{ gram}$$

One grain of rice is the same as 0.0176 gram of rice. It is convenient to express this fact as a fraction:

$$\frac{1\,\cancel{\text{grain}}}{1\,\cancel{\text{grain}}} = \frac{0.0176 \text{ gram}}{1 \text{ grain}}$$

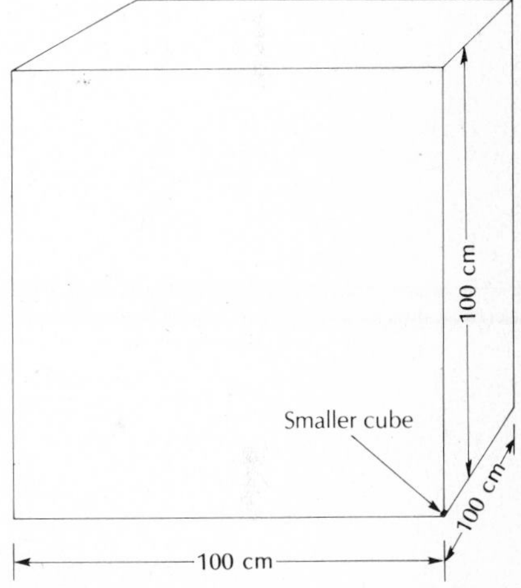

Figure 1.3
This large cube contains 1 million of the smaller cubes.

Step 3: Now that I knew the weight of 1 grain of rice, I could find the weight of any number of grains. So I calculated the weight of 1 million grains of rice:

$$1{,}000{,}000\,\cancel{\text{grains}} \times \frac{0.0176 \text{ gram}}{1\,\cancel{\text{grain}}} = 17{,}600 \text{ grams}$$

Step 4: Since you are more familiar with pounds than grams, I expressed the amount of rice that contains 1 million grains in units of pounds.

$$1 \text{ pound} = 454 \text{ grams}$$

Using this information, I found the weight of 1 million grains of rice expressed in pounds:

$$17{,}600\,\cancel{\text{grams}} \times \frac{1\text{ pound}}{454\,\cancel{\text{grams}}} = 38.8\text{ pounds}$$

One million grains of rice would weigh about 40 pounds.

In performing the calculations in this section, I have done some things that are new to many of you.

1. I was careful to always *include the units of measurement* along with the numbers that I wrote down.
2. I expressed the weight of the rice in the *metric unit* of grams.
3. When I did the calculations, I used a fraction known as a *unit-factor.*
4. I treated the units like an *x* or *y* in an algebra problem. The units "canceled."

These steps are common when we do calculations in science. You will be expected to do the same when you do calculations in this course. The following sections tell you why these steps are important and teach you how to use them. After you have learned these new skills, you will have an opportunity to use them as we continue to discuss numbers.

1.4 WHY ARE UNITS IMPORTANT?

It is amazing to teachers that so many students fail to see the importance of including units when they write a number that describes some quantity. If you were asked how much money you have, would you simply reply "eighty-two"? Of course not! Nobody would know whether you had eighty-two *cents* or eighty-two *dollars*. You would include the proper units in your statement; otherwise, it would not make sense. This is true of *any* quantity. *You must always include the proper units when you record a number that has units.*

There are many units in common use. Just wandering around the house, I found the following references to units. In the kitchen I opened a cookbook to a recipe entitled "Chicken Pilaf." It called for a 4-*pound* chicken, 1 *teaspoon* of curry powder mixed with 1 *cup* of water and 4 *tablespoons* of fat. The recipe says to cook the dish for ½ *hour* in an oven set at 350°*F* before draining the drippings. Rice is cooked in the drippings for 20 *minutes.*

Also in the kitchen is a bottle of diet cola boasting "less than 1 *calorie* per bottle," a bottle of vitamin C containing 250-*milligram* tablets, and an 8-*ounce* bottle of cough syrup. One must keep in mind that "oz" on the cough syrup stands for "fluid ounces" and does not mean the same as the "oz" on the bottle of oregano in the spice rack. That bottle contains ½ *oz* (meaning *avoirdupois* ounces, one of three "ounce-units" of mass). As indicated on the bottle, this ½ oz is equivalent to 14.1 *g*.

Standing in the corner of the kitchen is the .177 *caliber* air rifle that I bought my wife last Christmas. Caliber is not actually a unit, I suppose. It refers to the diameter of the gun's bore and is measured in hundredths or thousandths of an inch in this country. In Europe, the caliber of a gun is measured in *millimeters.*

When I went to the bathroom to shave, I noted that my electric shaver is designed to operate at 110–125 *volts* and 60 *hertz.* It requires a current of 0.07 *ampere* and has a power rating of 10 *watts.* As I poured SaniFlush® into the sluggish toilet bowl, I noticed that it is 75 *percent* sodium bisulfate, has a "net weight of 48 *oz* (3 *lb*) or 1.36 *kg*," and it cost me 91 *cents.*

Returning to the typewriter, it occurred to me that I buy paper by the *ream,* pencils by the *gross,* and eggs by the *dozen.* My bottle of correction fluid contains 18 *mL,* and my typewriter ribbon measures $\frac{1}{2}$ *in.* × 12 *yd.* In addition, I buy firewood by the *cord,* fruit by the *peck,* and diamonds—if I can afford them—by the *carat.*

I found all of these *units* without really trying. You can find just as many where you live. Keeping units straight in science and knowing just what they mean can be a real chore. If scientists did not limit the kinds of units that they use, communication among scientists throughout the world would be seriously impaired. For that reason, scientists have adopted standard units of measurement. The accepted system is a modification of the metric system: *Le Système international d'unités,* more commonly referred to as **SI.** The SI units that you will see most often are given in Table 1.1. Prefixes commonly used with SI units are given in Table 1.2.

Table 1.1. Commonly Used Metric (SI) Units

Quantity	*Common Units*	*Symbol*
length	kilometer	km
	meter	m
	centimeter	cm
	millimeter	mm
area	square meter	m^2
	square centimeter	cm^2
volume	cubic meter	m^3
	cubic decimeter	dm^3
	cubic centimeter	cm^3
mass	kilogram	kg
	gram	g
pressure	kilopascal	kPa
energy, work, or quantity of heat	joule	J
	kilowatt-hour	kW-h
temperature	kelvin	K
	degree Celsius	°C

Table 1.2. Commonly Used Metric (SI) Prefixes

Multiplication Factor	*Prefix*	*Symbol*	*Meaning (in USA)*
1,000,000 = 10^6	mega-	M	one million times
1,000 = 10^3	kilo-	k	one thousand times
0.01 = 10^{-2}	centi-	c	one hundredth of
0.001 = 10^{-3}	milli-	m	one thousandth of
0.000001 = 10^{-6}	micro-	μ	one millionth of

1.5 WHY USE METRIC (SI) UNITS?

There are many reasons to "go metric": one is that the United States will soon be the only industralized country in the world that uses English units. This causes problems in world trade and communication. It is to our advantage to get in step, and there is every reason to believe that we will do so in your lifetime. But why have other countries decided to use metric units?

People who are familiar with metric units realize that they are much easier to use than English units, because the system has the same base (ten) as ordinary numbers. (If this is not clear, read the review of decimal numbers found in Appendix B.)

The metric system was designed so that we can convert from one unit to another by multiplying or dividing by 10, 100, 1,000, or some power of 10. If you don't know the simple way to do that arithmetic, this is of little advantage, but once you *do* know the simple method described in Appendix B, it saves a tremendous amount of time and energy. You will begin to see the value of a decimal system of measurement later in this chapter, but first you need to learn to use units as a guide to calculation.

1.6 HOW TO SOLVE PROBLEMS WITH FACTOR-LABEL

Experience has shown the need for a quick, logical, and orderly process for solving mathematical problems. Many of these problems involve proportions, but the common language of proportions is difficult for many students. When problems grow complex, even the best students may become confused. We all need a systematic procedure to guide us in solving mathematical problems.

The procedure I am going to describe goes by many names, such as *factor-label, unit-factor,* and *dimensional analysis.* It involves using factors (numbers that multiply other numbers) that have a value of 1 (hence **unit-factor**) or represent a proportion. These factors carry *units* of measurement as a *label* (hence **factor-label**), which is used as a guide to calculation. Let's illustrate the procedure with an example.

Suppose that you are told that a desk is 60 in. long and that you want to know its length in feet. Knowing that 1 ft = 12 in., you can solve the problem using proportions like this:

$$60:12::x:1$$

This expression is read, "Sixty is to twelve as the unknown (x) is to one." In equation form, it should be written

$$\frac{60}{12} = \frac{x}{1}$$

Solving the equation for x, we get 5 as the correct answer and must recall that it has units of feet. Using factor-label, the problem would be set up like this:

$$? \text{ ft} = 60\,\cancel{\text{in.}} \times \frac{1 \text{ ft}}{12\,\cancel{\text{in.}}} = 5 \text{ ft}$$

Using this example, let us focus on what we have done:

1. We begin by writing down the given value that we want to convert to some other form. In this case we are given a length, 60 in., and want to convert it to feet. We write down

   60 in.

   When problems get more involved, it helps to write down what we want to find as an answer and set that equal to what we are given, like this:

   ? ft = 60 in.

   This expression can be read, "How many feet equal sixty inches?" or "How many feet are there in sixty inches?"
2. Note that we always include the units along with the numbers. They tell us what the numbers represent.
3. We then multiply the value we start with by a fraction that is equal to 1. In this case, we multiply by the unit-factor 1 ft/12 in. How do you know that this fraction is equal to 1? Well, we know that

   1 ft = 12 in.

If we now divide both sides of this equality by 12 in., we get

$$\frac{1\ \text{ft}}{12\ \text{in.}} = \frac{\overset{1}{\cancel{12\ \text{in.}}}}{\underset{1}{\cancel{12\ \text{in.}}}}$$

The numerator and denominator are the same on the right, so

$$\frac{1\ \text{ft}}{12\ \text{in.}} = 1$$

We could have divided both sides by 1 ft to get a different unit-factor, as follows:

$$\frac{\overset{1}{\cancel{1\ \text{ft}}}}{\underset{1}{\cancel{1\ \text{ft}}}} = \frac{12\ \text{in.}}{1\ \text{ft}} = 1$$

In this problem, it is the first unit-factor that we want.

4. When a problem is worked using factor-label, *all units cancel except the units that belong with the answer.* This is how we know which unit-factor to use. Note that when we multiply 60 in. by 1 ft/12 in., inches appear in both the numerator and the denominator and therefore cancel. The only unit left is feet, which is the unit we want in the answer.

$$?\ \text{ft} = 60\,\cancel{\text{in.}} \times \frac{1\ \text{ft}}{12\,\cancel{\text{in.}}} = 5\ \text{ft}$$

If we made the mistake of multiplying by the wrong unit-factor, the units would not cancel.

$$?\ \text{ft} = 60\,\text{in.} \times \frac{12\,\text{in.}}{1\ \text{ft}} = \frac{720\ \text{in}^2}{\text{ft}}$$

This answer doesn't make sense, and we know that it is incorrect.

As we consider the next problem, you will need to get involved. (If your math is rusty, you may want to do the basic math review in Appendix A.) You can't learn much by reading my solutions to problems. You need to think the problems through yourself.

Get a pencil and paper so that you can work the next example. Now place the paper over the book and move it down to the first dotted line. That's where you will stop reading and begin working the problem. After you answer the first part of the problem, you can move the paper down to the next dotted line. When you do, you will be able to compare your answer to mine. That way you can be sure you are on the right track. Continue in this way, step by step, until the problem is solved.

Please don't skip this exercise. It is easier to look ahead at my answer before you try to work the problem, but you won't learn as much. You must learn to think problems through and that means working them step by step.

Example 1.1

Convert 2,640 feet to yards using factor-label.

What is the first thing you do when solving a problem using factor-label?

Write down what you are given and set it equal to what you want to find.

Do that now.

$$? \text{ yd} = 2{,}640 \text{ ft}$$

Be sure to include the units along with the numbers. You will use the units as a guide to calculation.

The next step is to multiply what you are given by a unit-factor that represents a relationship between the units you are given and the units you want in your answer. What relationship exists between feet and yards?

$$3 \text{ ft} = 1 \text{ yd}$$

Now use this relationship to write *two* unit-factors.

$$\frac{3 \text{ ft}}{1 \text{ yd}} \quad \text{and} \quad \frac{1 \text{ yd}}{3 \text{ ft}}$$

Select the proper unit-factor and finish working the problem.

$$? \text{ yd} = 2{,}640 \cancel{\text{ft}} \times \frac{1 \text{ yd}}{3 \cancel{\text{ft}}} = 880 \text{ yd}$$

(If you have ever run track, you recognize this distance as half a mile.)

How did you know which unit-factor was the right one to use?

By using the units as a guide.

When you use the proper unit-factor, all units cancel except those that should appear in the answer. If this doesn't happen, either you are not finished with the problem or you have selected the wrong unit-factor.

Factor-label allows you to apply rules to get correct answers without going through the logic of the problem in detail. It is still a good idea to ask, "What did I actually do in that calculation?" Explain in an English sentence what you did to solve this problem.

I described it like this:

I wanted to know how many yards were equivalent to 2,640 ft, so I divided the number of feet I was given by the number of feet in 1 yd. I performed this division by multiplying by the unit-factor 1 yd/3 ft.

So far, the unit-factors that we have used are easily recognized as replacements for the number 1. Since 12 in. and 1 ft are different names for the same length, 12 in./1 ft and 1 ft/12 in. represent fractions that are equivalent to 1. The numerator and denominator are the same length, even though that length is expressed in different units. The same can be said for 3 ft/1 yd and 1 yd/3 ft.

1.7 RATIOS AS UNIT-FACTORS

Sometimes we use unit-factors in which the numerator and denominator are logically related but do not represent different names for the same quantity. This kind of unit-factor is illustrated in the next example. Again, cover each step of the problem with a sheet of paper until you have formulated your answers. Do this for all similar ("dotted line") examples in the text.

Example 1.2

If oranges cost 60¢ a dozen, how much would 4 oranges cost?

Before working the problem using factor-label, solve the problem any way you like—*but write down your reasoning.* In other words, tell how you go about solving the problem.

- -

The answer is 20¢. Many people can get the answer but can't explain their reasoning. Here is the reasoning used by other students.

"Let's see. Oranges cost 60¢ a dozen.

That means that 12 oranges cost 60¢, so 1 orange would cost 60¢ divided by 12, or 5¢.

Now if 1 orange costs 5¢, then 4 oranges would cost 5¢ times 4, or 20¢."

The arithmetic corresponding to this reasoning would look something like this:

$$\frac{60¢}{12} \times 4 = 20¢$$

Other students go about the problem differently:

"Let's see. Oranges cost 60¢ a dozen.

That means that:

12 oranges correspond to 60¢
6 oranges correspond to 30¢
3 oranges correspond to 15¢
1 orange corresponds to 5¢
4 oranges correspond to 20¢."

In this case the student has recognized a kind of logical equivalence between 12 oranges and 60¢. Of course, 12 oranges do not *equal* 60¢. Money and oranges just aren't the same thing. However, a kind of logical equivalence does exist. The grocer has essentially said that 12 oranges are *equivalent in value* to 60¢. This kind of correspondence is called a **ratio** and is represented like this:

$$12 \text{ oranges} \propto 60¢$$

Note that the student used this correspondence to get the next one by *dividing* both sides by 2.

$$6 \text{ oranges} \propto 30¢$$

Once he got to the point where he could say

$$1 \text{ orange} \propto 5¢$$

he got the next (and final) correspondence by *multiplying* both sides of the relationship by 4.

$$4 \text{ oranges} \propto 20¢$$

By multiplying or dividing both sides of this kind of correspondence relation by the same number, we can always produce a new correspondence relation that is equally true.

$$\begin{aligned} 12 \text{ oranges} &\propto 60¢ \\ 6 \text{ oranges} &\propto 30¢ \\ 24 \text{ oranges} &\propto 120¢ \\ 36 \text{ oranges} &\propto 180¢ \\ 1 \text{ orange} &\propto 5¢ \end{aligned}$$

All of these correspondences represent "fair trade" agreements if oranges sell for 60¢ a dozen. All were obtained by dividing or multiplying the original correspondence by some number.

Correspondences of this kind represent two quantities that are proportional and can be used to produce a "unit-factor" that is not equal to 1 in a strict mathematical sense but can be used in the same way as other unit-factors.

By dividing both sides of the correspondence by the same quantity, make two unit-factors from 12 oranges ∝ 60¢.

$$\frac{12 \text{ oranges}}{60¢} \text{ and } \frac{60¢}{12 \text{ oranges}} \text{ are the two factors.}$$

These may be read, "twelve oranges per sixty cents" and "sixty cents per twelve oranges."

Now show how you would use one of these factors to find the cost of 4 oranges.

$$?¢ = 4 \cancel{\text{oranges}} \times \frac{60¢}{12 \cancel{\text{oranges}}} = 20¢$$

Since the value we were given (4 oranges) has units of oranges, and the value we want to calculate has units of cents, we need the unit-factor that has cents in the numerator and oranges in the denominator.

Now that we have some experience in using units in calculations, let's look at the specific units we find in the metric system of measurement.

1.8 METRIC (SI) UNITS OF MASS

One million grains of rice weigh 17,600 grams. The **gram** (g) is the basic unit of mass in the metric system, and the kilogram (kg) is the basic unit in SI, but there are other units as well—just as pounds, ounces, and tons are all units of mass in the English system. Unlike units in the English system, however, units in the metric or SI system are always 10 times the next smaller unit. Table 1.3 lists units of mass in the metric (SI) and English systems.

Memorize! Memorize! Memorize! You *must* know those SI units that are underlined, their abbreviations, and the meaning of the prefixes used in Table 1.3. There is no way to learn this but to memorize. Close the book and drill until you can remember them.

Table 1.3. Metric (SI) and English Units of Mass

Metric	*English*
1 kilogram (kg) = 1,000 grams	1 long ton (tn) = 2,240 pounds
1 hectogram (hg) = 100 grams	1 short ton (tn) = 2,000 pounds
1 decagram (dag) = 10 grams	1 long hundredweight (cwt) = 112 pounds
1 gram (g) = 1 gram	1 short hundredweight (cwt) = 100 pounds
1 decigram (dg) = 0.1 gram	1 pound (lb) = 1 pound
1 centigram (cg) = 0.01 gram	1 ounce (oz) = 0.0625 pounds
1 milligram (mg) = 0.001 gram	1 dram (dr) = 0.00390625 pounds
	1 grain (gr) = 0.0001428125 pounds

The table contains several units that are seldom used in practice. The most common units in both systems are underlined. These units are sufficient to express the weights we encounter in most situations.

1.9 CONVERTING FROM ONE UNIT TO ANOTHER

To convert from one unit of measurement to another, you can use unit-factors. For example, a mass as large as 17,600 g would normally be expressed in kilograms. We see from Table 1.3 that 1 kg = 1,000 g. We can convert 17,600 g to kilograms like this:

$$? \text{ kg} = 17{,}600 \cancel{\text{g}} \times \frac{1 \text{ kg}}{1{,}000 \cancel{\text{g}}} = 17.6 \text{ kg}$$

Note that we selected the unit-factor that places grams in the denominator so that the units of grams will cancel to leave kilograms as the units in the answer. Also note that since the arithmetic involves only division by 1,000, the problem is solved by simply moving the decimal point three places to the left. Some people would not even write the problem down. The main purpose of writing this problem down is to ensure that the units are correct and that you divide by 1,000 rather than multiply by 1,000. If units don't cancel, you know that something is wrong.

It is often convenient to make conversions from one unit to another in two or more steps. Suppose, for example, that you know a person weighs 73 kg and you want to know how many milligrams this is. You can use two of the facts given in Table 1.3 to make this conversion. First, knowing that 1 kg = 1,000 g, you can convert kilograms to grams. Then, knowing that 1 mg = 0.001 g, you can convert grams to milligrams. This can be done in two separate steps:

$$? \text{ g} = 73 \cancel{\text{kg}} \times \frac{1{,}000 \text{ g}}{1 \cancel{\text{kg}}} = 73{,}000 \text{ g}$$

$$? \text{ mg} = 73{,}000 \cancel{\text{g}} \times \frac{1 \text{ mg}}{0.001 \cancel{\text{g}}} = 73{,}000{,}000 \text{ mg}$$

or combining the two steps:

$$? \text{ mg} = 73 \cancel{\text{kg}} \times \frac{1{,}000 \cancel{\text{g}}}{1 \cancel{\text{kg}}} \times \frac{1 \text{ mg}}{0.001 \cancel{\text{g}}} = 73{,}000{,}000 \text{ mg}$$

Note that, in setting up the problem, we start with what is given: 73 kg. We then multiply by a conversion factor that has kilograms in the denominator and some unit that brings us closer to the desired answer in the numerator. If necessary, we select other conversion factors until we arrive at the desired units in the numerator.

As this problem is stated, some of you may not follow the arithmetic. You aren't sure how to divide by 0.001. If this is a problem, first multiply the numerator and denominator of the second conversion factor by 1,000 to get

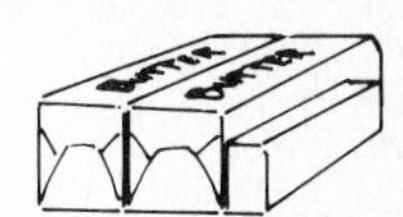

A kilogram of butter is a little over 2 lb.

A nickel weighs about 5 grams.

An ordinary flashlight battery weighs about 80 grams.

The ink in the period at the end of this sentence weighs about 1 milligram.

Figure 1.4
The masses of common objects in metric units.

1 in the denominator. (Remember, the value of a fraction is not changed as long as we multiply the numerator and denominator by the same number.) The following calculations illustrate this.

$$? \text{ mg} = 73 \cancel{\text{kg}} \times \frac{1000 \cancel{\text{g}}}{1 \cancel{\text{kg}}} \times \frac{1 \text{ mg}}{0.001 \cancel{\text{g}}} \times \frac{1000}{1000} =$$

$$? \text{ mg} = 73 \cancel{\text{kg}} \times \frac{1000 \cancel{\text{g}}}{1 \cancel{\text{kg}}} \times \frac{1000 \text{ mg}}{1 \cancel{\text{g}}} = 73{,}000{,}000 \text{ mg}$$

Now you should see that the answer is obtained simply by moving the decimal six places to the right—three places for the first thousand and three places for the second thousand.

Unless you travel in a country that uses metric measurement, you are unlikely to convert measurements from one system of measurement to another. However, we will ask you to do a few of these conversions early in this course just to help you develop a better understanding of the size of metric units. For example, the 73 kg mentioned in the last problem may mean very little to you, because you do not have a good idea of how large a kilogram is. Is this person fat, skinny, or just about average? The best way to get an idea of his or her size is to convert the 73 kg to the more familiar unit of pounds. In order to do this, you must have some conversion factor that gives you the relationship between metric and English units.

If you take a 1-lb weight and place it on the balance in the chemistry laboratory, you will find that it weighs 453.6 g. (This is only an approximate measurement. That is, it is correct only to the nearest tenth of a gram. With care, we could find the mass more precisely, but this will be sufficient for our purposes.) We can say that 1 lb = 453.6 g, and we can use this fact to find the mass of our 73-kg person in pounds.

$$? \text{ lb} = 73 \cancel{\text{kg}} \times \frac{1000 \cancel{\text{g}}}{1 \cancel{\text{kg}}} \times \frac{1 \text{ lb}}{453.6 \cancel{\text{g}}} = 160.9347443 \text{ lb, or about } 161 \text{ lb}$$

If this is a man of average height, his weight is also about average. If it is a woman of average height, she is heavier than average.

The arithmetic in the last problem raises an interesting question. If you did it by hand, you know that the answer does not come out "even." Some decision must be made about how far to carry the calculation.

In Chapter 2 you will learn a procedure that will help you decide how to record an answer to a calculation such as this one. For now, you need to practice using factor-label to convert from one unit to another.

From the equivalencies given in Table 1.3, construct the appropriate unit-factor and do the conversions indicated in the following problems. If you have difficulty, look at the solution to the problem at the end of the chapter.

1-1. ? tn = 24,560 lb	1-3. ? oz = 1.36 lb	1-5. ? kg = 354 g
1-2. ? dag = 127 g	1-4. ? g = 268 mg	

You probably found these problems easy. Only one unit-factor was needed to find the answer. The following problems can be worked in the same way, but you will need to use more than one unit-factor. You may work the problem in two or more steps *or* by using more than one unit-factor in a single step.

1-6. ? oz = 0.215 tn	1-8. ? cg = 2.34 dag	1-10. ? dr = 0.098 cwt
1-7. ? mg = 0.0013 kg	1-9. ? kg = 2,456,000 cg	

You will find questions like these throughout this text. *Don't skip them.* Research has shown (and good common sense confirms) that you learn better if you practice a skill when it is first introduced. Putting off the practice makes the problems more difficult and will make the next section harder to understand.

1.10 HOW SMALL IS A MILLIONTH?

We began our look at numbers by asking you to identify a million of something. At this point, we want to go the other way and ask you to find a millionth of something. Now I know that you could do this by simply reversing what you did when you counted a million. For example, I could get a 50-lb bag of rice, take out 1.28 grains and announce, "I have one-millionth of a 50-lb bag of rice." That's cheating. I want you to get one-millionth of a single object such as a candy bar, a spool of thread, a football field, or a watermelon. Could you do it? After you read the following paragraphs, give it a try. Right now, see whether you can draw a square around one-millionth of a sheet of typing paper.

Would it be this big?
or this big?
or this big?

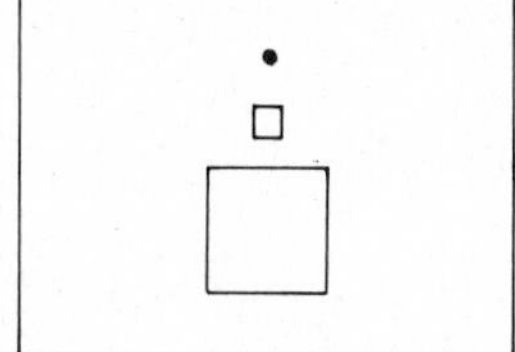

I will eventually tell you how to do it, but before I spoil the fun, stop reading for a minute or two and try to decide how *you* would solve the problem. When I have presented this problem to my classes, I have received several suggestions. One is to cut the paper in half, cut one of the halves in half, and so on until you get one-millionth of the sheet of paper. It is an interesting suggestion, but it would never yield exactly one-millionth. Another suggestion that I have received is to weigh the page on a balance, divide by 1 million, and then cut a piece of paper that weighs that much. This would work, but there would be a great deal of trial and error before the job was done.

The solution that I like best is to measure the length and width of the page, use these measurements to calculate its *area,* divide by 1 million to find the area of one-millionth of the sheet of paper, and then take the *square root* of that area to get the length of a side of the square that we want to draw. Did you follow that? Let's do it in steps.

1.11 METRIC (SI) UNITS OF LENGTH

Measuring the length of a sheet of paper provides an excellent opportunity to introduce more metric units. You have learned that the basic unit of mass in the metric system is the gram. The basic unit of length is the **meter.** Other units of length are derived by adding the prefixes given in Table 1.2 in the same way that they were used with gram to derive other mass units. Table 1.4 lists metric and English units of length. The most commonly used units are underlined.

Table 1.4 Metric (SI) and English Units of Length

Metric	*English*
1 kilometer (km) = 1,000 meter	1 mile (mi) = 5,280 feet
1 hectometer (hm) = 100 meter	1 rod (rd) = 16.5 feet
1 decameter (dam) = 10 meter	1 yard (yd) = 3 feet
1 meter (m) = 1 meter	1 foot (ft) = 1 foot
1 decimeter (dm) = 0.1 meter	1 inch (in) = $\frac{1}{12}$ foot
1 centimeter (cm) = 0.01 meter	
1 millimeter (mm) = 0.001 meter	

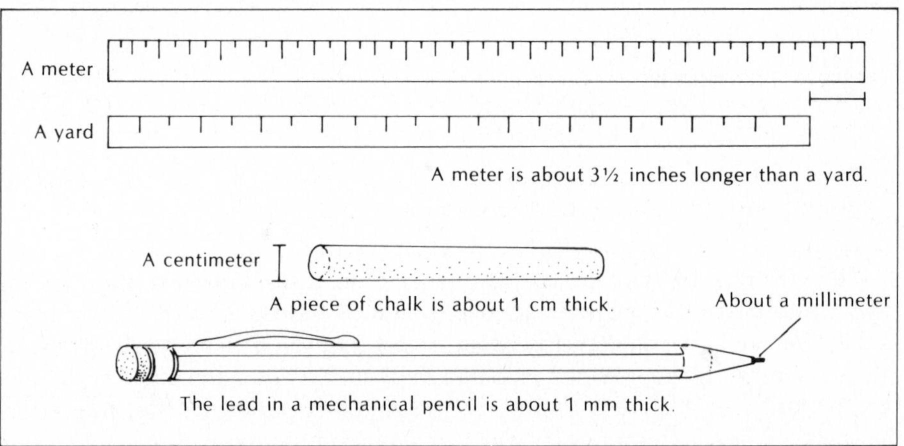

Figure 1.5 *The lengths of common objects in metric units.*

Using a ruler that is graduated in both centimeters and inches, measure the length of a sheet of typing paper in both centimeters and inches, and use the information to calculate the number of centimeters in 1 in.

1-11. length of page in centimeters =
1-12. length of page in inches =
1-13. number of centimeters in 1 in. =

If you had any difficulty in measuring the length of the paper, it is probably because you do not know how to read the scale on the ruler. The scale shown in Figure 1.6 should help you.

On the metric scale, each number represents the number of centimeters from the end of the scale. Note that the distance between the numbers is divided into 10 spaces.

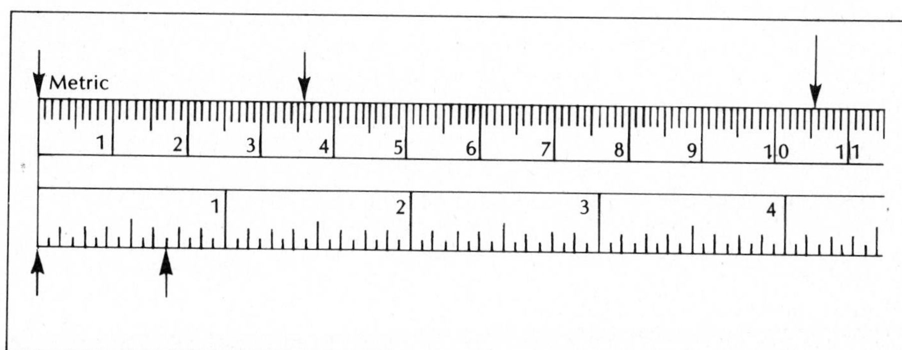

Figure 1.6 *Comparing inches and centimeters.*

1-14. What is the distance between two of the smallest divisions on the metric scale?

On the English scale, each number represents the number of inches from the end of the scale.

1-15. What is the distance between two of the smallest divisions on the English scale?

Since the metric scale is always divided into tenths, it is very easy to represent metric measurements as decimal numbers.

1-16. What is the distance between the short arrows shown on the metric scale in Figure 1.6? (Express your answer as a decimal.)
1-17. What is the distance between the arrows shown on the English scale in Figure 1.6? (Express your answer as a decimal.)

Note that you can express the answer to Question 1-16 by reading the scale directly, because the distance between numbers is divided into 10 equal parts. However, in order to express the answer to Question 1-17 as a decimal, it was necessary to convert a fraction to a decimal. This is one of the reasons why metric measurements are more convenient to use.

In using any measuring device (a ruler, for example), you should try to make the measurement as precisely as possible. Since the length that you want to measure may not correspond exactly to a mark on the scale, you will often have to estimate the distance between marks on the scale. Note that the long arrow in Figure 1.6 does not fall on one of the scale divisions.

1-18. What is the distance from the end of the metric scale to the long arrow shown in Figure 1.6?

In order to answer Question 1-18, imagine that the distance between marks is divided into 10 equal spaces and estimate the number of those spaces from the last mark on the scale to the arrow. Since this is an estimation, you may not get the same value as your neighbor, but you should be close. My estimate of the answer to Question 1-18 is 10.55 cm. Your answer should be no less than 10.53 cm and no more than 10.57 cm. In other words, our answers should differ by no more than 2 in the last digit.

Now that you know how to use the metric scale, measure the length and width of the page again and calculate its area.

1-19. What is the area of the sheet of typing paper?

The answer that you get to Question 1-19 should be recorded in units of square centimeters, which is usually abbreviated as cm^2. You find the **area** by multiplying the length (27.94 cm) by the width (21.59 cm). You should multiply both the numbers *and* the units. When you multiply the units, you get "cm × cm," which is the same as cm^2. The 2 is an exponent and indicates the number of times that centimeter is taken as a factor. (Exponents are discussed in more detail in Chapter 2 and Appendix C.)

I am sure that you have calculated areas before. Surprisingly, many people would not understand what we meant if we said that the area of a room is 200 square feet (ft^2) or that the area of the gray portion of this page is 357 square centimeters (cm^2). Do you?

A square centimeter is a square that measures 1 cm on a side. It is the size of each of the squares covering the gray portion of this page. When we say that the area of the gray portion of this page is 357 cm^2, we mean that it would take 357 squares that measure 1 cm on a side to cover it. Count them. When we say that the area of a room is 200 ft^2, we mean that it would take 200 squares that measure 1 ft on a side to cover the floor of the room.

1.12 ONE-MILLIONTH OF A PAGE

We began this exercise in measurement because we wanted to see whether it is possible to draw a square around one-millionth of a sheet of typing paper. Now that you have calculated the area, you should be able to find out.

1-20. What is the area of one-millionth of a sheet of typing paper?

Knowing that one-millionth of the page has an area of 0.0006032 cm^2 probably doesn't tell you whether you can draw a square that small. You need to know the length of each side of the square. The area of a square is found by multiplying the length of the side by itself.

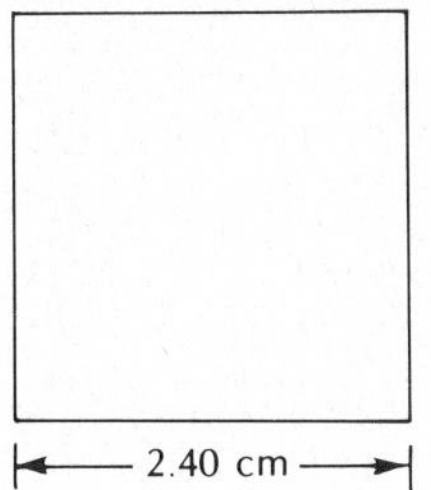

Figure 1.7

1-21. What is the area of the square shown in Figure 1.7?

You found the area by multiplying 2.40 cm by itself.

In our problem we know the *area* and need to find the *length of the side*. We need to ask, "What number multiplied by itself will give us 0.0006032?" This process is called finding the **square root.**

1-22. What is the square root of 0.0006032?

Finding the square root by long-hand calculation is rather difficult. Most people in this day and age use a calculator with a square root function, use math tables to look up the answer, or calculate an approximate answer by trial and error. Using any one of these procedures, you should have found that the square root of 0.0006032 is 0.02456. A square that measures 0.02456 cm on a side has an area of 0.0006032 cm^2. Use your metric ruler and see whether you can draw a square that small.

As you can see, one-millionth of a sheet of paper is a rather small area—smaller than the period at the end of this sentence. It is too small to draw accurately. However, it would be possible to draw a square around the amount of paper that represents one-millionth of a ream of paper. A ream contains 500 pages.

1-23. Calculate one-millionth of the area covered by a ream of paper, and draw a square around that area.

1.13 SUMMARY

We have said a lot about numbers in this chapter. You should have a better idea of how large a million is and how small a millionth is. You should know more about units, too. First of all, you should know that units are important and that we always include them with numbers if the number represents some quantity.

Perhaps the most important thing in this chapter is the technique of using units as a guide to calculations. We called that technique factor-label, but it goes by a number of other names. As the problems get more difficult, factor-label will become an increasingly important tool, and it is essential that you learn to use it now.

Finally, you should have learned SI units for mass and length. If you learned the prefixes given in Table 1.2, you only have to remember the basic units of gram and meter. SI units for other quantities, such as volume, force, and pressure, use the same prefixes. Learning them now will make it easy later on.

We will have more to say about numbers and measurement in Chapter 2, but you are already well on your way to understanding numbers, large and small.

Questions and Problems

You have worked several problems as you read through this chapter. The answers are given at the end of this section so that you can check your work. Some students find that working the problems in the chapter provides enough practice. Others need more practice. "Questions and Problems" are for those who need additional practice. The answers to these problems are not included in the text, but your instructor can provide you with answers if he or she wants you to have them.

1-24. Convert the following metric measurements to their equivalents in the indicated units. Write mass, length, volume, or area to indicate what the units measure.

a. 22.4 cm = ______ km d. 32 mm^2 = ______ m^2

b. 1.6 kg = ______ mg e. 256 mg = ______ g

c. 16 cm^3 = ______ mm^3 f. 0.0021 km = ______ mm

1-25. Use the conversion factors 2.54 cm = 1 in. and 1 kg = 2.2 lb to make the following conversions between English and metric units.

a. 1 yd = ______ cm d. 232 g = ______ lb

b. 30.5 cm = ______ in. e. 1 mi = ______ km

c. 115 lb = ______ kg f. 1 lb = ______ mg

The following problems are more involved, because they often require several steps that are not specifically described in the problem. You must decide how the problem can be solved logically and then develop the necessary unit-factors. The first problem is worked out in detail.

Example 1.3

A sprinter runs 100 yd in 10.0 sec. What is the speed in miles per hour?

(a) We are told that the sprinter runs 100 yd in 10.0 sec, so we can express his speed as 100 yd/10.0 sec. We want his speed in miles per hour, so we can restate the problem as

$$? \frac{\text{mi}}{\text{hr}} = \frac{100 \text{ yd}}{10.0 \text{ sec}}$$

(b) Now we look for unit-factors that will allow us to convert yards to miles and seconds to hours. We can do this one step at a time. Since 1,760 yd = 1 mi, we can write

$$? \frac{\text{mi}}{\text{hr}} = \frac{100 \cancel{\text{yd}}}{10.0 \text{ sec}} \times \frac{1 \text{ mi}}{1{,}760 \cancel{\text{yd}}} = \frac{1 \text{ mi}}{176 \text{ sec}}$$

$$? \frac{\text{mi}}{\text{hr}} = \frac{1 \text{ mi}}{176 \text{ sec}}$$

Now if we convert seconds to hours, we are done. Since 60 sec = 1 min and 60 min = 1 hr

$$? \frac{\text{mi}}{\text{hr}} = \frac{1 \text{ mi}}{176 \cancel{\text{sec}}} \times \frac{60 \cancel{\text{sec}}}{1 \text{ min}} = \frac{60 \text{ mi}}{176 \text{ min}}$$

$$? \frac{\text{mi}}{\text{hr}} = \frac{60 \text{ mi}}{176 \cancel{\text{min}}} \times \frac{60 \cancel{\text{min}}}{1 \text{ hr}} = 20.5 \text{ mi/hr}$$

(Note that 1 min/60 sec and 1 hr/60 min are unit-factors just like 60 sec/1 min and 60 min/1 hr, but if we had used the incorrect unit-factors the units would not have canceled to produce an answer with units of miles per hour.)

1-26. The density of iron in the metric system is 7.80 g/cm³ (grams per cubic centimeter). What is its density in pounds per cubic foot?

1-27. How many pounds of iron ore that contains 70 percent iron are needed to get 10,000 lb of iron?
(Hint: Percent means "parts per 100." Since the ore is 70 percent iron, 100 lb of ore contain 70 lb of iron. We establish a logical equivalence: 100 lb of ore = 70 lb of iron, from which we get unit factors of $\frac{70 \text{ lb iron}}{100 \text{ lb ore}}$ or $\frac{100 \text{ lb ore}}{70 \text{ lb iron}}$.)

1-28. How many seconds are there in one leap year?

1-29. If the burning of 1.0 lb of coal produces 14,000 BTU, how many calories could be produced from 1.0 kg of coal? (1 BTU = 252 cal.)

Answers to Questions in Chapter 1

1-1. $? \text{ tn} = 24{,}560 \cancel{\text{lb}} \times \dfrac{1 \text{ tn}}{2{,}000 \cancel{\text{lb}}}$
$= 12.28 \text{ tn}$ (short tons)

$? \text{ tn} = 24{,}560 \cancel{\text{lb}} \times \dfrac{1 \text{ tn}}{2{,}240 \cancel{\text{lb}}}$
$= 10.96 \text{ tn}$ (long tons)

1-2. $? \text{ dag} = 127 \cancel{\text{g}} \times \dfrac{1 \text{ dag}}{10 \cancel{\text{g}}} = 12.7 \cancel{\text{g}}$

1-3. $? \text{ oz} = 1.36 \cancel{\text{lb}} \times \dfrac{16 \text{ oz}}{1 \cancel{\text{lb}}} = 21.76 \text{ oz}$

1-4. $? \text{ g} = 268 \cancel{\text{mg}} \times \dfrac{0.001 \text{ g}}{1 \cancel{\text{mg}}} = 0.268 \text{ g}$

1-5. $? \text{ kg} = 354 \cancel{\text{g}} \times \dfrac{1 \text{ kg}}{1{,}000 \cancel{\text{g}}} = 0.354 \text{ kg}$

1-6. $? \text{ oz} = 0.215 \cancel{\text{tn}} \times \dfrac{2{,}000 \cancel{\text{lb}}}{1 \cancel{\text{tn}}} \times \dfrac{16 \text{ oz}}{1 \cancel{\text{lb}}}$
$= 6{,}880 \text{ oz}$ (short tons assumed)

1-7. $? \text{ mg} = 0.0013 \cancel{\text{kg}} \times \dfrac{1{,}000 \cancel{\text{g}}}{1 \cancel{\text{kg}}} \times \dfrac{1{,}000 \text{ mg}}{1 \cancel{\text{g}}}$
$= 1{,}300 \text{ mg}$

1-8. $? \text{ cg} = 2.34 \cancel{\text{dag}} \times \dfrac{10 \cancel{\text{g}}}{1 \cancel{\text{dag}}} \times \dfrac{1 \text{ cg}}{0.01 \cancel{\text{g}}}$
$= 2{,}340 \text{ cg}$

1-9. $? \text{ kg} = 2{,}456{,}000 \cancel{\text{cg}} \times \dfrac{0.01 \cancel{\text{g}}}{1 \cancel{\text{cg}}} \times \dfrac{1 \text{ kg}}{1{,}000 \cancel{\text{g}}}$
$= 24.56 \text{ kg}$

1-10. $? \text{ dr} = 0.098 \cancel{\text{cwt}} \times \dfrac{100 \cancel{\text{lb}}}{1 \cancel{\text{cwt}}} \times \dfrac{1 \text{ dr}}{0.0039 \cancel{\text{lb}}}$
$= 2509 \text{ dr}$ (short hundredweight assumed; rounded to nearest dram)

1-11. I got 27.92 cm. The figure you got may be slightly more or less than this.
1-12. I got 11.0 in. Your answer should be close to this.
1-13. 2.54 cm = 1 in. Your answer should be close to this.
1-14. 1 mm
1-15. $\frac{1}{16}$ in.
1-16. 3.6 cm
1-17. 0.6875 in.
1-18. 10.55 cm
1-19. $27.94 \text{ cm} \times 21.59 \text{ cm} = 603.2 \text{ cm}^2$
1-20. $603.2 \text{ cm}^2 \div 1{,}000{,}000 = 0.0006032 \text{ cm}^2$
1-21. 5.76 cm^2
1-22. 0.02456
1-23. A square measuring 0.5492 cm on a side

READING

This book is preparation for chemistry. It focuses on skills that you need when you encounter chemistry in a textbook or in a formal lecture. But we don't want you to think that these are the only places where you encounter chemistry. Chemistry is all around you, and you can learn a great deal about it by reading things that aren't textbooks. We encourage you to do that.

The reading that follows and other readings that you will find throughout this book are taken from current periodicals. They vary from popular treatments of scientific topics that appear in such magazines as *Newsweek, Popular Science,* and *The Smithsonian* to the more technical discussions found in *BioScience, Chemistry,* and the *Journal of Chemical Education.*

There are several reasons for including these readings. They show how ideas presented in this course are applied in other fields. They show how chemistry affects our everyday lives. They provide samples of the technical literature that is available to help you keep your science education up to date after college is over. I hope you find the articles interesting and informative. If you do, there is a treasury of other articles in the library. Try mining there.

Some students get the notion that math skills are useful only in science and math classes. Not so. This article shows how factor-label is used to solve problems in nursing. In spite of the authors' claim, factor-label won't work in *all* problems, but it certainly works for most of them.

HOW TO SOLVE DOSAGE PROBLEMS IN ONE EASY LESSON

Joseph J. Carr, Norman L. McElroy, Bonita L. Carr

Scene: A nurses' station on a medical-surgical unit; a registered nurse is speaking to the omnipotent giver of wisdom—the instructor.

NURSE: The ability to solve simple drug math problems is necessary to administer drugs correctly and to avoid making medication errors. Why, oh wise one, do many otherwise competent nurses have so much trouble with these problems?

INSTRUCTOR: The answer to your question is not simple and straightforward; several aspects must be considered. On the one hand, we can point to a general decline in secondary school standards which seems to have produced a generation of high school graduates who show a marked lack of arithmetic ability. Nursing educators, though, must guard against using this as an excuse for their own shortcomings. We have, for example, known students with relatively high college-board math scores who are inept problem solvers. Those who fail a pharmacy-math quiz will range from the bottom to the top of the class!

The crux of the problem is lack of *problem-solving ability*. It is unfortunate that most people, even many educators, assume that arithmetic ability and problem-solving ability are one and the same. They are not. The ability to work nicely laid out arithmetic problems is considerably different from the ability to solve the kind of practical problems encountered by the working nurse or the nursing student.

NURSE: . . . But when I was in nursing school we were taught all the methods and formulas we needed for solving. . . .

INSTRUCTOR: Reflect on your own words for a moment. You were given "all of the. . . ." What you were given was a collection of formulas and procedures, many of which you promptly and nearly completely forgot once the BIG TEST was over. I dare say that you were cut loose with a few words about ratio and proportion and the assumption that it was sufficient. Now, several years after graduation, when you have to apply that knowledge professionally, you become confused and have difficulty performing. What would you say if I told you that there is a single, unified, technique with which you could solve *all* your nursing math problems? Furthermore, this technique does not use ratios and proportions and is so absurdly easy that you will wonder why you didn't think of it yourself.

NURSE: I'd say that you were crazy!

INSTRUCTOR: I don't suppose you could be blamed for that, but believe me, there is such a technique, and it has been used by engineers, chemists, and physicists for at least a century.

NURSE: What could they possibly know about nursing?

INSTRUCTOR: Probably nothing, but they are experts in problem solving and that is what we are talking about, not nursing. In fact, it is the lack of that insight that causes so many to intermingle math instruction with pharmacy or nursing-skills instruction. There is no essential difference between many of the problems which physics students and nursing students must learn to master. The only real difference is that physics students aren't cut adrift to fend for themselves but, rather, are given a logical method for working them.

NURSE: What is this magical technique called?

INSTRUCTOR: Some people call it "units conversion" but that name tends to limit its application. I prefer *dimensional analysis* because it implies far more power.

NURSE: Will it help me determine I.V. drip rates?

INSTRUCTOR: Yes.

NURSE: Will it allow me to calculate how many cc. of a liquid drug to draw up to get an ordered amount of milligrams?

INSTRUCTOR: Yes.

NURSE: Will it allow me to convert units such as a kilogram to pounds, or *vice versa?*

Source: *American Journal of Nursing,* December 1976, pp. 1934–1937.

INSTRUCTOR: Of course.

NURSE: What about the dopamine problem? I bet that's beyond your method!

INSTRUCTOR: Duck soup. That problem is the same as any other, only more steps.

NURSE: Is it too complicated for one, quick lesson?

INSTRUCTOR: Not at all. First, let us review a few concepts which you already know. For example, consider a conversion factor such as the well-known

$$1 \text{ kg.} = 2.2 \text{ lb.}$$

This means that a platform scale will balance exactly if a 2.2-lb. weight were placed on one dish and a 1-kg. weight were placed on the other. That is, both expressions represent the identical amount of weight.

But what would happen if we were to divide both sides by 2.2 lbs.? This is a legal arithmetic move, because we do the same thing to both sides. Let's see.

$$\frac{1 \text{ kg.}}{2.2 \text{ lb.}} = \frac{\overset{1}{\cancel{2.2 \text{ lb.}}}}{\underset{1}{\cancel{2.2 \text{ lb.}}}}$$

Notice that the 2.2 lb. cancelled out leaving the right side equal to 1/1 or, simply, 1. Similarly, consider what would have happened if we had divided both sides by 1 kg. instead. We would now have

$$\frac{\overset{1}{1 \cancel{\text{kg.}}}}{\underset{1}{1 \cancel{\text{kg.}}}} = \frac{2.2 \text{ lb.}}{1 \text{ kg.}}$$

The left side is now made equal to 1. So we have two expressions that are both equal to 1.

NURSE: Doesn't that make them equal to each other?

INSTRUCTOR: Yes, and that proves that mathematically we are allowed to substitute one for the other without fear of changing the problem and making a mistake. We then can write

$$\frac{1 \text{ kg.}}{2.2 \text{ lb.}} = \frac{2.2 \text{ lb.}}{1 \text{ kg.}} = 1$$

We also know that we can multiply any number by 1 without changing its value. For example:

$$2 \times 1 = 2$$
$$9 \times 1 = 9$$
$$A \times 1 = A$$
$$4 \times \frac{2}{2} = 4$$

NURSE: That seems to be a useless fact!

INSTRUCTOR: Be patient. The fact that multiplication by a factor equivalent to 1 does not change the value of a quantity permits us to solve problems. In fact, it is probably the single most important tool the expert problem solver has.

Perhaps it would be clearer to you if we worked a sample problem. For instance, how many milligrams are there in 3 grains? We know two facts:

a. There are 3 gr.
b. That 60 mg. = 1 gr.

The second fact comes from table of conversion factors, which you may have or have memorized at one time. The fundamental trick in dimensional analysis is to multiply given amounts, such as in a. above, by conversion factors equivalent to 1 until *only the desired units are left uncancelled in the answer.*

In this particular problem, we have a conversion factor involving mg. and gr., a specified number of gr., and want only mg. left in the final answer. Conversion factor b. can be written in two forms.

$$\frac{60 \text{ mg.}}{1 \text{ gr.}} = 1$$

or

$$\frac{1 \text{ gr.}}{60 \text{ mg.}} = 1$$

We select the one which will cancel the gr. unit in the numerator. A unit is cancelled only by the same unit placed so that the two divide out. That is to say, a gr. in the numerator can only be cancelled by a gr. in the denominator. This requires that we use the b. version which has gr. in the denominator, so we write

$$3 \cancel{\text{gr.}} \times \frac{60 \text{ mg.}}{1 \cancel{\text{gr.}}}$$

The gr. term cancels out leaving us with . . .

$$3 \times 60 \text{ mg.} = 180 \text{ mg.}$$

NURSE: But how can you be sure that it is the correct answer? I have been sure on many examinations only to find out that I had made a serious procedural error.

INSTRUCTOR: In that case, you really didn't know what you were doing, and your procedures left you with no method for checking your approach to solving the problem. In dimensional analysis, we have a three-point check:

a. Is the conversion factor correct? (Look it up. Does 1 gr. = 60 mg.?)
b. Is the arithmetic correct? (Does 3 × 60 = 180?)
c. *Are the units that are desired (in this case, mg.) the only ones left uncancelled in the answer?*

If you can answer "yes" to all three points, then there is absolutely no chance that there is a mistake. . . . The answer is correct!

INSTRUCTOR: Here are several types of nursing problems worked out by the method of dimensional analysis. Would you look them over?

NURSE: May I take a few minutes to read them now?

INSTRUCTOR: Of course, take your time and study them closely.

NURSE (*reading silently*): Problem 1—The doctor ordered 5 mg. of a certain drug. The vial is labeled "1 mg. in 2 cc." How many cc. are required? We want an answer in cc.'s. "1 mg. in 2 cc." is the same as saying "1 mg. = 2 cc.", so we can say

$$5\,\cancel{\text{mg.}} \times \frac{2\text{ cc.}}{1\,\cancel{\text{mg}}} = \frac{5 \times 2}{1}\text{ cc.} = 10\text{ cc.}$$

Problem 2—A patient is to receive 50 mg. of a certain drug. The container is labeled "10 mg. per cc." How many cc. are required? We want an answer in cc.'s. "10 mg. per cc." is the same as saying 10 mg. = 1 cc., so we may say:

$$50\,\cancel{\text{mg.}} \times \frac{1\text{ cc.}}{10\,\cancel{\text{mg.}}} = \frac{50}{10}\text{cc.} = 5\text{ cc.}$$

In problems 1 and 2, it was necessary to recognize the meaning of something we read in the problem: The words "in" and "per" are used as verbal equal signs. For example.

1 mg. *in* 2 cc.

is the same as

1 mg. = 2 cc.

and,

10 mg. *per* cc.

is the same as if you wrote that

10 mg. = 1 cc.

Therefore,

$$\frac{10\text{ mg.}}{1\text{ cc.}} = \frac{1\text{ cc.}}{10\text{ mg.}} = 1$$

Problem 3—A patient is to receive by I.V. 3,000 cc. of a solution over 24 hours. The I.V. set has a drop factor of 15 gtt./cc. What drip rate is required?

We want an answer in: gtt./min.

$$\frac{3{,}000\,\cancel{\text{cc.}}}{24\,\cancel{\text{hr.}}} \times \frac{1\,\cancel{\text{hr.}}}{60\text{ min.}} \times \frac{15\text{ gtt.}}{1\,\cancel{\text{cc.}}}$$

$$= \left[\frac{3{,}000 \times 15}{24 \times 60}\right] \frac{\text{gtt.}}{\text{min.}} = ?$$

Only the *desired units* are left uncancelled in the answer so we may now carry out the arithmetic to get a numerical answer:

$$= 31\,\frac{\text{gtt.}}{\text{min.}}$$

Notice how we left the arithmetic unworked until the very last step. Always work with the units and forget the arithmetic until all factors have been included and only the desired units are left uncancelled.

Problem 4—A patient's order is for dopamine at a rate of 5 micrograms (μg.) per kilogram patient body weight (kg.) per minute (min.) The drug is supplied from the pharmacy mixed 400 mg. in 500 cc. D_5W and the patient weighs 90 kg. The I.V. set is a microdrop type with a drop factor of 60 microdrops/cc. What rate is required?

We want an answer in microdrops/min.

What do you know about this problem?

a. Dose: $\frac{5\ \mu g.}{kg.\text{-}min.}$
b. Supply: 400 mg. = 500 cc.
c. Patient weight: 90 kg.
d. Drop factor: 60 microdrops/cc.
e. 1,000 μg. = 1 mg.

We start off with 5 μg./(kg.-min.) but want to eliminate μg. and kg. and add microdrops.

$$\frac{5\ \cancel{\mu g.}}{\cancel{kg.}\text{-min.}} \times 90\ \cancel{kg.} \times \frac{1\ \cancel{mg.}}{1{,}000\ \cancel{\mu g.}} \times \frac{500\ \cancel{cc.}}{400\ \cancel{mg.}} \times \frac{60\ \text{microdrops}}{\cancel{cc.}}$$

Only microdrops remains uncancelled in the numerator. Our units, then, are gtt./min., which is what we want in the final answer. This tells us, then, that it is time to work the arithmetic to find the numerical part of the answer:

$$\left[\frac{5 \times 90 \times 1 \times 500 \times 60}{1{,}000 \times 400}\right] \frac{\text{microdrops}}{\text{min.}}$$

$$= 34 \text{ microdrops/min.}$$

Notice that the same method solved all three types of problems. The only thing more difficult about problem 4 is the extra steps. The idea is to do whatever legal arithmetic steps are necessary to cancel out units not required in the answer or to obtain units that are required. Only those units required by the answer may be left uncancelled in the final result.

NURSE: It seems so easy. . . .

INSTRUCTOR: Yes, it is easy. And dimensional analysis offers several advantages:

1. It is easy to learn.
2. It gives you a check method which shows absolutely whether or not you made a mistake.
3. It allows you to figure out which way to go in solving the problem, rather than forcing you to depend upon memory.
4. *It allows you to solve problems which you do not yet understand.* You may then restudy your solution to gain an understanding.

NURSE: But what about ratios and proportions?

INSTRUCTOR: They have their place, I suppose, but that method is too difficult to use in this type of problem. They require an insight of what to put where in the equation. Guess how you can tell that . . . by dimensional analysis! If you have to use dimensional analysis in that situation, why not use it totally? It saves steps and does not require a knowledge of simple algebra.

NURSE: One thing bothers me. Dimensional analysis is obviously very useful, yet it is also ridiculously easy to learn. Why aren't we taught that method?

INSTRUCTOR (*wearily*): Well, . . . I've been wondering about that, too!

2

Numbers from Measurement

OBJECTIVES

Uncertainty and Significant Digits

No measurement is certain. In this chapter you will learn what is meant by uncertainty and how to write numbers to show the amount of their uncertainty. The skills associated with this goal are subtle, and it may take weeks of practice before you feel comfortable with significant digits. However, you will be well on your way when you are able to do the following:

1. Using a measurement that you have made, describe what is meant by uncertainty of measurement.
2. Given a number that represents a measurement, identify the digits that are certain and the one that is uncertain. (See Problem 2-76.)
3. Determine the number of significant digits in a number resulting from measurement. (See Problem 2-77.)
4. Identify those zero digits that are significant and those that are not. (See Problem 2-77.)
5. Round the answer to a multiplication or division problem to the proper number of significant digits. (See Problem 2-78.)
6. Round the answer to an addition or subtraction problem to the proper number of significant digits. (See Problem 2-78.)
7. Identify numbers that have an unlimited number of significant digits and explain why they do. (See Problem 2-79.)
8. Using an example, illustrate the difference between the precision and the accuracy of a measurement.

Units of Measurement

You learned SI units for length and mass in Chapter 1. As you make other measurements, you need other units. The following objectives cover units for volume, temperature, heat, and the derived quantity of density.

9. Name the basic metric units for volume, temperature, and heat and give the correct abbreviation for each.
10. Convert from volume expressed in cubic measure (m^3, cm^3, and so on) to volume expressed in container sizes (L, mL, and so on). (See Problem 2-80.)
11. Convert from °C to K and vice versa. (See Problem 2-81.)
12. Given the mass and volume of an object, calculate its density. (See Problem 2-82.)
13. Given density and mass, calculate volume. (See Problem 2-82.)
14. Given density and volume, calculate mass. (See Problem 2-82.)

Scientific Notation

When numbers are very large or very small, ordinary decimal notation is difficult to use. You will be much better able to handle large and small numbers when you can do the following:

15. Convert any number from decimal to scientific (exponential) notation and vice versa.
16. Multiply or divide any number by a power of 10 by moving the decimal the proper number of places.
17. Add, subtract, multiply, or divide numbers written in scientific notation.

Words You Should Know

Building your vocabulary is important. When you finish this chapter, you should be able to use these words correctly.

uncertainty
accuracy
precision
parallax
significant digit
volume
liter
pipet
measuring pipet
volumetric flask
graduated cylinder
buret
meniscus
temperature
extensive property
intensive property
Celsius
Kelvin
Fahrenheit
energy
joule
heat
calorie
Calorie
density
proportional
scientific notation
exponential notation

2.1 UNCERTAINTY IN NUMBERS

In Chapter 1 we began looking at numbers. We saw how large numbers can be and how small, and we began to study the measurements used to produce numbers.

Most numbers come from some kind of measurement, and unless you are careful, they can be misleading. Take the distance from my house to Chicago as an example. In Chapter 1, I told you that it is 120 miles. Now just what does that mean? Is it *exactly* 120 miles? Is it *exactly* 633,600 feet? Is it *exactly* 7,603,200 inches? To *where* in Chicago? To the Sears Tower downtown? To the city limits? To O'Hare Airport? By highway or as the crow flies?

The point is that numbers resulting from measurement have some **uncertainty** associated with them. Some of this uncertainty is due to the way the measuring is done. Travelers are sometimes bothered because the mileage shown on a road map is different from the distance they measure with the odometer on their car. The discrepancy may be due to the fact that map mileages are measured from center city to center city, while the traveler is more likely to measure from city limit to city limit.

There is another kind of uncertainty associated with how close we are able to make the measurement. I don't know the distance from my house to Chicago to the nearest inch or the nearest foot. It would be virtually impossible to make the measurement that close. It is even doubtful that the number is correct to the nearest mile, or even to the nearest ten miles. I wouldn't bet my life that the distance is closer to 120 miles than it is to 110 or 130. Although the distance was measured with the odometer on my car, I don't know how accurately my odomoter records distances.

For most purposes, all of this fuss about whether I know the distance to Chicago to the nearest inch or nearest 10 mi would not matter. The figure of 120 mi is close enough. However, if I were a contractor building a highway, an uncertainty of 10 mi could be disastrous when I make my bid. Similar problems arise in science and, like the contractor, we must worry about uncertainty.

2.2 ACCURACY AND PRECISION

Two words often seen in a discussion of uncertainty are "precision" and "accuracy." Although they are related, they do not have the same meaning. **Precision** refers to the uncertainty of a measurement. It is usually related to the nature of the instrument that is used to make the measurement. For example, there are two types of balances in the chemical laboratory. One is called a triple beam or centigram balance, and the other is referred to as an analytical balance. The analytical balance is more *precise* than the triple beam balance. When I weighed my pencil on an analytical balance, I got a reading of 10.7282 g. When I weighed the same pencil on a triple beam balance, I got a reading of 10.71 g. In both cases the last digit is uncertain, but with the analytical balance, I am certain of the weight of the pencil to the nearest thousandth of a gram and have an estimate to the nearest ten-thousandth of a gram. With the triple beam balance, I am certain only to the nearest tenth of a gram, and have an estimate to the nearest hundredth. The analytical balance is 100 times more precise. I can weigh much closer. I can get an answer correct to more digits.

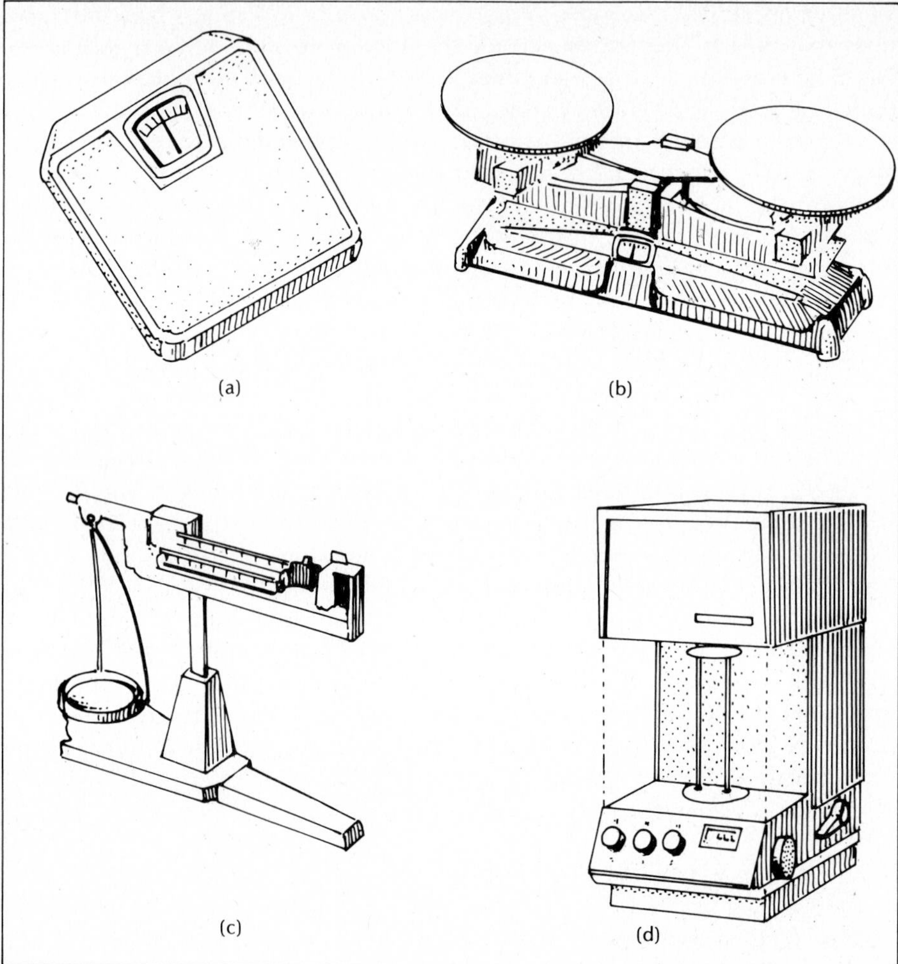

Figure 2.1
Some common balances. (a) A bathroom scale measures large masses but only to the nearest pound. (b) An inexpensive trip scale weighs up to 5 kg but only to the nearest gram. (c) This triple beam balance weighs up to 300 g with a precision of about 0.01 g. (d) A single-pan, electronic balance has a capacity of 200 g or less but has a precision of ±0.0005 g.

The **accuracy** of a measurement refers to how close it comes to the truth. The more precise measurement is usually the more accurate one, but this is not necessarily so. Take the case of the two measurements of the weight of the pencil. It is possible that when I weighed the pencil on the analytical balance, it had a piece of lint caught on the eraser and the reading I obtained was actually the weight of the pencil *and* the lint. In that case, the weight shown on the triple beam balance might be closer to the truth—more *accurate*—even though the weight shown on the analytical balance was more precise.

We can usually estimate the precision of a measurement on the basis of what we know about our measuring instruments. Since many factors that we are not aware of can affect accuracy, we can estimate the accuracy of a measurement only if we are able to compare our value to some known or accepted value. Unfortunately, the only way to get a known or accepted value is to make measurements, and that takes us right back to the problem of precision. We usually gain confidence in the accuracy of a value if the same value is obtained from many different experiments.

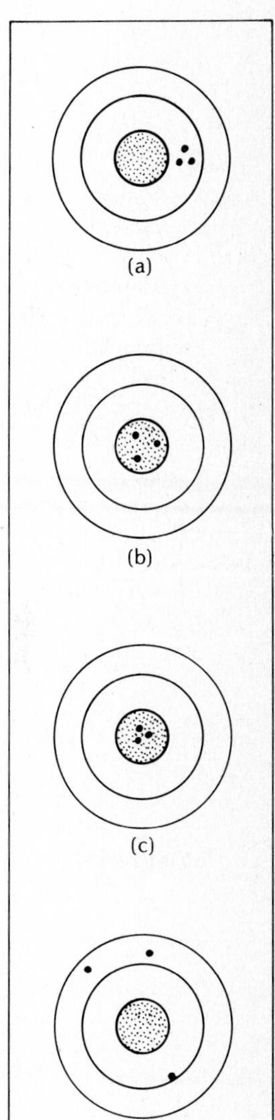

Figure 2.2
Accuracy and precision are illustrated by the holes formed in four targets. (a) The three shots fell close together. The shooting was very precise. The shots did not hit the center of the target, however, so the accuracy was low. (b) Accuracy has improved. The three shots are grouped near the center of the target. However, the shots show more scatter. The precision is less than that shown on the first target. (c) Shots that were very precise and very accurate. (d) Shots that were neither accurate nor precise.

2.3 SOME REASONS FOR UNCERTAINTY

In Chapter 1 you were asked to measure a sheet of typing paper. I found the length of the paper to be 27.92 cm. The number you got was probably different. When students in my class measured the same paper, they got values ranging from 27.85 cm to 27.96 cm. Several things could account for the different values. Some of them are:

1. Not having the end of the paper exactly on the zero point of the scale. (Always check to be sure the end is properly aligned.)

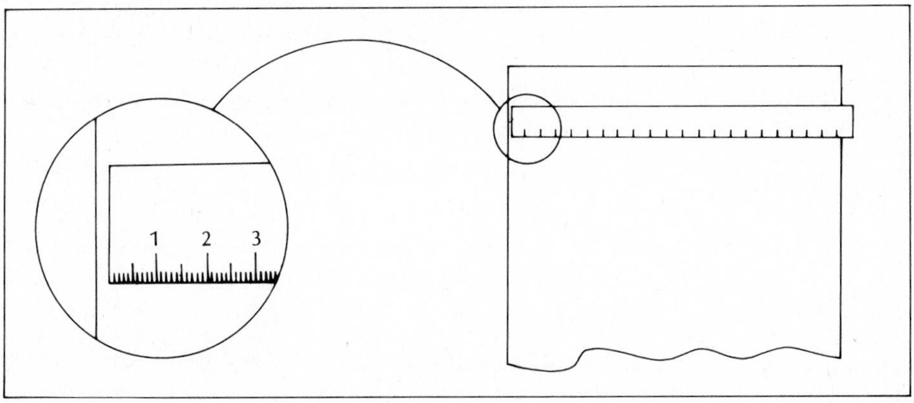

Figure 2.3

2. Using the end of the rule as the zero point when the end has been worn away with age. (It is usually better to use some other starting point on the rule—the 1-cm mark, for example—because the ends are often battered and don't give accurate results.)

Figure 2.4

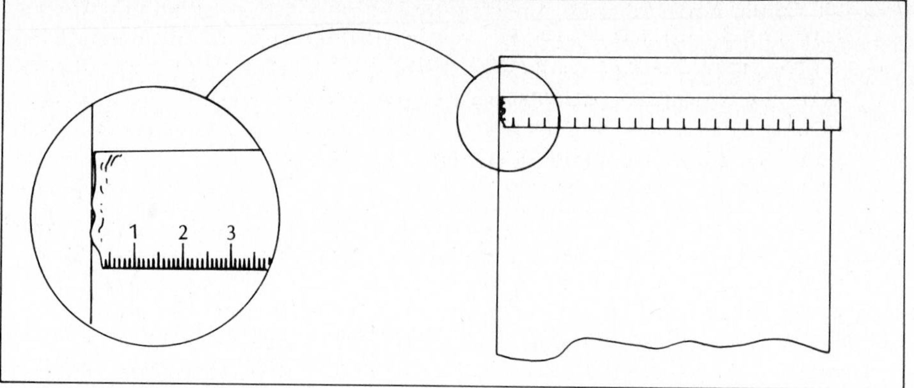

3. Placing the ruler on the paper at a slight angle rather than straight across.

Figure 2.5

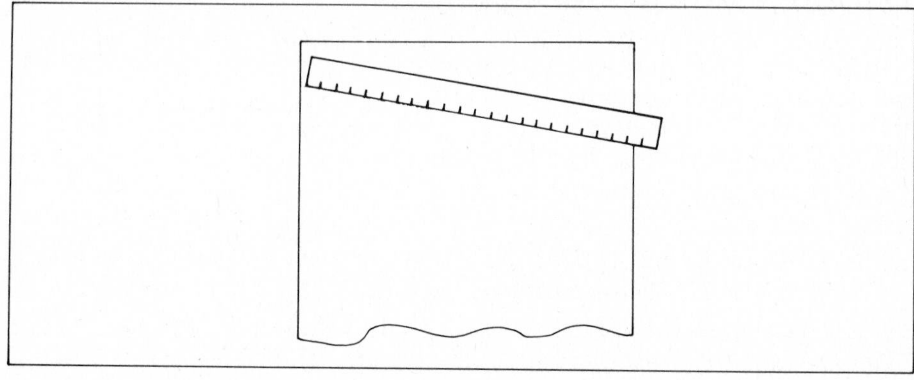

4. Not placing the paper on a flat surface when making the measurement.

Figure 2.6

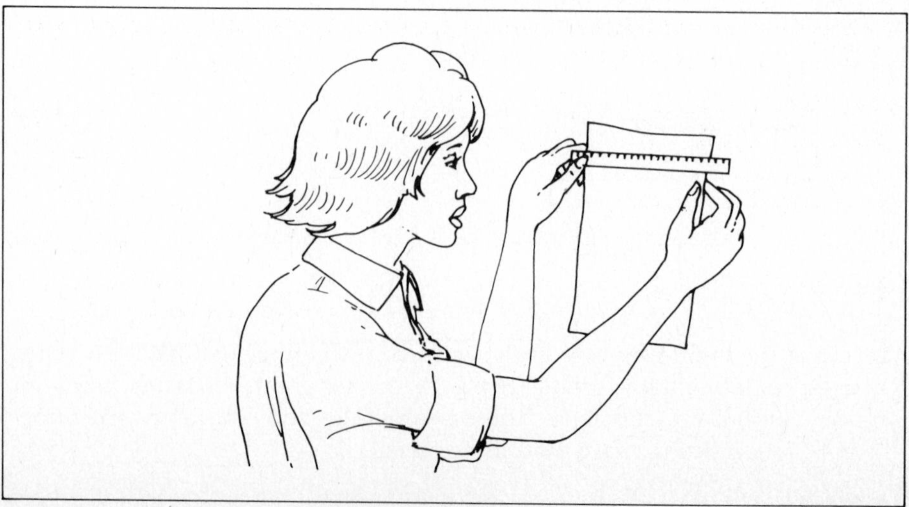

5. Parallax. Figure 2.7 shows three pictures of the same ruler. In the first picture, the view is from the left; in the second, it is from directly above; in the third, it is from the right. Only the picture taken from directly above shows the correct reading on the scale. The incorrect readings shown in the other pictures result from the angle from which the ruler is viewed. This phenomenon is called **parallax.**

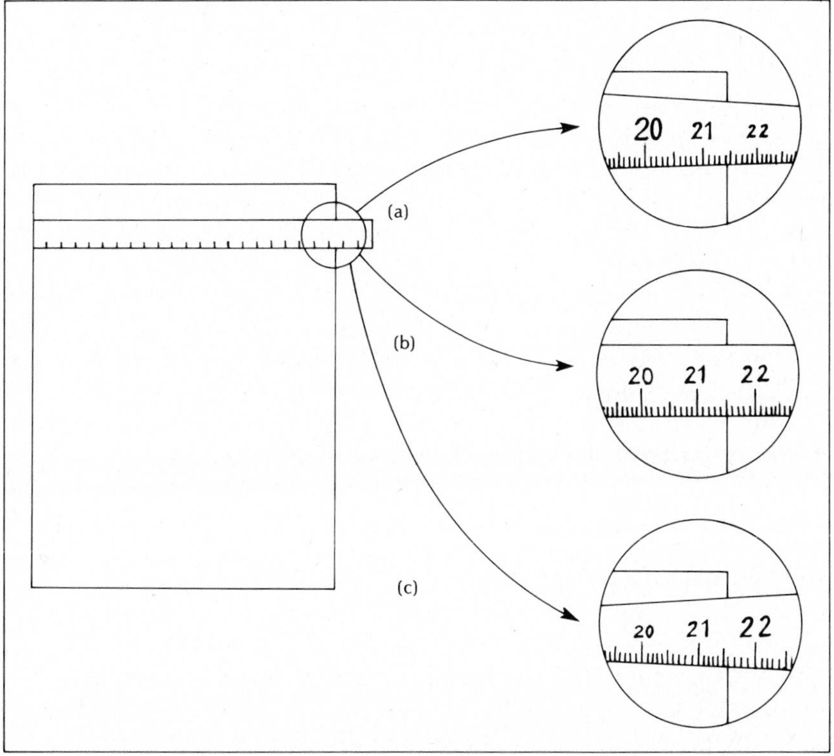

Figure 2.7

6. Uncertainty in estimating between marks on a scale. Careful technique can solve all the problems except this one. You must learn to live with it.

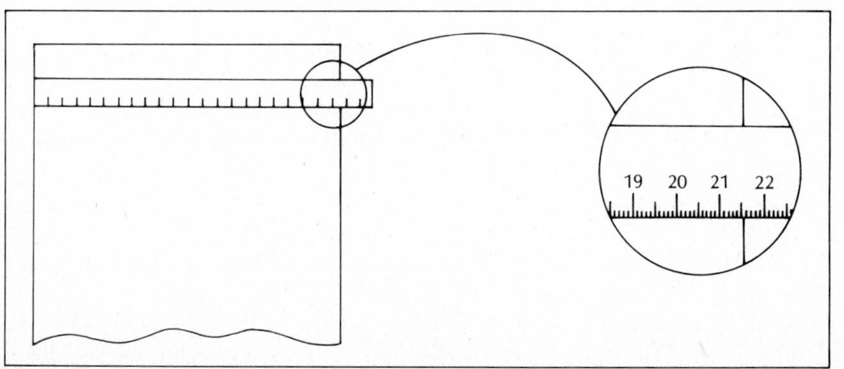

Figure 2.8

All of these "errors" combine to produce uncertainty in the measurement that is made. I cannot be sure that the length of the page of typing paper is exactly 27.92 cm, and students who got a value of 27.85 cm cannot be certain either. However, careful work can eliminate most uncertainty other than what results from limitations in the measuring instrument. Avoiding the problems we have listed, measure the following line:

I hope you agree that the line is more than 9.4 cm long but less than 9.5 cm long. How did you record the measurement? Did you say something like "about 9.5 cm" or "a little over 9.4 cm"? I hope not. Even though these values would be "close enough" for most purposes, scientists measure as precisely as possible. You can measure the line to the nearest hundredth of a centimeter by estimating where the line falls between the smallest scale divisions. This is shown in Figure 2.8.

My estimate for the length of the line is 9.46 cm. If you measured the line or if I measured it again, the value we obtained would probably be slightly different—perhaps 9.44 cm or 9.47 cm. Obviously there is some uncertainty about that last digit. Even so, I am pretty sure that the length of the line is closer to 9.46 cm than it is to 9.40 cm. By including this uncertain digit, I have a better estimate of the length of the line than I would have if I left it off.

2.4 SIGNIFICANT DIGITS

In recording measurements, it is customary to show all digits that are certain plus a final digit that is an estimate. These certain digits plus the final uncertain digit are known as significant digits. Every measurement should be recorded to the proper number of significant digits.

In normal usage, significant means "having meaning" or "important." As used here, the meaning is more restricted. Significant digits are those that would "have meaning" in reporting a measurement. The **significant digits** consist of all digits that we are certain about plus one final digit that represents our best estimate.

I am sure you agree that it would be foolish to record the length of the line you just measured as 9.462 cm. Since you had to estimate in order to record the length to the nearest hundredth of a centimeter (9.46 cm), including the final 2 has no meaning. It is not significant. Actually, recording the length as 9.462 cm is equivalent to lying. It implies that we know the length to the nearest thousandth of a centimeter, and that isn't possible. Nobody can read an ordinary ruler to the nearest thousandth of a centimeter.

Just as it is customary to record measurements to show all certain digits plus one uncertain digit, it is considered improper to record *more* than one uncertain digit. Such additional digits would have no meaning. They would not be significant and would probably be misleading.

Numbers recorded to the proper number of significant digits tell a great deal about how a measurement was made. Consider the following value for the mass of a book.

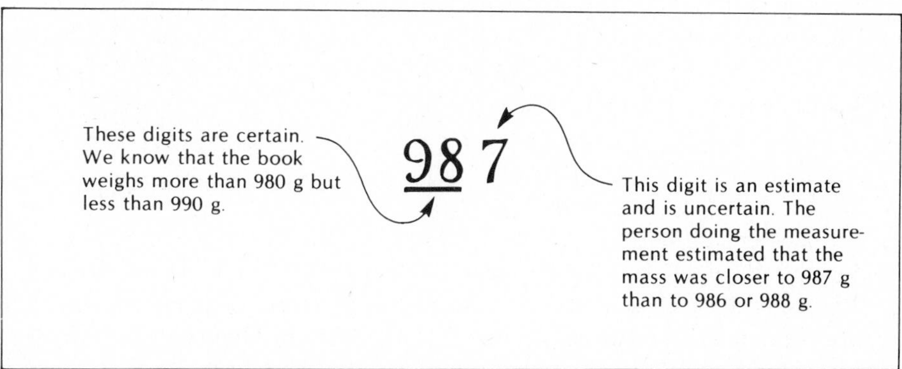

2-1. Which of the balances shown in Figure 2.1 was used to weigh the book?

If the measurement was recorded properly, the book must have been weighed on the trip scale shown in part (b) of Figure 2.1. The bathroom scale has an uncertainty of about 500 g. It could not be used to weigh the book to the nearest gram. The last two balances weigh to a fraction of a gram. If they had been used, the mass would be known to within less than a gram. However, these balances are not made to weigh objects as heavy as a book. Knowing that the last digit in the measurement is uncertain is our clue that the trip scale was used.

To be sure that you understand what we have said about significant digits so far, answer the following questions before you go on.

2-2. In recording a measurement, how many *certain* digits should be recorded?

2-3. How many uncertain digits should be recorded?

2-4. If I told you that the length of a room is 4.32 m, which digits in the number would you assume I am certain about? Which digit would you assume is just an estimate and therefore uncertain?

2-5. If you measured the same room and got a value of 4.31 m, would you assume that you had made an error? If not, why is your value different from mine?

2-6. If you measured the room and got a value of 4.23 m, would you assume that you (or I) had made an error? Why?

2.5 CALCULATIONS WITH UNCERTAIN MEASURES

Significant digits are usually no problem, as long as you are only recording a measurement that you have made. You are unlikely to record more than one uncertain digit. Problems can arise when you use those measurements to do a calculation, however. To illustrate the problem, we will return to a calculation you did earlier.

In Chapter 1 you measured the width of a sheet of typing paper in both centimeters and inches. You then divided the width in centimeters by the width in inches to find the number of centimeters in 1 in. Using my measurements, the calculation is as follows:

$$\frac{21.50 \text{ cm}}{8.50 \text{ in.}} = 2.529411765 \text{ cm/in.}$$

The division doesn't come out even. My calculator produces an answer to 10 digits, and I have recorded all of them. As the result is presently written, it appears that I have calculated the number of centimeters in 1 in. to the nearest 0.000000001 cm—a billionth of a centimeter. You have seen how small a millionth is, so you probably agree that my answer is misleading. Many of the digits recorded have no practical meaning.

When you make a measurement, you record one and only one uncertain digit. When you do a calculation, you want one and only one uncertain digit in the answer. The trick is telling which digit in the answer is the first uncertain digit. Fortunately, we can follow simple rules in order to round an answer to the proper number of significant digits.

Rule 1: *In any calculation involving multiplication or division, the answer should be rounded to the same number of digits as are found in the least precise number used in the calculation.*

When we follow this rule, nearly all our calculations result in an answer with one and only one uncertain digit. In the following example, the rule is applied to the calculation done to find the number of centimeters in an inch.

Example 2.1

$$\frac{21.50 \text{ cm}}{8.50 \text{ in.}} = 2.529411765 \text{ cm/in.}$$

According to the rule, where should the answer be rounded? The rule says to retain "the same number of digits as are found in the least precise number used in the calculation." The least precise number is the number with the least number of significant digits.

21.50 has *four* significant digits
8.50 has *three* significant digits

Then the least precise number is 8.50, with *three* significant digits. The answer should be rounded to three digits. Starting from the left, we count three digits:

2.529411765

The answer should be rounded to the *nearest* hundredth. The number is between 2.52 and 2.53. It is closer to 2.53. The answer should be recorded as 2.53 cm/in.

For the following problems, apply the rule for multiplication and division to round the answers to the proper number of significant digits. *Record the proper units for the answer.*

2-7. 21.3 cm × 1.3 cm = 27.69
2-8. 21.3 cm ÷ 1.3 cm = 16.384615
2-9. 6.34 cm^2 × 1.2 cm ÷ 1.217 cm^2 = 6.2514379
2-10. 13.21 m × 61.5 m = 812.415
2-11. 12.43 m × 2.35 m^2 = 29.2105

The rule that you have just practiced is a quick and convenient way to determine the number of digits to leave in an answer. You should not forget why you are using the rule, however. Our purpose is to express all numbers with only one uncertain digit. To prove that the rule accomplishes this, we will work the last problem that you did in the following example to show the uncertain digits.

Example 2.2

$$12.43 \text{ m} \times 2.35 \text{ m}^2 = ?$$

We assume that these numbers represent measurements and that the last digit of each is uncertain. We will draw a bar over these uncertain digits and any other uncertain digits that occur during the multiplication.

$12.4\bar{3}$	
$\times\ 2.3\bar{5}$	
$\overline{6215}$	Because the 5 we multiply by is uncertain, the digits that result from the multiplication are uncertain.
$372\bar{9}$	The 9 and the 6 are uncertain because the 3 in 12.43 is uncertain.
$248\bar{6}$	
$29.\overline{2105}$	The 2105 in the answer represents uncertain digits because uncertain digits were added.
$29.\bar{2}$	This is the correct answer, because we want to retain only one uncertain digit. Note that this is the same answer we would get by applying the rule we have stated.

So far we have talked only about multiplication and division. What about addition and subtraction? How would you round the answer to the following addition problem in order to keep only one uncertain digit in the answer?

2-12. 52.6 + 84.32 = 136.92

If you answered 137, you probably applied the rule for multiplication and division to this addition problem. It doesn't work. The following example shows why.

Example 2.3

Once again, a bar is drawn over each uncertain digit.

$$\begin{array}{r} 52.\overline{6} \\ +\ 84.3\overline{2} \\ \hline 136.\overline{92} \end{array}$$

Since the 6 in 52.6 is uncertain, adding it to any other number produces an uncertain number. Both the 9 and the 2 in the answer must be uncertain.

$$136.\overline{9}$$

This is the correct answer, because it contains only one uncertain digit.

You can see that any column containing an uncertain digit has an uncertain digit in the answer. Try to write a rule to express this idea. Use the rule to practice rounding the answers to the problems that follow.

2-13. State a rule (call it Rule 2) for rounding answers to addition and subtraction problems.
2-14. 63.43 + 34.5 = 97.93
2-15. 124 − 87.2 = 36.8
2-16. 27.35 − 21.3 = 6.05
2-17. 3217 + 13.2 + 1.30 = 3231.50

2.6 ROUNDING FIVES

The last two questions may have caused you to pause. Our rule is to round so that the answer is the *nearest* number with the proper number of significant digits. Both the 0 and the 5 in 6.05 are uncertain, so we must round to the nearest tenth, but what is the nearest tenth? The value 6.05 is half-way between 6.0 and 6.1. How shall we round?

The answer is that one choice is just as good as the other. However, there are circumstances that work out better if we round to the higher number about as often as we round to the lower number. Any rule that accomplishes this would be a good one. Here are two possibilities:

Rule 3a: *Flip a coin. If it is heads, round to the higher value; if tails, round to the lower value.*

Rule 3b: *Round to the higher number if it is even; round to the lower number if it is even. In other words, round to make the uncertain digit even.*

2.7 INFINITE SIGNIFICANT DIGITS

It is important to remember that significant digits have meaning only in relation to some kind of measurement or a number calculated from a measurement. They are a method of representing uncertainty.

There are numbers that have *no* uncertainty. How many eggs are in a dozen? About 12? No, *exactly* 12. The number of eggs in a dozen is established by definition rather than measurement. The same is true of the number of centimeters in a meter and the number of grams in a kilogram. Such numbers may be considered as having an infinite number of significant digits. If we tried to write the number of centimeters in a meter to include the first uncertain digit, we would write 100.0000 . . . , adding zeros forever. There is no uncertainty associated with such numbers. They do not affect the number of significant digits retained in an answer to a calculation. Similarly, numbers obtained by counting are usually known exactly; they have no uncertainty.

Example 2.4

We found that a sheet of typing paper is 21.50 cm wide. What is the width in meters?

$$? \text{ m} = 21.50\,\cancel{\text{cm}} \times \frac{1 \text{ m}}{100\,\cancel{\text{cm}}} = .2150 \text{ m}$$

Both 1 m and 100 cm in the conversion factor are exact numbers. They do not affect the number of significant digits in the answer. Since 21.50 has four significant digits, so does the answer.

2.8 WHEN ARE ZEROS SIGNIFICANT?

When you see a number like 27.3, you assume that there is one and only one uncertain digit. In other words, you assume that all three digits are significant. However, if you see a number like 4,500,000,000—the approximate human population of the earth—it is difficult to know how many digits are significant. Does this number tell us the population to the nearest 100 million, to the nearest 10 million, to the nearest 1 million, or what? Because zeros are needed to show the magnitude of the number, there is no way of knowing.

If you are told that a glass contains 100 mL of water, you can make several assumptions. You can assume that the water was measured to the nearest milliliter and that the volume is known to *three* significant digits, or you can assume that the volume was measured to the nearest 100 milliliters (mL) and that the volume is known to only *one* significant digit. Without additional information, it is impossible to be sure how many significant digits are indicated by a whole number with zeros at the end.

Unfortunately, some scientists and science writers assume that a number like 100 represents a measurement to the nearest hundred and contains one significant digit. Other scientists and science writers assume that a number like 100 represents a measurement to the nearest unit and contains three significant digits. During your study of science, you will undoubtedly encounter both conventions. In this book we will adopt the latter convention.

Rule 4: *Assume that all trailing zeros are significant.*

Note that the rule specifies *trailing* zeros. When a small number is written, zeros must be used in front to locate the decimal. A number like 0.00025 is assumed to have *two* significant digits. The leading zeros are there to locate the decimal.

Rule 5: *Assume that leading zeros are not significant.*

Using these rules, indicate the number of significant digits represented by the following numbers.

______	2-18. 207		______	2-22. 0.012
______	2-19. 13.502		______	2-23. 0.020
______	2-20. 320		______	2-24. 10001
______	2-21. 2.50		______	2-25. 1.00020

One problem is created by Rule 4. There are times when zeros must be used to locate a decimal at the end of a number and should not be considered significant. To illustrate the problem, convert the following measurements to the units indicated.

2-26. ? g = $0.\underline{45}$ kg

2-27. ? μm = $\underline{21.50}$ cm

(Underline the significant digits in your answer, as shown here.)

Converting a measurement from one unit to another does not change the uncertainty of the measurement. Question 2-26 gives the metric equivalent of 1 lb to the nearest hundredth of a kilogram. Converting to grams, you get an answer of 450 g, which is correct only to the nearest 10 g. (To the nearest gram, 1 lb is equal to 454 g.) Clearly, the answer of 450 g represents only two significant digits, even though Rule 4 would lead us to say that is has three.

The width of a sheet of typing paper to the nearest hundredth of a centimeter is 21.50 cm. This measurement has four significant digits. Converting it to micrometers does not change the uncertainty of the measurement, so our answer of 215,000 μm has *four* significant digits rather than the six suggested by Rule 4. The only way to avoid this problem is to represent these numbers some other way. The most common way is to represent the number in exponential notation. Exponential notation is discussed at the end of this chapter and in Appendix C.

2.9 MORE UNITS OF MEASUREMENT: VOLUME

Volume is a term that you have used for years. You know that a cup, a quart, and a gallon are all different measures of volume. However, many people are confused by other units of volume, such as cubic foot (usually abbreviated as ft^3) or cubic centimeter (abbreviated as cm^3).* Since cubic foot and cubic centimeter are the kinds of volume units normally used in science, you must be sure that you know what they mean.

The floor in my office is 15 ft long and 10 ft wide. The ceiling is 8 ft above the floor. Using the formula I learned in math class, I can find the volume of my office by multiplying the length by the width by the height.

$$15 \text{ ft} \times 10 \text{ ft} \times 8 \text{ ft} = \text{volume of my office}$$

Multiplying the numbers, I get 1,200, and multiplying the units, I get ft^3. The volume of my office is approximately 1,200 cubic feet (ft^3). What does this mean?

"Cubic foot" simply means a cube that measures 1 foot on a side, just as "square foot" means a square that measures 1 foot on a side. When I say that my office has a volume of 1,200 ft^3, I am saying that I could fill all the space in my office with 1,200 cubes that measure 1 foot on a side.

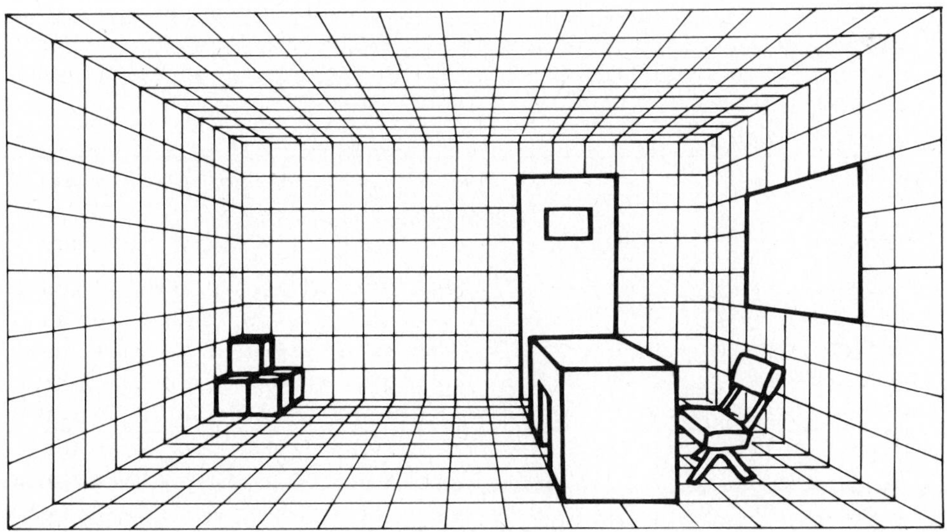

Figure 2.9
An office with a volume of 1,200 ft^3 could be filled by 1,200 cubes measuring 1 ft on each side.

* It is a shame that authors do not agree on abbreviations. Unfortunately, they do not. Scientists abbreviate cubic centimeters as cm^3, but "cc" is the abbreviation commonly used in medicine, and "cu cm" is used in some other fields.

In the English system of measurement, volume is expressed in units of cubic feet or cubic inches as well as in "container sizes" such as cups, quarts, and gallons. The basic SI unit of volume is the cubic meter (m^3). Cubic centimeters (cm^3) and a "container size" called the **liter** are also used.

The liter container is directly related to linear measurement. The original liter was made to contain the same volume as a cube that measures a decimeter (10 cm) on a side. The liter is the same as a cubic decimeter (dm^3).

2-28. How many cubic centimeters are there in a liter? (Hint: A cubic decimeter measures 10 cm × 10 cm × 10 cm.)

Since there is a simple relationship between the "container size" of liter and the commonly used volume units of cm^3 or m^3, it is easy to convert from liters to cubic centimeters. It is *not* easy to convert from "container sizes" in the English system to cubic feet or cubic inches, because there is no simple relationship between container sizes such as quarts or cups and the units based on linear measurement.

Table 2.1 shows several volume units in the metric system. The table is incomplete, however. By completing it, you can see how well you recall some of the facts you have already learned.

In the first column of Table 2.1, units based on the basic container size of liter are given (liter is abbreviated as L). The same standard prefixes introduced in Chapter 1 are used with liter to name other volume units. Without looking back to Chapter 1, see whether you can supply the missing prefixes.

Table 2.1 SI Units of Volume

Common Units	*Equivalent in cm^3*	*Equivalent in m^3*
2–29. 1 _____ liter (kL) = 1,000 L	1,000,000 cm^3	2–36. _____ m^3
2–30. 1 _____ liter (hL) = 100 L	2–37. _____ cm^3	0.1 m^3
2–31. 1 _____ liter (daL) = 10 L	10,000 cm^3	2–38. _____ m^3
1 liter (L) = 1 L	2–39. _____ cm^3	0.001 m^3
2–32. 1 _____ liter (dL) = 0.1 L	100 cm^3	2–40. _____ m^3
2–33. 1 _____ liter (cL) = 0.01 L	2–41. _____ cm^3	0.00001 m^3
2–34. 1 _____ liter (mL) = 0.001 L	1 cm^3	2–42. _____ m^3

Columns 2 and 3 of Table 2-1 give the equivalent of the container volumes in cm^3 and m^3. The relationship between liters and cubic centimeters is simple, because 1 L is 1,000 cm^3. However, converting from cubic centimeters to cubic meters will cause problems if you do not stop and think. Before you fill in the blanks in columns 2 and 3, be sure you can answer the following question.

2-35. How many cubic centimeters are there in 1 cubic meter? (Hint: A cubic meter is a cube that measures 1 meter on a side. What is the length of each side in centimeters? Then how many cubes measuring 1 cm on a side will fill the same space as a cube measuring 1 m on a side?)

USING CONTAINERS TO MEASURE VOLUME

If the volume of a cube or a rectangular solid is to be determined, a meter stick can be used to measure the length of each side and the volume can be calculated. This is not a convenient way to measure volumes of fluids. For this purpose, containers still work best.

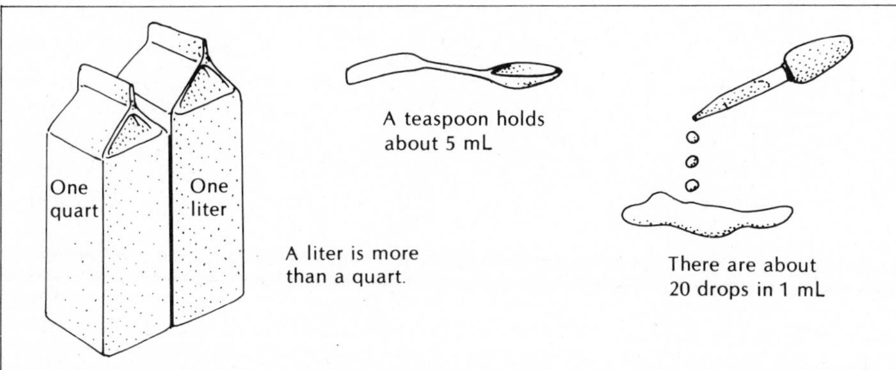

Figure 2.10 *The volumes of some common objects in metric units.*

Several types of containers are used to measure the volume of liquids. Fixed volumes can be measured accurately in a long-necked bottle called a **volumetric flask** or in a tube called a **pipet.** Look at the volumetric flask and pipet pictured in Figure 2.11. Note that there is a line scratched on the neck.

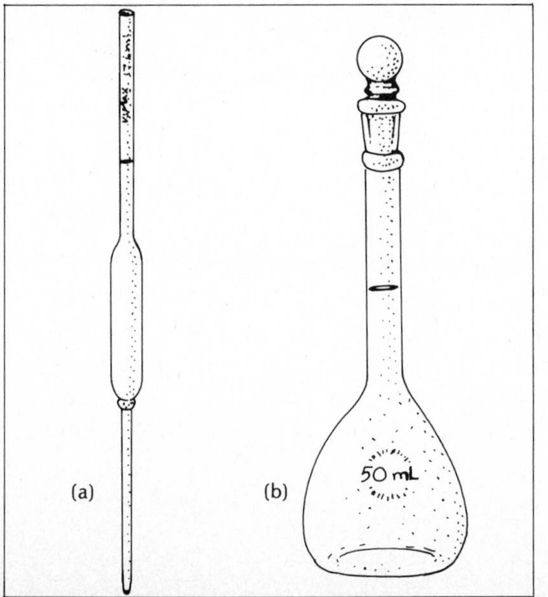

Figure 2.11 *(a) Transfer pipet. (b) Volumetric flask.*

When the flask or pipet is filled to that line, it contains the volume written on the container.

When a volume that is not fixed is to be measured, various glass cylinders with a scale etched on the wall are used. A **graduated cylinder,** a **buret,** and a **measuring pipet** are shown in Figure 2.12. The scale etched on the side is used in the same way as the scale on a measuring cup in the kitchen.

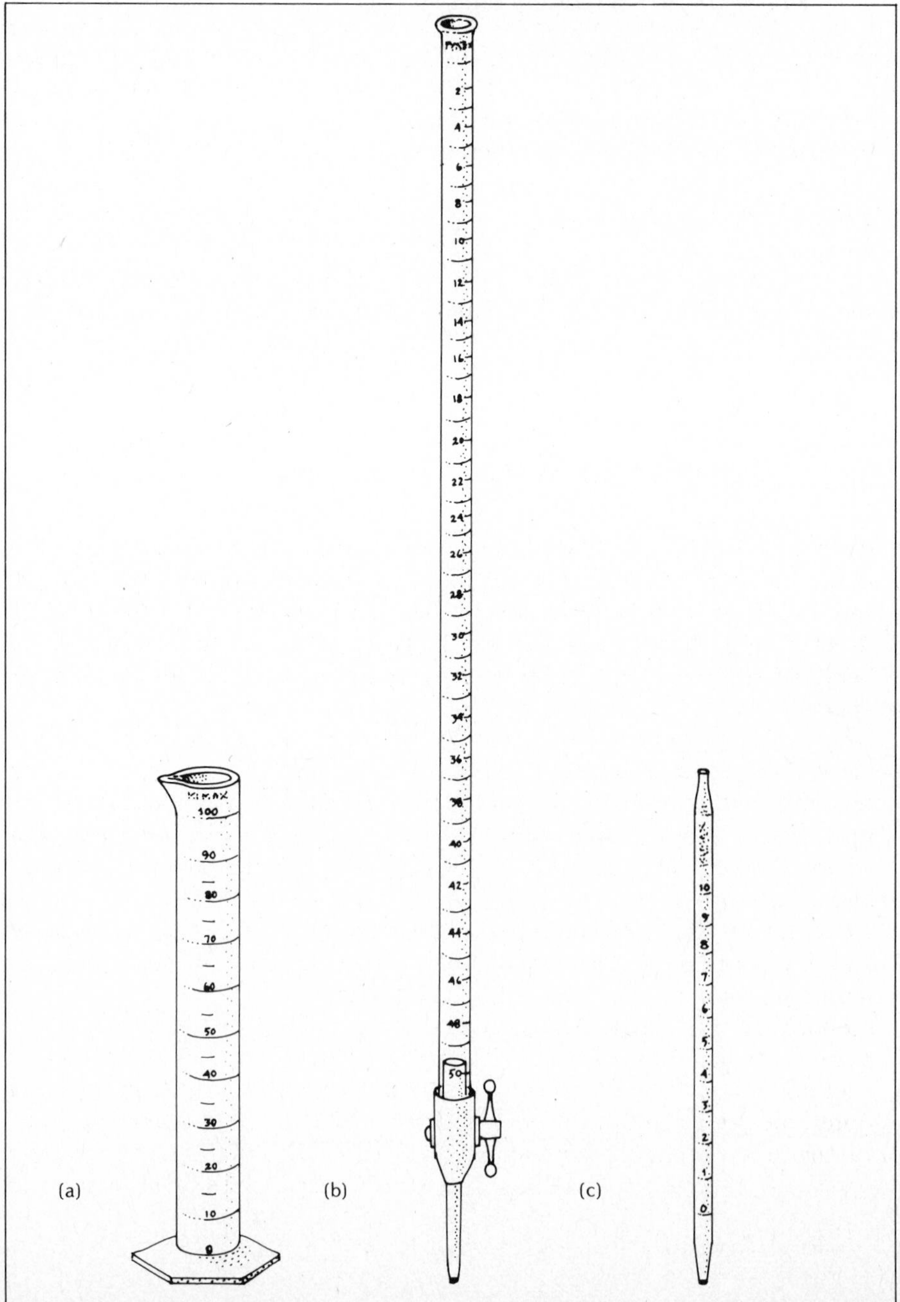

Figure 2.12
(a) Graduated cylinder. (b) Buret. (c) Measuring pipet.

Since we want to get the least amount of uncertainty when we measure in science, we do two things when we use a graduated cylinder or buret that we do not normally do when we use a measuring cup. First, we always estimate between divisions on the scale to give one more significant digit in the measurement. Second, we worry about the *meniscus*. When water is placed in a container, it doesn't lie flat. In glass, water hugs the glass and actually moves up the sides a bit. The surface looks like that shown in Figure 2.13(a). This curved surface is called the **meniscus.** ("Meniscus" means moon- or crescent-shaped.) To get an accurate reading, we must ask, "What point do we call the top of the water?" Most of the meniscus is empty. There is less uncertainty when we read the bottom of the meniscus.

However, other liquids and other containers might be read differently. When mercury is in a glass container or when water is in a plastic container, the meniscus looks like that shown in Figure 2.13(b). In this case, most of the meniscus is filled with the liquid and there is less uncertainty when we read the *top* of the meniscus. The idea is to reduce uncertainty rather than follow a rule. When you are reading a volume, you should look at the meniscus and decide how to read the volume to get the least uncertainty in your measurement.

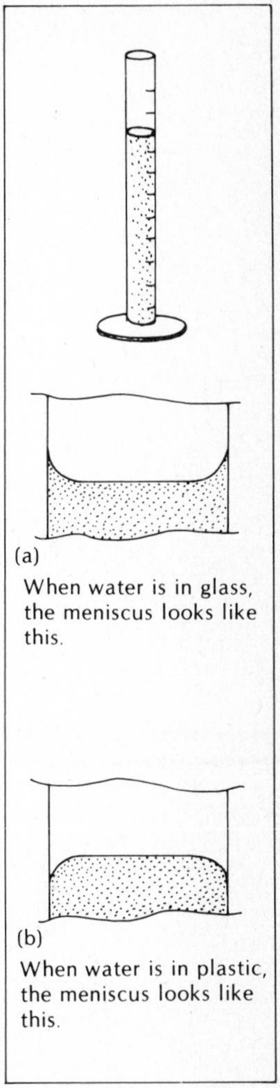

(a)
When water is in glass, the meniscus looks like this.

(b)
When water is in plastic, the meniscus looks like this.

Figure 2.13
The shape of the meniscus depends on the fluid and the material of which the container is made.

2.10 TEMPERATURE: AN INTENSIVE PROPERTY

We have talked about measuring length, volume, and mass. All of these measurements tell us something about the amount of material measured. For instance, 2 m of cloth contain more material than 1 m; 5 L of milk is more than 1 L; 1 kg of butter is a larger pile than 0.5 kg. Measurements that depend on the amount of matter present are said to measure **extensive properties.***

We make some measurements that do not depend on the amount of matter present. **Temperature** is one example. A thermometer in a *cup* of boiling water and a thermometer in a *gallon* of boiling water read the same. The temperature of a tiny chip of ice is the same as the temperature of the iceberg it is broken from. Temperature does not tell us anything about the amount of matter present.

We might think of temperature as a measure of "heat intensity." As such, it is an **intensive property.** Qualitatively, it is the amount of heat present in a given amount of matter. An example will help.

Identical burners on a kitchen stove produce about the same amount of heat in a given period of time. If you place a small pan of water on one burner and a large pan of water on another burner and heat them for a few minutes, the same amount of heat should be added to each pan of water. The same amount of energy has been added to each pan of water, but the temperature change in the two pans is *not* the same. The same amount of

* A property is an essential or distinctive attribute of something.

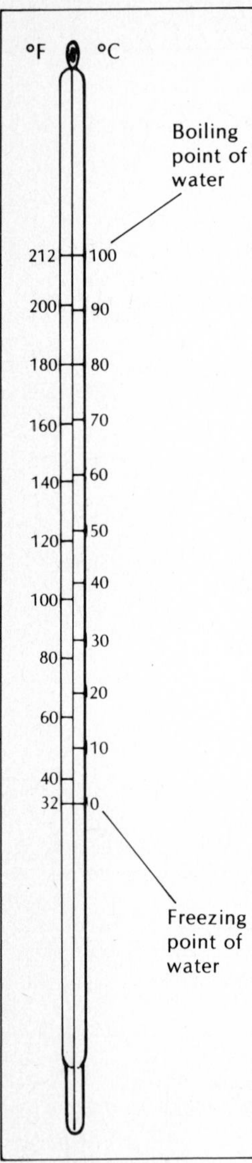

Figure 2.15
A thermometer that shows temperature in both Fahrenheit and Celsius.

Figure 2.14
When the same amount of heat is added to different amounts of water, the temperatures differ in the two pans.

energy has been added to different amounts of matter. The "heat intensity" or temperature of the water in the small pan will be greater than that in the large pan.

There are several scales for measuring temperature. You are accustomed to temperatures recorded in degrees **Fahrenheit** (F). The more commonly used scale is the **Celsius** (C) or centigrade scale. These two scales are compared in Figure 2.15. Scientists use the Celsius scale for most purposes, even though the official SI unit for temperature is the **Kelvin** (K). Kelvin temperature will be discussed in a later chapter.

2.11 HEAT: A FORM OF ENERGY

Energy is an idea, like love. And, like love, it is easier to recognize than to define. Most people recognize that the sun gives off energy in the form of light, that a stove gives off energy in the form of heat, that a bolt of lightning has electrical energy, and that a boulder rolling down a mountain has energy due to its motion. We might ask what these things have in common.

Physicists define **energy** as the ability to do work, but without a formal definition of work, this isn't much help. Most of us use more energy to play than to work!

I like to think of energy as what is required to do something to matter—to change it in some way. Light from the sun certainly can change matter. It causes green plants to grow and produce food. It causes your skin to redden and the pigment to turn brown. It causes the drapes to fade. Heat can change

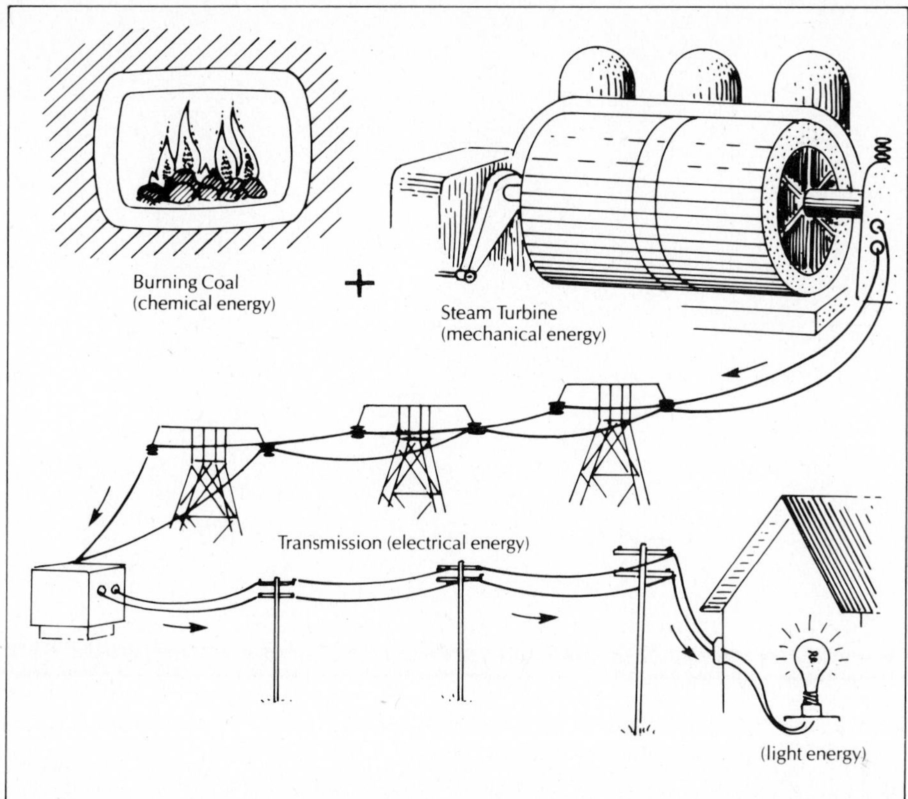

Figure 2.16 *Energy takes many forms. Here* chemical energy *in coal is changed to* heat energy *when the coal burns. The hot steam produces* mechanical energy *when it spins the turbine. The turbine spins a generator that produces* electrical energy. *The electrical energy can be transformed into other forms of energy, such as* light *from an electric lamp.*

matter too. It can burn toast, cook eggs, boil water, warm your body, or ignite paper. Lightning can split a tree, kill a cow, burn out your TV, or catch your house on fire. A rolling boulder can crush your car, move water, and, believe it or not, heat the water a little.

Within limits, it is possible to change one form of energy into another. Coal can be burned to produce heat, which boils water to form steam, which turns a turbine to produce electricity, which can be used to produce light. These forms of energy are measured in different units. The heat produced by the coal is normally measured in BTU (British thermal units) or *calories.* Mechanical energy is measured in *joules,* and electrical energy in *kilowatt-hours.* Energies expressed in one of these units can be converted to other units in the same way that centimeters can be converted to inches.

In the SI system, **joule** (J) is the accepted unit of energy, and it is recommended that all energies be expressed in this unit. A joule is not a great deal of energy. It is approximately the amount of energy lost when a flashlight battery falls a distance of 1 meter. It is the energy used by a 100-watt light bulb in 0.01 second—less time than it takes to flip the switch on and off. It is the amount of energy released when about 0.00002 gram of natural gas burns—less mass than can be detected by a good laboratory balance.

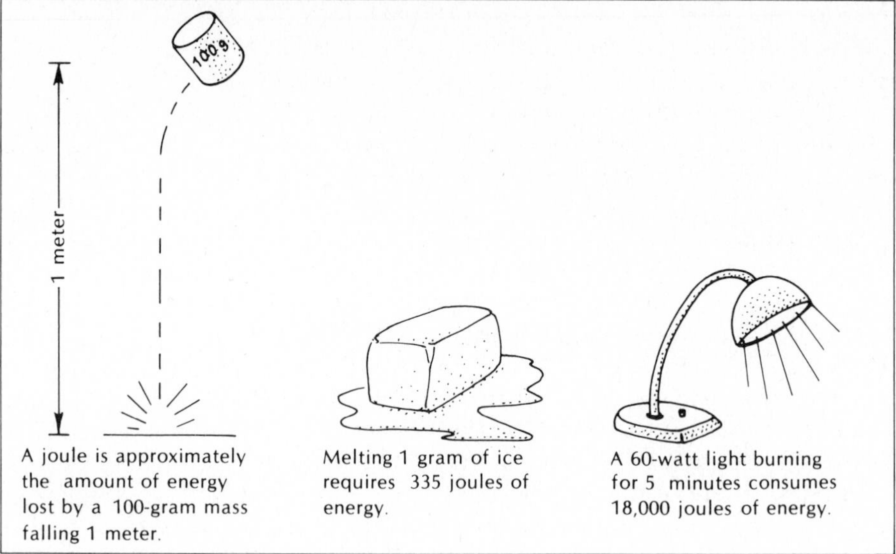

Figure 2.17 *The joules of energy involved in a few ordinary events.*

Traditionally, heat has been measured in units of calories. A **calorie** is 4.184 joule (J) and is the heat required to raise the temperature of 1 gram (g) of water 1 degree Celsius (C). Most people are more familiar with a kilocalorie. A kilocalorie or **Calorie** (spelled with a capital C) is the unit used when people "count calories" in an effort to lose weight.

Heat and temperature will be involved in many measurements that you make later on. We will not concern ourselves with those measurements now, but it is important that you understand the concepts of heat and temperature. Answering the following questions and checking your answers with those given at the end of the chapter should help you do that.

2-43. What is a comfortable room temperature?

2-44. What is normal body temperature?

2-45. An average teacup holds about 200 g of water. Make a rough guess at the energy required to heat water for a cup of tea.

2-46. People who "count calories" are trying to get an estimate of the energy released when food is "burned" in the body. A small peanut produces about 3 Calories. About how many joules of heat is that?

2-47. Which would be at a higher temperature, a cup of tap water or a tub full of tap water?

2-48. Which has more heat, ice or tap water?

2-49. Some liquids produce heat when they are mixed. When 25 mL of two liquids are poured together, the temperature rises 5 degrees. How much would the temperature rise when 50 mL of the two liquids were mixed?

2.12 DENSITY: ANOTHER INTENSIVE PROPERTY

Plastic and glass may look very much alike. Some plastics even feel a lot like glass, making it difficult to tell them apart. How can you tell whether a pair of eyeglasses are made of glass or plastic? Putting them into a fire to see if they burn would probably work, but it is hard on the glasses. Hitting them with a hammer or trying to scratch the lens with a nail might also work, but these techniques are equally bad.

We are often interested in finding out what kind of material an article is made of without ruining the article. One of the oldest techniques for doing this is to check the *density* of the material. You would probably express the notion of density by saying something like, "Plastic is lighter than glass." When you stop and think about it, this is a rather absurd statement. How heavy or how light something is depends on how much of it you have. (Mass is an extensive property.) A very large piece of plastic would certainly be heavier than a very small piece of glass. When you say that plastic is lighter than glass, you have in mind pieces that are of the same size. This notion of how heavy pieces of the same size would be, is what scientists call density. The easiest way to think about density is in terms of the mass of an item of a given size. **Density** is defined as the mass per unit volume of the material.

Density can be defined in almost any units we choose. For example, we could talk about the mass of one spoonful of sugar. If we measured the mass in grams, we could report the density in "grams per spoonful" (abbreviated as g/sf). If we measured the mass in pounds, we could report the density in "pounds per spoonful" (abbreviated as lb/sf). But just as mass could be measured in either grams or pounds, volume could be measured in units other than spoonfuls—cups, for example. If so, density might be represented as "grams per cup" (g/c) or "pounds per cup" (lb/c). Clearly, the number calculated for the density of an object depends on the units selected to measure the mass and volume. There will be hopeless confusion unless we agree on some particular units to use. To illustrate the need for common units, think about the following situation.

A student found a piece of metal. In an effort to identify it, she decided to check its density. She measured the volume and found that it was 10 cm^3. The metal weighed 27 g. Since the 10-cm^3 piece weighed 27 g, she knew that she could find the weight of 1 cm^3 by division.

2-50. What should the student calculate for the weight of 1 cm^3 of the metal?

On the basis of this simple calculation, the student reported that the metal had a density of 2.7 g/cm^3.

Another student found some metal in the same area and decided to calculate its density. However, he used different units. He measured volume in cubic inches and mass in ounces. The density of his metal was 1.424 oz/in³. Are the metals made of the same kind of matter? If their densities are the same, they may be, but since different units were used, the densities can't be compared directly.

2-51. Convert 1.424 oz/in³ to units of grams per cubic centimeter and see whether the two metals have the same density (1 oz = 31.1 g and 1 in³ = 16.4 cm³).

2-52. Measure the volume and mass of several pennies or nickels and find the density of the metal that these coins are made of.

You will notice that the density does not depend on how many pennies or nickels you use. It is an intensive property. Because intensive properties do not depend on the amount of matter present, they are often used to identify substances. Some of the intensive properties used to identify matter are density, melting point, boiling point, and color. Color is the least reliable, because it changes when some materials are melted or ground into a fine powder.

2.13 USING RELATIONSHIPS

You have seen that density can be useful in talking about matter. The old saw, "Which is heavier, a pound of lead or a pound of feathers?" would catch nobody off guard if density were used in everyday speech. Everyone recognizes that lead is more dense than feathers, while a pound is a pound is a pound. The trouble is that we talk about "heaviness" in everyday speech when we should be saying "density."

Density is useful for purposes other than description. It defines a relationship between two measures and can be used to find one measure when the other is known. Using relationships in this way is not new to you. Earlier in this chapter, we discussed the relationship between length measured in inches and length measured in centimeters. We expressed that relationship in several ways:

2.54 cm = 1 in. which is read, "2.54 centimeters are equal to 1 inch."

$$\frac{2.54\text{ cm}}{1\text{ in.}}$$ which is read, "2.54 centimeters per inch" and means that there are 2.54 centimeters for every inch, or that 2.54 cm corresponds to 1 in.

$$\frac{1\text{ in.}}{2.54\text{ cm}}$$ which is read, "1 inch per 2.54 centimeters" and still means that 1 in. corresponds to 2.54 cm.

Figure 2.18
A pound of lead and a pound of feathers weigh the same, but they don't take up the same amount of space. Lead is more dense than feathers.

Relationships of this kind can be used to change from a quantity that we know to an equivalent quantity in other units. A length given in inches can be changed to an equivalent length in centimeters, or vice versa.

Prices represent similar relationships. A price of $1.70 per pound of hamburger describes a proportional relationship between dollars and pounds. It can be expressed in the same way that we expressed the relationship between centimeters and inches.

$\$1.70 = 1 \text{ lb}$ which is not a true equality (dollars and pounds are not the same thing) but a logical "trade" equivalence. It says that, for trading purposes, $1.70 is equivalent to 1 lb of this meat. Hence it is the unit price.

$\frac{\$1.70}{1 \text{ lb}}$ which is read, "one dollar and seventy cents per pound" and means that you will pay $1.70 for every pound of meat you buy.

$\frac{1 \text{ lb}}{\$1.70}$ which is read, "one pound per one dollar and seventy cents," and still means that you get one pound of meat for each $1.70 you pay.

If you know the price of meat and the weight of a package, you can find the cost of the package. Just to review how it is done, work the following problems.

2-53. What would you pay for 3.50 lb of meat that is priced at $1.70/lb?
2-54. Steak is priced at $2.65/lb and roast sells at $1.95/lb. How much would a 2.43-lb roast cost?

The relationship between cost and amount of a product can be used in other ways, too. If you know the unit price and the total cost, you can find the amount in a package. This is illustrated in the following problems. If you have difficulty, turn to the solutions given in the answer section for this chapter.

2-55. How many pounds of steak can you buy for $10.00 if it sells for $2.65/lb?
2-56. A package of roast selling at $1.95/lb has a total cost of $7.84. A package of steak costs $11.36, and the steak sells for $2.65/lb. How much does the roast weigh?

Finally, you can find the unit cost if you know the total price and the weight. This is what people do when they do comparison shopping. Find the prices in the following problems.

2-57. If a 16-oz can of fruit sells for 72¢, how much does it cost per ounce?
2-58. A 17-oz can of fruit cocktail sells for 40¢, and a 29-oz can of peaches costs 60¢. Which fruit has the lower cost per ounce?

Just as prices at a store represent a relationship between two quantities, density represents a relationship between two quantities. The following questions illustrate that similarity. Work through them carefully and you will begin to see how useful proportional relationships such as density can be.

2-59. Mercury has a density of 13.6 g/cm^3. List three ways of expressing this relationship, and write down an English translation of what the mathematical statement means. (See the expressions for the relationship between centimeters and inches on p. 54 and the expressions for unit prices on p. 55.)
2-60. You know that 2.5 cm^3 of glass weigh 4.25 g. What is the density of the glass?
2-61. Ordinary table salt has a density of 2.16 g/cm^3. How much does 20.0 cm^3 of table salt weigh?
2-62. Borax has a density of 1.73 g/cm^3. What is the mass of a cube of borax measuring 2.54 cm on a side? (The answer will be the mass of 1 in^3 of borax and will represent the density of borax in units of grams per cubic inch.)
2-63. Which is heavier, 40.0 cm^3 of table salt or 20.0 in^3 of borax?

2-64. What volume of table salt weighs 15.4 g?

2-65. At normal room temperature and normal atmospheric pressure, air has a density of 0.001184 g/cm³. How much does the air in a laboratory weigh if the room measures 7 m × 10 m × 3 m?

2-66. What is the volume of 1.0 g of air?

2-67. Gold has a density of 19.3 g/cm³. A cube of metal that looks like gold measures 2.00 cm on a side and weighs 125 g. Is it gold?

2-68. What volume is occupied by 1.00 kg of gold?

2.14 NONPROPORTIONAL RELATIONSHIPS

The relationships we discussed in the last section are all **proportional** relationships—that is, they are related by multiplication. If any of these relationships is multiplied or divided by another number, the new relationship is just as true as the old one. Since we know that 2.54 cm = 1 in.,

we can say 5.08 cm = 2 in. by multiplying by 2,

or we can say 12.7 cm = 5 in. by multiplying by 5.

Not all relationships are proportional. There is a relationship between my age and my son's age, but it is not a proportional one. In 1980 I was 44 and my son was 22. I was twice as old as my son then, but I have not always been twice his age. In 1960 I was 24 and he was only 2. Then I was 12 times his age. There is a constant difference between our ages, so there is an *additive* relationship between our ages rather than a multiplicative one. I will always be 22 years older than my son—my age at the time he was born.

You must be sure that you know the kind of relationship between two quantities, or you will make needless errors.

2.54 cm = 1 in.	We can multiply both sides of this equality by 2 and get a new relationship that is equally true.
5.08 cm = 2 in.	This new relationship can be obtained by multiplication *only* because the relationship is proportional. Try it with a relationship that is *not* proportional and you get into trouble.

Consider the following equivalent temperatures:

100°C = 373 K	Both of these temperatures represent the normal boiling point of water. But if we multiply both sides by 2, we get a new relationship that is *not* true.
200°C ≠ 746 K	Celsius and Kelvin temperatures are *not* proportional. They are related by *addition* rather than multiplication: 200°C = 473 K.

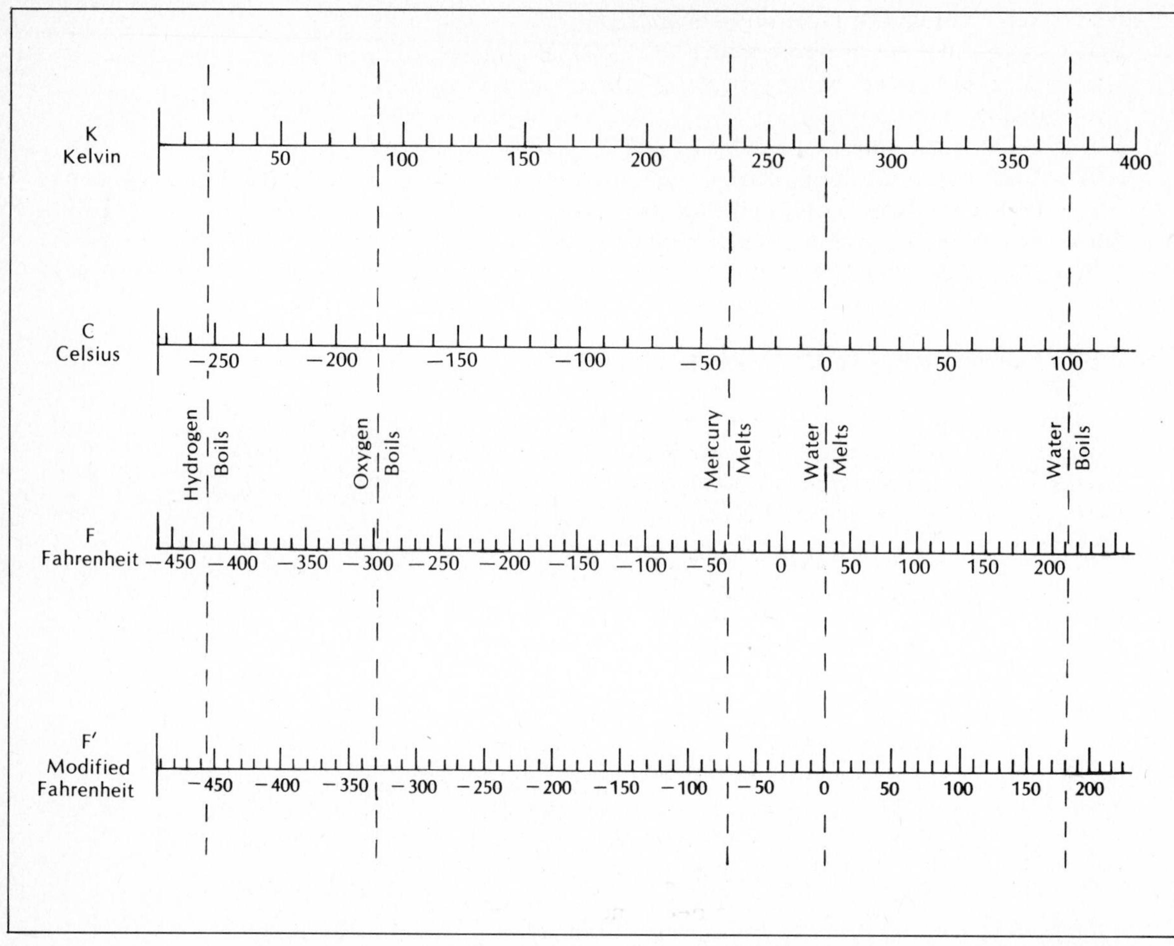

Figure 2.19
A comparison of several temperature scales. The melting point (or freezing point) of water is 273 K, 0°C, and 32°F. Note that the divisions on the Kelvin scale and the Celsius scale are the same size. A change of 1°C and a change of 1 K represent the same change in temperature. Also note that the divisions on the F scale and the F′ scale are the same size.

Figure 2.19 shows the scales used to measure temperatures in Kelvin, Celsius, and Fahrenheit. Note that the sizes of the Celsius and Kelvin degrees are identical. The only difference between them is the point on the scale that is called zero. To convert between Celsius and Kelvin, we either add or subtract 273.*

* The difference between Celsius and Kelvin temperatures is 273 to the nearest degree. More precisely, the difference is 273.16.

Now let's compare the Celsius and Fahrenheit scales. Between these scales there are *both* an additive relationship and a proportional relationship. The size of the degrees is not the same on the two scales, and neither is the temperature described as zero. On the Celsius scale, the freezing point of pure water is called zero. On the Fahrenheit scale, the freezing point of a salt and water mixture is called zero.

A fourth scale labeled F′ is shown in Figure 2.19. It is identical to the Fahrenheit scale except that 32 has been subtracted from all Fahrenheit values so that the freezing point of water is now labeled zero. This removes the additive relationship between Celsius and Fahrenheit; the new F′ scale is proportional to the Celsius scale.

2-69. Pick a temperature on the Celsius scale of Figure 2.19 and set it equal to the corresponding temperature on the F′ scale. Now multiply by 2 or some other small number. Is the new "equality" true? Is the C scale proportional to the F′ scale? Try the same thing with the C and F scales.

2-70. What is the relationship between degrees on the C scale and degrees on the F′ scale? In other words, how many Fahrenheit degrees correspond to a Celsius degree?

2.15 STILL LARGER AND SMALLER NUMBERS

Chapter 1 began with a discussion of large numbers and small numbers. You counted a million and found a millionth. In the process, you learned to make several kinds of measurements, and you have now learned to put some of those measurements together to define new properties such as density. We now return to numbers large and small.

One million is a rather large number and a millionth (0.000001) is a rather small number, but you will encounter numbers much larger and much smaller in the study of science. Chapter 1 mentioned a light year, about 6,000,000,000,000 mi, and the size of a single atom, 0.000000001 in. in diameter. These numbers are much larger than a million and much smaller than a millionth. You need some idea of just how much larger or smaller.

In naming numbers, a billion is the name that follows a million. Let's do some simple calculations to get a better idea of how much larger it is. To do this, go back to the calculation that you did in counting a million of something and extend your work to find the mass or the space occupied by a billion of those objects. If a million grains of sugar weigh 5 lb, how much will a billion grains weigh? If there are a million bricks in a single building on campus, how many college campuses would it take to have a billion bricks in all their buildings? As an example, I will do the calculation for the rice.

The weight of 1 million grains of rice is 17,600 g. The weight of 1 billion grains is

$$\frac{17{,}600\ \text{g}^*}{1{,}000{,}000\ \cancel{\text{grains}}} \times 1{,}000{,}000{,}000\ \cancel{\text{grains}} = 17{,}600{,}000\ \text{g}$$

By comparison, the weight of an automobile is about 1,800,000 g. Hence 1,000,000,000 grains of rice would weigh about the same as 10 automobiles—not an inconceivable number, but clearly much more than the 38.8 lb we calculated for the weight of 1 million grains.

Now let's make a huge jump in size. Repeat your calculation to find the mass or the space occupied by 602,000,000,000,000,000,000,000 of the objects that you counted. Why *this* number? Later you will see that this is a number of major importance in chemistry. You need some conception of just how enormous this number is. Once again, I will do some calculations for this many grains of rice.

First, how many piles of 1 million grains would we have?

$$602{,}000{,}000{,}000{,}000{,}000{,}000{,}000\ \cancel{\text{grains}} \times \frac{1\ \text{pile}}{1{,}000{,}000\ \cancel{\text{grains}}} = 602{,}000{,}000{,}000{,}000{,}000\ \text{piles}$$

And how much would this weigh?

$$\frac{17{,}600\ \text{g}}{1\ \cancel{\text{pile}}} \times 602{,}000{,}000{,}000{,}000{,}000\ \cancel{\text{piles}} = 10{,}600{,}000{,}000{,}000{,}000{,}000{,}000\ \text{g}$$

Without some comparisons, this mass is inconceivable. How many automobiles would weigh this much?

$$10{,}600{,}000{,}000{,}000{,}000{,}000{,}000\ \cancel{\text{g}} \times \frac{1\ \text{auto}}{1{,}800{,}000\ \cancel{\text{g}}} = 5{,}900{,}000{,}000{,}000{,}000\ \text{autos}$$

This doesn't help either, but knowing that there are about 4,000,000,000 people on the face of the earth, we can calculate the number of automobiles that would mean for each person.

* The units are actually

$$\frac{17{,}600\ \text{grams of rice}}{1{,}000{,}000\ \text{grains of rice}}$$

However, lengthy units are awkward. So long as we keep in mind that we are talking about rice, the shortened units are less confusing.

$$\frac{5{,}900{,}000{,}000{,}000{,}000 \text{ autos}}{4{,}000{,}000{,}000 \text{ people}} = 1{,}500{,}000 \text{ autos/person}$$

Not knowing what 1.5 million cars would look like in my garage, I still do not find this figure very helpful. Let us try another comparison. In 1976 the annual rice production was 320,000,000 metric tonn or 320,000,000,000,000 g. At that rate of production, how many years would it take to grow all that rice?

$$10{,}600{,}000{,}000{,}000{,}000{,}000{,}000 \not{\text{g}} \times \frac{1 \text{ yr}}{320{,}000{,}000{,}000{,}000 \not{\text{g}}} = 33{,}000{,}000 \text{ yr}$$

In other words, the amount of rice in 602,000,000,000,000,000,000,000 grains is about the same as *all* the rice that could be produced in about *33 million years* if production were at the level of 1976. It seems safe to say that we are talking about more rice than has ever been produced on the face of the earth!

Let us do one more calculation. How much space would be occupied by 602,000,000,000,000,000,000,000 grains of rice? To answer this question, we need to know the space occupied by 1 g of rice. I don't know, but since I know that rice is primarily starch, we should get a good estimate if we assume that 1 g of rice occupies the same amount of space as 1 g of starch. This is about 0.654 cm^3. Then we find that

$$10{,}600{,}000{,}000{,}000{,}000{,}000{,}000 \not{\text{g}} \times \frac{0.654 \text{ cm}^3}{1 \not{\text{g}}} = 6{,}930{,}000{,}000{,}000{,}000{,}000{,}000 \text{ cm}^3$$

Once again, we have difficulty visualizing a volume this great. Imagine a huge cube, measuring 19,100,000 cm on each side. (This number is the cube root of 6,930,000,000,000,000,000,000.) Perhaps it would help if we converted this length to more familiar English units.

$$19{,}100{,}000 \not{\text{cm}} \times \frac{1 \not{\text{in.}}}{2.54 \not{\text{cm}}} \times \frac{1 \not{\text{ft}}}{12 \not{\text{in.}}} \times \frac{1 \text{ mi}}{5280 \not{\text{ft}}} = 119 \text{ mi}$$

This helps a little. The cube would have an edge about as long as the road from my home in Lafayette, Indiana, to Chicago, Illinois. If this much rice were spread over the continental United States, it would cover the surface of the country to a depth of *over half a mile!*

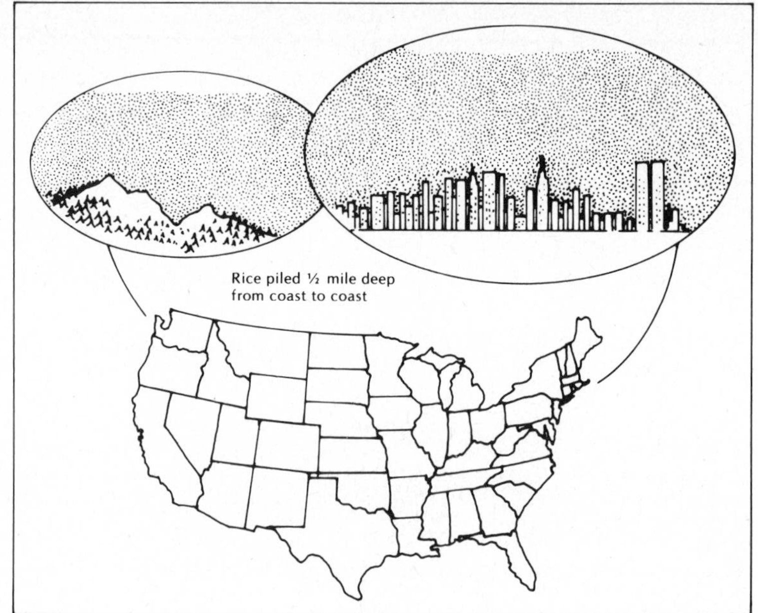

Figure 2.20
If 602,000,000,000,-000,000,000,000 grains of rice were spread over the continental United States, the pile would be over half a mile deep!

This exercise in arithmetic was intended to convince you of two things. First, even though 1 million is a rather large number, it is extremely small compared to some of the numbers you will encounter in chemistry. Second, it is very awkward to do arithmetic with such large numbers. We can't do a thing about the first problem, but we *can* do something about the second.

2.16 USING SCIENTIFIC NOTATION

If you did the arithmetic suggested in the last section for the objects that you counted, you must have grown tired of writing zeros. If you did not use a calculator and went through the entire process without losing a zero or two, you were very fortunate. It is difficult to work with very large or very small numbers when they are written in the ordinary way. Since most of the numbers that we use are known only to three or four significant digits, it isn't necessary to write these long numbers. We can express them in scientific notation.

Scientific notation is a special form of **exponential notation** in which a number is written as the product of a number between 1 and 10 times some power of 10. Some of you are familiar with exponential notation. Those who are not should *study* Appendix C before going any further. If you have difficulty with this section, *review* Appendix C even if you have studied exponents before.

We will work through many of the calculations carried out in the last section. The only difference is that we will express numbers in scientific notation. As you have done before, you should work the problems with me.

Get a pencil and paper. Move the paper down the page to the first dotted line. Read up to that point and follow the instructions before moving the paper to the next dotted line. In this way, you can practice calculations in scientific notation and be sure that you know how to do them.

Example 2.5

The first calculation in the last section was to learn the weight of 1 billion grains of rice. It was set up like this:

$$\frac{17{,}600 \text{ g}}{1{,}000{,}000 \text{ grains}} \times 1{,}000{,}000{,}000 \text{ grains} =$$

Rewrite the problem, expressing all numbers in scientific notation.

- -

$$\frac{1.76 \times 10^4 \text{ g}}{10^6 \text{ grains}} \times 10^9 \text{ grains} = ?$$

(You could have written 1×10^6 and 1×10^9, but the 1 is not needed.) Now solve the problem.

- -

$$1.76 \times 10^7 \text{ g}$$

It is easiest to deal with the exponents first. (Exponents are added together in order to multiply and subtracted in order to divide.) I began by multiplying the 10^4 and 10^9 found in the numerator to get 10^{13}. Next, I divided 10^{13} by the 10^6 found in the denominator to get 10^7. This is still multiplied by 1.76, as indicated by the final answer of 1.76×10^7. The number is left in this form.

The next problem worked in the last section was to find the number of piles containing 1 million grains that could be obtained from 602,000,000,000,000,000,000,000 grains of rice.

$$602{,}000{,}000{,}000{,}000{,}000{,}000{,}000 \text{ grains} \times \frac{1 \text{ pile}}{1{,}000{,}000 \text{ grains}} = ?$$

Express the numbers in scientific notation and then work the problem.

- -

$$6.02 \times 10^{23} \cancel{\text{grains}} \times \frac{1 \text{ pile}}{10^6 \cancel{\text{grains}}} = 6.02 \times 10^{17} \text{ piles}$$

In the next calculation, we found the mass of 6.02×10^{23} grains of rice.

$$\frac{1.76 \times 10^4 \text{ g}}{1 \text{ pile}} \times 6.02 \times 10^{17} \text{ piles} = ?$$

$$1.06 \times 10^{22}\ \text{g}$$

I worked with the exponents first. Multiplying 10^4 by 10^{17}, I got 10^{21}. Next I multiplied 1.76 by 6.02 to get 10.6. Then the answer is 10.6×10^{21}. This answer is correct, but it is standard practice to write the number as a product of a number between 1 and 10 and some power of 10. Converting to scientific notation, we get an answer of 1.06×10^{22}.

The final calculation that you were asked to do was to show the volume of 6.02×10^{23} grains of rice.

$$1.06 \times 10^{22}\ \text{g} \times \frac{0.654\ \text{cm}^3}{1\ \text{g}} = ?$$

$$6.93 \times 10^{21}\ \text{cm}^3$$

The only arithmetic here is multiplying 1.06 by 0.654 to get 0.693. When the answer of 0.693×10^{22} is converted to scientific notation, it reads 6.93×10^{21}.

Some students miss the last part of the example. They ask, "Why did you multiply the 1.06 by 0.654 but not multiply 10^{22} by 0.654?" If this is a question in your mind, work through Example 2.6 with me.

Example 2.6

Do the following multiplication:

$$(2 \times 3) \times 4 = ?$$

24

Most people would work this problem by multiplying 2 by 3 to get 6 and then multiplying 6 by 4 to get 24. However, you could multiply *either* 2 *or* 3 by the 4 first. You would *not* multiply both the 2 *and* the 3 by 4. If you did, you would get an incorrect answer.

The arithmetic is the same in the following problem:

$$(2 \times 10^2) \times 4 = ?$$

$$8 \times 10^2$$

Although the order of multiplication does not matter here any more than in the foregoing problem, the simple procedure is to multiply 2 by 4 and then multiply the 8 by 10^2.

Students sometimes have difficulty *adding* exponential numbers. Do the following example with me to see whether you have this problem. If you do, study the last part of Appendix C.

Example 2.7

Add 2×10^3 and 3×10^4

$$3.2 \times 10^4$$

Unless a calculator is used for this problem, the addition cannot be done until one of the numbers is changed to the same power of 10 as the other number. The following solutions are equivalent.

$$(0.2 \times 10^4) + (3 \times 10^4) = 3.2 \times 10^4$$

or

$$(2 \times 10^3) + (30 \times 10^3) = 32 \times 10^3 = 3.2 \times 10^4$$

Addition of exponential numbers is discussed in Appendix C.

2.17 REDUCING UNCERTAINTY

In examining large and small numbers, we have introduced skills you need to do careful, scientific work. Many of these are basic math skills, while others deal with measurement. We have tried to present these skills so that it is clear *why* the skills are necessary. Contrary to what some students believe, skills that are taught in chemistry are not presented in an effort to confuse them or make life difficult. Unfortunately, the need for the skill isn't always clear, and the student leaves the beginning course wondering, "Why did they make me do all those weird things?" Even now we have done some things that we have not explained. Perhaps it would help to discuss them.

When I tried to find the amount of rice that would contain 1 million grains, I counted out 300 grains and weighed them. Why? Why didn't I just weigh one grain? I certainly could have. Most of you know at least one of the reasons. Not all grains of rice are the same size. If I weighed only one grain and it was a very large one, my estimate for the weight of 1 million grains would have been too large. If I weighed a very small grain, my estimate for the weight of a million grains would have been too small. The magnitude of the possible errors can be seen in Table 2.2.

The difference between the mass of the smallest grain and the mass of the largest grain is 0.0057 g, or about 30 percent of the average mass. In selecting 300 grains of rice, weighing them, and dividing by 300 to get the average mass of one grain, I made an assumption. I assumed that the 300 grains selected would be representative of all grains of rice. If I was careful

Table 2.2 Mass of Different Grains of Rice

Mass	
0.0171 g	
0.0186 g	
0.0154 g	
0.0151 g	
0.0186 g	Average mass = 0.0179 g
0.0149 g	
0.0206 g	
0.0206 g	
0.0174 g	
0.0203 g	

in my selection, it was probably a reasonable assumption. If I was not careful, I am probably in error. The point is that, although I wanted to say something about *all* grains of rice, I did my measurements on only 300. Unless the grains in my pile of 300 were like the grains of rice in the larger pile from which they were taken, my calculations were misleading.

Sampling is an important consideration. If you do an experiment in the laboratory, you want to say something about what will happen when *any* similar experiment is done. But because you are human and make human errors, you may have inadvertently done something to affect the result of the experiment. If you draw conclusions based on that one experiment, your conclusions may be wrong. For this reason, scientists like to repeat their experiments several times and have several people do the same experiment. We believe that unintentional errors will "average out" and help us to draw better conclusions. As you do experiments in the laboratory or make measurements, you would be wise to follow the same procedure.

Many students ask, "How many times must I repeat an experiment to have confidence in the result?" or "How many grains of rice are enough to count?" There is no simple answer. Students get the impression that science is always exact. It is not. Students believe that scientists know things exactly. They do not. Students believe that scientists never need to make value judgments. This is not true. In counting 300 grains of rice, I made a judgment that 300 would be enough to give a good estimate of the average weight of one grain. I could have counted more, but doing so would have increased the chance of miscounting. I could have taken 10 grains of rice, weighed them individually, and calculated the average, as shown in Table 2.2. The resulting estimate would have been a good one, but it would not be identical to the one I obtained by weighing 300 grains. The average weight of the 10 grains was 0.0179 g; the average of the 300 grains was 0.0176 g. Had I counted 1,000 grains, I would probably have found a slightly different average. Still, the values would almost certainly disagree only in the third significant digit, the one known to be uncertain.

We sometimes face another problem of measurement: limitations in our instruments. For example, if I tried to weigh a single grain of rice on the triple beam balance shown in Figure 2.1, I would find it impossible. The balance is accurate only to about plus or minus 0.02 g, and that is all one grain of rice weighs! By weighing 10 grains at a time, I get an average value of 0.016 g—not really very good, but much better than I could get by trying to weigh 1 grain at a time. If I weigh as many as 300 grains of rice on the triple beam balance, I should be able to get an estimate that is certain in the first two places.

When I weighed 300 grains of rice on a triple beam balance, I got a reading of 5.35 g. (The same 300 grains weighed 5.28 g on an analytical balance. The difference between the readings on the two balances is much greater than the 0.02 g expected. This is not too surprising. The balances used in general chemistry receive heavy use *and* abuse. Rough use causes them to provide less precision than they are designed for.) Still, when we divide this value by 300, the answer is 0.0178 g, which is very close to our original estimate of 0.0176 g. This technique of taking large samples and finding an average value often enables us to get very accurate estimates of measurements that cannot be made directly with great precision.

If you try to measure the thickness of this page using a centimeter rule, your answer will be neither precise nor accurate. However, if you measure the thickness of all pages in the book and divide by the number of pages, your estimate should be very precise. Try it. Record your answer to the proper number of significant digits and compare it with the answers obtained by others in your class. Then answer the following questions.

2-71. What is the thickness of this page in meters? Express the answer in scientific notation and to the correct number of significant digits.

2-72. Compare your answer to the answers obtained by others in the class. Do they all agree except for the last (uncertain) digit?

2-73. If the answers seem to contain more uncertainty than indicated by the significant digits recorded, what could have caused that uncertainty? (You might want to ask others just how they did the measurement. Differences in procedure often make a difference in results.)

2-74. Which would you consider to be the best estimate of the thickness of this page, your value or an average of all values calculated by the class? Why?

2-75. Did you see any values that you would like to ignore because they are obviously wrong? If so, how did you decide they were wrong?

2.18 SUMMARY

Much of this chapter has dealt with the fact that we are unable to make exact measurements; the measurements are *uncertain.* You have learned some of the reasons for uncertainty in measurements and you have learned to use *significant digits* to express the uncertainty of a measurement.

In addition to learning that measurements are uncertain, you have studied some new units of measurement. *Heat,* a form of energy, was discussed and both *calories* and *joules* were introduced as units of heat. The amount of heat in a substance depends on the amount of substance present; it is an *extensive* property. In contrast, you saw that *temperature* does not depend on the amount of substance present and is an *intensive* property.

Another property discussed in the chapter is *volume,* an extensive property that can be measured in "cubic units" such as a cubic centimeter (cm^3) or "container units" such as a liter (L).

Volume can be combined with mass to define another intensive property, *density.* Density, you learned, can be used like a unit-factor to obtain the mass of an object when you know its volume, or to find the volume of an object when you know its mass. Density is also used to identify unknown substances.

Finally, as we continued to explore still larger and smaller numbers, *scientific notation* was introduced as a means of expressing very large or very small numbers. By now we have developed most of the tools needed to observe matter carefully. In the following chapter you will begin to make those observations and see what chemistry is all about.

Questions and Problems

2-76. The following numbers represent measurements made in the laboratory and properly recorded to indicate the uncertainty of the measurement. Circle the uncertain digit(s) in each number.

a. 21.35 cm
b. 8.705 g
c. 121.2000 g
d. 0.000823 kg
e. 27,000 mm
f. 0.0910 m
g. 30 mg
h. 38,002 cm
i. 2.50 in.
j. 96,000,000 mi

2-77. For each of the numbers in Problem 2-76, indicate the number of significant digits represented.

2-78. The following problems involve calculations with numbers resulting from measurements recorded to the proper number of significant digits. Round the answers to the proper number of significant digits.

a. $2.86 \times 1.824 = 5.21664$
b. $21 \div 8 = 2.625$
c. $3.7 + 1.86 + 0.0024 = 5.5624$
d. $98.0 \times 1.22 = 119.56$
e. $2.1 + 1.0 = 3.1$
f. $56 - 3.47 = 52.53$
g. $4.6215 \div 0.0015 = 3081$
h. $10.00 \div 2.000 = 5$
i. $6.00 - 1.000 = 5$
j. $87.1 \times 21.6457 \times 0.020 = 37.7068094$

2-79. How many significant digits are in the number underlined? If there is an infinite number, explain why.

a. 1 gross = $\underline{144}$
b. my height = $\underline{69.7}$ in.
c. 1 km = $\underline{1000}$ m
d. 1 yr = $\underline{365}$ days
e. 1 yd = $\underline{36}$ in.
f. body temperature = $\underline{37}$°C
g. $\sqrt{2} = \underline{1.414}$
h. $\underline{\sqrt{2}} = 1.414$

2-80. Convert the following measurements to the indicated units.

a. ? mL = 2.13 L
b. ? mL = 2.13 m^3
c. ? cm^3 = 256 mL
d. ? L = 356 cm^3
e. ? m^3 = 498 L

2-81. Convert the following measurements to the indicated units.

a. ? K = 27°C
b. ? °C = 140 K
c. ? °C = 0 K
d. ? K = 100°C

2-82. The following table lists the mass, volume, and density of several objects. One of the three measures is missing for each object. Supply the missing values.

	Mass	*Volume*	*Density*
a.	45.2 g	8.4 cm^3	______
b.	______	13.3 mL	4.3 g/cm^3
c.	2,356 g	______	4.3 g/cm^3
d.	1.5 kg	0.283 m^3	______
e.	1.5 kg	______	13.6 g/cm^3

2-83. In Problem 2-82 and throughout this chapter, we have calculated density in units of grams per cubic centimeter. These are the units commonly used by chemists to express the density of common solids and liquids, because the numerical value turns out to be a convenient size—for most substances, it is between 1 and 10. In spite of this, the units for density in the SI system are kilograms per cubic meter. Convert the densities in Problem 2-82 to the correct SI units. How do the numerical values in units of kilograms per cubic meter compare to the values in units of grams per cubic centimeter?

2-84. Convert *all* numbers shown in Problem 2-78 to exponential notation.

2-85. Do the following calculations, and record your answer in *both* exponential and decimal notation. Record the answer to the proper number of significant digits.

a. $(7.24 \times 10^5) \times (1.62 \times 10^{-3}) =$
b. $(1.87 \times 10^5) - (2.75 \times 10^4) =$
c. $(3.456 \times 10^4) + (5.46 \times 10^4) =$
d. $(6.02 \times 10^{23}) \div (3 \times 10^{19}) =$
e. $(9.99 \times 10^{-3}) \div (3.33 \times 10^{-5}) =$
f. $(5.25 \times 10^{64}) \times (5 \times 10^{-62}) =$

More Challenging Problems

2-86. Convert the following measurements to the indicated units.

a. ? m^3 = 1.75×10^9 μL
b. ? g/cm^3 = 6.74 lb/ft^3
c. $tonn/m^3$ = 1.32 tn/yd^3

2-87. Derive a formula that could be used to convert from Kelvin to Fahrenheit temperature. Use it to find the Fahrenheit equivalent of 0 K.

2-88. Cement has a density of about 3 g/cm^3 when it has set. How many pounds would a cubic yard of cement weigh? How many cubic feet of cement would you be able to lift?

2-89. A rock with a volume of 1.2×10^{-4} ft^3 has a mass of 2.4×10^{-2} lb. Could this rock be diamond?

2-90. Three different people weighed the same rock on three different balances. The values they obtained were 34.2 g, 34.18 g and 34.1694 g. What is the best value to use for the mass of the rock? Why?

2-91. Three people weighed the same rock and obtained values of 34.2 g, 35.0 g, and 34.6 g. What is the best estimate for the weight of the rock?

Answers to Questions in Chapter 2

2-1. The trip scale, which weighs to the nearest gram

2-2. All that you know

2-3. Only one

2-4. The 4 and the 3 are certain. The 2 is uncertain.

2-5. No. The values 4.32 and 4.31 differ only in the digit that is estimated. We can't expect everyone to make the same estimate.

2-6. Yes. The answers differ in the tenths place, which is not an estimate. It should be a certain digit. Any two people who make the measurement should agree on the number of tenths in the answer.

2-7. 28 cm^2 (We round to two digits because 1.3 has only two digits. We round to 28 because 27.69 is closer to 28 than to 27.)

2-8. 16 (The answer has no units because cm ÷ cm gives 1.) 16 × 1 = 16

2-9. 6.3 cm (6.2514379 is closer to 6.3 than to 6.2. The units are cm because (cm^2 × cm)/cm^2 = cm.)

2-10. 812 m^2

2-11. 29.2 m^3

2-12. 136.9

2-13. You may say this in several ways. I say it like this: In addition or subtraction, round the answer to the largest place value (first column from the left) that contains an uncertain digit.

2-14. 97.9 (Both the tenths place and the hundredths place are uncertain. Therefore, we round to the nearest tenth.)

2-15. 37 (The units place is uncertain in 124.)

2-16. 6.0 (The tenths place is uncertain. I drop a 5 after an even digit. Zero is considered even.) 6.1 is acceptable.

2-17. 3232 (The units place is uncertain. I round up when a 5 follows an odd number.) 3231 is acceptable.

2-18. 3

2-19. 5

2-20. 3 (The final zero is assumed to be significant.)

2-21. 3

2-22. 2 (Leading zeros are used to place the decimal and are not considered significant.)

2-23. 2 (The initial zeros are used to place the decimal, but the final zero is not. The final zero is a significant digit.)

2-24. 5 (All zeros located between nonzero digits are significant.)

2-25. 6

2-26. $? \text{ g} = 0.45 \cancel{\text{kg}} \times 1000 \text{ g}/1 \cancel{\text{kg}} = 450 \text{ g}$

2-27. $? \ \mu\text{m} = 21.50 \cancel{\text{cm}} \times \dfrac{1 \cancel{\text{m}}}{100 \cancel{\text{cm}}} \times \dfrac{1{,}000{,}000 \ \mu\text{m}}{1 \cancel{\text{m}}}$

$= 215{,}000 \ \mu\text{m}$

2-28. $? \text{ cm}^3 = 1 \cancel{\text{L}} \times \dfrac{1 \cancel{\text{dm}^3}}{1 \cancel{\text{L}}} \times \dfrac{10 \text{ cm} \times 10 \text{ cm} \times 10 \text{ cm}}{1 \cancel{\text{dm}^3}}$

$= 1{,}000 \text{ cm}^3$

2-29. kiloliter

2-30. hectoliter

2-31. decaliter

2-32. deciliter

2-33. centiliter

2-34. milliliter

2-35. $? \text{ cm}^3 = 1 \text{ m}^3 \times \dfrac{100 \text{ cm} \times 100 \text{ cm} \times 100 \text{ cm}}{1 \text{ m} \times 1 \text{ m} \times 1 \text{ m}}$

$= 1{,}000{,}000 \text{ cm}^3$

2-36. 1

2-37. 100,000

2-38. 0.01

2-39. 1,000

2-40. 0.0001

2-41. 10

2-42. 0.000001 (a millionth)

2-43. Most people consider a temperature of 18°C to 24°C comfortable. If you don't know what this is in Fahrenheit degrees, look at Figure 2.15, where the Celsius and Fahrenheit scales are compared.

2-44. 37°C

2-45. About 70,000 J. This assumes that tap water is at a temperature of 16°C and that it is heated to 100°C. The temperature change would be 84 Celsius degrees and it takes 4.184 J to raise the temperature of each gram of water 1 Celsius degree. The calculations are

$$? \text{ J} = 200 \cancel{\text{g}} \times \frac{4.184 \text{ J}}{\cancel{\text{g}} \cancel{\text{deg}}} \times 84 \cancel{\text{deg}} = 70{,}000 \text{ J}$$

2-46. 10,000 J. Remember that Calorie spelled with a capital C refers to the "food Calorie," which is actually a kilocalorie. Three kilocalories would be 12,552 J. Rounded to one significant digit, the answer is 10,000 J.

2-47. They should be the same. Temperature is an intensive property and does not depend on the amount of matter present.

2-48. You really can't answer this question. Heat is an extensive property. It depends on the amount of matter present. There is *far* more heat in a huge iceberg than in a cup of boiling water, even though the temperature of the boiling water is higher. This seem strange to most people. When the temperature is low, we tend to think that there is no heat present. That is far from the truth. If your home is heated by a heat pump, you prove it each time you use the heat pump. A heat pump extracts heat from the cold air outside and puts it inside the warmer house to keep you warm.

2-49. 5 Celsius degrees. This is another tricky question. You may have said that the answer is 10 Celsius degrees. Twice as much liquid is mixed, so there should be twice as much heat produced. There is. But don't forget that this heat is now distributed throughout twice as much liquid. The change in heat intensity (temperature) will be the same when 50 mL of each liquid are mixed as when 25 mL of each liquid are mixed. There is twice as much heat distributed throughout twice as much volume.

2-50. 2.7 g

2-51. 2.70 g/cm³

2-52. Pennies have a density of 8.9 g/cm³, and nickels have a density of 8.3 g/cm³.

2-53. $? \$ = 3.50\,\cancel{lb} \times \frac{\$1.70}{1\,\cancel{lb}} = \$5.95$

2-54. $? \$ = 2.43\,\cancel{lb} \times \frac{\$1.95}{1\,\cancel{lb}} = \$4.74$

2-55. $? \text{ lb} = \$10.00 \times \frac{1 \text{ lb}}{\$2.65} = 3.77 \text{ lb}$

2-56. $? \text{ lb} = \$7.84 \times \frac{1 \text{ lb}}{\$1.95} = 4.02 \text{ lb}$

2-57. 4.5¢/oz

2-58. 40¢/17 oz = 2.4¢/oz, and 60¢/29 oz = 2.1¢/oz. The peaches are cheaper.

2-59. 13.6 g = 1 cm³ This means that 13.6 g of mercury is the same amount of mercury as 1 cm³ of mercury.

$\frac{13.6 \text{ g}}{1 \text{ cm}^3}$ This is read, "13.6 grams per cubic centimeter" and means that each cubic centimeter of mercury has a mass of 13.6 g.

$\frac{1 \text{ cm}^3}{13.6 \text{ g}}$ This is read, "1 cubic centimeter per 13.6 grams" and means that each 13.6 g of mercury has a volume of 1 cm³.

2-60. 1.7 g/cm³

2-61. $? \text{ g} = 20.0\,\cancel{cm^3} \times \frac{2.16 \text{ g}}{1\,\cancel{cm^3}} = 43.2 \text{ g}$

2-62. $? \text{ g} = 2.54 \text{ cm} \times 2.54 \text{ cm} \times 2.54 \text{ cm} = 16.39 \text{ cm}^3$

$$? \text{ g} = 16.39\,\cancel{cm^3} \times \frac{1.73 \text{ g}}{1\,\cancel{cm^3}} = 28.3 \text{ g}$$

2-63. $? \text{ g salt} = 40.0\,\cancel{cm^3} \text{ salt} \times \frac{2.16 \text{ g}}{1\,\cancel{cm^3}} = 86.4 \text{ g salt}$

$$? \text{ g borax} = 20.0\,\cancel{in^3} \text{ borax} \times \frac{28.3 \text{ g}}{1\,\cancel{in^3}} = 566 \text{ g borax}$$

The 20 in³ of borax are heavier.

2-64. $? \text{ cm}^3 \text{ salt} = 15.4\,\cancel{g} \text{ salt} \times \frac{1 \text{ cm}^3}{2.16\,\cancel{g}} = 7.13 \text{ cm}^3 \text{ salt}$

2-65. The volume of the room is 7 m × 10 m × 3 m = 200 m³ (one significant digit is assumed.)

$$? \text{ cm}^3 = 200\,\cancel{m^3} \times \frac{10^6 \text{ cm}^3}{1\,\cancel{m^3}} = 2 \times 10^8 \text{ cm}^3$$

$$? \text{ g air} = 2 \times 10^8\,\cancel{cm^3} \text{ air} \times \frac{1.184 \times 10^{-3} \text{ g}}{1\,\cancel{cm^3}}$$

$$= 2 \times 10^5 \text{ g air}$$

2-66. $? \text{ cm}^3 \text{ air} = 1.0\,\cancel{g} \text{ air} \times \frac{1 \text{ cm}^3}{1.184 \times 10^{-3}\,\cancel{g}}$

$= 8.4 \times 10^2 \text{ cm}^3 \text{ air}$

2-67. The volume of the metal is 8.00 cm³. Dividing the mass of the metal by the volume, we get a density of

$$\frac{125 \text{ g}}{8.00 \text{ cm}^3} = 15.6 \text{ g/cm}^3$$

Since the density of gold is 19.3 g/cm³, this is not gold.

2-68. $? \text{ cm}^3 \text{ gold} = 1.00\,\cancel{kg} \text{ gold} \times \frac{1{,}000\,\cancel{g}}{1\,\cancel{kg}} \times \frac{1 \text{ cm}^3}{19.3\,\cancel{g}} =$ 51.8 cm³ gold

2-69. Yes. F′ = 1.8C. If C is doubled, F′ is doubled. The scales are proportional.

2-70. 1.8 Fahrenheit degree = 1 Celsius degree

2-71. 8.0×10^{-5} m is the value I got.

2-72. Probably not.

2-73. Poor technique, how hard the pages are mashed together, not estimating between marks on the scale, dividing by twice as many pages as you should (each sheet consists of *two* numbered pages).

2-74. An average of several estimates almost always contains less uncertainty than any single estimate used to obtain the average. The assumption is that random errors make some values too high and some too low. When an average is taken, these low and high values "average out."

2-75. If several values are close together and another value is very different, we are inclined to think that the one that differs a great deal from the others represents a gross error and should be discarded. There are statistical techniques that can be used to decide when it is appropriate to disregard an experimental value, but these techniques will not be taught in this course.

READING Accurate measurement often depends on special instruments. This short article provides a historical account of efforts to measure the freezing point of mercury. Not only will you find it interesting, but you will be impressed by how far science has come since the birth of our nation.

EARLY RESEARCH ON THE FREEZING POINT OF MERCURY

K. M. Reese

Readers may recall the National Bureau of Standards' announcement that the redetermined triple-point temperature of mercury is −38.84168° C, and that the triple point of the element can serve as a reliable temperature reference point (C&EN, Jan. 3, page 14). The NBS work is "a major contribution to the science and art of accurate measurement," according to Arthur F. Scott of Reed College in Portland, Ore. Thus stimulated, Dr. Scott has set down the circumstances surrounding the first attempt to measure the constant, then known only as the freezing point of mercury.

By the end of the 17th century, natural philosophers were using thermometers containing either colored alcohol or mercury. Fahrenheit introduced his temperature scale in 1714 or thereabouts, and Celsius introduced his scale in 1742. These scales were fixed by reference to the freezing-point and boiling-point temperatures of water. Later in the 18th century, scientists decided that it would be desirable to standardize the methods for determining the fixed points of mercurial thermometers. At the suggestion of the English chemist Henry Cavendish, the Royal Society appointed a committee of seven to study the problem. Cavendish was a member and is said to have made a major contribution to the resulting report, which appeared in 1777, just 200 years ago.

The report was a landmark in the development of thermometry, Scott says. In part it specified means of estimating the correction for the emergent column of mercury in the thermometer. It defined standard atmospheric pressure as 29.8 inches and gave instructions for correcting the observed boiling point for different barometric pressures. Cavendish's work on the report, Scott says, was doubtless what caused him to take an interest in a then-recently discovered phenomenon, the freezing of mercury.

Source: *Chemical and Engineering News,* February 1977. Reprinted with permission of the American Chemical Society.

The freezing of mercury was first reported by J. A. Braun of St. Petersburg, who did it with a freezing mixture of snow and nitric acid. The product was a "solid, shining metallic mass, which extended under the strokes of a pestle; in hardness rather inferior to lead, and yielding a dull dead sound like that metal."

Braun's discovery aroused great curiosity in all quarters. At the request of the Royal Society, Gov. Thomas Hutchins of Albany Fort at Hudson Bay repeated Braun's experiment in 1775 and sought with indifferent success to determine the freezing-point temperature of mercury. Later, Hutchins did a second series of experiments with the remote guidance of Cavendish, who provided apparatus and thermometers of his own design. The English chemist believed that the erratic results of the first attempt had resulted from the partial freezing of the mercury in the thermometer, and he designed the apparatus for Hutchins so as to obviate this difficulty. Cavendish described the equipment as follows:

"This apparatus was intended to determine the precise degree of cold at which quicksilver freezes; it consisted of a small mercurial thermometer, the bulb of which reached about 2½ inches below the scale, and was inclosed in a glass cylinder swelled at the bottom into a ball, which, when used, was filled with quicksilver, so that the bulb of the thermometer was intirely surronded with it. If this cylinder is immersed in a freezing mixture till great part of the quicksilver in it is frozen, it is evident, that the degree shewn at that time by the inclosed thermometer is the precise point at which mercury freezes."

The apparatus and the several thermometers were made in 1776, but "for several reasons," Scott says, did not reach Hutchins until 1781. Hutchins made his measurements in the winter of 1781–82; his report of the results to the Royal Society appeared in the *Philosophical Transactions* in 1783. In the same volume Cavendish published his "Observations on

Mr. Hutchins's Experiments for determining the Degree of Cold at which Quicksilver freezes," in which he analyzed in great detail each of Hutchins' experiments. Cavendish reported that the thermometers used by Hutchins had been returned to London and there recalibrated, in the presence of an ad hoc committee, in the manner recommended in the [1777] report of the committee of the Royal Society. He summed up the results of the Hutchins experiments in these words:

"It follows, that all experiments agree in showing that the true point at which quicksilver freezes is $38\frac{2}{3}°$, or in whole numbers 39° below nothing."

Cavendish's figure, Dr. Scott notes, is in degrees Fahrenheit; on the Celsius scale it becomes −39.26° C.

3

Questions About Matter

OBJECTIVES

Properties

You know the difference between a fig tree and a fig newton. They have different traits or properties. In this chapter you will learn to use properties that you have already studied and some new ones to answer important questions about matter. Do you know what we mean by a property? You probably do if you can do the following:

1. Given any object, list at least five properties of that object. (See Problem 3-17.)
2. Distinguish between properties that are characteristic of an object (they would help you identify that object) and properties that are not characteristic of the object.
3. Given a list of properties, identify those that are intensive and those that are extensive. (See Problem 3-21.)
4. Name two properties each of solids, liquids, and gases.

Purity

In order to determine whether something is pure, we examine the properties of the material. You should be able to do the following:

5. Name properties that are particularly useful in determining whether something is pure or impure. (See Problem 3-24.)
6. Describe observations that would convince you that a substance is pure. (See Problem 3-24.)

7. Describe at least three ways in which a mixture can be separated, and name the property of the substances in the mixture that allows the separation. (See Problem 3-20.)

Compounds and Elements

There are two kinds of pure substances: compounds and elements. You will need to know the difference between them.

8. State evidence that would convince you that a substance is a compound, an element, or a mixture. (See Problems 3-22, 3-23, and 3-24.)

Chemical and Physical Changes

Some changes produce new substances and some do not. You need to be able to tell the difference. Can you do the following?

9. When shown some change in matter, indicate whether the change is physical or chemical and state the reasons for your conclusion. (See Problem 3-18.)
10. Name physical and chemical properties of some substance that you are familiar with. (See Problem 3-17.)

Words You Should Know

property	weight	chemical change
melting point	heat of fusion	distillation
freezing point	heat of vaporization	physical change
boiling point	homogeneous	element
mass	heterogeneous	compound

3.1 WHAT CHEMISTRY IS ABOUT

Come on! Really strike the match and look at it! It may seem dumb, but understanding chemistry requires making observations and thinking about them. Striking a match is an "armchair experiment" in the effects of heating matter.

Chemists are interested in matter. All of it. They are interested in dirt, air, and you. They want to know what matter is, how matter can be changed from one form to another, and what properties matter has that make it useful.

This is a chemistry book, so it deals with matter. We will be trying to find out more about matter. To do this, we will ask a number of questions and suggest ideas that will help explain why matter behaves as it does. To begin our study of matter, let's do something to matter and see what happens. As a result, we can raise some useful questions.

One of the things we can do to matter is to heat it. When we do, we find that it does not always behave the same. Got a match? If you have, strike it and observe what happens when the wood or paper stem of the match gets hot.

Several questions could be raised in connection with this exercise. The most obvious question is, "After heating do we still have the same material that we had before?"

3-1. Answer the question and state *why* you answered as you did.

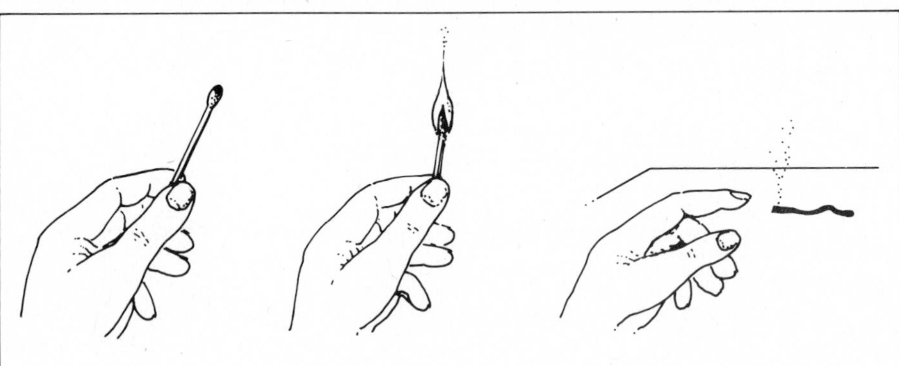

Figure 3.1
Is an unburned match made of the same kind of matter as a match that has burned?

In the case of the burning match, this question is easy to answer (negatively), but in other cases it is not. Still, the way we answer the question is always the same. We ask, "Are the properties of the material the same after heating as they were before?" If the properties are the same, we still have the same matter; if the properties are different, we do not. This idea is sometimes stated in a kind of rule: *Two substances are the same if their properties are the same.*

3.2 PROPERTIES

A **property** of a substance is some characteristic that can be observed. Color is a property. The match that you burned changed color. From that observation alone, you were inclined to say that the matter after burning was different from the matter in the original match. We might then ask, "What *is* this new material?" Once again, the way to decide is to compare the properties of this new matter with the properties of other matter that we already know about. The burnt match has the same color as coal, charcoal, and numerous other kinds of matter. We cannot identify matter on the basis of color alone.

Some properties have proven more useful than others in identifying substances. One of the most useful is the temperature at which a substance changes from a solid to a liquid—its **melting point**. The melting point of a pure substance is the same as its **freezing point**—the temperature at which the substance changes from a liquid to a solid. Another property often used for identification is the temperature at which a substance changes from a liquid to a gas—its **boiling point**.

You could test a colorless liquid to see whether it is water by measuring the temperature at which it changes from liquid to solid (freezing point) or the temperature at which it changes from liquid to gas (boiling point). Pure water freezes (or melts) at 0°C and boils at 100°C. By contrast, ethyl (grain) alcohol—another colorless liquid—melts at −177°C and boils at 78°C. Other colorless liquids have other freezing and boiling points.

Another property used to detect changes in matter is **mass** or **weight**.* Weighing a match before and after burning reveals that it decreases in mass when it burns. This indicates that some of the matter in the original match is not in the black stick that remains. Mass is a measure of the amount of matter present. If the mass of an object decreases, some of the matter is gone. This, of course, is why people watch their weight. When they lose weight, they know that they have lost some fat, excess water, muscle, or some part of the matter that makes up their body. If the loss is fat, they are usually happy.

Even though a loss or gain in mass is a good clue that a chunk of matter has *changed* in some way, it is not a good clue to the identity of a substance. Weighing two white solids wouldn't tell you whether they are both salt, both sugar, or some of each.

3-2. What intensive property that involves mass *could* be used to determine whether two solids are the same kind of matter?

3-3. How could you get a measure of this intensive property for the two chunks?

3.3 HEAT OF FUSION AND HEAT OF VAPORIZATION

Inexperienced cooks turn the stove on high and boil food at a rapid rate. The experienced cook turns the burner down when the pot begins to boil, knowing well that the food will cook no faster with the burner at a higher setting. How rapidly food cooks depends on the temperature, and once the liquid in the pot is boiling, the temperature won't change.

Put a pot of water on the stove and place a thermometer in it. Heat it to boiling and watch the temperature. The temperature will stay constant as long as there is liquid in the pot. This may seem odd at first. After all, heat is still being added to the pot. Why doesn't it continue to raise the temperature?

* Mass is the amount of a substance. Weight is the force that results from gravitational attraction. They are not the same idea. However, we normally determine the amount of substance (mass) by comparing the weight of the substance to the weight of a known mass. We call the process "weighing," even though we are actually trying to find the mass and record the result in units of mass. You can see why the terms get confused. The situation is just about hopeless! Almost everyone except physicists uses the terms interchangeably. It is a bad habit, but it's hard to break. When we use *mass* and *weight* in chemistry, we are usually talking about *mass*. May the physicists forgive us!

Figure 3.2 *Pure substances such as water boil at a constant temperature.*

The answer is that once the water boils, the heat energy is used to change the liquid water into gaseous water.

This energy is called the **heat of vaporization**. Heat of vaporization is an intensive property that is characteristic of the material being boiled. It takes over 2,200 J (540 cal) to change 1 g of water to a gas, but only about half that much (1,200 J) to change 1 g of methanol (wood alcohol) to a gas. Keep in mind that the temperature is not changed by this additional energy. All that is happening is a change from liquid to gas.

The energy required to change a liquid to a gas is large compared to the energy required to change the temperature of the liquid or gas. It takes more than *five times* as much energy to change water from a liquid (at 100°C) to a gas (at 100°C) as it takes to heat the water from its freezing point (0°C) to its boiling point (100°C).

Figure 3.3 shows a plot of data obtained when moth flakes (paradichlorobenzene) were melted and then cooled in a beaker of water. The line connecting the x's shows the temperature of the moth flakes over a period of time, and the line connecting the dots shows the temperature of the beaker of water that the flakes were in. Figure 3.4 shows how the data were collected.

During the first 15 minutes of cooling, everything looks normal. Since the air in the room is cooler than the water in the beaker, heat is transferred from the water to the room, and the water cools. In a similar fashion, the test tube filled with moth flakes gradually transfers heat to the cooler water. The curves for the water and the moth flakes remain parallel until the temperature falls to about 53°C. Then something strange happens.

The temperature of the water continues to drop as before, but the temperature of the moth flakes does not. After a momentary dip in temperature, the temperature of the flakes levels off at 53°C and remains there for several minutes.

If you do this experiment (and I hope that you will), you will find that moth flakes begin to change from a liquid to a solid at about 53°C, while the

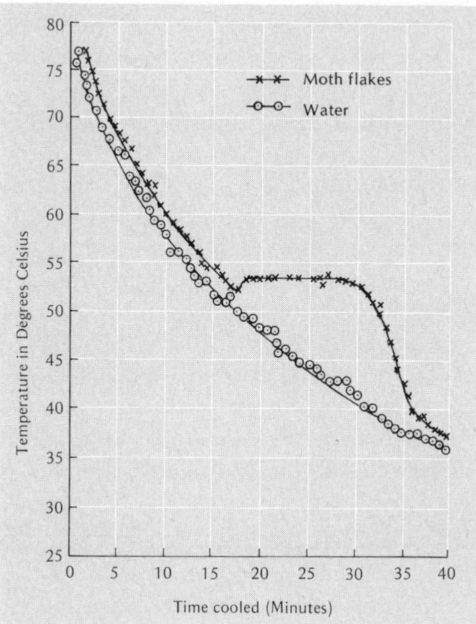

Figure 3.3
Cooling curves for water and moth flakes.

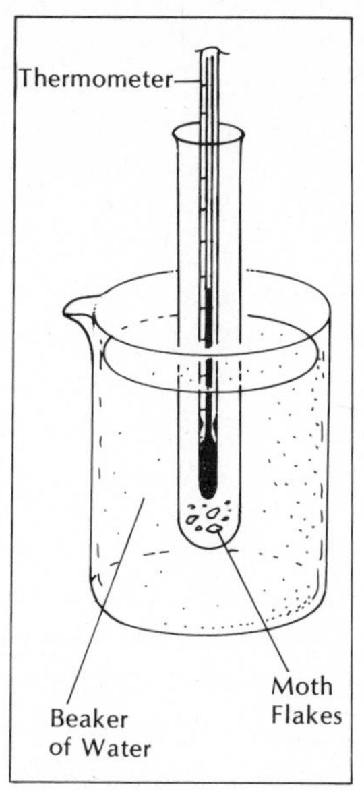

Figure 3.4
Apparatus for measuring the temperature of moth flakes as they cool.

water remains liquid. The temperature of the moth flakes remains at 53°C until all of the liquid has changed to solid. *The freezing point (or melting point) of any pure substance is constant.* Even though heat is still being transferred from the moth flakes to the cooler water, this loss of heat does *not* result in a change in temperature. Rather, it results in a change of state from liquid to solid. This heat is called the **heat of fusion**. (To fuse is to melt. Heat of fusion literally means "the heat required to melt something.") Once all of the moth flakes have frozen, the temperature drops rapidly until it is just a little higher than the water temperature again.

Like heat of vaporization, heat of fusion is an intensive property that is characteristic of a substance. Ice is very effective at cooling your soft drink or tea, because it absorbs a great deal of energy when it melts. It takes more than 300 J to change a gram of solid water to liquid water, and that is without changing the temperature from the melting point of 0°C.

Nature sometimes plays tricks. The dip recorded in Figure 3.3 just before the moth flakes begin to solidify is a result of a phenomenon sometimes seen when materials freeze. It is called supercooling. The liquid cools below the temperature at which it normally freeezes. Once solid does appear, the material freezes quickly, releasing enough heat to raise the temperature back to the normal freezing point, where it stays until the change in state is complete. Supercooling is observed often in some materials and rarely in others.

We have mentioned only a few of the properties that may be used to identify substances. Properties that you are sure to see in later studies are surface tension, index of refraction, electrical conductivity, thermal conductivity, heat of formation, heat of combustion, crystal structure, specific rotation of polarized light, solubility, specific heat, dielectric constant, and viscosity. There are many others.

Our purpose here is not to tell you about all of the properties that you might observe, but to remind you that the only way to identify matter is to observe such properties. If you want to determine whether two substances that look alike are the same kind of material, observe and test the two substances to see whether their properties differ. If you want to know whether something has been changed to a new kind of matter by heating, beating, or just sitting, observe its properties "before and after." If the properties differ, the substances differ.

Answering the following questions will give you a better understanding of how we use properties and should clarify the meaning of those properties that were discussed in this section.

3-4. Sticking a thermometer into a glass of clear liquid will not reveal what the liquid is. Under what conditions could you measure the temperature of the material to get useful information about its identity?

3-5. The crystal on my watch and the glass in my window look alike. The watch crystal weighs 3.5 g. The window pane weighs 832 g. Why can't I be sure that they are made of different materials, on the basis of this information alone?

3-6. What measurement could be made on the watch crystal and window pane and combined with the mass to give information that would suggest whether they are made of the same material?

3-7. It takes about 400 J to melt a gram of aluminum at its normal melting point of 659°C. It takes only 110 J to melt a gram of silver at its normal melting point of 961°C. What is this heat needed for melting called? What units would be appropriate to express this intensive property?

3-8. The heat of vaporization for gold is 1580 J/g, and the heat of fusion for gold is 64 J/g. What does this tell you about the energy required to change gold from a liquid to a gas compared to the energy required to melt it?

3.4 WHAT DO WE MEAN BY PURITY?

Much is said these days about purity. We worry about pure air, pure water, and pure foods. It must be something worth talking about. We shall.

What is meant by purity is not always obvious. When one speaks of pure water, for example, it would seem that one means 100 percent water with absolutely nothing else in it. However, this isn't the case. Oh, it is possible to get 100 percent water. The distilled or deionized water used in laboratories is essentially 100 percent water, but the water that you drink isn't. Most water has oxygen and carbon dioxide dissolved in it. (Without oxygen dissolved in the water, fish would quickly die by suffocation.) Most water also has various solids dissolved in it. The dissolved materials are responsible for the taste of water. Water with impurities that are not harmful to plants or animals is generally considered "pure." In normal usage, pure water means water that is safe to use for drinking, cooking, washing clothes, and the many other uses to which we put it. It is not considered pure by chemists, because it is not 100 percent water.

Pure air is even more of a problem than pure water. By volume, the purest air that you can breathe on the face of the earth is a mixture of about 21 percent oxygen, 78 percent nitrogen, 1 percent argon, and traces of carbon dioxide, water, and a number of other gases. As long as the amount of these other gases is not too high, the air is considered pure by even the most concerned environmentalist. Once again, pure air usually means air that is free from any materials that harm plant or animal life.

A chemist who describes something as pure means that there is only one kind of substance present. This is important, because the chemist is interested in learning why changes occur in matter. If we place some solid in a beaker of water and there is evidence of a chemical reaction—a color change, formation of a new solid, or evolution of a gas, for example—we want to be able to attribute that change to some reaction with the water. If the water used is not 100 percent pure, we cannot be sure that this is the case. It should be clear why distilled or deionized water is always used in chemical and medical laboratories.

If purity of substances is important, we need to know how to determine whether a substance is pure. Sometimes the task is easy; at other times it is not. Some examples will help.

Pick up a handful of dirt (outside or in your room; it doesn't really matter). Look at it. You can see that the dirt is not pure. Some specks are darker than others. With luck, you will see specks that crawl around! There are likely to be pieces of grass or bits of leaves mixed in with the dirt. Not all of the parts are alike. The material is **heterogeneous** (from the Latin or Middle French *heter,* meaning "different," and the Greek *genos,* meaning "kind"). *Any material that is heterogeneous is obviously not pure.*

It is tempting to say that those things that appear **homogeneous** (from two Greek words, *hom,* which means "the same or alike," and *genos*) *are* pure. Not so. Many materials that *appear* homogeneous are made up of two or more kinds of matter. Milk, coke, water from the tap, butter, and axle grease all appear to be homogeneous; none is pure.

To determine whether a homogeneous material is pure, one frequently tests it to see whether it can be separated into components. If butter is melted in a pan, it separates into an oil (called clarified butter) and a solid material that floats to the top. If a pan of tap water is boiled on the stove until all of the water is gone, a small amount of white solid remains in the pan, indicating that the original water was not pure. Milk can be heated in a vacuum chamber to drive away the water and leave powdered milk. A homogeneous mixture of powdered glass and salt can be separated by placing it in water so that the salt dissolves, leaving the glass as a solid. A mixture of powdered sugar and salt could *not* be separated by this technique. If you relied on it, you might mistakenly assume that you had a pure white solid, when in fact you had a mixture. The particular tests that separate mixtures vary from mixture to mixture, and purification of materials can be a tedious and time-consuming process.

3.5 CHECKING ON PURITY

It is easy to determine that heterogenous materials are mixtures. Nobody would look at a bowl of party mix and claim that the assortment of cereals and nuts represented a single substance. It is easy to separate the components into separate piles. But what if you pick out a single nut? It is homogeneous in appearance. How would you know whether it is a single substance or several different substances mixed together so thoroughly that they *look* like a single substance? It is not clear that it can be separated into components. And what about the glass container that the party mix is in? Is glass a single substance or a mixture of several different substances? How can we test homogeneous materials such as these to see whether they are pure?

One of the most common techniques for testing the purity of a homogeneous substance is to check melting point or boiling point. In the section on properties, we mentioned that pure substances have a characteristic melting point and boiling point. Furthermore, the melting and boiling behavior of pure substances differ from the behavior of mixtures.

Figure 3.5 shows two heating curves. The lower curve with o's marking the data points shows the temperature of a beaker of pure methanol as it is heated over a period of 23 min.* At first the temperature of the methanol rises steadily, but when the methanol begins to boil, the temperature remains constant for the rest of the time. The boiling point of a pure liquid is constant and is characteristic of the material boiled.

Now look at the curve formed with the x's. These data show the temperature of a beaker containing 75 mL of water and 25 mL of methanol. Once again, the temperature rises gradually until the material begins to boil, and then it levels off. However, the temperature does *not* remain constant. *Mixtures ordinarily do not have a constant boiling point.*

It is possible to make mixtures that have constant boiling points, but they are rare and require a specific composition. Such mixtures are called *azeotropic* mixtures, and they are important in industries where mixtures are separated by distillation. One such mixture is 95.5 percent ethanol (the alcohol in

* Methanol (wood alcohol) is toxic to humans. Inhaling the vapors can cause blindness or even death. If this experiment is done, it must be done so that the fumes do not escape into the room.

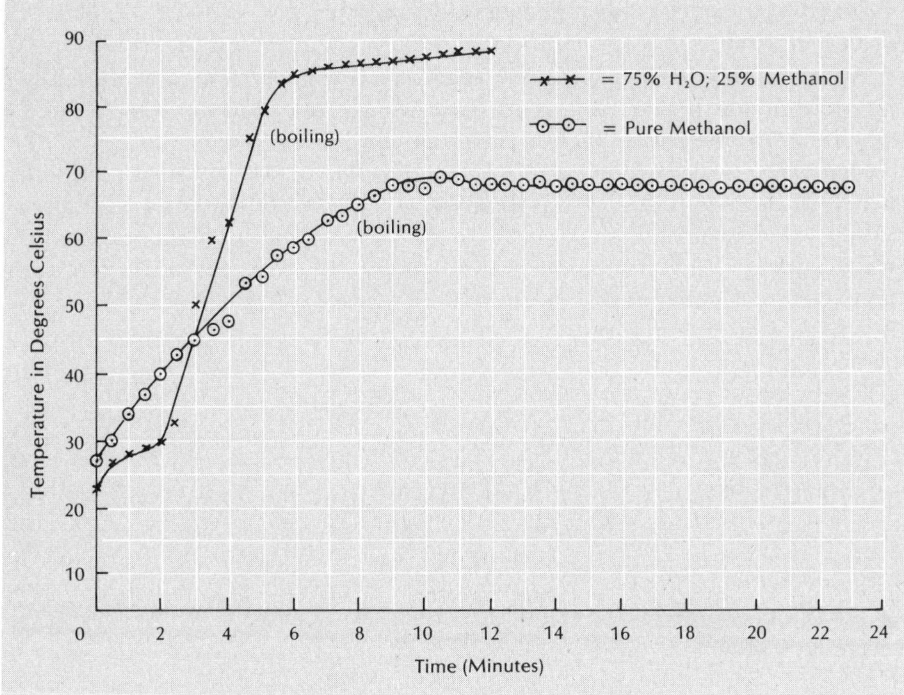

Figure 3.5
Heating curves showing the change in temperature as pure methanol and a mixture of methanol and water are heated to boiling.

alcoholic beverages) and 4.5 percent water. This mixture has a constant boiling point of 78.1° C. Any other mixture of ethanol and water would show a change in temperature as the mixture boils.

Just as mixtures do not have constant boiling points, they do not have constant freezing points. Any mixture of water and antifreeze (ethylene glycol) will freeze if the temperature is lowered enough. However, the temperature does not remain constant as freezing takes place (unlike the temperature of the pure moth flakes plotted in Figure 3.3). Rather, the temperature decreases slowly during freezing and then falls more rapidly after all of the mixture has become solid.

3-9. Sketch a graph similar to Figure 3.3 showing the temperature change of a *mixture* during the time when it is freezing.

Just as there are mixtures of special composition that have a constant boiling point, there are mixtures of special composition that have a constant melting point. Such mixtures are called *eutectic* mixtures. It would be very unusual to find that the composition that produces an azeotropic mixture also produces a eutectic mixture. Consequently, if one finds that *both* the boiling point and the melting point of a substance are constant, it is almost certain that the substance is not a mixture.

3.6 DESTRUCTIVE DISTILLATION OF WOOD

We began this chapter by asking you to do an experiment. You were asked to strike a match and look at it. Now we ask that you consider a similar experiment done under slightly different conditions.

One of the ancient chemical arts is making charcoal. If you struck a wooden match at the beginning of this chapter, you probably made some charcoal. When wood is heated, charcoal is formed. When the job is done right, the wood is heated in a container that keeps out air. The apparatus shown in Figure 3.6 is a convenient one for making charcoal in the laboratory.

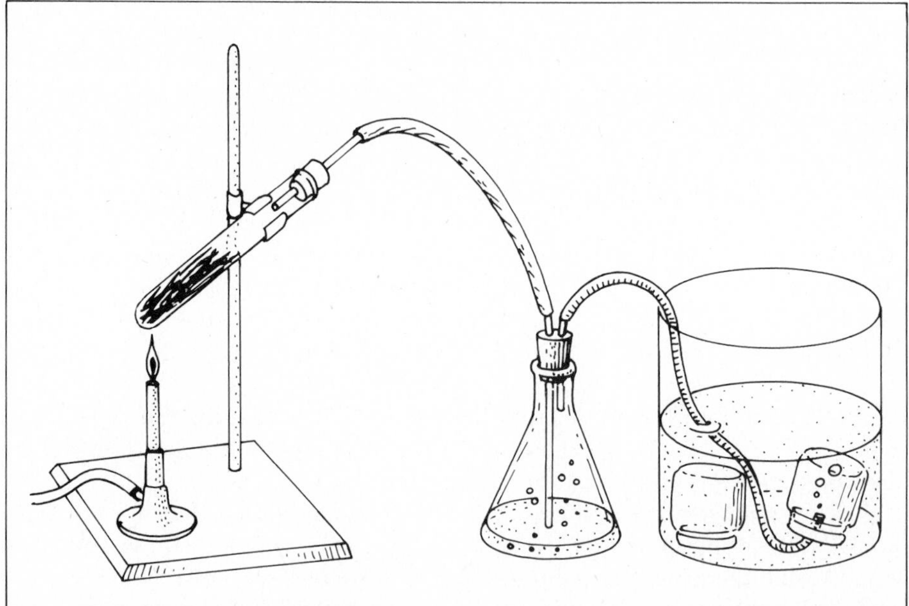

Figure 3.6 *Apparatus for the destructive distillation of wood.*

Sticks of wood are placed in the test tube shown on the left and heated with a burner. That's all that is required to make the charcoal. However, bad-smelling smoke is produced in the process and pollutes the air if it is allowed to escape. The rest of the apparatus is there to trap the matter in the smoke. As the smoke enters the center bottle and cools, a brown liquid forms in the bottle. The gas that doesn't condense to form this brown liquid passes through the exit tube and finally bubbles through a pan of water and is collected as it forces water out of the inverted bottle in the pan. Thus all products of the reaction are collected. Charcoal is left in the original container, a brown liquid is collected in the center container, and a colorless, stinking gas is collected in the final bottle.

This chemical process, called destructive distillation, has been known for centuries, but it is still useful to observe and talk about. To begin the discussion, consider the following questions.

3-10. The destructive distillation of wood has just been described as a *chemical* process. This means that some new kind of matter was formed in the process. What evidence exists that this is true?

3-11. Name at least two observations to indicate that the material collected in the center bottle is *not* the same as the material collected in the final bottle.

It should be clear from the properties of the original wood and the properties of the final products that the change that occurred when the wood was heated produced new kinds of matter. Such changes are called **chemical changes.** Let us consider the materials produced by this chemical change.

What can we learn about the liquid produced? Is it a pure substance like water, or is it a mixture of several substances like crude oil? We can examine its properties to find out. You know that the boiling point of a pure substance is constant, whereas the boiling point of a mixture is not. We may be able to find out whether the liquid is pure by checking its boiling point.

Any time you don't know what a substance is, consider it poisonous and treat it accordingly. To prevent noxious vapors from filling the room, the gas produced by boiling can be passed through a cooled piece of glass (a condenser) to convert the material back to a liquid. A simple apparatus for doing that is shown in Figure 3.7. Boiling and reliquefying a substance in this way is called **distillation.** Table 3.1 (p. 86) shows the temperature of the liquid at 15-sec intervals as it was heated over a 5-min period.

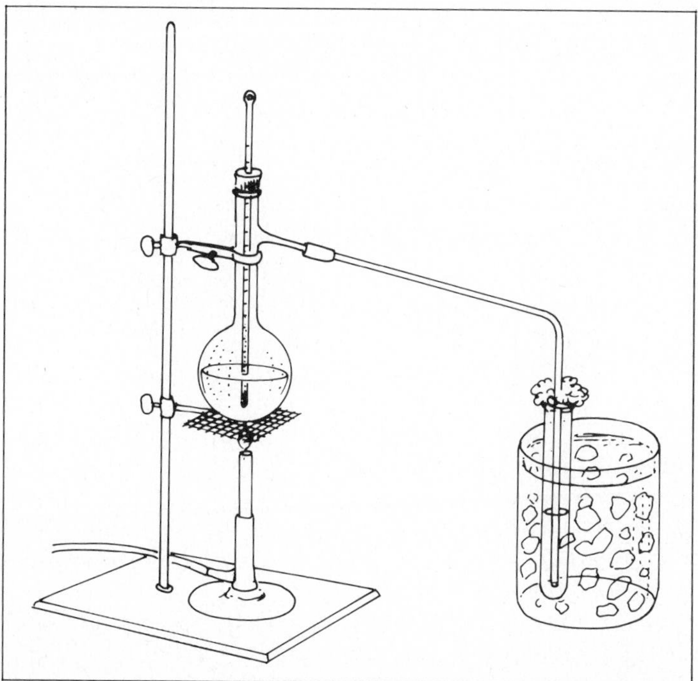

Figure 3.7
Apparatus for distillation of a liquid.

Table 3.1. Time and Temperature Data for the Distillation of an Unknown Liquid

Time (sec)	*Temperature (°C)*	*Time (sec)*	*Temperature (°C)*
0	24	165	100.5
15	24	180	101.2
30	24	195	101.9
45	24	210	102.1
60	31	225	102.4
75	46	240	103.0
90	81	255	103.0
105	97.5	270	103.3
120	99.0	285	104.0
135	99.8	300	104.2
150	100.0		

3-12. Graph the data given in Table 3.1. (Appendix E explains how to construct a graph.)
3-13. From the graph, what can you say about the purity of the liquid?
3-14. If you have observed this distillation, describe other observations that support your conclusions.

When we boil the light brown liquid to see whether the boiling point is constant, the liquid boiling over in the beginning is almost colorless. The liquid remaining in the boiling flask is dark brown. It appears that the boiling process results in the separation of a colorless liquid from a dark brown liquid. We seem to be separating a mixture. But how can we be sure? What we have observed can be explained in either of two ways:

1. The light brown *mixture* is made up of two components. When it is boiled, the colorless component boils away, leaving the dark brown component behind.
2. The light brown liquid is a *pure substance.* When it is heated, this pure material changes chemically to form at least two new substances. One of these substances is colorless and boils away.

3.7 CHEMICAL AND PHYSICAL CHANGES

We have already said that a chemical change is one that produces some new kind of matter. A **physical change** is a change that makes things appear different without really forming anything new.

When we melt butter or ice, we don't think that we get anything new. We just have liquid butter and liquid water instead of solid butter and solid water. Melting and freezing are considered physical changes. So is boiling.

Another example of a physical change is the separation of a party mix into piles of nuts, candies, and other tidbits. Nothing new is made, but it looks different because it is no longer mixed.

Distillation—the process used on the brown liquid—is much like separating the party mix, except that the pieces are too small to see. The light brown liquid obtained from the destructive distillation of wood looks different from the colorless and dark brown liquids produced when we distill it. Nevertheless, we can imagine that the light brown liquid is only the other two mixed together.

As the two explanations at the end of the previous section indicate, it isn't always easy to tell whether a change is physical or chemical. There are clues we can use, but no set rules.

One clue to the nature of a change is *the energy involved.* If a change requires a lot of energy to make it take place, or if a lot of energy is released when it takes place, it is usually a chemical change. It takes a lot of heat to break down wood into charcoal and smoke. That suggests that the destructive distillation of wood is a chemical change. It is. However, it also takes a lot of heat to boil water, and that is considered a physical change.

Another clue to the nature of a change is *how easy it is to reverse the change.* If you cool the steam from boiling water, you get liquid water again. If the temperature of melted ice is lowered, ice forms once more. The separated party mix can be reformed by simply pouring the ingredients together and mixing them. The brown liquid and the colorless liquid we obtained from distillation can be mixed to form the light brown liquid that we started with. These are physical changes. By contrast, mixing the charcoal, the light brown liquid, and the colorless gas obtained when the wood was heated will never produce wood again. Nobody knows how to get wood from these products. Most chemical changes are not easily reversed.

3.8 IS CHARCOAL PURE?

The black solid left when wood pieces are heated looks homogeneous. Is it pure? How could you find out? Checking the melting point of the charcoal to see if it is constant is inviting. However, since the material was heated as much as possible with a laboratory burner when the destructive distillation of wood was done, it is unlikely that it can be melted easily. Checking the boiling point is also out. We might try solubility. Some mixtures—salt and sand, for example—can be detected by mixing them in water. The salt dissolves; the sand does not. The charcoal could be ground to a powder and its solubility checked. I did this and saw no evidence of anything dissolving. What shall we do?

Many problems in chemistry are like this one. The measurements we need to answer our question cannot be made. We are left to guess or dream up a new strategy.

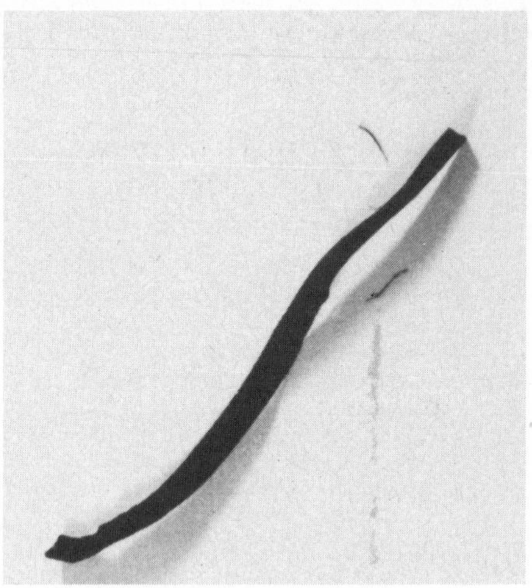

Figure 3.8 *Charcoal before and after burning.* (*Photo by Dudley Herron*)

Let us try a new strategy. I know that charcoal will burn in air to produce heat and some new material. (Remember that the distillation that produced the charcoal was carried out in the absence of air.) Figure 3.8 shows some of the charcoal from the wood distillation and the white ash that is left when the charcoal burns. A colorless, odorless gas is also produced, but it is difficult to detect.

3-15. Would you describe the burning of charcoal as a chemical or a physical change? Why?
3-16. From your experience with charcoal, what evidence indicates that the white ash is not the only product of the burning?

The fact that more than one product is formed when charcoal burns might lead you to think that charcoal is a mixture. If it is a single substance, why would it produce two (or more) different things when it burns? Inviting as this logic may be, it is dangerous. To see why, we need to look more carefully at pure substances.

3.9 ELEMENTS AND COMPOUNDS

Table 3.2 gives the names, melting point, boiling point, and density of several pure substances.

All of the substances that are listed in the table are pure. They have constant physical and chemical properties. Nevertheless, there are differences in the nature of these pure substances.

You will note from the table that the first two substances, sugar and baking soda, do not have a melting point or boiling point recorded. The reason is simple. If you have ever heated sugar on the stove, you know that it does not melt very well. In the process of melting, it turns dark brown, indicating that

Table 3.2. Representative Properties of Some Pure Substances

Name	*Melting Point* (°C)	*Boiling Point* (°C)	*Density* (g/cm^3)
Sucrose (common sugar)	—	—	1.58
Sodium bicarbonate (baking soda)	—	—	2.16
Sodium chloride (table salt)	801	1413	2.16
Water	0	100	1.00
Iron	1535	3000	7.87
Carbon (graphite)	3550	4827	1.9–2.3
Copper	1083	2595	8.96

some chemical change has occurred. The chemical change may be a result of the sugar's combining with air or simply breaking down into simpler components. The change in color and other properties takes place even when we heat sugar in a container without air. If sugar is weighed before heating, and then all of the products are weighed after heating, the total mass does not change. This shows that the sugar did not combine with any other material. If it had, the products would weigh more than the original sugar. Still, we know that new materials are formed because the properties have changed. The sugar must have broken down.

When baking soda is heated, a white solid with properties different from those of the soda is produced. Water forms at the mouth of the test tube, and

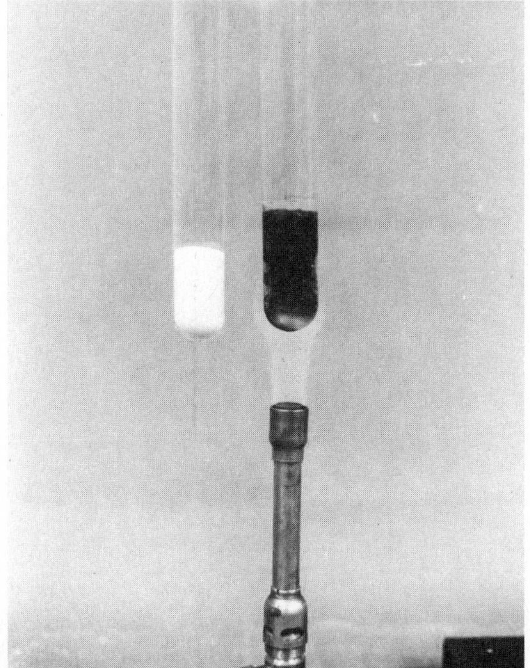

Figure 3.9
Sugar before and after heating.
(Photo by Tom Greenbowe)

a colorless, odorless gas is given off. The white solid weighs less than the original soda, but the total weight of the white solid, water, and colorless gas is the same as the weight of the original soda. This evidence indicates that the soda is breaking down into some kind of simpler components.

The next two substances in the table, salt and water, do not break down into simpler components when they are heated. However, if they are melted (a difficult job for the salt but no problem at all for the water) and an electric current is passed through them, new materials are formed. The weight of the new materials is exactly the same as the weight of the salt or water that disappears. Once again, the evidence indicates that these substances can be broken down into simpler materials.

The last three substances in Table 3.2 have defied all efforts to break them down into simpler components. Neither heat nor electricity will do the job; indeed, nothing has been found that will break them down into simpler materials. There seems to be something fundamental about them. They are called **elements**.

The first four substances in Table 3.2 *can* be broken down into elemental substances like the last three. (Note that I said it *can* be done; I didn't say it is easy.) It is as though sugar, salt, baking soda, and water are somehow "compounded" or made from the simple, elemental substances. We call such pure substances **compounds**.

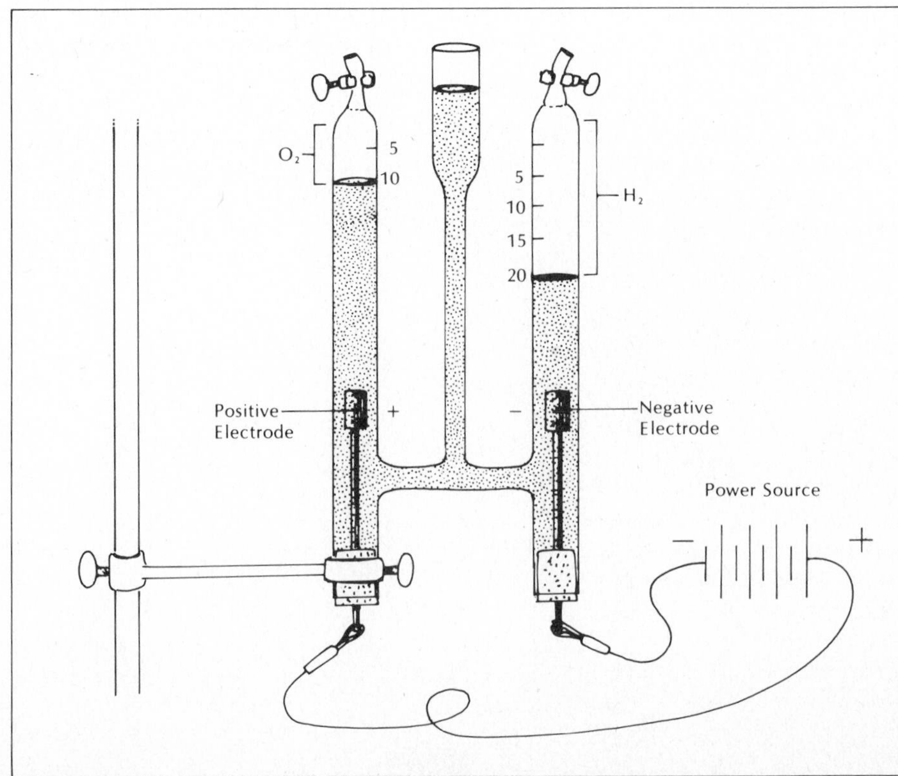

Figure 3.10 *Water can be decomposed into hydrogen and oxygen gas by passing an electric current through the water.*

It is often difficult to determine whether a pure substance is a compound composed of simpler elements or an element that cannot be broken down into simpler pure substances. For many years water was considered an element, because nobody had found a way to break it down into the elements hydrogen and oxygen.

There are only about 100 known elements, and many of those are rare. The names of the elements are found in Chapter 5 and on the inside cover of this text.

3.10 MIXTURES AND COMPOUNDS

We have said that compounds can be decomposed into two or more elements. Then isn't a compound just a mixture?

It *is* possible to make a mixture of elements, but such mixtures are quite different from compounds made from those same elements. A mixture of oxygen and hydrogen, for instance, would look like air. Water, a compound made from oxygen and hydrogen, doesn't look like air at all.

One of the most important differences between mixtures of elements and compounds of elements is that *compounds have a definite composition* and mixtures do not. All water on the face of the earth contains eight times as much oxygen as hydrogen by weight. In other words, there are 8 g of oxygen for every 1 g of hydrogen in water. No other proportions of oxygen and hydrogen are found in any sample of water.

Mixtures do not have a definite composition. You can mix 1 g of hydrogen with 1 g of oxygen, 1 g of hydrogen with 5 g of oxygen, or 1 g of hydrogen with 50 g of oxygen. In fact, any proportions are possible.

The reason why compounds have a definite composition will become clear as you begin to look at the microscopic nature of matter in Chapter 4.

3.11 SUMMARY

We began by saying that chemists are interested in matter. We then encouraged you to look carefully at some matter and some changes in matter to begin the perplexing task of explaining what matter is like and how it changes from one form to another.

We focused our attention on properties—the characteristics of matter that tell us there are different kinds of matter. We found that extensive properties such as mass and volume aren't much help in identifying substances, but intensive properties such as density, melting point, and boiling point can help a great deal.

At this point things became complex. We posed some questions about matter and found that the answers are sometimes difficult to obtain.

First, we asked whether something was pure or impure. Pure matter is composed of a single kind of substance; impure matter is a mixture. Matter that appears heterogeneous is almost certain to be impure, but many kinds of matter that appear to be homogeneous really aren't.

Deciding whether homogeneous material is pure is the hard part. However, observing the behavior of pure and impure materials when they boil or melt is one useful test. Pure substances don't change temperature during boiling or melting. Mixtures do. We pointed out that there are exceptions to this rule. But azeotropic and eutectic mixtures—those that boil or freeze at constant temperature—require special compositions and need not concern us at this point in our study.

The second question we asked was whether some observed change produced a new kind of matter. Changes that produce a new kind of matter were described as chemical changes, and those that do not produce a new kind of matter were described as physical changes. The trick is to decide whether something new is produced. Solids, liquids, and gases certainly don't look the same, but changes from solid to liquid or from liquid to gas are considered physical changes. At least this kind of change would be physical so long as it could easily be reversed by lowering or raising the temperature. Ice, water, and steam are all the same substance in three different forms. Changing the shape of an object and breaking it into smaller pieces are also physical changes.

It is usually easy to determine that something new is formed. Burned matches don't look like unburned matches, and when wood is heated, the products are nothing like the original wood. The properties are different. We are quite confident that the substances are different.

However, it can be difficult to be certain whether an apparent change in a homogeneous material is due to a chemical change that produces new materials or to a physical change that separates substances mixed together. We found that the brown liquid obtained from distillation of wood does not have a constant boiling point. That told us it is not pure. However, it was difficult to rule out the possibility that it started out pure and changed to produce several different substances when it boiled. The best evidence that it was impure in the beginning was that the colorless liquid that boiled away could be poured back into the original flask to produce a mixture with the same properties as the original liquid. Chemical changes are not usually reversed so easily.

Finally, we pointed out that there are two kinds of pure matter. One kind can be broken down into simpler kinds of matter by heat or electricity. This kind of matter is called a compound. Substances that cannot be broken down into simpler kinds of matter are called elements.

Figure 3.11 summarizes the way we have classified matter. Study the diagram and see whether you can decide how to make the decisions called for at the forks. For example, how can you decide whether something is an element or a compound?

This introduction to matter is complicated by the fact that the important differences among elements, compounds, and mixtures are invisible. Before we can go far in understanding matter, we must invent a way to talk about

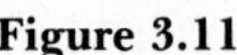

Figure 3.11

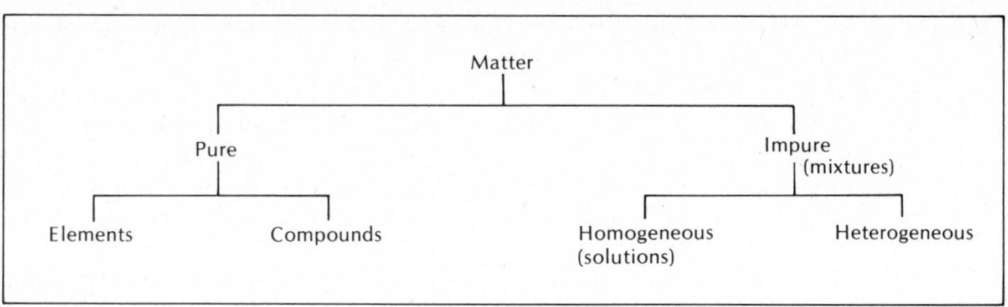

the invisible world of atoms and molecules. It is the subject of the next chapter. Before taking that step in the story of chemistry, you may want to read the end-of-chapter article about some of the things discussed in this chapter.

Questions and Problems

3-17. Read "The Wonder of Water" and make a list of chemical properties and a list of physical properties of water.

3-18. Indicate whether each of the following changes is chemical or physical, and explain why you answered as you did.
   a. Leaves changing color in the fall
   b. Crushing a dry leaf in your hand
   c. An aspirin dissolving in your mouth or stomach
   d. An aspirin acting on the body to relieve a headache
   e. Water boiling
   f. Beans cooking
   g. Grass growing
   h. Cutting grass
   i. Bleaching clothes
   j. Using Drano® to unstop a sink

3-19. From the following list of observations, indicate whether the change is more likely to represent an element combining to form a compound or a compound decomposing into simpler compounds or into elements. Give the reasons for your decision.
   a. When a white solid is heated in a closed container, a gas appears to be produced, and the solid that remains after heating weighs less than the original solid.
   b. When a metal is heated in air, it changes color and gains in weight.
   c. A log is weighed and left to rot. The rotted log weighs less than it did before rotting.
   d. A black powder is heated with oxygen. The only product is a colorless gas that weighs the same as the original powder and oxygen combined.
   e. A colorless solid is heated in oxygen. Three gaseous products are identified. The three products weigh the same as the original solid and the oxygen gas.

3-20. Describe a way to separate the following mixtures.
   a. Salt dissolved in water
   b. Alcohol dissolved in water
   c. Pieces of iron and wood
   d. Sugar and powdered glass
   e. Large pieces of aluminum and zinc that look identical

3-21. Which of the following are intensive properties and which are extensive properties? Explain how you decided.
a. Temperature
b. Mass
c. Weight
d. Density
e. Melting point
f. Volume
g. Length
h. An hourly wage (such as $2.80 an hour)
i. Yearly earnings
j. Heat of vaporization

3-22. Throughout human history, it has been difficult to interpret observations made of nature. Because water did not break down when it was heated and all other efforts failed, people concluded that water was an element. This proved to be wrong. In similar fashion, students might heat a sample of copper in air, note the black solid that forms and conclude, "Aha! we have broken down the copper into a black solid and a colorless gas, just as the wood was broken down when it was heated!" They would be wrong. How could you prove that the black solid is actually the result of something combining with the copper rather than the copper breaking down into simpler substances?

3-23. When iron is left in air, the iron changes to a reddish-brown solid. How could you prove that the reddish-brown solid (rust) is iron combined with something else rather than a decomposition product?

3-24. We did not obtain any evidence that would tell us whether the charcoal obtained from the destructive distillation of wood was pure or a mixture. We *do* have evidence from the burning that indicates that it is not a pure element. How do you know that the charcoal is not a pure element? Could the charcoal be a compound? Explain. Could the charcoal be a mixture? Explain.

Answers to Questions in Chapter 3

3-1. No. You probably said something like, "It doesn't look the same."

3-2. Density

3-3. Determine the mass, determine the volume, and divide the mass by the volume. The only difficulty would be measuring the volume. You could do this by putting some liquid that will not dissolve the solids in a large graduate. By measuring the volume of the liquid before the solid is placed in it and then measuring the volume after the solid is placed in it, you could find the volume of the solid.

3-4. The boiling and melting points of a substance are characteristic of the material. Measure the temperature when the material is boiling or melting.

3-5. They could be different-sized pieces of glass or plastic.

3-6. If I weighed pieces of the same size or measured the volume of each and calculated the density, I could make a comparison. If the densities are different, pieces are not made of the same material. If the densities are the same, the pieces *may* be made of the same material.

3-7. Heat of fusion. Several units could be used to express this, but the units must be in terms of some amount of heat per some amount of material. Possibilities are J/g, J/m^3, and cal/g.

3-8. It requires over twenty-four times as much energy to boil gold as it requires to melt it. Changing from a liquid to a gas always requires more energy than changing from a solid to a liquid.

3-9. See Figure 3.12 (facing page).

3-10. The color has changed; there are a liquid and a gas present at room temperature, in addition to a solid like the original wood. The solid that remains, unlike the wood, is soft and easy to break. The odor is different. *Many* properties have changed. There must be new kinds of matter.

3-11. At the same temperature (the temperature of the room), the material in the center bottle is a liquid; the material in the final bottle is a gas. The liquid in the center bottle is light brown; the gas is colorless. The odor of the two materials is not the same.

3-12. See Figure 3.13 (facing page).

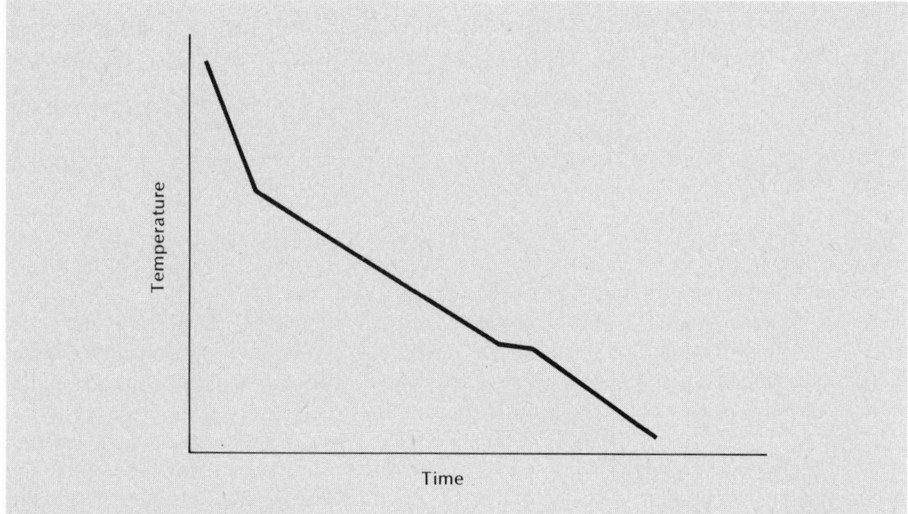

Figure 3.12

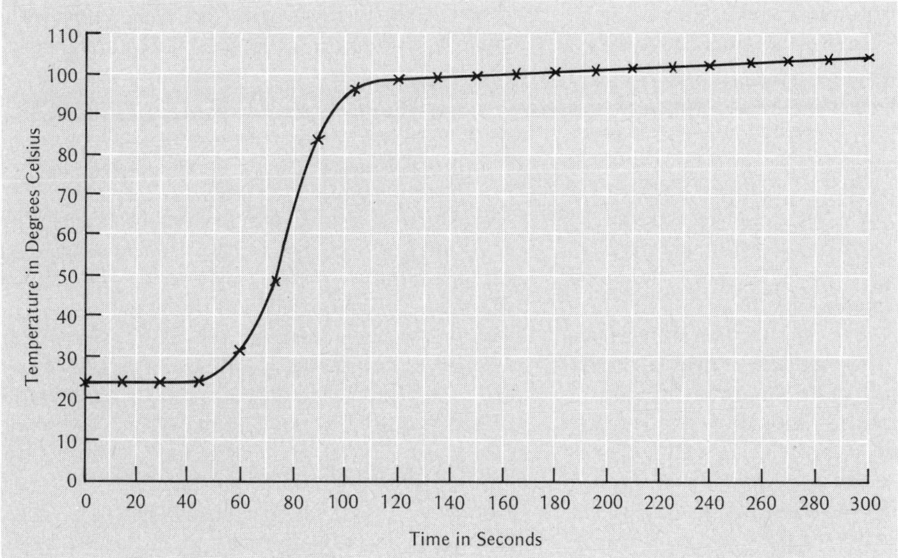

Figure 3.13 *Temperature of wood distillate as it is heated over a period of 300 seconds.*

3-13. It is impure. (It *could* have been pure before heating, but after it was heated to boiling, it was a mixture. The boiling point was not constant.)

3-14. The first liquid that boils over is almost colorless. As it continues to boil, it takes on a slight yellow color. Meanwhile, the liquid that remains becomes darker brown in color.

3-15. Chemical. What we have after burning has different properties from the charcoal that we start with.

3-16. The ash weighs less than the charcoal that you start with. Something else must have been produced to account for this loss in mass.

READING No substance on earth is more important than water. Our lives and more depend on it. This article describes some of the properties of water that make it so important.

THE WONDER OF WATER

T. A. Boyd

Life on earth has evolved either around or in water. This event must mean that this seemingly simple compound of hydrogen and oxygen has special and even wondrous properties. And indeed it has. In many respects water is unique among the countless compounds in nature. Once in a wry remark to fellow chemists, Nobel prize winner in chemistry, Peter Debye, said: "We are just beginning to know what water is, although we have been calling it H_2O for more than a century."

According to one theory, evolution of life began in the sea. And the highest evolution of all, the human body, is about 70% water—not just plain water, though, but a water solution similar in salinity to that of the sea. The blood, which is largely water, carries food to the cells as well as oxygen to burn the food. As the blood returns it bears the ashes of the body fires, part of which is water itself, formed by burning hydrogen in food. About 0.5 litre of water is produced in the body each day. Some desert animals do not drink water at all but rather obtain what they need by burning hydrogen in their food.

Water performs another vital function in our bodies—it controls body temperature. When a person gets too warm, water oozes from a million pores in the skin, and cools by its evaporation, thus keeping the body temperature normal. Compared to other liquids the cooling effect of evaporating water is exceptionally large: 539 calories per gram, or 1100 Btu (British thermal units) per pound.

Water is twice blessed, said E. E. Slosson. It bestows this blessing as it comes and as it goes. We appreciate the coolness of a glass of cold water when we drink it, but it is many times more cooling later as it oozes from pores in the skin and evaporates. Even a cup of hot tea can be cooling, because as it evaporates from the skin it carries away about 10 times as much heat as was taken in with the tea.

Water is equally vital to plants. Absorbed from the soil by the roots, it rises and circulates throughout the plant, bearing with it minerals dissolved from the soil. Then under the influence of sunlight and chlorophyll of the leaves, it unites chemically with carbon dioxide from the air to build tissues of the plant, and the unused portion transpires from the leaf surfaces into the surrounding atmosphere. Growing a bushel of corn is said to require about 2500 gallons of water.

For the convenience of cooks, water boils at a usable temperature of 100° C which, compared to related compounds, is extraordinarily high. If its boiling point were too low for cooking at normal atmospheric pressures, cooks would certainly be vexed because pressure vessels would be required. Most of our food contains water.

The freezing point of water is extraordinarily high also, a fortunate peculiarity that allows water to exist as a liquid rather than a solid within the normal temperature range on much of our planet.

Another distinctive property of water is that it boils away very slowly. It takes nearly seven times as much heat, called heat of vaporization, to boil off a pint of water as it does to warm it from 21° C to its boiling point of 100° C. This fortunate property causes water to evaporate slowly from rivers, lakes, and seas.

Strangely, water vapor is lighter than air. Thus, evaporation at the Earth's surface causes the lower layers of the atmosphere to be lighter than those above. Then, by various means, including convection and wind action, the moist air is uplifted into the higher atmosphere where it expands under the lower pressures in the upper regions. Expansion causes cooling and some of the water vapor condenses to form clouds from which falls rain that makes life possible on land.

For a compound having its vapor lighter than air, the boiling point of water is quite extraordinary. If its

Source: *Chemistry*, Vol. 47, No. 6 (June 1974), pp. 6–8. Reprinted with permission of the copyright owner, THE AMERICAN CHEMICAL SOCIETY.

molecular aggregation in liquid form were the same as that in vapor form, its boiling point would be many degrees below zero. Liquid ammonia, the molecular weight of which is close to that of water (ammonia, 17; water, 18) boils at 33° C below zero. And this boiling point is higher than that of many similar liquids.

Another uncommon and fortunate property of water is that, when frozen, it is lighter than the liquid. If ice were heavier, it would sink to the bottom in winter and, as a frozen mass, settle there. It would not melt in summer because the water above it is a poor conductor of heat and little warmth would reach the ice. Thus, with each succeeding winter in temperate and frigid zones, our lakes and rivers, and even our seas, could gradually accumulate ice at their bottoms, until finally they would be almost a mass of solid ice (*Chemistry,* September 1973, page 8).

This property of being lighter as a solid than as a liquid is most unusual, indeed. Nearly all liquids contract as they freeze, and become heavier instead of lighter. And as it cools even water contracts until it reaches a temperature of about 4° C. But as it is cooled further, it reverses its behavior completely and begins to expand until its volume has increased by about 9%. In nature this behavior is extremely rare.

Another important singularity in the freezing of water is that ice forms slowly, even on extremely cold nights. The process is greatly retarded by the heat of fusion given off. To convert a kilogram of water into a kilogram of ice requires four times as much refrigeration as it does to cool the water from 21° C to its freezing point. This heat-evolving property slows the formation of ice and thus prevents bodies of water from freezing solid in winter. It also helps to moderate temperatures of cold nights.

Just as the freezing of water is slowed by the heat evolved, melting is slowed by a reverse process where heat is absorbed. This helps to lessen or even to prevent floods in the springtime. It also enables snow and ice in glaciers to be stored in the mountains, and melt slowly in summer. Thus, throughout the warm season water is fed gradually to lower ground areas where it can be used by vegetation. Sometimes, the steady flow also spins hydroelectric power turbines.

The specific heat, which, except for ammonia, is higher than that of any other known liquid or solid in nature, enables water to absorb and store great quantities of heat from the sun and the air. So the oceans and lakes serve as immense storehouses of solar energy which regulate climate in many areas and moderate extremes in others. The Gulf Stream flowing from the Gulf of Mexico across the Atlantic carries warm water from the tropics to temper the climate of Northern Europe. This flow is called the heating apparatus of Western Europe.

Water is a medium of transportation. Early man swam in it and traveled over it by means of logs, canoes, boats, and ships. And water is admirable for such activities. It is dense enough to buoy up heavy objects and at the same time is mobile enough to flow with great freedom. If water were as light as gasoline, men or animals could not swim. A wreck at sea would be a more terrible calamity than it is now, because falling into the sea would mean almost certain death. Even fish would have to be much lighter.

Still other properties of water are extremely vital to the life on Earth. Especially important is that water dissolves more substances than any other liquid. Because it dissolves life-sustaining oxygen, as well as many minerals and other substances, populations of the world's waters is perhaps larger than those on land. And, because everyone must drink water and its vapor is everywhere, it is fortunate indeed that water has neither taste nor odor. It does not burn, puts out fires, and is not destroyed in the environment. In winter it does not fall from the sky as large frozen pellets like hail, but floats gently down as snow flakes which spread a warming blanket over the freezing ground.

Furthermore, the Earth's supply of water does not diminish with time. Each hydrogen-containing compound that burns adds to the water supply of the world. Except for nuclear power plants, water vapor pours from the stack of every powerhouse and from the chimney of every house that burns fuel. Every automobile engine manufactures about a gallon (3.78 litres) of water for each gallon of gasoline burned. Today the automobiles of the United States alone increase the amount of water in the world by nearly 100 billion gallons (378 billion litres) each year. Thus, as the world's supply of combustibles decreases, its water reserves increase.

Although water is one of the most amazing compounds in nature, it is composed of just two common elements, hydrogen and oxygen. If 1244 volumes of hydrogen are mixed with 622 volumes of oxygen and the mixture is ignited, for example by a spark, and if the vessel is cooled, one volume of liquid water is obtained. And the force of attraction between the hydrogen and oxygen atoms in a molecule of water is very, very high. Hermann Helmholtz, eminent German physicist, reported in 1881 that such attraction is 21 billion times the force of gravity.

Also amazing is the utter immensity of the volume of water on Earth—over 325 million cubic miles. That is 18 times the volume of all land above sea level. Water now covers about three fourths of the Earth's surface, but if all ups and downs of the surface were somehow ironed out, the seas would then blanket the whole Earth to a depth of about 8000 feet or 2.4 kilometres.

Should fossil fuels and sources of atomic energy, such as uranium and thorium, become exhausted, the unlimited waters of the seas might furnish fuel. That fuel would be hydrogen, which in some respects is ideal because of its high heat of combustion and its combustion product is water which in turn can furnish more fuel.

The production of hydrogen from water might be done somewhat as follows: About 0.02% of hydrogen in water is deuterium which is a hydrogen isotope having twice the mass of ordinary hydrogen. It is present in natural water as deuterium oxide, sometimes called heavy water. At high temperatures and pressures, deuterium nuclei combine in a process called fusion and release an immense amount of energy. That energy, converted to electricity, could then be used for the electrolysis of water to produce the hydrogen needed for fuel. However, the economical production of energy by controlled fusion is a problem yet to be solved, and years will pass before this method of producing hydrogen will be needed. But already deuterium is being extracted from the sea. And, as Alvin Weinberg, former director of Oak Ridge National Laboratory said, "It would be premature to say that economical fusion processes cannot be developed."

Indeed, this simple compound of hydrogen and oxygen, so vital to us and all-pervading in our world, is a wondrous liquid.

4

Atoms and Molecules

Making sense of the things we see sometimes depends on our ability to think in a new way. When chemists began to think in terms of atoms, the pace of discovery and understanding increased significantly. Today the subject of chemistry and a discussion of atoms and molecules go hand in hand. Until you are able to think about matter in terms of atoms and molecules, you can't go far in chemistry. The importance of this short chapter cannot be overemphasized.

OBJECTIVES

1. Draw pictures to represent an element and a compound. (See Problems 4-16 and 4-17.)
2. Represent and state the difference between a molecule of an element and a molecule of a compound. (See Problems 4-16 and 4-17.)
3. Draw pictures to represent the difference between a mixture and a pure substance. (See Problem 4-18.)
4. Using a microscopic model, describe how a solid, a liquid, and a gas differ. Describe the changes that occur when a substance goes from one physical state to another. (See Problem 4-19.)
5. Use pictures of atoms and molecules to describe what happens when a specified chemical change occurs. (See Problem 4-20.)
6. State how you could determine whether a pure substance is an element or a compound.

Words You Should Know

macroscopic	bond	law of definite composition
microscopic	molecule	
atom	decompose	electrolysis

4.1 MACROSCOPIC OBSERVATIONS AND MICROSCOPIC MODELS

Up to this point, you have been able to see or feel or smell the things we have discussed. We have been talking about **macroscopic** observations (*macro* means "large"; *scopic* means "viewing or observing"). Melting point, boiling point, heat of fusion, temperature, and mass are all properties of large chunks of matter and are called macroscopic properties. We can learn a lot about matter from such properties. But we are still curious. We wonder, "What makes matter behave as it does?" If we could see deep inside a mixture, how would it differ from a pure substance? Is there something about the inner structure of matter that explains why a pure substance has a definite boiling point but a mixture does not? Is there some way to explain why sugar and salt can be broken down into simpler substances but iron and aluminum cannot?

"Explain" is a devious word. When we explain, we often think that we are talking about things as they really are, when in fact we are only talking about things as they might be. Some scientific explanations simply provide a way to think about why things happen. Sometimes they are accurate descriptions of what happens, but at other times they are not. We must realize that explanations in science may be intelligent guesses rather than established fact.

In this chapter we will begin to develop a **microscopic** model to explain the behavior of matter (*micro* is a prefix meaning "small"; when used with SI units, it means one-millionth). We will not attempt to explain everything completely at this point. As you continue to study chemistry, the model will take on more detail, enabling you to explain more things about matter.

4.2 ATOMS

We have spoken as though matter is composed of some kind of small pieces. Is it? Everybody seems to think so. Atoms and molecules are discussed in the popular literature as though they were as well known as toads and calluses.

People have not always believed that matter is made of atoms. It was not until the early 1800s that the notion became popular. The original idea of atoms was developed from a kind of armchair argument. If one takes an iron nail (or any other matter) and breaks it and breaks it again and again, it seems reasonable that one will eventually obtain some "smallest possible piece" that can still be called iron. The word **atom** means this "smallest possible piece" of something. We begin our microscopic description of matter with this simple notion.

We believe that *all* ordinary matter is made up of atoms. Different kinds of matter differ in the kinds of atoms present and the way in which the atoms are arranged. Mixtures as well as pure substances must be made up of atoms, and we will discuss how the atoms may be arranged differently in such substances.

Since we can't see real atoms and it is easier to talk about things that can be seen, atoms are normally represented with spheres of various sizes and colors. A sphere of a particular size and color is used to represent atoms of one kind, and a sphere of a different size or color is used to represent atoms of a different kind. The size of the spheres are usually selected to suggest the relative sizes of the actual atoms. Atoms that we think are very small are represented by small spheres, and atoms that we think are much larger are represented by larger spheres. It is difficult to keep everything to scale, and you should not assume that the spheres are exact representations of real atoms. We simply use the spheres to help us imagine what is happening in the microscopic world that we cannot see.

In the following sections, we will use sketches of such spheres to describe what is happening when wood is heated, a mixture is distilled, or a pure solid is melted.

4.3 THE NATURE OF ELEMENTS

Chapter 3 ended with the observation that there are two kinds of pure substances: compounds, which can be broken apart into simpler substances, and elements, which cannot. We'll begin our discussion of the microscopic world by describing elements.

The differences in the behavior of elements and compounds can be explained by the fact that elements contain only one kind of atom and compounds contain more than one kind of atom.

Copper is an element. If we imagine a tiny piece of copper from a penny being magnified to enormous size, it might look something like Figure 4.1. Each sphere represents an atom of copper. These atoms are held together by some kind of force or **bond.** We will not worry about bonds until later. The bonds that hold atoms together in a copper penny are strong enough for the penny to hold its shape, but not so strong that they cannot be broken.

Figure 4.1
If the copper in a penny were pure, we think that it would contain identical atoms in an ordered array, as shown here. The atoms are in fixed positions.

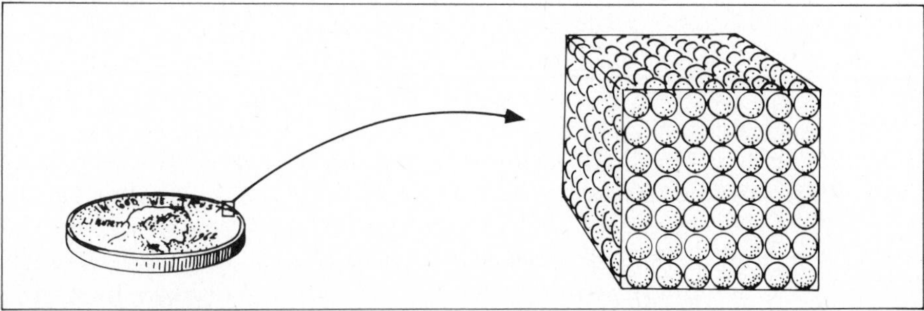

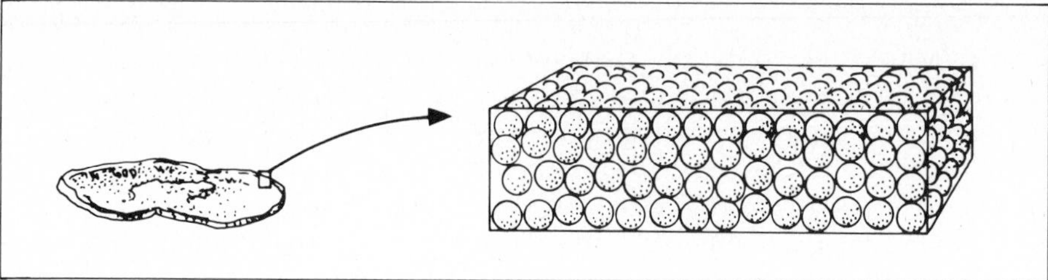

Figure 4.2
When a metal like copper is hammered, the atoms are moved around in the solid, but they are not changed. The atoms remain fixed in their new positions.

You know that copper is malleable. It can be hammered flat, as shown in Figure 4.2. The energy of the falling hammer breaks some of the bonds between atoms and moves them before new bonds form to hold the atoms in a new position. This ability of the atoms to form new bonds in almost any new position is characteristic of metals, but it is not characteristic of all solids.

When copper or another metal is heated to a high temperature, it melts. In the liquid state, it does not hold its shape and can be poured from one container to another. This means that the bonds holding the atoms in fixed positions must have been broken or severely weakened. Otherwise, the atoms could not flow past one another as the liquid is poured. That's where the heat goes when melting takes place. Instead of raising the temperature of the material, the heat of fusion goes to work breaking the bonds between atoms.

4.4 HOW TEMPERATURE AFFECTS ATOMS

The mention of temperature is a reminder that the model depicted in Figure 4.2 lacks something. How do a copper penny at 20°C and one at 30°C differ? To increase the temperature, energy must be added. What does the energy do to the atoms?

We believe that temperature is a measure of the motion of atoms in a substance. Figures 4.1 and 4.2 show the atoms motionless. This is misleading. Like students seated in a lecture hall, atoms are in fixed positions, but they can still wiggle a lot. The hotter it gets, the more they wiggle. At some point, the wiggle of the atoms is so great that the bonds holding them in place are overcome and the solid is said to melt. How much the atoms can wiggle before the bonds break, and how much energy is required to overcome the bonds, depend on the kind of atoms in the solid. Thus both the melting temperature and the heat of fusion are characteristic of the kind of matter present.

Once copper or any other element has melted, it might look something like Figure 4.3. Although the position of each atom is no longer fixed, the atoms must still be close together. We believe this is so, because neither solids nor

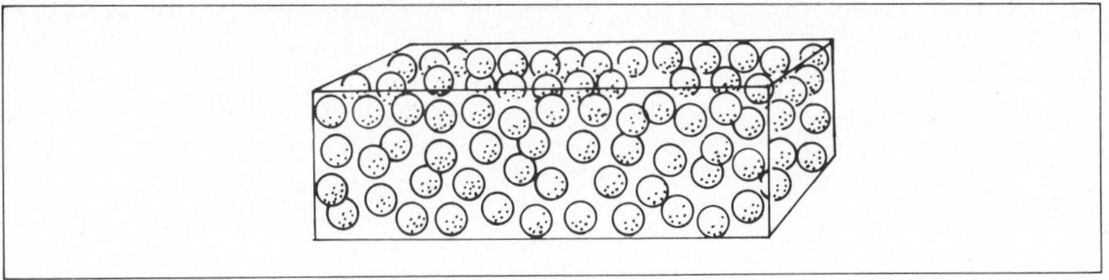

Figure 4.3
Atoms in a liquid are close together, but they are free to move past one another or "flow." For most materials, the atoms require slightly more space in the liquid than they did in the solid state.

liquids can be compressed very much. No matter how hard we push, we can't get atoms in a solid or a liquid much closer together.

When liquid copper is heated to a very high temperature (2,567°C), it finally boils. When this happens, the atoms move very far apart and the gaseous copper must have *much* more room to move about in. Figure 4.4 represents our model of an element in the gaseous state, except for two details. First, the spheres should be moving about in a random manner. Second, the spheres should be shown much farther apart.

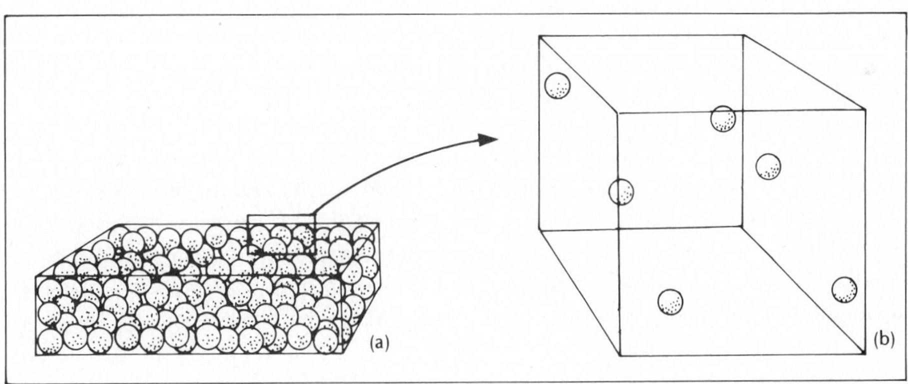

Figure 4.4
Atoms in a gas are very far apart and are moving about in constant, random motion.

4.5 ATOMS AND MOLECULES

Figure 4.4 shows that elements in the gas phase exist as individual atoms moving about independently of one another. This is normally what happens with elements, but not always. Several elements exist as particles composed of two atoms hooked together. At least one element, phosphorus, exists as four atoms connected in a single particle, and sulfur exists as a ring of eight atoms hooked together. Several of these particles are shown in Figure 4.5.

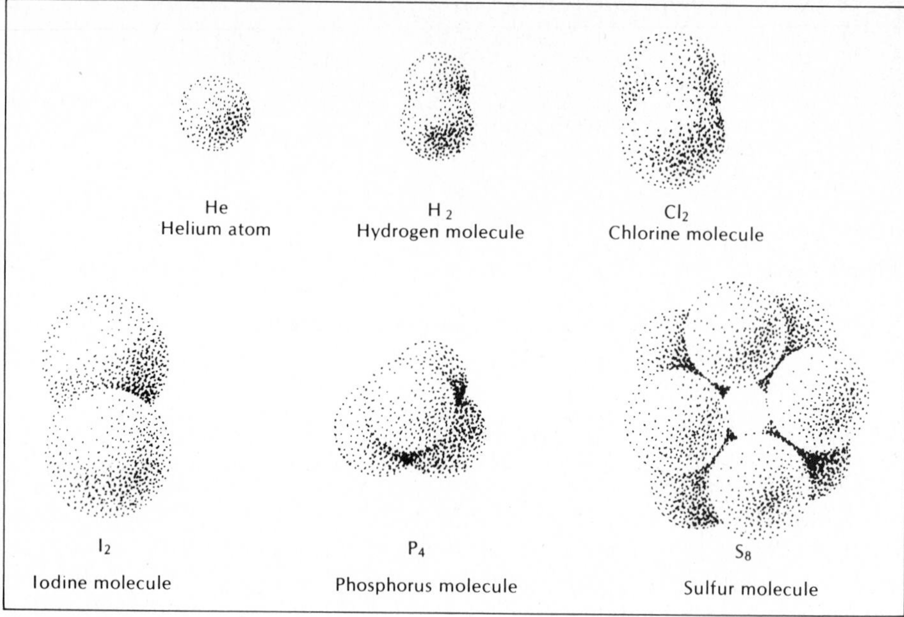

Figure 4.5
Most elements exist as isolated atoms in the gas phase. Helium is an example of such an element. Some elements exist as diatomic molecules, however. Hydrogen, chlorine, and iodine are examples of such elements. A few elements exist as larger molecules. Phosphorus can exist as a molecule of four atoms, and sulfur can exist as a molecule of eight atoms.

Particles made up of more than one atom are called **molecules.** The particles shown in Figure 4.5 are molecules of various elements. There is no simple way to tell which elements exist as individual atoms and which exist as molecules. However, when we look at elements more systematically in the next chapter, you will learn about common elements that exist as diatomic molecules (*di* means "two"; diatomic molecules have two atoms).

4.6 OTHER MOLECULES: COMPOUNDS

Only a few of the elements exist as molecules, but all compounds exist as molecules. Compounds are "compounded" from elements. When two or more elements combine to form a compound, identical particles made up of several atoms are formed. Some of the molecules produced when wood is heated are shown in Figure 4.6.

Compare Figure 4.6 and Figure 4.5. In both cases the particles shown are molecules, because there are two or more atoms stuck together in a single unit. In Figure 4.5, however, all atoms in the molecule are alike, so we have molecules of an *element.* In Figure 4.6, particles consist of at least two kinds

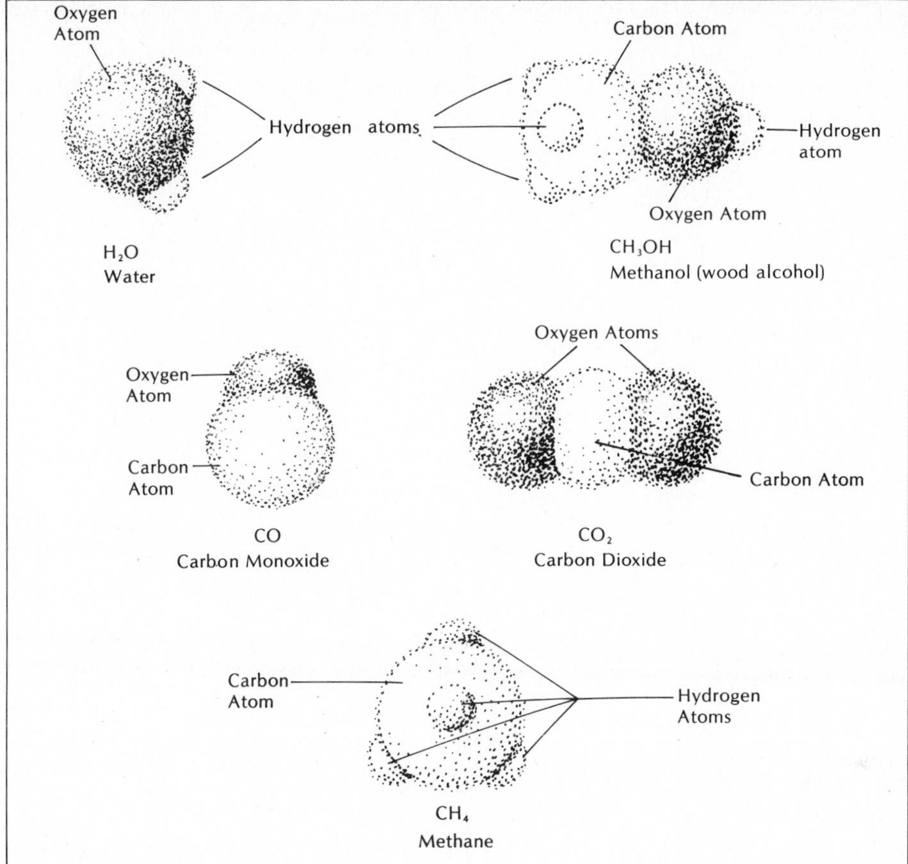

Figure 4.6 *Some molecules that are formed when wood is heated.*

of atoms stuck together. These particles are molecules of *compounds.* In a pure compound, all of the molecules are identical, and we say that *compounds are made up of a single kind of particle.* But you must keep in mind that the particle we speak of is the molecule. Even though these *molecules* are all identical, the *atoms within* the molecule are different.

Before going on, answer the following questions about Figure 4.7, page 106, to be sure you understand what we have said about atoms and molecules.

4-1. Does diagram (a) represent an element or a compound?

4-2. Diagram (b) shows two kinds of atoms. Does it represent a compound? Explain your answer.

4-3. There are two diagrams that show only molecules. Which ones are they, and how are they different?

4-4. All of the diagrams represent substances in a single physical state. Do they represent solids, liquids, or gases? How do you know?

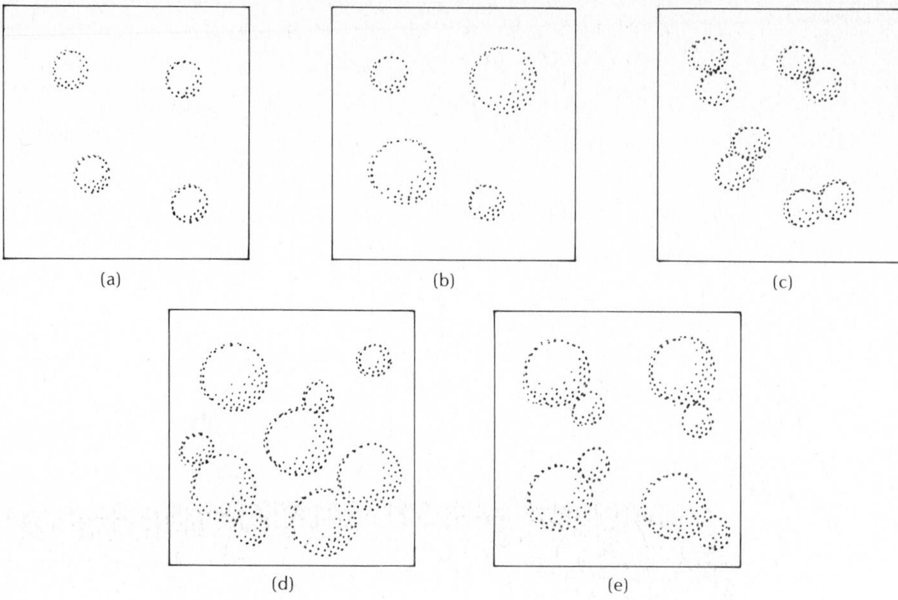

Figure 4.7

4.7 CHEMICAL AND PHYSICAL CHANGES

Now that you know the difference between an element and a compound, we can use the notion of atoms to explain the difference between chemical and physical changes. In a physical change, the same kind of matter assumes a different form. Figures 4.1, 4.3, and 4.4 (pp. 101–103) show copper metal as a solid, a liquid, and a gas. In all three pictures, the copper is represented by the same kind of particle. Each sphere represents an atom of copper, and in all three figures, the only particles represented are atoms of copper. Although there are changes in how orderly the arrangement of the atoms is and how far apart the atoms are, the particles are the same. We describe such changes as physical changes.

It might help to illustrate these same physical changes with a compound. Figure 4.8 shows water as a solid, a liquid, and a gas. As before, the particles in the solid are shown in a fixed position. In the liquid, the particles are free to move past one another and are arranged randomly. In the gas, the particles are far apart and move randomly.

Two molecules have been shaded in Figure 4.8 to emphasize that the same kind of particle is present in all three phases. In physical changes such as this, the kind of particle making up the material does not change, even though the arrangement of the particles may.

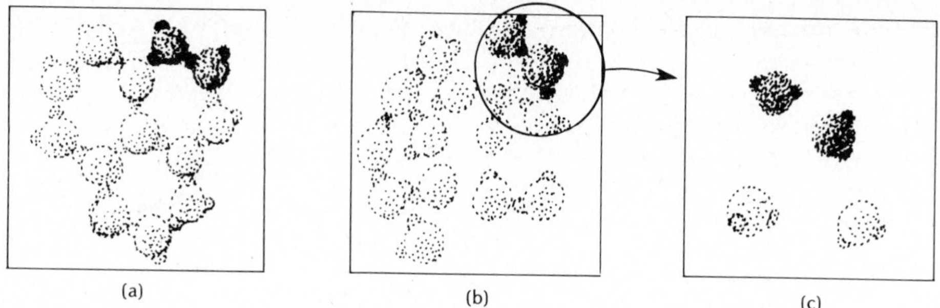

Figure 4.8
Water as a solid (ice), a liquid, and a gas (steam or water vapor). (a) In a solid compound, all of the atoms are stuck together in fixed positions, and atoms of one molecule are bonded to atoms of another molecule. However, some of the bonds are stronger than others. (b) When the solid melts, it breaks apart into molecules rather than separate atoms. (c) When the liquid evaporates, the same molecular particles persist.

4-5. Which diagram(s) in Figure 4.9 could represent a solid compound?
4-6. Which diagram(s) in Figure 4.9 could represent a gaseous element?
4-7. Which diagram(s) could represent neither a pure element nor a pure compound?

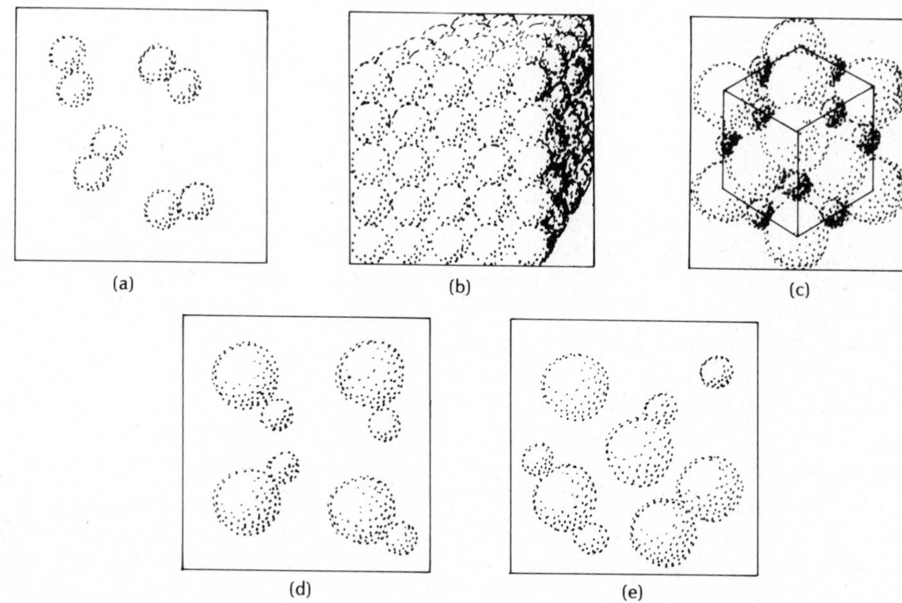

Figure 4.9

4-8. A carbon dioxide fire extinguisher contains liquid carbon dioxide under pressure. When this pressure is removed, the carbon dioxide is released as a gas. In the process, the escaping gas gets very cold and some of the carbon dioxide forms a solid powder—dry ice. Fill in the following drawing to show what the carbon dioxide might look like in these three forms. One molecule of carbon dioxide looks like this:

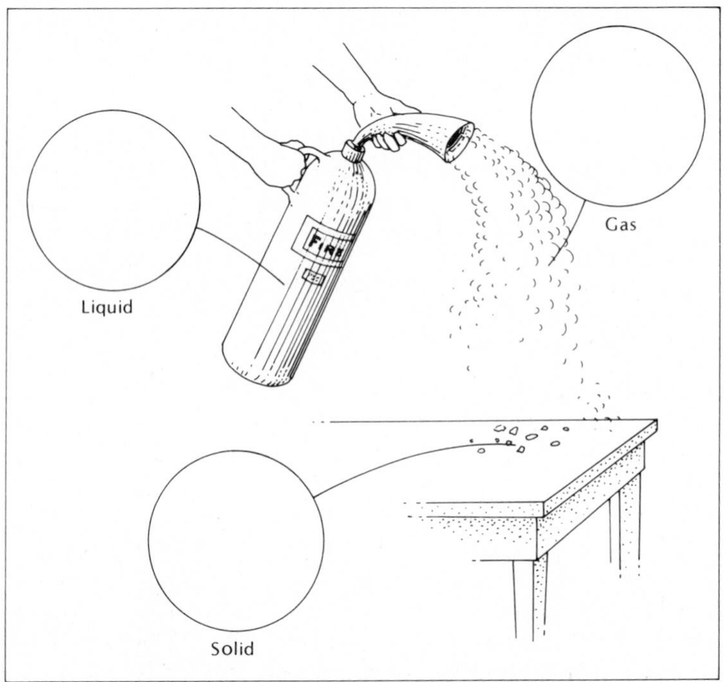

Now let us extend our discussion to show a chemical change. In Chapter 3 we mentioned the **electrolysis** of water. When an electric current is passed through water, two gases are formed. One of the gases is lighter than air and will burn with a pale blue flame. It has properties identical to hydrogen gas. It *is* hydrogen gas. From the properties of the other gas, we know that it is oxygen. These are new substances with new properties, so a chemical change must have taken place. What does this chemical change look like at the microscopic level? Figure 4.10 (facing page) gives the answer.

Note that the particles in the liquid are molecules of water. The particles in the gas are either molecules of hydrogen or molecules of oxygen. The particles have changed. They do not look the same. They have different properties. There has been a chemical change. In this particular chemical change, a compound has **decomposed** into its elements.

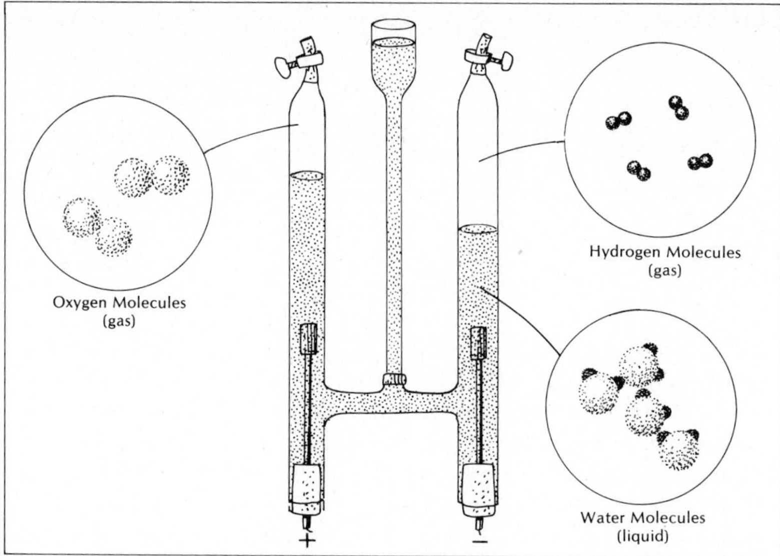

Figure 4.10
Microscopic changes during the electrolysis of water. The same atoms are present before and after the electrolysis, but the atoms are connected to make different kinds of particles. The water is made up of particles containing one oxygen atom and two hydrogen atoms. The hydrogen gas contains only hydrogen atoms joined together to form diatomic molecules. Oxygen gas also exists as diatomic molecules of oxygen atoms.

4-9. Diamond is pure carbon. At very high temperatures, a diamond burns. It combines with oxygen gas in the air to produce carbon dioxide gas. Draw pictures to illustrate the changes that take place at the microscopic level when a diamond burns.

4-10. Is the burning of a diamond a chemical change or a physical change? Why do you think so?

4.8 MIXTURES OF ATOMS AND MOLECULES OF ATOMS

Diagram (e) in Figure 4.9 represents a mixture. There are several kinds of particles in the diagram. Just as heterogeneous mixtures are made of large particles that are not all alike, homogeneous mixtures are made of smaller particles that are not all alike.

There can be confusion about what is meant by a particle. Look at Figure 4.11. How many "particles" are there in the picture? If particle means "bike," the answer is clearly two. This is what most people would answer. But look

Figure 4.11 *Two bicycles would normally be considered two particles.*

at Figure 4.12. Here the two bikes have been taken apart, and it is clear that each bike is made of many particles. The problem is similar to the one we face when we talk about atoms and molecules. If atoms are stuck together in a molecule, we usually think of the molecule as a single particle, just as we think of a bike as a single particle when it is fully assembled. When the molecule decomposes to form separate atoms, we think of the atoms as separate particles—just as we think of the bicycle parts as particles when the bicycle is disassembled.

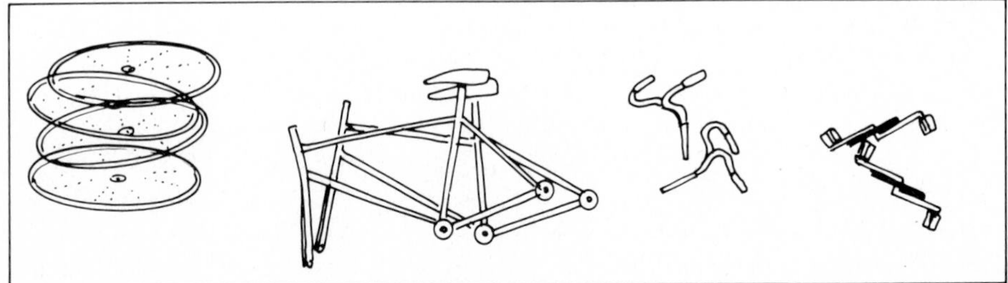

Figure 4.12 *When the bicycles are taken apart, we say that there are many particles. The parts are no longer connected.*

If you keep this idea of particles in mind, you will not confuse pictures representing pure compounds with pictures representing mixtures. Diagrams (c) and (d) in Figure 4.9 (p. 107) represent a pure compound. The only difference between the two is that diagram (c) represents a solid and diagram (d) represents a gas. Even though there are two kinds of atoms in these diagrams, we would not say that there are two kinds of particles. The atoms are stuck together to make molecules containing one atom of one element and one atom of another element. Note that each molecule has the same composition. The elements are always present in the same ratio. This characteristic of compounds is known as the *law of definite composition*. A particular compound has molecules that are all alike. There is a definite composition to the molecules.

In the mixtures shown in diagram (e) of Figure 4.9, there are different kinds of particles that are not joined together. One particle is a single small atom. Another particle is a single large atom. Still other particles are made up of a single small atom and a single large atom stuck together. The particles of this substance have no definite composition. Furthermore, there would be no definite composition to the mixture as a whole. You could make a mixture with many more small atoms or one with many more molecules. Mixtures have a variable composition. Compounds have a fixed composition.

4.9 MICROSCOPIC CHANGES DURING MELTING AND DISTILLATION

In Chapter 3 we found that the properties of mixtures are not the same as the properties of pure substances. Let us see whether the atomic model can explain such differences. You should be able to explain some of the behavior yourself. As before, get a sheet of paper and place it at the first dotted line. After you have read to that point, try to write your own explanation before moving ahead to read mine.

Example 4.1

Pure solids are held together in a rigid mass by some kind of force. Since the particles in a pure solid are all identical, the forces between these particles should also be identical (or nearly so). As a solid is heated, the heat makes the particles in the solid wiggle or vibrate until, finally, some of the bonds between the particles break and the particles become free to move about. When this happens, we say that the solid is melting. In a pure substance, the temperature does not change while melting is taking place. Why? Draw a picture to illustrate the melting of a pure compound like water.

- -

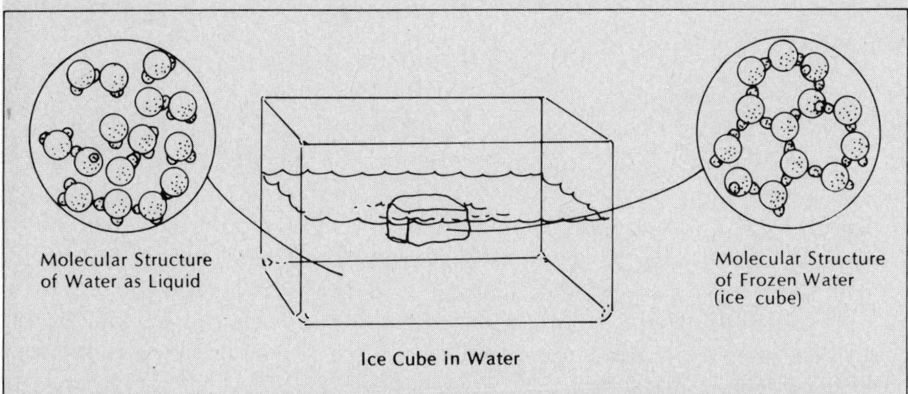

Figure 4.13 *When ice melts, two atoms of hydrogen are still connected to one atom of oxygen to form a single particle of water. However, these molecules of water are no longer bonded to one another.*

Since all of the particles are alike in a pure solid, the bonds that hold the particles together must also be alike. If the temperature is high enough to cause vibration which breaks some bonds, that vibration is intense enough to break any of the bonds. As heat is added during the melting process, it can break more bonds, but it can't cause particles that are still stuck together to vibrate more violently than those that have already come unglued. Therefore, there is a period of time during the melting of pure substance when the heat being added is all being used to overcome the forces that hold the particles together in the solid. The temperature of the substance cannot rise until all of the solid melts.

Unlike a pure substance, mixtures are not made up of identical particles. Would you expect the forces holding particles together in a solid mixture to be identical?

Probably not. If all of the people in your class joined hands, there would be stronger "bonds" between some people than between others. As in the childhood game of Red Rover, some connections would be easier to break than others. Using this idea, can you explain why mixtures change temperature as they melt? Draw a picture to illustrate a solid mixture of two elements melting.

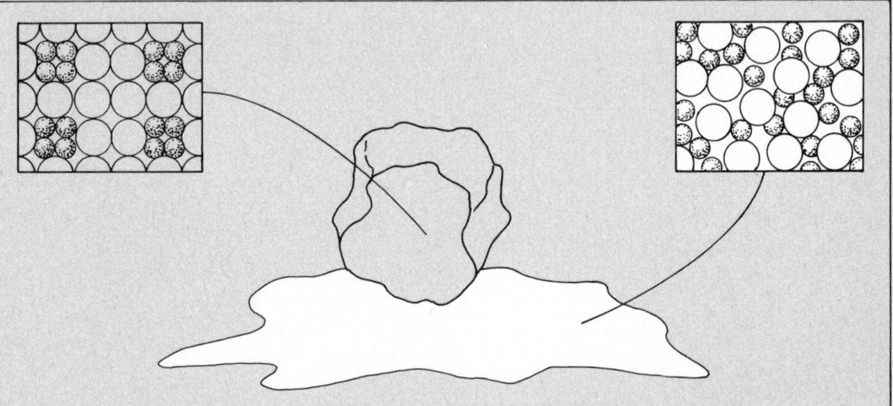

Figure 4.14
Mixtures are composed of many different kinds of particles assembled in a random way. The bonds between some particles are weaker than the bonds between other particles. Consequently, mixtures melt over a wide temperature range.

How you said it doesn't matter, but you should have conveyed the idea that, as the temperature is raised, the weaker forces between particles begin to break first, freeing some particles to move as a liquid. However, since forces between other particles are stronger, the temperature must be higher before these particles come apart. As the solid melts, the temperature slowly rises until, finally, even those particles held together with the strongest forces are shaken apart by the increasingly turbulent vibration of the molecules.

In Chapter 3 we discussed the separation of a liquid mixture by distillation. Complete separation of liquid mixtures by distillation is difficult, but mixtures like salt and water are easily separated this way. What would you do to separate salt and water by distillation? Draw a picture to illustrate what happens, using to illustrate salt and to illustrate water.

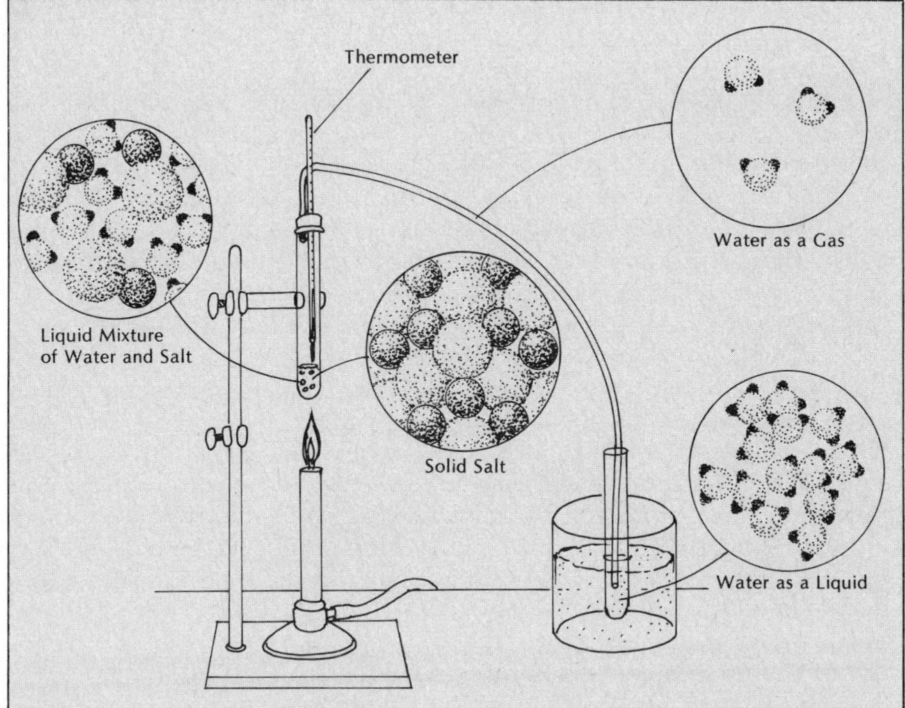

Figure 4.15
The figure shows the microscopic nature of the salt and water as they are separated by distillation.

Distillation is separation by boiling. If salt water is boiled, the water changes to gas before the salt. The gaseous water can be passed through a tube and cooled until it returns to the liquid phase. The salt is left behind in the original container. Can you describe this process on a microscopic level?

The energy required to make salt behave as a gas is much higher than the energy required to make water behave as a gas. As a mixture of salt and water is heated, energy is added and the molecules move faster and faster. At some point, the water molecules have enough energy to overcome the forces that hold the molecules together in the liquid mixture. They speed away as a gas. However, since the energy needed to change salt particles to gas is much greater, no salt is changed to gas at this temperature. As more heat is added, more water boils away. All of the energy added is used to overcome the forces holding the water molecules in the mixture. There is only a very gradual change in temperature so long as water is present. Once all the water is gone, the heat raises the temperature of the salt. If the temperature is raised high enough, the salt finally melts and then boils. However, this temperature is so high that it is seldom reached. Salt cannot be boiled with an ordinary burner.

4.10 CHEMICAL CHANGES OF ELEMENTS

Some changes in matter are easier to understand when viewed at the microscopic level of atoms than when viewed at the macroscopic level of everyday observation. In Chapter 3 we pointed out that it is difficult to determine whether a pure substance is an element or a compound. We know that a compound can be broken down into simpler substances and that an element cannot, but how do we know when a chemical change involves breaking down rather than building up? When an element reacts chemically, it must always produce a substance that weighs more than the original piece.

When a piece of copper is heated in air, the copper color disappears and a dull, black solid appears in its place. Does the copper decompose, or does it react with something in the air?

The clue to what happens comes from the observation that the black solid weighs *more* than the original copper. In fact, 100 g of copper produce about 125 g of the black solid. You already know that oxygen in the air combines with many substances. It will come as no surprise that oxygen combines with copper to form the black solid known as copper oxide. We can represent the experimental facts like this:

100 g copper + 25 g oxygen → 125 g copper oxide

Figure 4.16

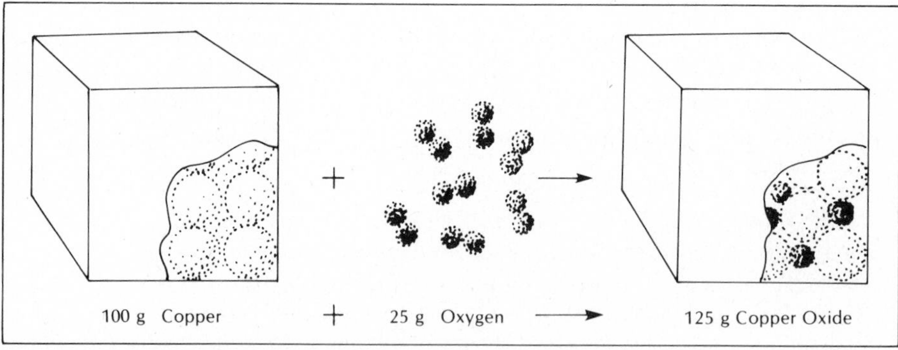

Figure 4.16 shows what we think is happening at the microscopic level. Copper oxide consists of alternating atoms of copper and oxygen in a black crystal. It has a definite composition and is a compound rather than a mixture. This compound always has 25 g of oxygen for every 100 g of copper in it.

4-11. Using the idea of atoms and the fact that copper oxide has a definite composition (one atom of copper for each atom of oxygen), explain why copper oxide always has 25 g of oxygen for every 100 g of copper.

4-12. Which weighs more, an atom of copper or an atom of oxygen? How do you know?

4.11 CHEMICAL CHANGES OF COMPOUNDS

The only way for an element to change chemically is to combine with some other element or compound to form a new molecule. This isn't true of compounds. Since compounds are composed of different atoms hooked together, they can change chemically by decomposing to form elements, decomposing to form other compounds, or combining with some element or compound to form a more complex compound. An example of each of these changes is described in the following exercises. Try to draw pictures to represent the starting materials and the final products of the changes described. If you can, you are well on your way toward thinking about matter in terms of atoms and molecules.

4-13. Ordinary table salt is a compound of two elements, sodium and chlorine. Solid salt is represented nicely by diagram (c) in Figure 4.9. The large spheres represent the chlorine, and the small spheres represent the sodium. When salt is melted and an electric current is passed through it, the compound decomposes into the elements in much the same way that water decomposes into hydrogen and oxygen (see Figure 4.10). Draw pictures to represent the change:

table salt → sodium metal + chlorine gas

4-14. Carbon monoxide gas is a compound composed of one atom of carbon and one atom of oxygen. It will combine with oxygen gas to form molecules of carbon dioxide gas. Carbon dioxide has two atoms of oxygen and one atom of carbon in each molecule. Draw pictures to illustrate the change:

carbon monoxide gas + oxygen gas → carbon dioxide gas

4-15. Ordinary baking soda is a compound composed of four elements: sodium, hydrogen, carbon, and oxygen. A molecule of the compound might be represented like this:

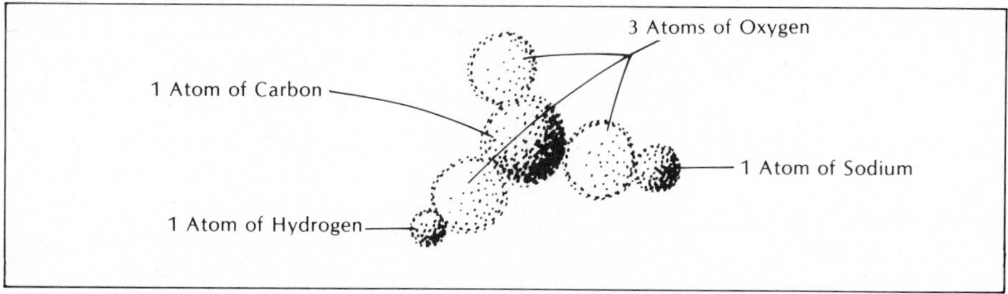

When baking soda is heated to a high temperature, the molecules break apart into three new molecules. One is water, another is carbon dioxide gas, and the third is sodium oxide, a molecule composed of two sodium atoms and one oxygen atom. Draw pictures to represent the change:

baking soda → sodium oxide + water + carbon dioxide gas

You probably had some trouble drawing the pictures. As molecules get larger and several atoms are included in each molecule, it becomes difficult to represent the molecule with pictures. And the molecules that have been discussed here are among the simplest that exist! Many of the molecules making up your body contain millions of atoms, and many of the simpler molecules have 100 atoms. Even though we must continue to think in terms of atoms and molecules, we can't go on drawing pictures. In the next chapter, you will begin to learn a language to describe what happens to atoms and molecules during chemical changes.

4.12 SUMMARY

Although it may not seem so, we have come a long way in our discussion of matter. Before going on, let us review a few points that we have made.

We began by saying that chemists are interested in matter. As chemists, we want to know how matter behaves and why. In attempting to explain observations about matter, we have noted that matter may be pure or impure. If it is pure, it is composed of only one kind of particle and its properties are uniform throughout. Some mixtures *appear* to be uniform throughout, but it is possible to separate these mixtures into components and then put the components back together again to form the original mixture. In most cases, this is fairly easy to do.

We have also seen that some pure substances—those that we call compounds—can be separated into components, but this separation is usually much more difficult to perform and requires more energy. The kind of particle that makes up compounds is called a molecule. Molecules are particles composed of more than one atom. When elements form molecules, the atoms are all alike. In compounds, the atoms in the molecule are not all alike. The atoms in a molecule are bound together as a single unit, and the properties of the compound are due to the properties of the molecule rather than to the properties of the individual atoms.

It takes a lot of energy to separate the atoms in a compound, but it can be done. When we do separate compounds into components, each component being composed of a single kind of atom, we call the components elements. Elements are the "elemental" building blocks of all matter, and, as mentioned earlier, there are only about 100 of them in all the universe. (See Chapter 5 or the inside cover of the text for the names of all known elements.)

When compounds are broken down into elements or other compounds, or when elements combine to form compounds, new kinds of particles are formed. We call this kind of change a chemical change. When mixtures are separated into components or when elements or compounds are changed from solid to liquid to gas, there is no change in the kind of particle that exists. We call this kind of change a physical change. Physical changes usually involve much less energy than chemical changes.

We have used spheres of various sizes and colors to represent atoms, one kind of sphere for each element. We have represented the molecules of compounds by showing these spheres hooked together to form a larger particle in much the same way that the parts of a bicycle go together to make a single unit. These are far from perfect representations, but they can help

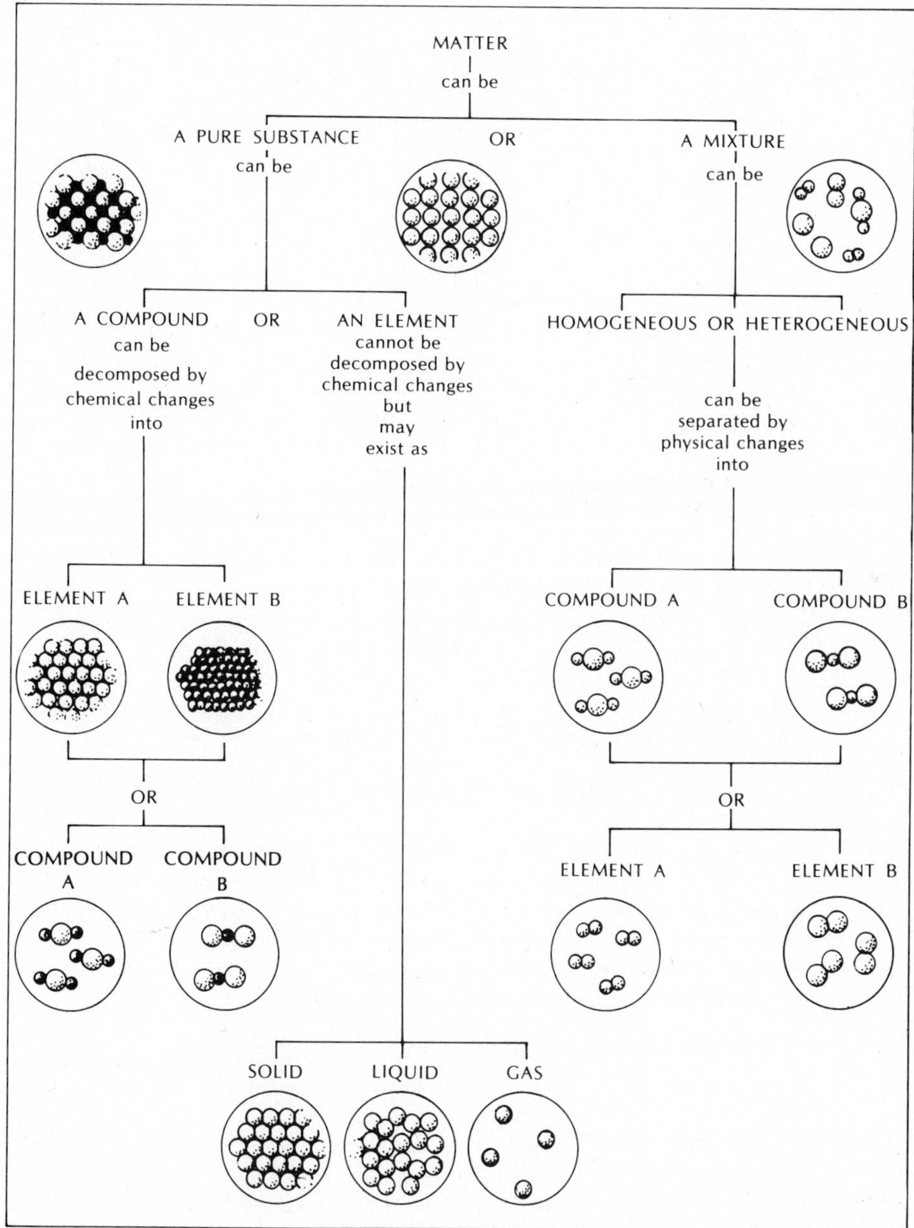

Figure 4.17 *Much of what has been said about matter is summarized here. Study the figure carefully, paying particular attention to the nature of the microscopic particles that make up each kind of matter.*

us imagine how atoms and molecules differ. And with them we can explain most of what we know about the behavior of matter.

Other models can be used to explain how matter behaves, and they are better than our simple one for some purposes.

Questions and Problems

4-16. Different elements may exist at room temperature in slightly different forms. The condition of several elements at room temperature and normal pressure is described in the following list. Draw a picture to represent the microscopic nature of each element.

a. Nitrogen, the major component of air, is a diatomic gas.
b. Neon, the element in neon lights, is a gas that exists as individual atoms.
c. Tin is a solid metal.
d. Mercury is a metal that is liquid at normal temperatures.
e. Bromine is a liquid consisting of diatomic molecules.

4-17. Draw a picture to represent a single molecule of each of the following compounds.

a. Hydrogen sulfide is the odor from rotten eggs. The molecule has two hydrogen atoms stuck to a single sulfur atom.
b. Two simple molecules that pollute the air are nitrogen monoxide and nitrogen dioxide. Both molecules have an atom of nitrogen. The monoxide has a single oxygen atom connected to the nitrogen atom; the dioxide has two oxygen atoms.
c. Sulfur dioxide is the most serious air pollutant released from burning coal. It contains one sulfur atom and two oxygen atoms.
d. Ammonia is used for cleaning, for fertilizer, and as a raw material for several chemical processes. It is a simple molecule composed of three hydrogen atoms connected to a single nitrogen atom.

4-18. Draw pictures to represent the following different arrangements of atoms.

a. A mixture of hydrogen and oxygen gas, and a compound formed from hydrogen and oxygen
b. Two oxygen atoms, and one oxygen molecule
c. A mixture containing two elements, and a compound formed from those elements
d. A mixture of two compounds

4-19. Draw a picture to represent some substance as a solid, as a liquid, and as a gas. After drawing the pictures, describe what would happen to the atoms or molecules as the solid is slowly heated until it becomes a gas.

4-20. Draw pictures to illustrate what might happen in the following reaction to produce iron from iron ore.

$$\text{iron ore} + \text{coke} \rightarrow \text{iron metal} + \text{carbon dioxide gas}$$

Assume that iron ore is a pure compound containing one atom of iron for each atom of oxygen, and assume that coke is pure carbon.

4-21. Questions 3-19, 3-20, 3-23, and 3-24 describe a number of changes. Try to explain these changes in terms of atoms and molecules. Share your explanation with a friend. Get her or him to tell you why it is a good explanation or a bad one.

Answers to Questions in Chapter 4

4-1. An element

4-2. No. The atoms are not joined together to form a molecule.

4-3. Diagrams (c) and (e) show molecules. Diagram (c) shows molecules of an element. (Both atoms in the molecule are alike.) Diagram (e) shows molecules of a compound. (The two atoms in the molecule are different.)

4-4. All the diagrams represent gases. The molecules (or atoms) are shown rather far apart. In both solids and liquids, the particles would be touching.

4-5. Diagram (c) could represent a solid compound. The atoms are touching, there are two kinds of atoms, and the atoms are arranged in an ordered array.

4-6. Diagram (a)

4-7. Diagram (e)

4-8.

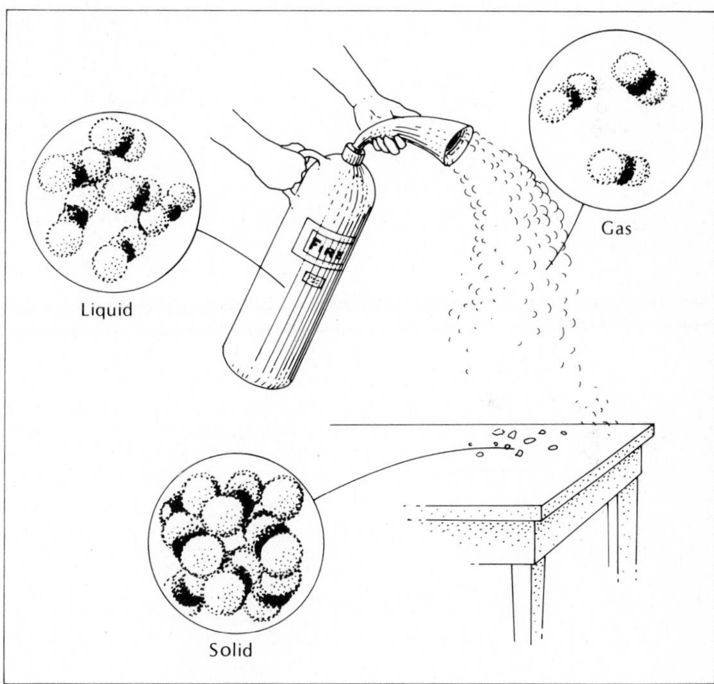

4-9.

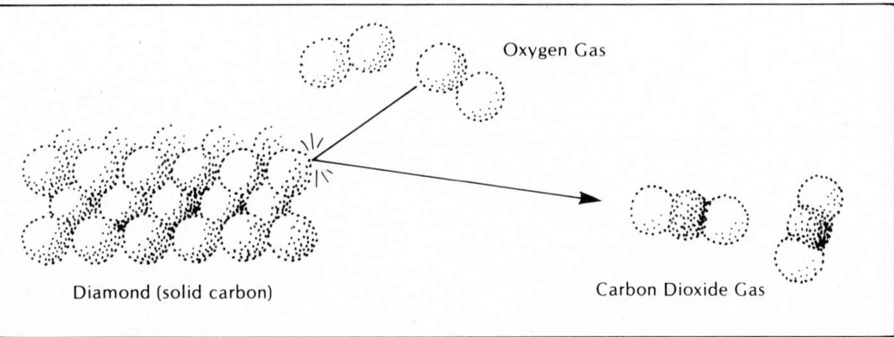

4-10. It is a chemical change. A different substance (carbon dioxide) with a different kind of molecule and different properties is formed.

4-11. If all copper atoms weigh the same and all oxygen atoms weigh the same, a fixed ratio of copper and oxygen atoms would result in a fixed ratio of copper and oxygen mass in the compound. (This relationship will be discussed at length in a later chapter.)

4-12. Copper. In copper oxide, the number of copper atoms is equal to the number of oxygen atoms. If the copper weighs more, copper atoms weigh more.

4-13.

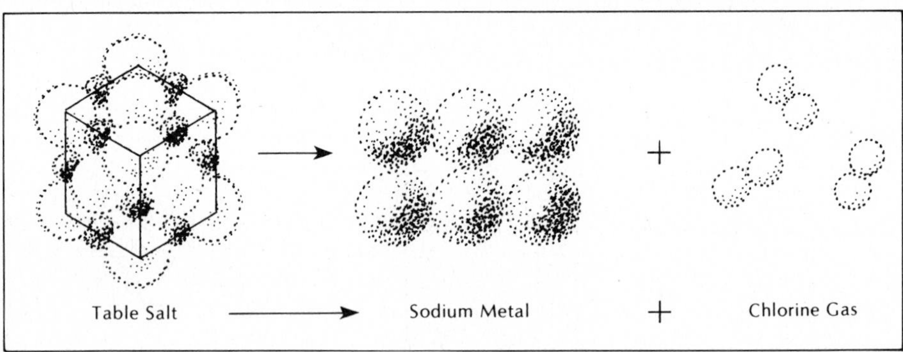

4-14.

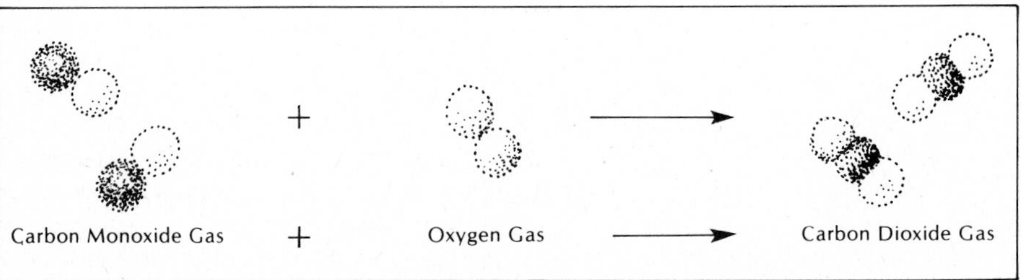

4-15.

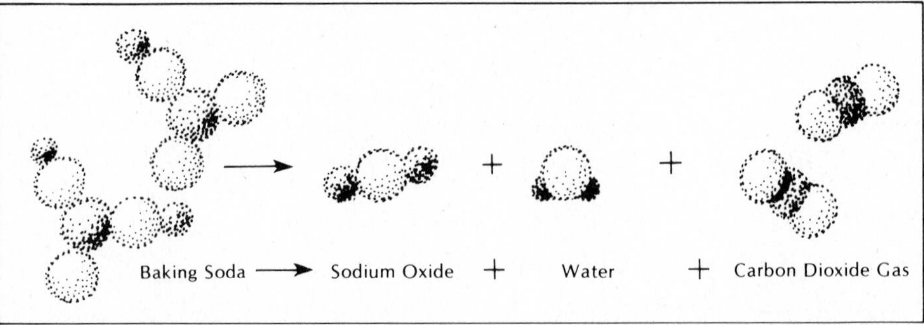

READING The law of constant composition, which was introduced in this chapter, was a very important development in chemistry. The chemist largely responsible for this law was Louis Joseph Proust. He is recognized as a great experimentalist. However, even the best scientists make mistakes. The following article describes Proust's discovery of a new element—and his subsequent discovery that he was wrong.

"EIN NEUES METALL"

William H. Waggoner

Louis Joseph Proust (1754–1826) is most generally remembered as the French chemist responsible for the law of constant composition. Less well known is the fact that most of his professional career was spent in Spain. There he was associated with the seminary at Vergara, the artillery school at Segovia, and the university of Salamanca, before being made director of the Royal Laboratory in Madrid, a splendid laboratory in which, it was said, "even the common apparatus was of platinum." In 1808, while Proust was on leave in France, the laboratory was wrecked by the citizens of Madrid during the siege of that city by Napoleon's army.

Widely regarded as an analyst, Proust published numerous memoirs on the compositions of minerals and metallic compounds. It was he who developed the use of hydrogen sulfide as a precipitant for heavy metals and laid the groundwork for the classical scheme of qualitative analysis. In 1802, Proust wrote to Jean Claude Delametherie (1743–1817), editor of the *Journal de Physique,* about "Silene, a new metal which I have found in an Hungarian lead ore."

The alleged new element had not been obtained in the metallic state and Proust thought that the reduction would be difficult to accomplish, "as it retains oxygen with great force." Two oxidation states were recognized. In the maximum state, the oxide, its solutions, and gases were yellow; in the minimum state, they were green. A continuing investigation was implied.

In a second letter written later that same year, Proust disposed of his alleged discovery in a single sentence: "The new metal has been found to be nothing but uranium." He requested that his work be made known, however, as it supplemented that published by Martin Heinrich Klaproth (1743–1817), the discoverer of that element. Proust's two letters were abstracted by Ludwig Wilhelm Gilbert (1769–1824) and published in 1803 in the German journal which he edited. Following the nomenclature used by Proust, Gilbert referred to "ein neues Metall, le Silene." In the collective index to the *Annalen* which appeared in 1826, however, this single reference was listed under the title Silenium, a name remarkably similar to that of still another element.

Although Proust's discovery proved wrong, his descriptions of uranium's properties were remarkably exact, even prophetic. Klaproth is credited with the discovery in pitchblende of a new metal for which he suggested in 1789 the name uranit, in honor of the planet Uranus, discovered eight years previously. The following year, he changed this name to uranium in order that the spelling might conform to the form adopted generally for other metals. The element was reportedly isolated as a black powder by the reduction of the yellow oxide, and Berzelius characterized uranium as one of the metals easiest to obtain.

Actually, neither Klaproth nor Berzelius did obtain the metal. Fifty years after the original discovery, Eugene Melchoir Peligot (1811–1890) isolated the element by reducing uranium tetrachloride (UCl_4) with potassium and demonstrated that the material considered previously to be the metal was actually the dioxide. Melting point of metallic uranium was not determined accurately until 1942 when scientists of the Manhattan Project lowered the accepted value by some 700°C.

One final note concerning nomenclature of ura-

Source: *Chemistry,* Vol. 48, No. 9 (October 1975), p. 27.

nium. Klaproth favored the use of mythological sources for the names of elements, but other chemists preferred to relate names to the person associated with their discovery. At about the same time that Klaproth's announcement was made, Johann Gottfried Leonhardi (1746–1823) objected to the suggestion that the new metal be named after the planet honoring the Greek sky god, Uranus. He proposed instead it be called klaprothium in honor of the discoverer. Needless to say, this proposal was not accepted, and modern chemists can name transuranium elements after mythological beings without having to consider what those names might have been, had that series become known as the transsilenium or transklaprothium elements.

Suggested Reading

(1) Peligot, E. M., *Compt. Rend.*, **1841**, 12, 735; 13, 417.
(2) Proust, L. J., *J. Phys.*, **1802**, 55, 297 and 457.
(3) Seaborg, G. T., "Man-Made Transuranium Elements," **1963**, Prentice-Hall, Englewood Cliffs, NJ.

5

Introduction to Chemical Language

As you come to understand more about matter, you need a better language to describe what you know. You know that there are elements and compounds, but you haven't learned to name them. You know that pictures of molecules are awkward, but you haven't learned a substitute. In this chapter you will find the names and symbols for the elements, and you will learn to use symbols to represent the formula for a compound. In many cases, these formulas can take the place of the awkward pictures.

OBJECTIVES

Names and Symbols

1. Given the name of an element, write the correct symbol.
2. Given the symbol for an element, write the correct name.

Formulas

3. Write the formula that represents a compound, and tell what each part of the formula represents.
4. From the formula of a compound, state the number of atoms of each element in one molecule of the compound.
5. From the formula of a compound, construct a model or draw a picture of a model to represent a molecule of the compound.
6. Indicate whether a model or picture of a model represents an element, a compound, or a mixture.
7. Given a model or picture of a model representing a compound, write the formula of the compound.

Naming Compounds

8. Given the formula of a binary compound, name the compound.
9. Given the name of a binary compound, write the formula.

Combining Number

10. Predict the combining number of an element from its position in the periodic table.
11. Given the combining number of two elements that combine, write the formula for the compound.
12. Given the formula of a binary compound and the combining number of one of the elements in the compound, determine the combining number of the other element in the compound.

Words You Should Know

In addition to the following terms, you should know the names and symbols for all common elements.

chemical symbol	nonmetal	stereo-
periodic table	formula	-ide
atomic radius	binary compound	nomenclature
metal	multiple proportions	combining number
		group

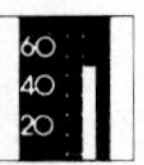

5.1 CHEMICAL SYMBOLS

The models we used in Chapter 4 are useful, but they are awkward. It takes a long time to draw pictures. It would help to have some way to represent atoms and molecules in symbols that could be written easily on paper. We usually do this by using abbreviations for the names of the elements. Such abbreviations are called **chemical symbols.** The simplest way to abbreviate is to use the first letter of the name of the element as its symbol. Since the names of several elements begin with the same letter, it is often necessary to use two letters to produce a different symbol for each element. Table 5.1 lists alphabetically the names and symbols for all known elements and gives some information about the origins of the symbols.

Twelve elements have just one letter for the symbol. All others use two letters. Current practice is to use two letters for the symbol of any new element. The first letter of the symbol is always capitalized.

Note that some symbols are not letters taken from the name of the element. That is because we do not use the original name for that element. Many elements discovered long ago were given Latin or Greek names, and some elements were named by people who speak languages other than English. These names are the basis for the chemical symbol. The original name for the element that we call sodium was *natrium,* and the symbol for sodium (Na) comes from that Latin name. The symbol for iron (Fe) comes from the Latin name *ferrum.* What we know as tungsten gets its symbol (W) from the German name for the element, *wolfram.* You will find other examples in Table 5.1.

Since symbols are used in place of the name of an element, you must learn to recognize the element by its symbol. However, you need not memorize all of them.

The thirty-nine elements marked with an asterisk in Table 5.1 are ones mentioned most often in introductory courses. You must know them. You may want to learn the symbols for other elements that you have heard of, or your instructor may advise you to learn symbols for a few other elements. As indicated in Table 5.1, thirteen of the elements do not occur in nature. The only one of these artificial elements that you are likely to hear much about is plutonium, an element produced in nuclear reactors and a raw material for nuclear weapons.

Table 5.1. Names and Symbols for All Known Elements

Name	*Symbol*	*Origin of Symbol*
actinium	Ac	Gr.** *Aktis, Aktinos,* beam or ray
aluminum*	Al	L. *alumen, alum*
americium†	Am	The Americas
antimony*	Sb	L. *stibium,* mark
argon*	Ar	Gr. *argos,* inactive
arsenic*	As	L. *arsenicum,* Gr. *arsenikon*
astatine	At	Gr. *astatos,* unstable
barium*	Ba	Gr. *barys,* heavy
berkelium†	Bk	Berkeley, California
beryllium*	Be	Gr. *berryllos, beryl*
bismuth	Bi	Ger. *Bisemutum*
boron*	B	Ar. *Buraq,* Pers. *Burah*
bromine*	Br	Gr. *bromos,* stench
cadmium	Cd	L. *cadmia*
calcium*	Ca	L. *calx,* lime
californium†	Cf	California
carbon*	C	L. *carbo,* charcoal
cerium	Ce	The asteroid Ceres
cesium	Cs	L. *caesius,* sky blue
chlorine*	Cl	Gr. *chloros,* greenish-yellow
chromium*	Cr	Gr. *chroma,* color
cobalt*	Co	Ger. *Kobold,* goblin
copper*	Cu	L. *cuprium,* from the island of Cyprus
curium†	Cm	Pierre and Marie Curie
dysprosium	Dy	Gr. *dysprositos,* hard to get at
einsteinium†	Es	Albert Einstein
erbium	Er	Ytterby,‡ a town in Sweden
europium	Eu	Europe
fermium†	Fm	Enrico Fermi

Table 5.1. (*Continued*)

fluorine*	F	L. & F. *fluerre,* flow or flux
francium	Fr	France
gadolinium	Gd	Gadolinite, a material named for a Finnish chemist
gallium	Ga	L. *Gallia,* France
germanium	Ge	L. *Germania,* Germany
gold*	Au	L. *aurum,* shining dawn
hafnium	Hf	L. *Hafnia,* Copenhagen
helium*	He	Gr. *helios,* the sun
holmium	Ho	L. *Holmia,* Stockholm
hydrogen*	H	Gr. *hydro,* water, and *genes,* forming
indium	In	Indigo
iodine*	I	Gr. *iodes,* violet
iridium	Ir	L. *iris,* rainbow
iron*	Fe	L. *ferrum*
krypton	Kr	Gr. *kryptos,* hidden
lanthanum	La	Gr. *lanthanein,* to be hidden
lawrencium†	Lr	Ernest O. Lawrence
lead*	Pb	L. *Plumbum*
lithium*	Li	Gr. *lithos,* stone
lutetium	Lu	Lutetia, an ancient name for Paris
magnesium*	Mg	Magnesia, a district in Greece
manganese*	Mn	L. *magnes,* magnet
mendelevium†	Md	Dmitri Mendeleev
mercury*	Hg	Hydragyrum, liquid silver
molybdenum	Mo	Gr. *molybdos,* lead
neodymium	Nd	Gr. *neos,* new, and *didymos,* twin
neon*	Ne	Gr. *neos,* new
neptunium†	Np	The planet Neptune
nickel*	Ni	Ger. *nickel,* Satan
niobium	Nb	Niobe, daughter of Tantalus
nitrogen*	N	L. *nitrum,* Gr. *nitron*
nobelium†	No	Alfred Nobel
osmium	Os	Gr. *osme,* a smell
oxygen*	O	Gr. *oxys,* sharp, acid, and *genes,* forming
palladium	Pd	The asteroid Pallas
phosphorus*	P	Gr. *phosphoros,* light-bearing
platinum	Pt	Sp. *platina,* silver
plutonium†	Pu	The planet Pluto
polonium	Po	Poland
potassium*	K	L. *kalium*
praseodymium	Pr	Gr. *prasios,* green, and *didymos,* twin
promethium†	Pm	Prometheus
protactinium	Pa	Gr. *protos,* first
radium	Ra	L. *radius,* ray
radon	Rn	From radium
rhenium	Re	L. *Rhenus,* Rhine
rhodium	Rh	Gr. *rhodon,* rose
rubidium	Rb	L. *rubidius,* deepest red
ruthenium	Ru	L. *Ruthenia,* Russia
samarium	Sm	Samarskite, a mineral

Table 5.1. *(Continued)*

scandium	Sc	L. *Scandia,* Scandinavia
selenium	Se	Gr. *Selene,* moon
silicon*	Si	L. *silex, silicis,* flint
silver*	Ag	L. *argentum*
sodium*	Na	L. *natrium*
strontium	Sr	Strontian, a town in Scotland
sulfur*	S	L. *sulphurium*
tantalum	Ta	Gr. *Tantalos*
technetium†	Tc	Gr. *technetos,* artificial
tellurium	Te	L. *tellus,* earth
terbium	Tb	Ytterby, a town in Sweden
thallium	Tl	Gr. *thallos,* green shoot
thorium	Th	Thor
thulium	Tm	Thule, Scandinavia
tin*	Sn	L. *stannum*
titanium	Ti	L. *Titans*
tungsten*	W	Ger. *Wolfram*
uranium*	U	The planet Uranus
vanadium	V	Vanadis
xenon	Xe	Gr. *xenon,* stranger
ytterbium	Yb	Ytterby, a town in Sweden
yttrium	Y	Ytterby, a town in Sweden
zinc*	Zn	Ger. *Zink*
zirconium	Zr	Ar. *zargun,* gold color

* These thirty-nine elements are common, and you will encounter them often in this book. Your instructor may ask you to memorize other symbols as well, but you *must* know these.

† These thirteen elements have been made artificially. They have not been found in nature. Claims have been made for the production of three other artificial elements, but names for those elements have not been established.

** Gr., *Greek;* L., *Latin;* Ger., *German;* Ar., *Arabic;* Pers., *Persian;* F., *French;* Sp., *Spanish.*

‡ Ytterby is the site of a quarry that yielded many new elements.

5.2 THE PERIODIC TABLE

On the inside cover of this book, you will find the symbols for the elements arranged in what is known as the **periodic table.** There are many arrangements of the periodic table. All attempt to arrange the elements in such a way as to show similarities and trends in the properties of the elements.

The periodic table summarizes a great deal of information about the elements. It is one of the most versatile tools that we have for understanding chemistry. Throughout this book you will encounter trends that can be observed in the periodic table of the elements. Two of those trends are discussed here.

ATOMIC RADIUS

In Chapter 4 we used spheres to represent atoms. At that time, we mentioned that small spheres are used to represent small atoms and that

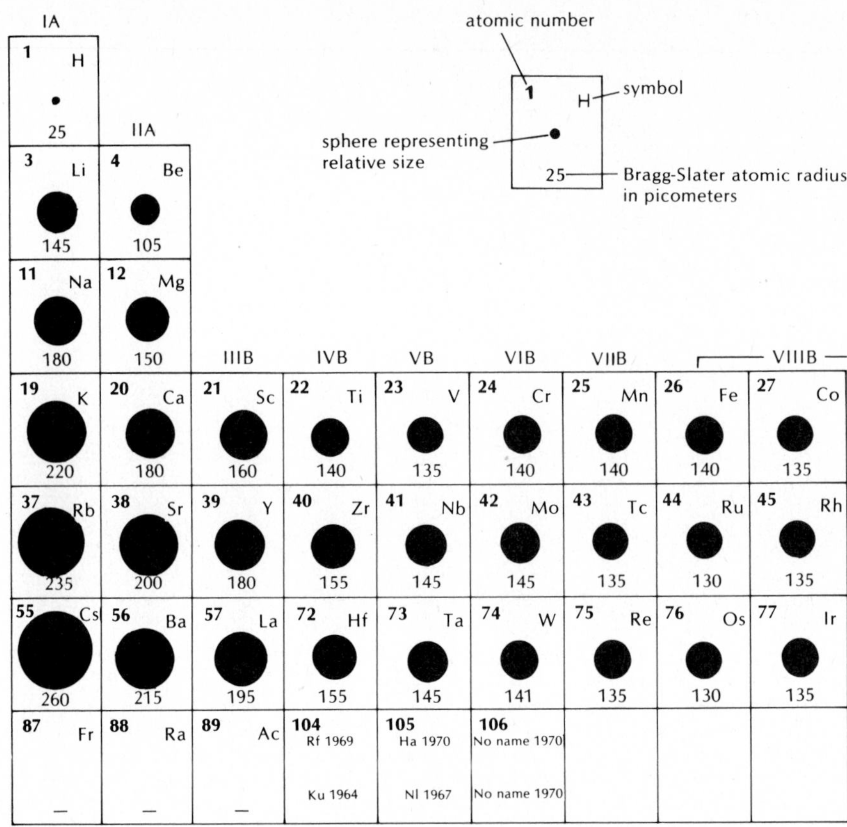

58 Ce	59 Pr	60 Nd	61 Pm	62 Sm
181	182	182	—	181
90 Th	91 Pa	92 U	93 Np	94 Pu
—	—	—	—	—

Figure 5.1
The relative size of atoms. Each block shows the atomic number and chemical symbol for the element. The circle in each block shows the relative size of the atom. The number under the circle gives the Bragg–Slater atomic radius in picometers. (A picometer is 10^{-12} m.)

larger spheres are used to represent larger atoms. We did not mention how large atoms are or how their sizes compare. Figure 5.1 shows the relative sizes of atoms arranged in a periodic table.

First of all, you notice that atoms are very small. The **atomic radius** is expressed in picometers (pm), and 1 pm is 10^{-12} m. That is just about too small to imagine, but we might give it a try.

Recall that in Chapter 1 you found a millionth of a sheet of paper. That was 10^{-6} of the sheet of paper. Well, 10^{-12} is $10^{-6} \times 10^{-6}$, or a millionth of

	IB	IIB	IIIA	IVA	VA	VIA	VIIA	VIIIA
							1 H 25	2 He 50
			5 B 85	6 C 70	7 N 65	8 O 60	9 F 50	10 Ne 65
			13 Al 125	14 Si 110	15 P 100	16 S 100	17 Cl 100	18 Ar 95
28 Ni 135	29 Cu 135	30 Zn 135	31 Ga 130	32 Ge 125	33 As 115	34 Se 115	35 Br 115	36 Kr 110
46 Pd 140	47 Ag 160	48 Cd 155	49 In 155	50 Sn 145	51 Sb 145	52 Te 140	53 I 140	54 Xe 130
78 Pt 135	79 Au 135	80 Hg 150	81 Tl 190	82 Pb —	83 Bi 160	84 Po 190	85 At —	86 Rn 145

63 Eu 199	64 Gd 179	65 Tb 180	66 Dy 180	67 Ho 179	68 Er 178	69 Tm 177	70 Yb 194	71 Lu 175
95 Am —	96 Cm —	97 Bk —	98 Cf —	99 Es —	100 Fm —	101 Md —	102 No —	103 Lr —

a millionth. If you can imagine taking a millionth of a millionth of a page, you are close to imagining the size of atoms (sure enough, it *is* impossible).

Next note that the size of the atoms increases as you go down any column in Figure 5.1. Furthermore, there is a general trend toward smaller size as you go from left to right. The table is arranged to emphasize these regularities in the properties of the elements. However, you should remember that there are exceptions to most generalizations. As you can see in Figure 5.1, there are exceptions to the rules we have given for the size of atoms.

METALS AND NONMETALS

The heavy zig-zag line near the right of the periodic table divides the elements into **metals** and **nonmetals.** The majority of the elements are metals. As you know, metals reflect light when they are polished. They can be bent, beaten flat without shattering, and drawn into a wire. They have high melting points, so they are solids at normal temperature. They are good conductors of heat and electricity.

Once again, what we have said about metals is *generally* true, but we know of exceptions. Mercury is a liquid at room temperature, but it has most of the other characteristics of a metal. Aluminum tears when you try to flatten it and breaks rather easily when it is bent, but it is shiny and a good conductor of electricity. Both germanium and antimony are very brittle, and antimony is a poor conductor of heat and electricity, but both are regarded as metals.

It is no accident that three of the four metals mentioned as lacking some metallic property lie just to the left of the zig-zag line on the periodic table. Nor is it accidental that those elements just to the right of that line shine like metals and are semiconductors. (Semiconductors do not conduct electricity as well as most metals, but they are better conductors than glass and other substances known as insulators.)

Like size, metallic properties vary with position in the periodic table. In general, the more metallic elements are toward the left and toward the bottom of the periodic table. The less metallic elements are toward the upper right of the table. From its position in the periodic table, we would expect sodium to exhibit more properties of a metal than magnesium, and we would expect magnesium to be more metallic than aluminum. Judging by the trend from top to bottom of the table, we would predict that sodium has more metallic properties than lithium, which is above it, but fewer metallic properties than potassium, which lies below. It is a good idea to remember trends such as these. If you do, you will find the periodic table very useful as you study chemistry.

By the way, did you have any trouble locating the metals mentioned in the last paragraph? You were warned a little earlier that you would need to know the symbols for these common elements. If you did not recognize the symbols for these elements in the periodic table shown in Figure 5.1, you should get busy and learn them *now.* You will use them a great deal in the next section.

5.3 CHEMICAL FORMULAS

The symbols for elements are used in many ways. At times they are used to refer to any amount of the element. We might say, "The automobile is made of Fe." Here the symbol is used in place of the word "iron," and it represents an indefinite amount of the element. At other times we use the symbol to represent a single atom of the element. This is the way the symbol is used when we write chemical formulas.

It is common practice to represent a compound by a **formula** as well as a name. A chemical formula, such as H_2O, consists of the symbols of the elements in the compound combined with small numbers (subscripts) at the base and to the right of the symbols. The subscript indicates the number of atoms of that element in one molecule of the compound. Two rules for writing formulas are:

Rule 1: *Represent each kind of element in a compound with the correct symbol for that element.*

Rule 2: *Indicate by a subscript the number of atoms of each element in a molecule of that compound.* (If there is only one atom of a particular element in the molecule, no subscript is used.)

Applying these rules for a molecule composed of one atom of oxygen (O) and two atoms of hydrogen (H), the formula could be written OH_2. This formula says that a molecule of the compound is composed of O (oxygen) and H (hydrogen) and that there are two H atoms for each O atom.

Does OH_2 look familiar? Probably not, but H_2O *does* look familiar to many of you. Both represent a molecule consisting of two atoms of hydrogen and one atom of oxygen. In order to write formulas that are easily recognized by others, we need a third rule.

Rule 3: *Write the symbol for the more metallic element first.*

You have already seen that the more metallic elements are located toward the left of the periodic table and that the less metallic elements are located toward the right. An exception is hydrogen. Although hydrogen is located in the left column of the periodic table, it has none of the properties that we associate with metals. In naming compounds, however, we treat hydrogen like a metal. Its symbol appears first in most of its compounds.

Compounds can form between two nonmetals. Carbon dioxide is a common example. Neither carbon nor oxygen is a metal, but from its position in the periodic table, we would expect carbon to have more of the properties of metals than oxygen does. Carbon is the more metallic of the two elements in the compound, and the symbol for carbon is written first.

With these three rules, you can write the formula for any compound containing only two elements. Such compounds are called ***bi*nary compounds.** (*Bi* is a prefix meaning "two.") The rules are illustrated in Figure 5.2.

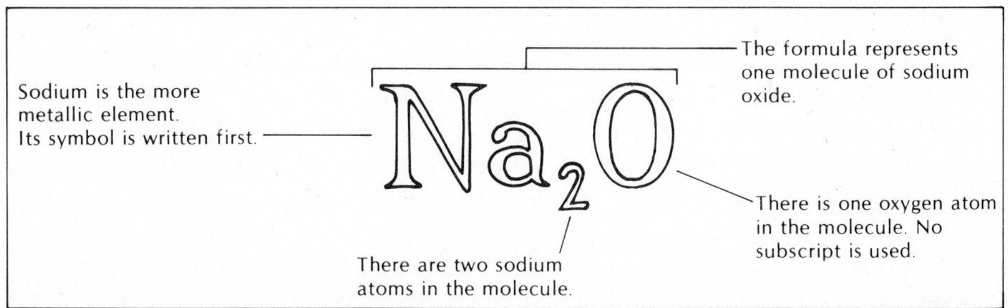

Figure 5.2 *The information conveyed by a chemical formula.*

To be sure that you understand what a formula represents, use spheres of different size and color or draw pictures to represent the molecules described by the following formulas. Use Figure 5.1 to draw the spheres about the right size.

5-1. CO
5-2. HCl
5-3. N_2O
5-4. CO_2
5-5. N_2H_4

Now see whether you can write formulas for molecules containing the following atoms.

5-6. one atom of carbon and four atoms of chlorine
5-7. one atom of chlorine and one atom of sodium
5-8. one atom of magnesium and one atom of oxygen
5-9. one atom of sulfur and two atoms of potassium
5-10. three atoms of fluorine and one atom of boron
5-11. one atom of copper and two atoms of iodine
5-12. one atom of iodine and one atom of copper

5.4 MULTIPLE PROPORTIONS

When you worked the last set of problems, you may have noticed that there were two compounds containing copper and iodine and two compounds containing carbon and oxygen. In each case, the two formulas represent different substances. The proportions of the two elements differ in the two compounds, and the properties of the two substances are very different. CO, for example, is carbon monoxide, a poisonous gas. CO_2 is carbon dioxide, which you exhale each time you breathe. It is not poisonous in low concentrations.

Elements can combine in several ratios or **multiple proportions.** Consider H_2O and H_2O_2. Both formulas represent compounds formed from hydrogen (H) and oxygen (O); but in the first compound there are two hydrogens for each oxygen atom, and in the second there is only one hydrogen for each oxygen atom. The proportions of the two elements differ in the two compounds. This makes a great deal of difference in the properties of the compound.

In Figure 5.3 you probably recognize H_2O as the formula for ordinary water. It is perfectly safe to drink. H_2O_2 is the formula for hydrogen peroxide, which is used to bleach hair. It is *not* safe to drink.

There are thousands of compounds that contain only carbon and hydrogen. Each of these hydrocarbons has different properties, and each is a unique substance. CH_4 (methane) is the major component of natural gas. C_2H_2 (acetylene) is used in acetylene torches for welding. C_3H_8 (propane) is the

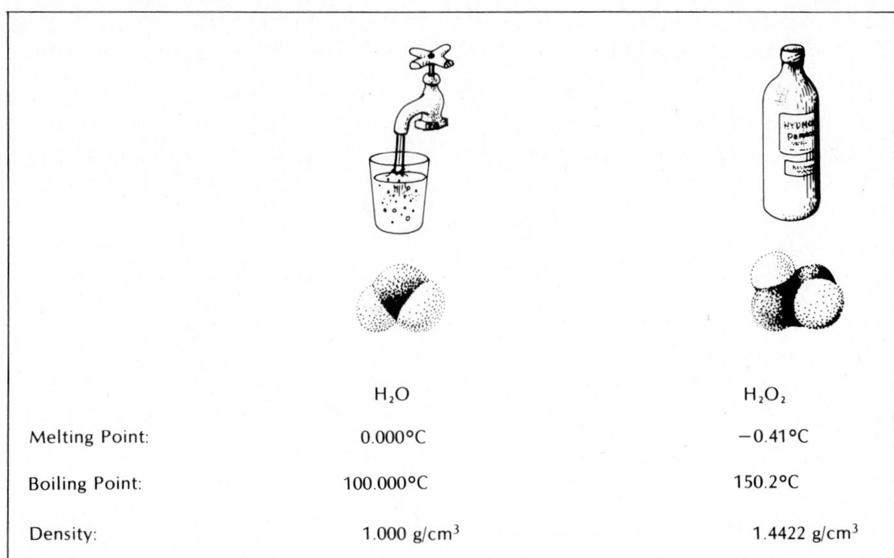

Figure 5.3 *H_2O and H_2O_2 contain the same elements but are very different substances, as these data indicate.*

major component of bottled gas, which is often used where natural gas is not available. C_4H_{10} (butane) is used in butane cigarette lighters. Note that all four of these compounds are composed of carbon and hydrogen. These compounds differ only in the proportion of each element in the different molecules. Still, these compounds have different melting points, boiling points, and densities, as shown in Table 5.2.

Table 5.2. Properties of Some Simple Hydrocarbons

Name	*Formula*	*Model*	*Melting Point (°C)*	*Boiling Point (°C)*	*Density† (g/cm³)*
methane	CH_4		−182.5	−161.5	0.555^{0}
acetylene	C_2H_2		−80.8	−84.0* (sublimes)	0.618^{-32}
propane	C_3H_8		−189.7	−42.1	0.501^{20}
butane	C_4H_{10}		−138.3	−0.5	0.578^{20} 0.601^{0}

Values taken from *Handbook of Chemistry and Physics*, 53rd Edition. Cleveland, Ohio: The Chemical Rubber Company (1972).

*Acetylene changes from a solid to a gas at this temperature. The process is called sublimation.

†Density changes with temperature. The temperature at which the densities were measured is indicated by the superscript following each value. Values for 20° and 0°C are given for butane.

5.5 THINGS TO COME

It is amazing that the millions of unique substances on earth and in the heavens are made from roughly 100 elements put together in different ways. At first it seems impossible. Now you know some of the reasons why it can happen. You know that: *The properties of a substance depend on WHAT atoms are in the molecule.*

H_2O

is water, which is essential for life. It is a liquid at room temperature and has no odor and no taste.

H_2S

is hydrogen sulfide. It gives rotten eggs their characteristic smell. It is a gas at room temperature and is very toxic.

The properties of a substance depend on HOW MANY of each kind of atom are in the molecule.

H_2O

is the familiar water. It has the same two elements as the compound on the right, but it does not have the same properties.

H_2O_2

is hydrogen peroxide. The extra oxygen atom in the molecule gives it very different properties. It is used as bleach and is very toxic.

In a later course, you will learn that *the properties of a substance depend on the ORDER in which atoms are connected.*

C_3H_7OH

is propanol, but there are two kinds:

```
  H  H  H
  |  |  |
H—C—C—C—O—H
  |  |  |
  H  H  H
```

is 1-propanol or normal propyl alcohol. It melts at $-126.5°C$, boils at 97.4°C, and has a density of 0.80 g/cm^3.

```
  H  H  H
  |  |  |
H—C—C—C—H
  |  |  |
  H  O  H
     |
     H
```

is 2-propanol or isopropyl alcohol. It is commonly used as rubbing alcohol. It melts at $-89.5°C$, boils at 82.4°C, and has a density of 0.79 g/cm^3.

Properties even change when the same atoms are connected in a way that produces different SHAPES.

$$C_6(H_2O)_6$$

is one way to write the formula for the common sugar, glucose.

```
 +----------O---------+
 |    H        H      |
 |    |        |      |
 | H  O   H    O      |    H
 | |  |   |    |      |    |
 +-C--C---C----C------C----C--O--H
   |  |   |    |      |    |
   O  H   O    H      H    H
   |      |
   H      H
```

shows which atoms are connected in glucose. There are *seven* other compounds with these *same* atoms connected in this *same* sequence. They are called allose, altrose, gulose, mannose, idose, galactose, and talose. These eight compounds have many properties that are the same, but there are differences in their chemical behavior that can be very important in living organisms.

Compounds of this kind are known as **stereoisomers.** *Stereo* is a prefix indicating three dimensions in space. Stereophonic sound gets its name from the fact that it seems to come from many directions rather than one. It is three-dimensional sound. As this prefix suggests, stereoisomers get their different properties from differences in the shape of the molecules. You need not worry about stereoisomers now. The subtle differences among these compounds go well beyond what you will study in this course. Right now, we need to learn how to name those simple compounds that we *will* study.

5.6 NOMENCLATURE

With the three rules given in Section 5.3, you can write the formula for a binary compound, provided that you know the number of each kind of atom in a molecule. Formulas are a convenient way to represent compounds, but there are times when we need names too. The time has come to put names with the formulas.

Nomenclature can get complicated. There are millions of compounds to name, and if it isn't done systematically, confusion reigns. To make matters worse, many compounds were known and their names were accepted before the need for some system became apparent. Furthermore, as more compounds were made or discovered in nature, a system that seemed fine in the beginning became awkward and had to be discarded. The problem is that the new system has not wholly replaced the old system in current usage. It is difficult to discard all of the books that use the older system.

In this book we will introduce the latest systematic nomenclature, or part of it. We will tell you how to name binary compounds now. In a later chapter, we will tell you how to name compounds with more than two elements. The principles introduced here can be expanded in later courses when you have to name still more complicated substances.

You already know the names for some compounds. You have known about carbon dioxide and carbon monoxide for years. Let's use these names to illustrate the rules of nomenclature.

5.7 NAMING COMPOUNDS FORMED FROM NONMETALS

To illustrate the principles of nomenclature, we will start with the formula for carbon dioxide and show, step by step, how the name is derived.

Figure 5.4

Figure 5.4 shows that carbon dioxide contains carbon and oxygen. We begin by naming those elements. As is customary, we write the name of the more metallic element first. Neither carbon nor oxygen is a metal in the strict sense. However, carbon is farther to the left in the periodic table and has more metallic properties than oxygen. Carbon is written first.

Figure 5.5

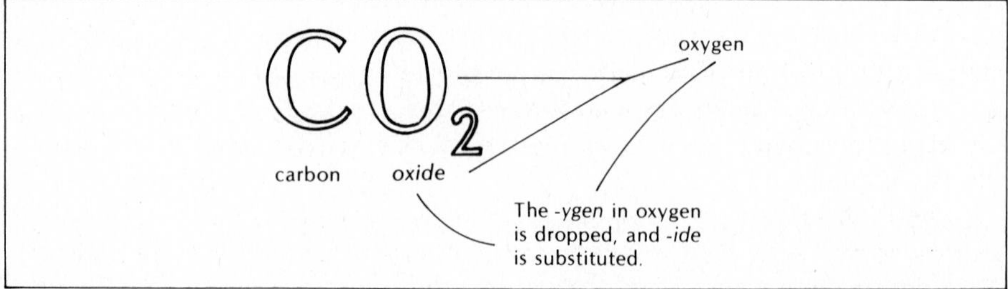

If the elements were merely named, there could be confusion about whether we were referring to the separated elements or a compound formed from those elements. As shown in Figure 5.5, the name for the nonmetal is modified when a compound is named. The last part (usually the final syllable) is dropped, and a suffix is added. For all binary compounds, the suffix added is *ide*.

Figure 5.6

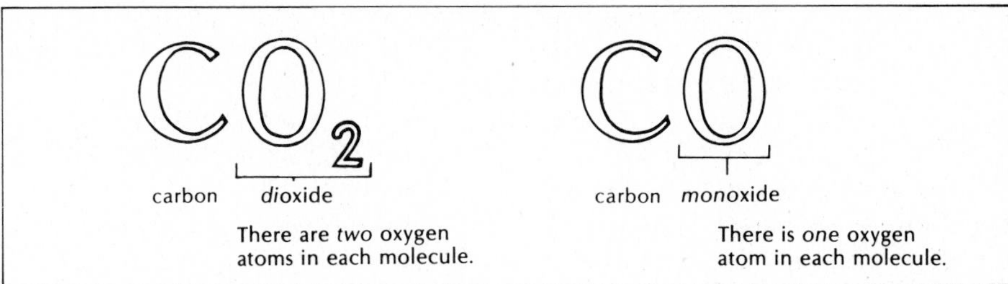

If there were only one compound containing carbon and oxygen, carbon oxide would be a satisfactory name. However, you know that carbon and oxygen form two compounds. As shown in Figure 5.6, prefixes are added to indicate the number of atoms of each element in the compound. *Di* is a prefix meaning "two," and *mono* is a prefix meaning "one." The *mono* prefix is frequently omitted, particularly for well-known substances. However, experts on nomenclature caution that this can be dangerous and suggest that it is better to include the *mono* prefix. Although you are unlikely to hear them in common use, Figure 5.7 gives the complete, systematic names for these compounds of carbon and oxygen.

Figure 5.7

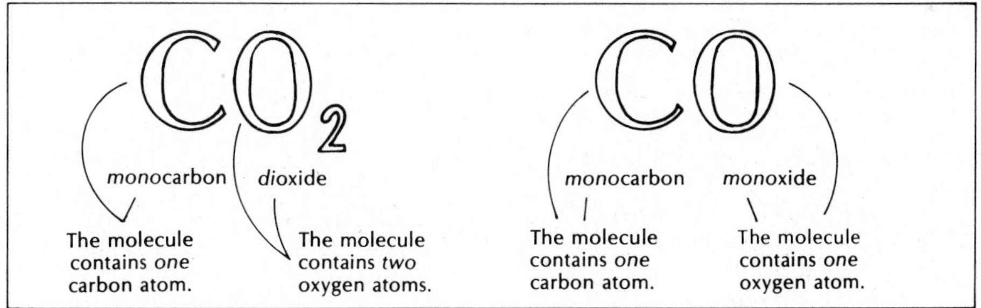

Table 5.3 lists the Greek prefixes that are used to indicate the number of atoms of an element in a compound. Using these prefixes, write names or formulas for the compounds in questions 5–13 through 5–25.

Table 5.3. Greek Prefixes

Prefix	*Number*	*Prefix*	*Number*
mono-	1	penta-	5
di-	2	hexa-	6
tri-	3	hepta-	7
tetra-	4	octa-	8

5-13. SiC	5-20. NO
5-14. CS_2	5-21. chlorine dioxide
5-15. SF_6	5-22. dichlorine monoxide
5-16. OF_2	5-23. iodine tribromide
5-17. SO_2	5-24. nitrogen triiodide
5-18. SO_3	5-25. diphosphorus tetraoxide
5-19. N_2O_5	

Nomenclature will become more involved as we go along. If you go through this section carefully and practice naming these compounds now, you will have little difficulty later. If you do the job poorly, however, you are likely to get confused when additional rules for naming compounds are added. If you miss any of these names, practice naming the compounds listed at the end of the chapter.

The system just described should always be used in naming compounds made up of nonmetals. It may be used to name compounds formed between metals and nonmetals, too, but it generally is not. This is quite evident from Table 5.4, which lists names of several binary compounds containing metals.

Table 5.4. Names and Formulas of Binary Compounds

Name	*Formula*
sodium chloride	NaCl
potassium iodide	KI
magnesium oxide	MgO
magnesium bromide	$MgBr_2$
calcium chloride	$CaCl_2$
aluminum chloride	$AlCl_3$
aluminum oxide	Al_2O_3
silver sulfide	Ag_2S
copper(II) oxide	CuO
copper(I) oxide	Cu_2O

All of the names in Table 5.4 end in *ide*. The *ide* ending is used for *all* binary compounds. The first three names in the table seem to be satisfactory. (The *mono* prefix hasn't been used, but this is acceptable.) However, as you move down the table, it is apparent that prefixes are not used for *any* of the names. This would *not* be acceptable.

For the last two names on the list, something new has been introduced. Roman numerals are part of the name, and it is clear that these numerals do *not* indicate number of atoms. What they *do* indicate is the *combining number* of the metallic element in the compound.

5.8 COMBINING POWER

The idea of combining power grew out of the observation that some elements combine with one atom to form a molecule, while other elements combine with many atoms to form a molecule. The combining power or **combining number** of an element is the number of hydrogen atoms that combine with one atom of the element. This is illustrated in Figure 5.8.

	IA	IIA	IIIB	IVB	IB	IIB	IIIA	IVA	VA	VIA	VIIA	0
1	HH											
2	LiH	BeH_2					B_2H_6 (B_4H_{10})	CH_4 (C_2H_6)	NH_3 (N_2H_4)	H_2O	HF	
3	NaH	MgH_2					AlH_3	SiH_4 (Si_2H_6)	PH_3 (P_2H_4)	H_2S	HCl	
4	KH	CaH_2				ZnH_2	Ga_2H_6	GeH_4 (Ge_2H_6)	AsH_3 (As_2H_4)	H_2Se	HBr	
5	RbH	SrH_2				CdH_2	InH_3	SnH_4	SbH_3	H_2Te	HI	
6	CsH	BaH_2							BiH_3			
Combining Number	1	2	Variable				3	Variable		2	1	0

Figure 5.8 *Formulas for compounds containing hydrogen.*

Figure 5.8 shows a part of the periodic table divided into **groups.** However, in place of the symbols for the elements, the formula for a compound formed between that element and hydrogen is shown. Look at the first column (known as Group IA) of Figure 5.8. Note that one hydrogen atom is combined with one atom of each element in Group IA of the periodic table. It appears that each element in Group IA has the power to combine with only one hydrogen atom. We say that *elements in Group IA have a combining number of 1.*

Now look at the second column (Group IIA) of Figure 5.8. Each element in Group IIA is combined with two atoms of hydrogen. *Elements in Group IIA of the periodic table have a combining number of 2.* The combining number of the elements in the "variable" columns varies from one compound to another.

Move across the table to Group VIIA of the periodic table. The formula for binary compounds of hydrogen and each element in Group VIIA indicates that these elements also combine with one hydrogen atom. *Elements in Group VIIA normally have a combining number of 1.* Figure 5.8 also suggests that *elements in Group VIA have a combining number of 2.*

Not all elements combine with hydrogen, but the combining power of elements that do combine with hydrogen can be used to assign combining numbers to other elements. From the formula H_2O, we would assign oxygen a combining number of 2. Then, if we assume that the combining power of oxygen is the same in Cu_2O as in H_2O, copper can be assigned a combining number of 1 in Cu_2O.

Cu_2O is not the only compound of copper and oxygen, as indicated in Table 5.4. The other compound has the formula CuO. If we still assume that oxygen has a combining power of 2, we can assign copper a combining number of 2 in this compound. Since one atom of copper is combined with one atom of oxygen, the two elements have the same combining power. The number indicating this combining power (the combining number) is also the same.

There are many elements like copper that change combining power from one compound to another. All of the "B" groups of the periodic table contain elements that change combining number as they form different compounds. As you can see from Figure 5.8, the elements in Groups IVA and VA also change combining number as various compounds are formed. It is not possible to predict the combining number of these elements from the position of the element in the periodic table.

The idea of combining power lies at the heart of all rules for writing chemical formulas. However, it is awkward to apply the concept directly. The following rules are based on what we know about the combining power of elements. They can be used to predict the formulas of many compounds. These rules seem somewhat arbitrary at this point, and in many ways they *are* arbitrary. However, after you have learned more about the nature of atoms, you will begin to see why the elements have the combining numbers indicated by these rules.

5.9 RULES FOR ASSIGNING COMBINING NUMBERS

Rule 1: *Any element in Group IA of the periodic table always has a combining number of* 1.

Rule 2: *Any element in Group IIA of the periodic table always has a combining number of* 2.

Rule 3: *The most common combining number of elements in Group IIIA is* 3. You may assume that aluminum always has a combining number of 3.

Rule 4: *Elements in Groups IVA through VIIA have combining numbers that vary from one compound to another.* However, the following rules about a few of these elements will enable you to write formulas and names for most compounds that you encounter.

a. Nitrogen has a combining number of 3 in nitrides. (Nitrides are binary compounds between nitrogen and a metal.)
b. Sulfur has a combining number of 2 in sulfides. (Sulfides are binary compounds between sulfur and a metal.)
c. Oxygen has a combining number of 2 in virtually all compounds. (In this course you may assume that the combining number of oxygen is always 2, unless you are specifically told otherwise.)
d. All halogens (elements in Group VIIA) have a combining number of 1 in halides. (Halides are binary compounds of a halogen and a metal.)

Rule 5: *The combining number of most other elements is variable and is determined from the formula or the name of the compound.*

5.10 PREDICTING FORMULAS FROM COMBINING NUMBERS

As you can see, the combining number of many elements can be determined by the position of the element in the periodic table. These combining numbers can be used to predict the correct formula for compounds. We will use calcium chloride to illustrate how it is done.

Since calcium is in Group IIA of the periodic table, we know that calcium *always* has a combining number of 2. Chlorine, the other element in calcium chloride, is in Group VIIA of the periodic table, so we know that it has a combining number of 1 in this binary compound with a metal. In order to write the correct formula for calcium chloride, we need to follow one simple rule:

The number of atoms of the metallic element times its combining number must be equal to the number of atoms of the nonmetallic element times its combining number. The rule is illustrated in Figure 5.9.

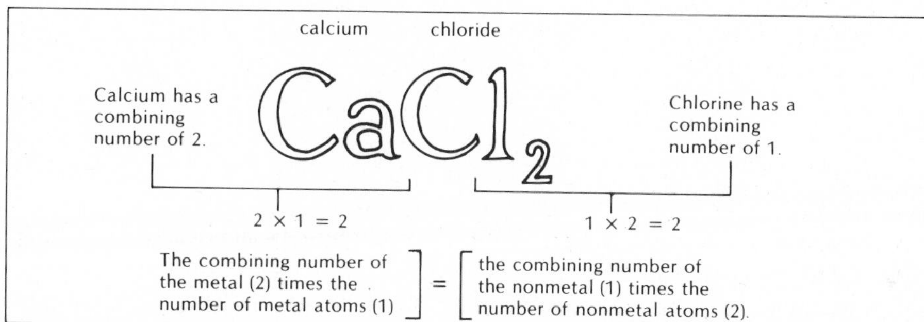

Figure 5.9

The correct formula for calcium chloride is $CaCl_2$. In that formula, the number of atoms of the metallic element (1) times its combining number (2) must be equal to the number of atoms of the nonmetallic element (2) times *its* combining number (1). We see that this is true. Since the formula for many compounds of metals can be predicted from combining numbers, the name does not include prefixes to indicate the number of atoms in the formula. Although monocalcium dichloride is a correct name for $CaCl_2$, it is never used. The simplified name calcium chloride is used.

Starting with one of these simplified names, we will figure out the formula for the compound. Going step by step, we will write the formula for sodium oxide.

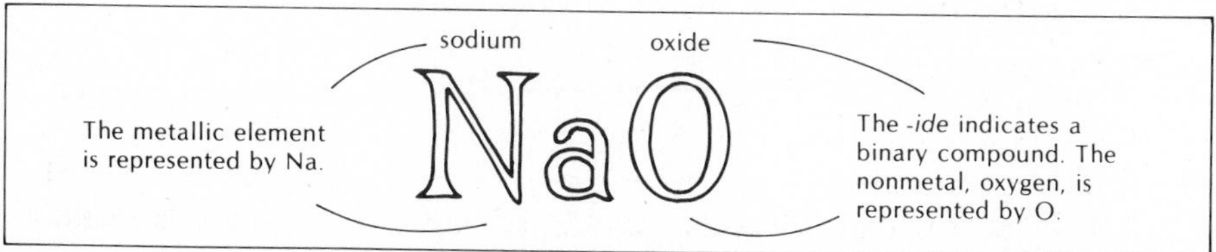

1. From the name, we know that we are talking about a compound of Na and O. We first write down these symbols in the correct order.

2. We need the combining numbers of Na and O. Since Na is in Group IA of the periodic table, we know that its combining number is 1. We also know that oxygen always has a combining number of 2.

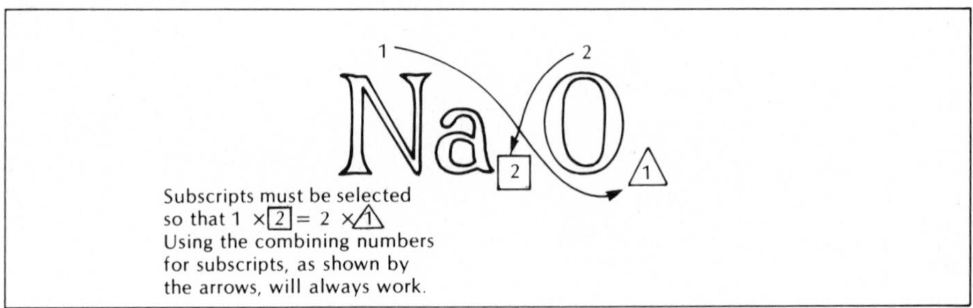

3. Now we decide how many atoms of each element are needed to make the product of combining number and number of atoms the same for the metal and the nonmetal. The simplest procedure is to use the combining number of one element as the subscript for the other.

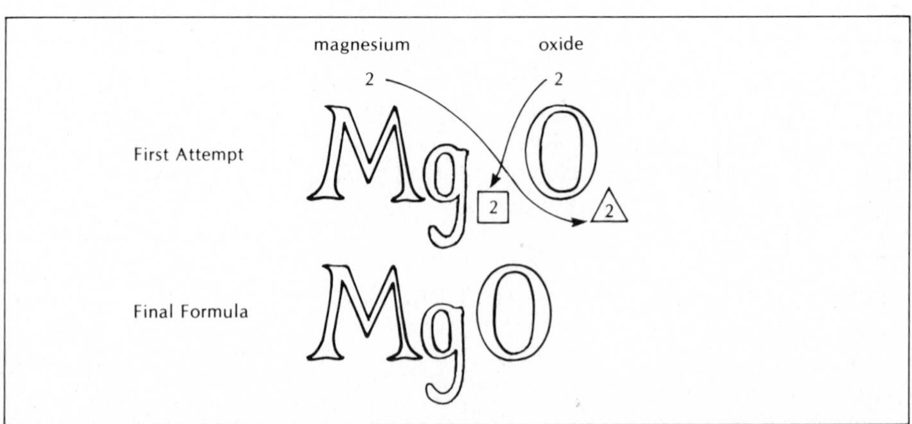

4. When this procedure is used, you may end up with formulas like Mg_2O_2. When both of the subscripts in the formula can be divided by some number to simplify the formula, you should do so, unless you *know* that the actual molecule is represented. (H_2O_2 is correct as written, because

we know from experimental evidence that the molecule has two atoms of hydrogen and two atoms of oxygen. Without such experimental evidence, the simplest possible formula is assumed. In our example, that formula is MgO.)

Practice these rules by writing formulas for the following compounds.

Practice is needed to develop any new skill. One way to practice is with a partner. Read your partner the names, and have her write down the correct formulas and then read the formulas back for you to give the names. Then reverse the procedure. Keep it up until you are both confident that you understand how to use the periodic table to find combining numbers and how to use the combining numbers to write correct formulas for binary compounds.

5-26. potassium bromide
5-27. magnesium chloride
5-28. calcium oxide
5-29. aluminum iodide
5-30. barium fluoride
5-31. sodium iodide
5-32. calcium bromide
5-33. potassium oxide
5-34. beryllium chloride
5-35. aluminum oxide

5.11 NAMING COMPOUNDS OF METALS

Since the combining number of many metals varies from one compound to another, it is necessary to indicate the combining number in the name of the compound. You have already seen how prefixes are used to clarify the formula for compounds formed from two nonmetals. For compounds of metals, a Roman numeral is used in the name to indicate the combining number of the metal.

Copper combines with oxygen to form two compounds, CuO and Cu_2O. Since you already know that the combining number of oxygen is 2, the combining number of Cu must be 2 in CuO. (Make sure that you understand this.)

5-36. What is the combining number of copper in Cu_2O?

The name for CuO is copper(II) oxide, and the name for Cu_2O is copper(I) oxide. The Roman numeral after the name of the metal tells you the combining number of the metal. (Note that the Roman numeral does *not* indicate the number of metal *atoms* in the formula.) Write the formula or name for each of the following compounds.

Go back and review Questions 5-13 through 5-25. Be sure you understand the difference between the rules used to name compounds composed of two nonmetals and the rules used here for naming compounds containing metals.

5-37. tin(IV) chloride
5-38. tin(II) bromide
5-39. tin(IV) oxide
5-40. iron(III) sulfide
5-41. iron(II) fluoride
5-42. mercury(II) oxide
5-43. cobalt(III) oxide
5-44. SnO
5-45. $SnBr_4$
5-46. FeS
5-47. $FeCl_3$
5-48. Hg_2Cl_2 (Experiments reveal that the molecule contains two atoms of each element.)
5-49. Cr_2O_3
5-50. $NiCl_2$

5.12 OLDER NAMES

If you use older reference books or read labels on bottles in the laboratory, you may encounter an older system for naming compounds of metals. Tin(IV) chloride may be called stannic chloride, tin(II) bromide may be called stannous bromide, iron(III) sulfide may be called ferric sulfide, and iron(II) fluoride may be called ferrous fluoride. This older system employs the Latin names for the elements and uses the suffixes *ous* and *ic* to indicate the lower or higher of two combining numbers. The system is awkward and confusing. Since it is being replaced by the system introduced in this book, it is not worth learning. However, you should be aware that it exists. You may have to recognize older names for a few common compounds.

5.13 SUMMARY

An important part of any subject is learning the language. In this chapter you have begun to learn the language of chemistry. First you were introduced to the symbols used to represent elements, and by now you should have memorized the symbols for the common ones. Next you saw how elements are arranged in the periodic table, and you examined some of the regularities that are apparent in that table: For example, you found that the size of atoms increases as you go from top to bottom within a column of the periodic table, and you found that there is a general trend toward decreasing size as you go across the table from left to right.

An important classification of the elements is that of metals and nonmetals. Although the elements are generally divided into these two classes, you saw that the boundary is not sharp. In general, elements become less metallic as you go from left to right in the periodic table and more metallic as you go from top to bottom.

The change in metallic character of the elements was used as a guide in writing formulas to represent compounds. Common practice is to write the more metallic element first in the formula and the name.

Naming compounds can be a problem, because there are two conventions in common use. In naming binary compounds formed from two nonmetallic elements, Greek prefixes are used to indicate the number of atoms of each element in the compound: N_2S_5 is *di*nitrogen *penta*sulfide. The *mono* prefix is frequently omitted.

Compounds between metals and nonmetals could be named in the same way as compounds between two nonmetals, but they generally are not named that way. When the combining numbers of the elements in the compound can be predicted from the periodic table or are common knowledge (Zn and Ag, for example), there is nothing in the name to indicate the combining number of the metal or the number of atoms of each element in the formula. Al_2O_3 is simply called aluminum oxide, because oxygen is always assumed to have a combining number of 2 and aluminum has a combining number of 3.

In the names of compounds containing metals that can have several combining numbers, the combining number of the metal is indicated by a Roman numeral following the name of the metal. Copper(I) oxide contains copper with a combining number of 1; copper(II) oxide contains copper with a combining number of 2.

Cu_2O and CuO serve as good examples of elements that form compounds in multiple proportions. The same two elements can form very different substances by combining in different proportions. Even the way the atoms are connected in a compound affects the properties of the substance formed, and the geometric shape of a molecule may determine important characteristics of the compound.

You have named several compounds and written many formulas in this chapter, but some of you will need more practice. To help you practice, answers are given to the supplementary questions at the end of the chapter.

Questions and Problems

To check your understanding of material in this chapter, work the problems in sets of ten *without referring to any section of the chapter.* If you miss *any* of the first ten, review the material in this chapter or discuss your errors with your instructor until you know where you made your mistakes. Then work ten more. Keep practicing until you are able to answer at least one set of ten questions without an error. (Answers are given at the end of this chapter.)

Set I

5-51. What element is represented by K?
5-52. What is the symbol for mercury?
5-53. What is the formula of a molecule containing one atom of zinc and two atoms of iodine?
5-54. What is the formula of a molecule containing one atom of oxygen and two atoms of silver?

In the following questions, write the formula if the name is given, and write the name if the formula is given.

5-55. aluminum chloride
5-56. Al_2S_3
5-57. HgO
5-58. cobalt(II) chloride
5-59. CS_2
5-60. dinitrogen tetroxide

Set II

5-61. What element is represented by Br?
5-62. What is the symbol for copper?
5-63. What is the formula of a molecule containing one atom of phosphorus and five atoms of chlorine?

5-64. What is the formula of a molecule containing three atoms of oxygen and one atom of sulfur?

In the following questions, write the formula if the name is given, and write the name if the formula is given.

5-65. N_2O_3
5-66. iron(II) bromide
5-67. magnesium nitride
5-68. Cr_2O_3
5-69. MnO_2
5-70. cobalt(III) fluoride

Set III

5-71. What is the symbol for lead?
5-72. What element is represented by Au?
5-73. What is the formula of a molecule containing one atom of boron and one atom of nickel?
5-74. What is the formula of a molecule containing one atom of iodine and one atom of lithium?

In the following questions, write the formula if the name is given, and write the name if the formula is given.

5-75. potassium bromide
5-76. SiC
5-77. sulfur dichloride
5-78. SnO_2
5-79. CCl_4
5-80. aluminum nitride

Set IV

In the following questions, write the formula if the name is given, and write the name if the formula is given.

5-81. barium nitride
5-82. CaC_2
5-83. chromium(III) silicide
5-84. CoO
5-85. copper(II) oxide
5-86. $AuBr_3$
5-87. iron(III) chloride
5-88. B_4C
5-89. nitrogen triiodide
5-90. Ni_5P_2

Answers to Questions in Chapter 5

CO	HCl	N_2O	CO_2	N_2H_4
(a)	(b)	(c)	(d)	(e)
5-1.	**5-2.**	**5-3.**	**5-4.**	**5-5.**

5-6. CCl_4
5-7. NaCl
5-8. MgO
5.9. K_2S
5-10. BF_3
5-11. CuI_2
5-12. CuI
5-13. silicon carbide (monosilicon monocarbide)
5-14. carbon disulfide (monocarbon disulfide)
5-15. sulfur hexafluoride (monosulfur hexafluoride)
5-16. oxygen difluoride (monoxygen difluoride)
5-17. sulfur dioxide (monosulfur dioxide)
5-18. sulfur trioxide (monosulfur trioxide)
5-19. dinitrogen pentoxide
5-20. nitrogen monoxide (mononitrogen monoxide)
5-21. ClO_2
5-22. Cl_2O
5-23. IBr_3
5-24. NI_3
5-25. P_2O_4
5-26. KBr
5-27. $MgCl_2$
5-28. CaO
5-29. AlI_3
5-30. BaF_2
5-31. NaI
5-32. $CaBr_2$
5-33. K_2O
5-34. $BeCl_2$
5-35. Al_2O_3
5-36. 1
5-37. $SnCl_4$
5-38. $SnBr_2$
5-39. SnO_2
5-40. Fe_2S_3
5-41. FeF_2
5-42. HgO
5-43. Co_2O_3
5-44. tin(II) oxide
5-45. tin(IV) bromide
5-46. iron(II) sulfide
5-47. iron(III) chloride
5-48. mercury(I) chloride
5-49. chromium(III) oxide
5-50. nickel(II) chloride
5-51. potassium
5-52. Hg
5-53. ZnI_2
5-54. Ag_2O
5-55. $AlCl_3$
5-56. aluminum sulfide
5-57. mercury(II) oxide
5-58. $CoCl_2$
5-59. carbon disulfide
5-60. N_2O_4
5-61. bromine
5-62. Cu
5-63. PCl_5
5-64. SO_3
5-65. dinitrogen trioxide
5-66. $FeBr_2$
5-67. Mg_3N_2
5-68. chromium(III) oxide
5-69. Manganese dioxide. Manganese(IV) oxide would be the name predicted from the rules given, and it is acceptable. However, this compound is normally named by the rules for naming nonmetals.
5-70. CoF_3
5-71. Pb
5-72. gold
5-73. NiB
5-74. LiI
5-75. KBr
5-76. silicon carbide (monosilicon monocarbide)
5-77. SCl_2
5-78. tin(IV) oxide
5-79. carbon tetrachloride (monocarbon tetrachloride)
5-80. AlN
5-81. Ba_3N_2
5-82. calcium carbide

5-83. Cr_4Si_3
5-84. cobalt(II) oxide
5-85. CuO
5-86. gold(III) bromide
5-87. $FeCl_3$
5-88. tetraboron carbide (tetraboron monocarbide)
5-89. NI_3
5-90. Pentanickel diphosphide. This is another example of a metallic compound named like a compound between two nonmetals. Since both nickel and phosphorus have variable combining numbers, this name is less ambiguous than the one derived by using other rules.

READING We mentioned that there are many arrangements of the periodic table. The arrangement described in the following article is as unusual as it is interesting. Because it shows the relative abundance of the elements on earth, it will give you an idea of why it is more important to learn the names and symbols of some elements than those of others.

PERIODIC TABLE WITH EMPHASIS

William F. Sheehan

Periodic tables have been devised in so many two- and three-dimensional forms (*1*) that one more can be justified only if it's fun. And for fun, here is one where two nearly universal features of all tables are modified.

One is equal treatment of all elements, except lanthanides and actinides—neglected in some tables. Although each element is unique and as privileged as any other, abundances, costs, and special properties lead chemists to deal with some more often than others.

The second feature modified here is the arrangement of elements by their idealized electronic structure, which is an unreal setting for chemical interaction because chemists deal with atoms as influenced by other atoms. Whatever success they have with periodic tables comes from reading into them wisdom and insight gained only by dealing with real situations. Only then do positions in a table take on the desired meaning for memory, prediction, and correlation.

The table presented here departs from the humdrum. The size of each element's niche is, in most cases, roughly proportional to its abundance on Earth. In addition, certain chemical similarities, such as between beryllium and aluminum, or between boron and silicon, are suggested by their placement as neighbors. Periodic tables based on elemental abundances would vary from planet to planet and comparing such tables would show how planets differ.

It is also fun to devise less distorted tables based on melting points, acidity or basicity of oxides in water, electronegativity, superconductivity (*2*), solubility of sulfides, date of discovery, crystal structure, ion size, compressibility, critical temperature (*3*), or other chemical or physical properties.

In each table niche size reflects quantitatively on that particular property of an element. For example, to emphasize ability of elements to form d^2sp^3 hybrid bonds, niches of transition elements near iron, nickel, and cobalt would be magnified more than silicon and selenium. Similarly, sodium, potassium, rubidium, cesium, and others would occupy quite small slots in the table.

Freedom of area size and shape introduces new dimensions in learning, usefulness, and fun. Adjustable areas also make a periodic table adaptable to a specific purpose that may be more chemical than electronic.

Source: *Chemistry,* Vol. 49, No. 3 (April 1976), pp. 17–18.

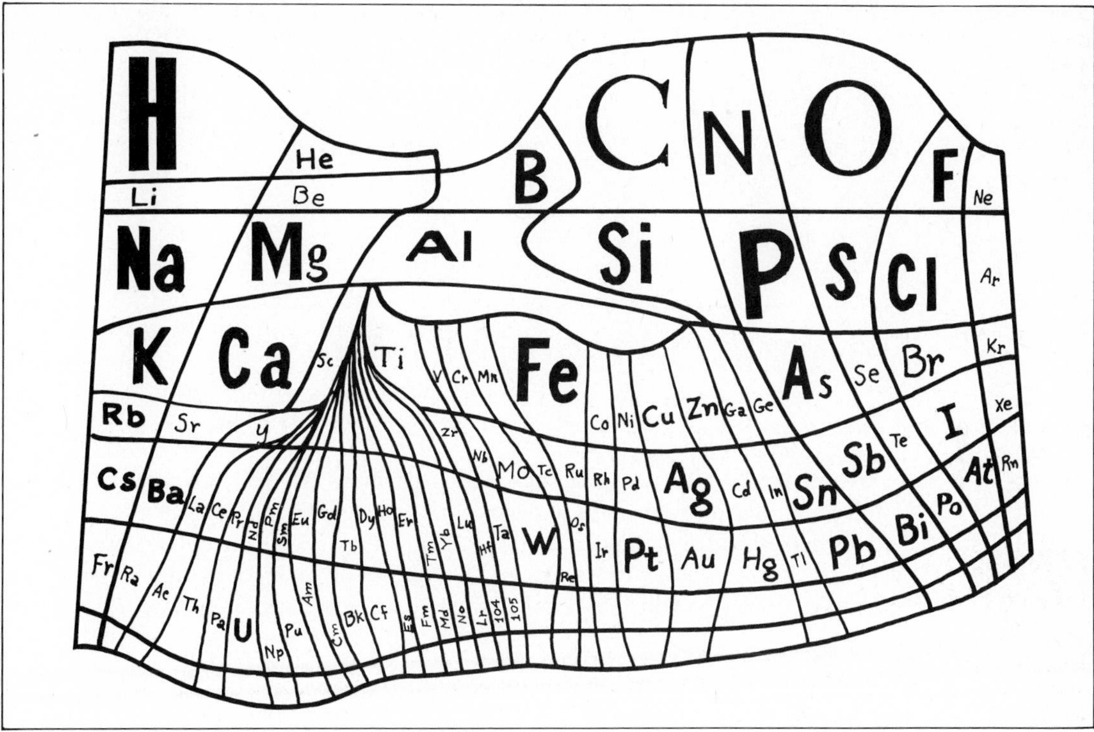

A periodic table where elemental abundances on Earth are indicated by area size; certain chemical similarities are also indicated. To accommodate all elements, some distortions are necessary. For example, some elements shown do not occur naturally. (William F. Sheehan © 1970. All Rights Reserved.)

REFERENCES

(1) Mazurs, E. G., "Graphic Representations of the Periodic System During One Hundred Years." The University of Alabama Press, University, Ala., 1974.

(2) Matthias, B. T., *Am. Sci.,* **58,** 80 (1970).

(3) Horvath, A. L., *J. Chem. Educ.,* **50,** 335 (1973).

(4) Mazurs, E. G., "Ups & Downs of the Periodic Table," *Chemistry,* **39,** 6 (July 1966).

6

Atomic Mass and Moles

Moles are small, furry animals that burrow in the ground, right? Wrong! Moles are pigmented spots on the skin, right? Wrong! Moles are . . . well, moles are what this chapter is about. It is a very important chapter, and you should be sure that you can meet the following objectives by the time you finish it. Doing so will make the rest of the course almost a snap.

OBJECTIVES

Atomic and Molecular Mass

An important characteristic of an atom is its atomic mass. This is a relative mass that is used to find the molecular mass of a compound and to count atoms and molecules. You are well on the way to understanding these ideas when you can:

1. State the meaning of relative mass.
2. Calculate the relative mass of several objects when you are given their actual mass and a relative mass for one of them. (See Problem 6-65.)
3. Given the formula (or name) and a periodic table, calculate the molecular mass of any compound. (See Problem 6-68.)
4. Identify the atomic mass of an element on the periodic table.

Mole and Mole Mass

Atomic mass and molecular mass are important, because the relative mass expressed in grams always contains 6.02×10^{23} atoms or molecules. Several relationships involve this number, which is called a mole. You need to understand them all.

5. Demonstrate that the relative mass of two models in tonns, kilograms, or any other mass unit contains the same number of particles. (See Problem 6-66.)

6. Find the number of moles in a pile of objects when you know the mass of the pile and the mass of one of the objects. (See Problem 6-67.)
7. State the number of particles in one mole.
8. Given the formula of a compound, find the mass of one mole of the compound. (See Problem 6-69.)
9. Given the formula and the mass of a sample of a compound, find the number of moles contained in the sample. (See Problem 6-72.)
10. Given the mass of a sample of an element, find the number of moles contained in the sample. (See Problem 6-71.)
11. Given the mass of a sample of an element, find the number of atoms in the sample. (See Problem 6-71.)
12. From the mass of a compound and its formula, find the number of molecules in the sample. (See Problem 6-73.)
13. Find the number of each kind of atom contained in a given mass of any compound. (See Problem 6-74.)
14. Given the mass of any number of molecules of a compound, find the molecular mass of the compound. (See Problem 6-70.)

Empirical Formula and Percent Composition

One of the most important uses of atomic mass and moles is to find the formula of a compound. You will know how to do this and how to express compositions in terms of percentages when you have met the following objectives:

15. Given the mass of each element in a sample of a compound, calculate the empirical formula of the compound. (See Problem 6-75.)
16. Given the mass of each element in a sample of a compound, calculate the percent composition of the compound. (See Example 6-75.)
17. Given the percent composition of a compound, calculate its empirical formula. (See Problem 6-75.)
18. Given the formula of a compound, calculate its percent composition. (See Problem 6-76.)

Words You Should Know

atomic mass
relative mass
mole
mole mass
molecular mass
empirical formula
percent composition

Writing the formula for a compound is a cinch—once you know the elements in the compound and the number of atoms of each kind in one molecule. But atoms are too small to see and so are molecules. How do we *know* how many atoms are in one molecule? That question required about 100 years to answer. It isn't easy! The key to the solution comes from atomic mass, so that is where we begin.

6.1 ATOMIC MASS*

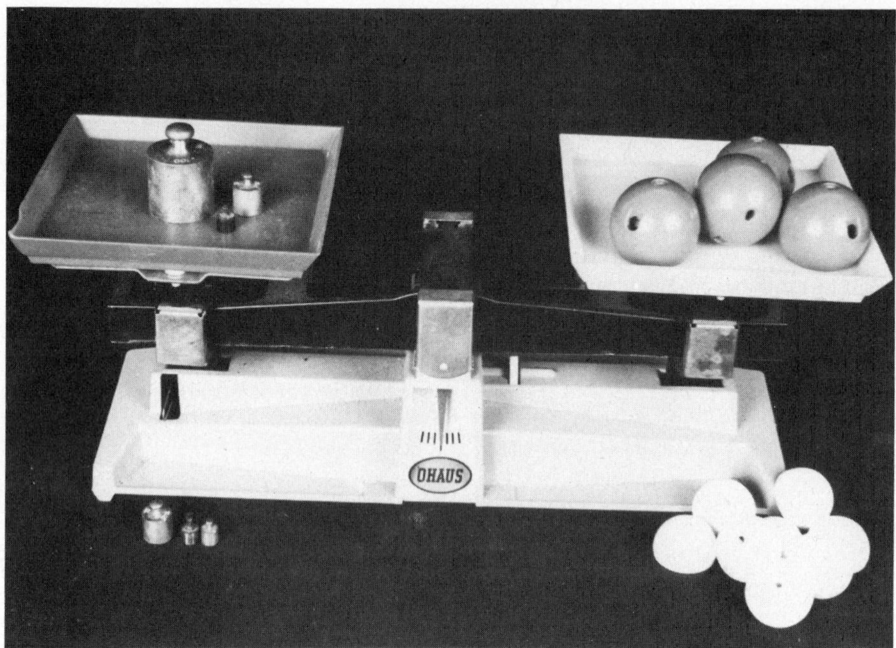

Figure 6.1 *Oxygen atom models being weighed on a balance.* (*Photo by Tom Greenbowe and Keith Herron*)

If real atoms were as big as the models used to represent them, it would be no problem to find the mass of one atom. The atoms could be weighed on a balance. That is exactly what was done with a set of large models representing atoms in order to derive the data recorded in Table 6.1. Table 6.1 shows the actual mass of these atom models.

The models in Table 6.1 were designed to be as nearly like the atoms that they represent as possible. Models that weigh as little as a single atom cannot be made, but the mass of these models is proportional to the mass of real atoms.

If you divide the mass of the carbon model by the mass of the hydrogen model, you will see that the carbon model is 12 times as heavy. Dividing the mass of the nitrogen model by the mass of the hydrogen model, you see that the nitrogen model is 14 times as heavy. In similar fashion, the oxygen model is 16 times as heavy as the hydrogen model, and the chlorine model is 35.5 times as heavy as the hydrogen model. These values are listed in Table 6.1 as the relative masses of the atom models.

* Again we face the problem of mass and weight. The term traditionally used by chemists is "atomic weight," even though mass is correct. Newer books are beginning to use "atomic mass," but you may encounter atomic weight in other references. It means the same thing.

6.2 RELATIVE MASS

There are two common ways to compare sizes. One way is to take a difference. This is what we do when we say, "Bill is 3 inches taller than John" or remark that "the price of gasoline has increased by 6 cents a gallon." Another way to compare is to say that one measure is so many times another. For example, we might say that 1 inch is 2.54 times as long as 1 centimeter or that 1 kilogram is 2.2 times as heavy as 1 pound. Relative mass is this kind of comparison.

Table 6.1. Mass of Several Atom Models

Atom Model	*Actual Mass of Model (g)*	*Relative Mass of Model*
hydrogen	1.72 g	1.00
carbon	20.64 g	12.0
nitrogen	24.08 g	14.0
oxygen	27.52 g	16.0
chlorine	61.06 g	35.5

Table 6.1 lists the relative mass of carbon as 12, and this means that the carbon model is 12 times as heavy as the hydrogen model. The relative mass of chlorine is given as 35.5, which means that 1 chlorine model weighs as much as 35.5 hydrogen models.

Look at the periodic table on the inside front cover. At the bottom of the square representing hydrogen, you see 1.01. In the square for carbon, you see 12.0. Nitrogen shows 14.0, oxygen 16.0, and chlorine 35.5. These numbers represent the relative mass of atoms of real elements, just as Table 6.1 shows the relative mass of the models we have built. (As we said earlier, the models were built to be as much like the real atoms as possible.)

The relative mass of an atom is called its **atomic mass.** It is *not* the actual mass of the atom in grams. It is the mass of that element *relative* to some element taken as a standard. In making our models, we compared everything to hydrogen. The carbon model was made 12 times as heavy as the hydrogen, the oxygen model was made 16 times as heavy, and so on. The original atomic mass scale had exactly the same meaning. Hydrogen was assigned a value of exactly 1.00. Since carbon was found to be 12 times as heavy, it was given a value of 12, and so on. Later, oxygen was used as a standard for atomic mass but was assigned a value of exactly 16. Still later, carbon was used as the standard. These changes in standard have resulted in little change in the relative masses indicated in the periodic table, and we can still think of the **relative mass** (or *atomic mass*) of an element as the number of hydrogen atoms that weigh the same as one atom of that element.

6.3 DETERMINING RELATIVE MASS

To be sure that you grasp the idea of relative mass, complete the following exercise. The actual masses of several objects are given. Cellophane tape has been selected as a standard and given a value of 1.0. Complete the table with the relative mass of each object. (If you have trouble getting started, check the answer to Question 6-1.)

	Object	*Actual Mass*	*Relative Mass*
	roll cello tape	28 g	1.0
6-1.	typewriter ribbon	42 g	________
6-2.	home stapler	115 g	________
6-3.	calculator	238 g	________

Although the calculations seem easier when an object is assigned a relative mass of 1.0, this isn't necessary. Find the relative mass of each object when the relative mass of the stapler is taken as 4.00.

	Object	*Actual Mass*	*Relative Mass*
6-4.	roll cello tape	28 g	________
6-5.	typewriter ribbon	42 g	________
	home stapler	115 g	4.00
6-6.	calculator	238 g	________

Now let's look at the kind of information that is used to find the relative mass of atoms. The reaction of copper metal with oxygen gas to produce black copper(II) oxide was discussed in Chapter 4. The experimental facts are that 125 g of CuO contain 100 g of Cu and 25 g of O. Although we don't know how many atoms of copper and oxygen are in 125 g of CuO, the formula indicates that there are the same number of Cu atoms and O atoms. Some number of Cu atoms weigh 100 g. *That same number* of O atoms weigh 25 g. The only way this can be true is for each Cu atom to weigh four times as much as an O atom. If oxygen has a relative mass of 16, copper must have a relative mass of (4×16) or 64. The periodic table gives the relative mass of copper as 63.5, which is 64 when rounded to the two significant digits that are justified by our calculations.

What we have just done illustrates the procedure used to determine relative masses. It may be clearer after you work through the following example. As before, use a sheet of paper to cover the page to the dotted line. Do what is asked, and then check your answer by moving the paper to the next dotted line.

Example 6.1

Using the models described in Table 6.1, we constructed several models of hydrogen chloride gas (HCl). These HCl models weighed 376.68 g. The H in all of the HCl models weighed 10.32 g. How much did the Cl in all of the models weigh?

Subtracting the mass of the hydrogen (10.32 g) from the total mass of the HCl (376.68 g), you find that the chlorine in the HCl models weighed 366.36 g. The mass relationship is summarized by the following sentence:

$$\underbrace{\text{10.32 g of H models}} + \underbrace{\text{366.36 g of Cl models}} \rightarrow \text{376.68 g of HCl models}$$

We want to compare the mass of the H models with the mass of the Cl models.

The Cl models are heavier than the H models. How many times as heavy are they?

Did you get 35.5?

We are asking what number times 10.32 is equal to 366.36:

$$(?)(10.32) = 366.36$$

Algebraically, we might write

$$10.32x = 366.36$$

$$x = \frac{366.36}{10.32}$$

$$x = 35.5$$

This is the relative mass of chlorine found in Table 6.1 Even though we don't know how many Cl models and how many H models were weighed, we do know that we weighed the same number of each. The pile of Cl models weighed 35.5 times as much as the pile of H models. Each Cl model must weigh 35.5 times as much as each H model.

It is important to understand that the relative mass of two elements does not depend on the number of atoms that are being weighed. To reinforce the point, various numbers of HCl models are pictured here. Use the data in Table 6.1 to calculate the actual mass of the models shown.

	Mass of H in the Models (g)	*Mass of Cl in the Models (g)*
	1 H =	1 Cl =
	2 H =	2 Cl =
	3 H =	3 Cl =

1 H = 1.72 g
2 H = 3.44 g
3 H = 5.16 g

1 Cl = 61.06 g
2 Cl = 122.12 g
3 Cl = 183.18 g

Now see what number you get when you divide the mass of the chlorine by the mass of the hydrogen.

In all three cases, you should have found that the chlorine is 35.5 times as heavy as the hydrogen. No matter how many models we are talking about, a pile of chlorine models weighs 35.5 times as much as a pile containing *the same number* of hydrogen models.

The point of this exercise is simply that you do not need to know the mass of one atom in order to find the relative mass. If you know the mass of *any number* of hydrogen models and the mass of *the same number* of some other atom models, you can find the relative mass of the other atom model.

Now do another exercise. When you finish, you should have discovered another important property of relative masses.

6-7. What is the relative mass of the hydrogen models?
6-8. How many hydrogen models weigh 1.00 tonn? (Hint: A tonn is 10^6 g, and each hydrogen model weighs 1.72 g.)
6-9. What is the relative mass of the carbon model?
6-10. How many carbon models weigh 12.0 tonn?
6-11. What is the relative mass of the oxygen model?
6-12. How many oxygen models weigh 16.0 tonn?
6-13. What is the relative mass of the chlorine model?
6-14. How many chlorine models weigh 35.5 tonn?
6-15. When the relative mass of *any* of the atom models is expressed in tonns, how many models does it take to weigh that amount?
6-16. It takes 581 oxygen models to weigh 16.0 kg. How many carbon models weigh 12.0 kg?

The questions that you just answered should convince you of an important fact. *If the relative masses of two elements are expressed in the same mass units, you have the same number of atoms of each element.*

In other words, 1.01 g of H, 12.0 g of C, 14.0 g of N, and 16.0 g of O contain the same number of atoms. It is also true that 1.01 oz of H, 12.0 oz of C, 14.0 oz of N, and 16.0 oz of O contain the same number of atoms. However, grams is the unit of mass normally used in science, so we will talk about the number of atoms in 1.01 g of hydrogen, 12.0 g of carbon, and so on. That number is called one mole.

6.4 HOW MANY IS A MOLE?

The mole concept is important. It is important because it is used to find the number of atoms in a sample of an element, the number of molecules in a sample of a compound, the mass of one element that will react with a given mass of another element, the molecular mass of a compound, and even the formula of a compound. Before we proceed to show how the idea of a mole can be used to do all these things, let us define the term.

A **mole** *is simply the amount of material that contains 6.02×10^{23} particles.* The particle can be anything. We usually talk about atoms, molecules, or electrons, but we could just as easily talk about a mole of apples or a mole of cars in the same way that we talk about a dozen donuts. A mole of apples is 6.02×10^{23} apples, and it would make quite a pile! In Chapter 1 we found that a mole of rice grains is enough to cover the United States to a depth of over half a mile. A mole of cars (6.02×10^{23} cars) would be much larger than the entire earth. You can see why we usually use moles only when we are talking about very small particles like atoms or molecules.

Example 6.2

Mole is a name for a number, much like dozen is a name for a number. When used as a unit, mole is abbreviated as mol. If you have 12 donuts, do you have a dozen donuts?

- - - - - - - - - - - - - - - - - - - -

Of course! It was a silly question. If you have 6.02×10^{23} donuts, do you have a mole of donuts?

- - - - - - - - - - - - - - - - - - - -

Yes. 6.02×10^{23} is the number in 1 mol, just as 12 is the number in a dozen.

How many dozen eggs do you have when you have 6 eggs?

- - - - - - - - - - - - - - - - - - - -

0.5 doz eggs. $$6 \text{ eggs} \times \frac{1 \text{ doz}}{12} = 0.5 \text{ doz eggs}$$

How many moles of eggs do you have when you have 3.01×10^{23} eggs?

- - - - - - - - - - - - - - - - - - - -

0.5 mol eggs. 3.01×10^{23} eggs $\times \dfrac{1 \text{ mol}}{6.02 \times 10^{23}} = 0.5$ mol eggs

Is 6.02×10^{24} pencils a mole of pencils? If not, how many moles is it?

---

10 mol pencils. 6.02×10^{24} pencils $\times \dfrac{1 \text{ mol}}{6.02 \times 10^{23}} = 10$ mol pencils

Confusion usually begins when other considerations come into play. If there are 24 bikes in a shop, how many dozen *wheels* are on the bikes?

The answer is 4 doz wheels. If you got 2 for the answer, you didn't read the question carefully. There are 2 wheels for each bike, so

$$? \text{ doz wheels} = 24 \cancel{\text{bikes}} \times \frac{2 \text{ wheels}}{1 \cancel{\text{bike}}} \times \frac{1 \text{ doz}}{12} = 4 \text{ doz wheels}$$

If a tank contains 3.01×10^{28} water molecules, how many moles of hydrogen atoms are in the tank?

10^5 mol H. The problem is like the one for the bike. There are 2 hydrogen atoms in each water molecule, so

$$? \text{ mol H} = 3.01 \times 10^{28} \cancel{H_2O} \times \frac{2 \text{ H}}{1 \cancel{H_2O}} \times \frac{1 \text{ mol}}{6.02 \times 10^{23}} = 10^5 \text{ mol H}$$

So far we have started with some number and calculated the number of moles, but we can also go the other way.

How many water molecules are there in 3.5 mol H_2O?

2.11×10^{24} molecules H_2O. (This time, you were asked about the entire molecule rather than one of its parts.)

$$? \text{ molecules } H_2O = 3.5 \cancel{\text{mol}} H_2O \times \frac{6.02 \times 10^{23} \text{ molecules}}{1 \cancel{\text{mol}}}$$

$$= 2.11 \times 10^{24} \text{ molecules } H_2O$$

How many H atoms are there in 3.5 mol H_2O?

4.21×10^{24} H atoms. (*Now* you are asked about one of the parts.)

$$? \text{ H atoms} = 3.5 \cancel{\text{mol } H_2O} \times \frac{6.02 \times 10^{23} \cancel{\text{molecules}}}{1 \cancel{\text{mol}}} \times \frac{2 \text{ H atoms}}{1 \cancel{H_2O \text{ molecule}}}$$

$$= 4.21 \times 10^{24} \text{ H atoms}$$

Now answer the following questions. In some of them, you will need to recall what you learned about naming compounds and writing formulas, as well as what is meant by a mole.

6-17. How many CO_2 molecules are there in 1.25 mol of CO_2?
6-18. How many moles of CO do you have when you have 1.2×10^{25} CO molecules?
6-19. How many oxygen atoms are there in 1.25 mol of sulfur dioxide?
6-20. How many moles of oxygen atoms do you have when you have 1.2×10^{25} dinitrogen pentoxide molecules?
6-21. How many molecules would be needed to make 5.33 mol of copper(II) chloride?
6-22. How many copper atoms are there in 5.33 mol of copper(II) chloride?
6-23. How many moles of chlorine atoms are there in 5.33 mol of copper(II) chloride?
6-24. How many moles of copper(II) chloride could be made with 1.2×10^{23} atoms of chlorine?
6-25. How many molecules of tin(IV) oxide are there in 3.15 mol of the compound?
6-26. How many oxygen atoms are there in 3.15 mol of tin(IV) oxide?

6.5 MOLE MASS

Even though a mole represents a given number of particles, it is important in chemistry because of its relationship to mass. Recall that a mole was introduced as the number of atoms in 1.01 g of H, 12.0 g of C, 16.0 g of O, or a pile of any element that weighs the same as its atomic mass in grams. The mass that contains 1 mol of particles is called the **mole mass.** This relationship to mass makes the mole important in chemistry.

We can't count atoms very conveniently because they are so small, but we can weigh large piles of atoms. If we know that the mass of 6.02×10^{23} atoms is the same as the atomic mass expressed in grams, we can "count" atoms of an element by weighing. The following example shows how.

Example 6.3

How many atoms are there in 26 g of carbon?

The atomic mass of carbon is 12.0, so 12.0 g of carbon is the mass of 6.02×10^{23} atoms. The mole mass of carbon is 12.0 g.

$$? \text{ atoms C} = 26 \cancel{\text{g C}} \times \frac{6.02 \times 10^{23} \text{ atoms C}}{12 \cancel{\text{g C}}} = 1.3 \times 10^{24} \text{ atoms C}$$

Usually we do not bother to find the number of atoms. We talk about moles in much the same way that we talk about dozens of eggs or reams of paper.

If the question had been "How many *moles* of carbon are there in 26 g?" we would proceed as follows:

$$? \text{ mol C} = 26 \cancel{\text{g C}} \times \frac{1 \text{ mol C}}{12 \cancel{\text{g C}}} = 2.2 \text{ mol C}$$

The logic is the same. We have merely changed the way we express the amount.

1 mol C is the same as 6.02×10^{23} atoms C, which is 12.0 g C.

2.2 mol C is the same as 1.3×10^{24} atoms C, which is 26 g C. Each of these two sentences describes the same amount of carbon in three different ways.

For each of the following amounts of the substances indicated, find the *number of atoms* represented and the *moles of atoms* represented.

6-27. 28 g of sodium
6-28. 122 g of Fe
6-29. 150 g of chlorine
6-30. 21 g of F
6-31. 5.0 g of zinc
6-32. 2.4 g of Ca

The following questions require nothing new, but you must read carefully to be sure you understand what you are to find.

6-33. How many moles of Na are there in 42 g of Na?
6-34. How many moles of O are there in 8.25 g of O?
6-35. How much do 2.18 mol of copper weigh?
6-36. How much do 0.28 mol of iron weigh?
6-37. How many atoms are there in 7.2 mol of chlorine?
6-38. How many moles are equal to 1.0×10^{9} atoms?
6-39. How many atoms are there in 36 g of bromine?
6-40. How much do 1.2×10^{25} atoms of sulfur weigh?

6.6 MOLECULAR MASS

We have introduced atomic mass and pointed out that the atomic mass of an element is the mass of an atom of that element relative to the mass of a hydrogen atom. The atomic mass of the element expressed in grams (the mole mass) is the mass that contains 6.02×10^{23} atoms. This number of atoms is 1 mol.

Just as we have talked about the relative mass of atoms, we can talk about the relative mass of molecules. To illustrate this, let's return to the models. Listed in Table 6.2 are the actual mass and relative mass of several molecular models. The actual mass was found by weighing, as shown in Figure 6.2.

The relative mass of these molecules was found in the same way that the relative mass of atoms was found in Table 6.1. The actual mass of the molecular model was divided by the actual mass of the hydrogen model.

The relative mass has the same meaning as before. When we say that the relative mass of CH_4 is 16, we mean that the CH_4 model weighs 16 times as

Table 6.2. Mass of Molecular Models

Formula	*Actual Mass of Model (g)*	*Relative Mass of Model (molecular mass)*
HCl	62.78	36.5
CH_4	27.52	16
CO_2	75.68	44
N_2	48.16	28
NH_3	29.24	17
O_2	55.04	32

much as the H model. The relative mass of a molecule is called the **molecular mass,** just as the relative mass of an atom is called the atomic mass.

We find the molecular mass by adding the atomic masses of all elements in the molecule. Table 6.1 shows that the atomic mass of hydrogen is 1 (in the example we have used a more exact value for hydrogen, so the molecular mass can be calculated to three places) and the atomic mass of chlorine is 35.5. Adding these together, you get 36.5, the molecular mass of HCl. You can find the molecular mass of any molecule in the same way.

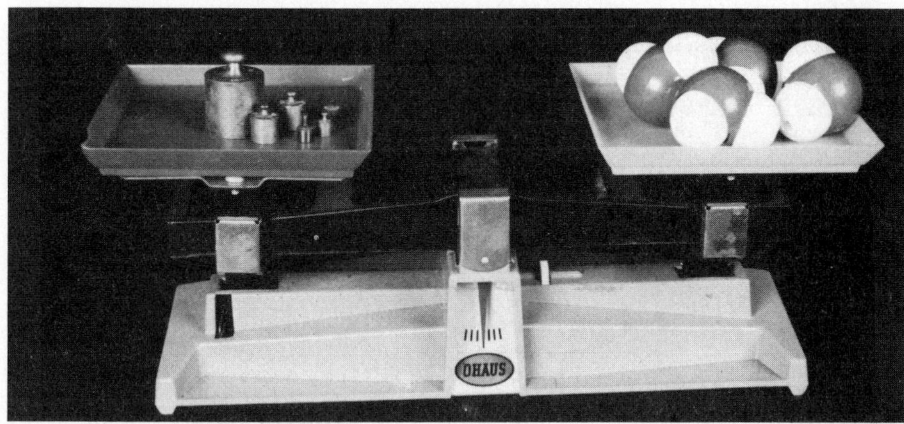

Figure 6.2 *Water molecule models being weighed on a balance.* (*Photo by Tom Greenbowe and Keith Herron*)

Example 6.4

What is the molecular mass of C_3H_8?

The molecular mass is the sum of all of the atomic masses in the molecules. For the carbon we have

$$C_3H_8$$

12.0	×	3	=	36.0
the atomic mass for carbon		the number of carbon atoms		the mass due to carbon

For the hydrogen we have

		C_3H_8		
1.01	×	8	=	8.08
the atomic mass for hydrogen		the number of hydrogen atoms		the mass due to hydrogen

Adding the mass of carbon and the mass of hydrogen, we get

36.0	+	8.08	=	44.1
the mass of the three carbon atoms		the mass of the eight hydrogen atoms		the molecular mass of C_3H_8

Using the atomic masses given in the periodic table, find the molecular mass for each of the following molecules to the nearest tenth of a unit.

6-41. carbon monoxide
6-42. C_2H_6
6-43. H_2SO_4
6-44. $CaCO_3$
6-45. $KMnO_4$
6-46. carbon tetrachloride
6-47. tin(IV) oxide
6-48. iron(III) oxide

6.7 MOLE MASS OF MOLECULES

You have learned that the atomic mass of an element expressed in grams is called the mole mass and contains 1 mol of atoms. It will come as no surprise that the molecular mass of a compound expressed in grams is also called the mole mass and contains 1 mol of molecules. The following argument should help you see why this is so.

1 H atom + 1 Cl atom → 1 HCl molecule

2 H atoms + 2 Cl atoms → 2 HCl molecules

We could continue with 5, 10, or 1,000 atoms of each kind combining to give 5, 10, or 1,000 molecules of HCl. Eventually,

(1) 6.02×10^{23} H atoms + 6.02×10^{23} Cl atoms → 6.02×10^{23} HCl molecules

Since 1 mol is the same as 6.02×10^{23}, we could write the last sentence in the following way:

(2) 1 mol H atoms + 1 mol Cl atoms → 1 mol HCl molecules

How much would this many atoms and molecules weigh? For atoms, we know that the mole mass is the atomic mass in grams. We can write

1.01 g H atoms + 35.5 g Cl atoms → ? g HCl molecules

Is the mass of 1 mol of HCl molecules the same as molecular mass (36.5) in grams? It must be. The HCl is formed by the H and Cl atoms sticking together. If the H atoms weigh 1.01 g and the Cl atoms weigh 35.5 g, the sum is 36.5 g.

(3) 1.01 g H + 35.5 g Cl → 36.5 g HCl

Look again at the chemical sentences marked (1), (2), and (3). All three say exactly the same thing in different ways. All three show the amount of hydrogen and chlorine that combine to give a mole of hydrogen chloride. All three sentences are talking about the same amount of the three substances.

During the course, you will have many occasions to express amounts of chemicals in these three ways: in terms of the *number* of particles, in terms of the *moles* of particles, and in terms of the *mass* of the particles.

A MNEMONIC FOR MOLE RELATIONSHIPS

Beginning students often have trouble remembering relationships illustrated by the chemical sentences we have just discussed. Someone has remarked that "the mole is at the heart of things." The diagram shown in Figure 6.3 may help you remember the importance of the mole as well as the many relationships involving it.

If you know the mass of an element in a sample, you can divide by the atomic mass to find the number of moles of that element in the sample. If you need to know the number of atoms in the sample, you can then multiply by 6.02×10^{23}. These operations are shown by the top arrows in Figure 6.3.

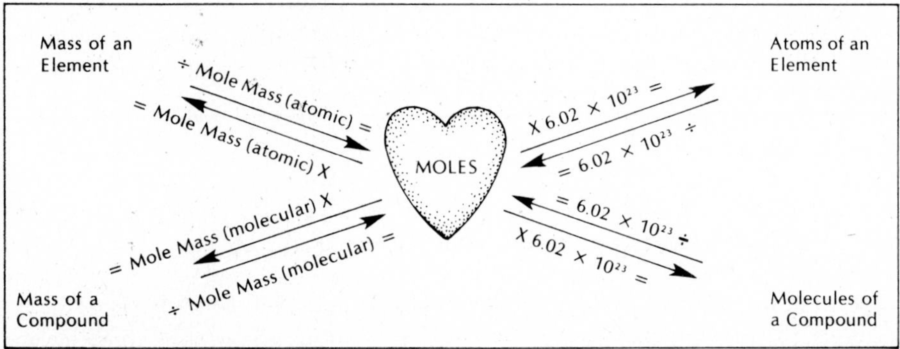

Figure 6.3 *Relationships involving the mole.*

If you know the number of atoms in a sample of an element and need to know the mass of the sample, you can follow the reverse procedure indicated by the upper arrows pointing toward the heart in Figure 6.3. The lower arrows in the figure represent similar computations for compounds.

If you find Figure 6.3 helpful, copy it on a card and use it as you work problems in this chapter. However, you should try to understand the relationships so that you do not need this crutch. Try to work each problem without referring to the card. If you can't, use the card to see whether it suggests a relationship that will help solve the problem. After some practice, you should be able to work the problems without using the card at all.

Now it is time for practice. Find the number of moles or the number of molecules in the indicated amounts of compound. (Hint: You can use the molecular masses that you calculated in Questions 6-35 through 6-42 to find the answers to the following questions.)

6-49. How many moles of CO molecules are there in 52 g of CO?

6-50. How many moles of C_2H_6 are there in 124 g? (In the language of chemistry, this is the way such questions are generally asked. It is understood that we want to find the number of moles of C_2H_6 *molecules* in 124 g of C_2H_6.)

6-51. How many moles of CCl_4 are there in 56 g?

6-52. How much do 2.50 mol of H_2SO_4 weigh?

6-53. How much do 0.25 mol of Fe_2O_3 weigh?

6-54. How many molecules (*not* moles of molecules) are there in 52 g of CO?

6-55. How many molecules are there in 22.4 g of tin(IV) oxide?

6-56. How many molecules are there in 116 g of carbon tetrachloride?

6-57. What is the mass of 3.01×10^{23} molecules of iron(III) oxide?

6-58. What is the mass of 1.2×10^{25} molecules of carbon monoxide?

Are you aware that we have not answered the question asked at the beginning of this chapter? We asked, "How do we *know* how many atoms are in one molecule?" We introduced atomic mass and molecular mass in an attempt to find an answer. Unfortunately, in order to find the atomic mass of various elements, we *assumed* the formula for some simple compounds. It seems that we cannot find atomic masses of elements unless we know the formula for some compounds, and we cannot determine the formula of a compound unless we know the atomic mass of the elements. That's a problem. In fact, the problem was so great that it was not solved for almost 100 years. The following account describes how the solution finally came about.

6.8 A SHORT HISTORY OF ATOMIC MASS AND MOLES

John Dalton was an Englishman born just 10 years before the United States declared its independence. In 1803 he proposed an atomic theory. He suggested that there were "elemental atoms" and "compound atoms," and he developed a table of atomic masses. He did this by making an assumption about molecules. He assumed that when only one compound of two elements is known, it must have molecules containing a single atom of each element.

Dalton knew that water was composed of hydrogen and oxygen, so he assumed that the formula for water was HO. This assumption is summarized as follows:

1 H atom + 1 O atom → 1 HO molecule

Dalton also knew from experiment that

$$1 \text{ g H} + 8 \text{ g O} \rightarrow 9 \text{ g water}$$

From this experimental fact and his assumption that water is HO, Dalton assigned hydrogen an atomic mass of 1 and oxygen an atomic mass of 8. Using these values and other compounds, Dalton found other values for atomic mass that were just as wrong. It was an honest mistake, but a mistake just the same. We now know that water is H_2O, making oxygen 16 times as heavy as hydrogen.

The data he needed to get the right answer were available to Dalton, but he did not interpret those data correctly. What were the data?

Gay-Lussac was a French chemist who lived from 1778 to 1850. He was a good experimentalist, and his work led him to investigate how gases combine to form gaseous compounds. It has been noted that, even though the proportions of elements in a given compound were always the same (law of definite composition), the same elements could combine to form different compounds with different ratios (law of multiple proportions). Much of Gay-Lussac's work was done on gases that formed different compounds. He noted that, in *all* cases, the ratio of the *volumes* of gases was a simple, whole-number ratio. His analysis of data for various oxides of nitrogen illustrate what he found (Table 6.3). The formulas shown in the last column of Table 6.3 are

Table 6.3. Combining Volumes of Gases

Compound Name (ancient name)	*Volume of Nitrogen*	*Volume of Oxygen*	*Accepted Formula*
dinitrogen oxide (nitrous oxide)	100	49.5	N_2O
nitrogen monoxide (nitrous gas)	100	108.9	NO
nitrogen dioxide (nitric acid)	100	204.7	NO_2

the formulas accepted today. They were *not* known at the time Gay-Lussac did his work.

Look at the volumes of nitrogen and oxygen that react to form each compound, and compare them to the modern-day formula for that compound. In dinitrogen oxide, the *volume* of nitrogen that combines is twice the *volume* of the oxygen. The modern-day formula for the gas is N_2O, indicating twice as many *atoms* of nitrogen as atoms of oxygen. For nitrogen monoxide, equal *volumes* of nitrogen and oxygen react to form the compound. The accepted formula indicates that equal numbers of *atoms* of nitrogen and oxygen combine to form the compound. In nitrogen dioxide, the *volume* of oxygen gas that combines is twice the *volume* of the nitrogen gas. The modern formula indicates two *atoms* of oxygen for one *atom* of nitrogen.

The volumes of gases that react to form a compound can be expressed as a simple ratio: in this case 2:1, 1:1, or 1:2. It is the same as the ratio of atoms in the molecule. This seemed too simple to Dalton and others of his day, and it went unnoticed by all but one man, Amadeo Avogadro.

Avogadro wrote a paper in 1811 in which he suggested that Gay-Lussac's data on the combining volumes of gases could be explained very simply, if one made two bold assumptions. The first assumption was that *equal volumes of gases contain equal numbers of particles.* The second assumption was that *elemental gases exist as diatomic molecules* rather than as individual atoms.

Neither assumption was accepted for fifty years! Why not? Because scientists are human and they tend to listen to individuals who have prestige and reputation. In the early 1800s, Dalton had the reputation and Avogadro did not. Dalton saw no reason to complicate matters by assuming that elemental gases were diatomic, and the idea that a little atom such as hydrogen would occupy the same volume in a gas as a much heavier atom such as oxygen seemed preposterous. Dalton's influence was too great. It was not until Cannizzaro revived Avogadro's strange hypotheses in 1860 that the idea was finally accepted. Avogadro died in 1856.

It is rather sad when a person's ideas are not appreciated until after his or her death, but it often happens that way. To make up for such disrespect, we frequently establish memorials to such people. This is what happened in the case of Avogadro. Since his ideas later enabled us to learn the number of particles in 1 mol, the number is named in his honor; 6.02×10^{23} is called Avogadro's number in honor of a man who never heard of it! Later in chemistry you will have an opportunity to learn some of the techniques that have been used to determine the value of that number.

The story of the development of early atomic theory is a fascinating one, and it is told rather clearly in the words of those scientists who took part in the development. Those of you who enjoy history might be interested in reading about Dalton, Gay-Lussac, Avogadro, and Cannizzaro in a book on the history of science. You can probably find any of the following as well as other good ones in your library:

Schwartz, George and Bishop, Philip W. *Moments of Discovery: The Origins of Science.* New York: Basic Books, 1958.

———. *Moments of Discovery: The Development of Modern Science.* New York: Basic Books, 1958.

Asimov, Isaac. *The New Intelligent Man's Guide to Science.* New York: Basic Books, 1965.

Knickerbocker, William S. *Classics of Modern Science.* Boston: Beacon Press, 1962.

Taton, René (Ed.). *History of Science: Science in the Nineteenth Century.* New York: Basic Books, 1965.

Newman, James R. *Science and Sensibility, Vols. 1 and 2.* New York: Simon and Schuster, 1961.

6.9 SOME CONFUSING POINTS

Several issues related to the mole concept tend to cause confusion among beginners. It is worth dwelling on some of these issues, starting with a very innocent question like those asked earlier in this chapter.

ATOMS OR MOLECULES

"How much do 1.5 mol of hydrogen weigh?" Now what's wrong with that? From the periodic table you can easily see that 1 mol of hydrogen weighs 1 g, so 1.5 mol of hydrogen should weigh 1.5 g. True, *if* we are talking about 1.5 mol of hydrogen *atoms.* Unfortunately, we use the word "hydrogen" when we refer to hydrogen atoms (H) or ordinary hydrogen gas (H_2). Hydrogen gas exists as diatomic molecules with the formula H_2. Clearly 1.5 mol of H_2 contain 3 mol of hydrogen *atoms* and weigh 3 g. Students (and teachers) should be careful when they speak and read to be sure that they know which is intended. The fact is that "1.5 mol of hydrogen" may mean 1.5 mol of hydrogen *atoms* or it may mean 1.5 mol of hydrogen *molecules.* Usually the meaning is clear from the context of the question. If it is not, you had better ask!

DOES THE COMPOUND FORM MOLECULES?

Another point of confusion is contained in the question, "How many molecules of salt are there in 59 g of NaCl?" The issue here is a little different. It revolves around the question, "What is a molecule of NaCl?"

Many compounds such as salt, baking soda, alum, washing soda, and lye do not form particles of a definite size that can correctly be described as molecules. A crystal of salt, for example, may be very small or several centimeters on a side, as shown in Figure 6.4 (p. 168). The crystal is made up of alternating atoms of sodium and chlorine in the ratio indicated by the formula NaCl (Figure 6.5, p. 168).

Figure 6.4
A crystal of salt can be any size. Compounds like salt do not form molecules of a definite size. (*Photo by Dudley Herron*)

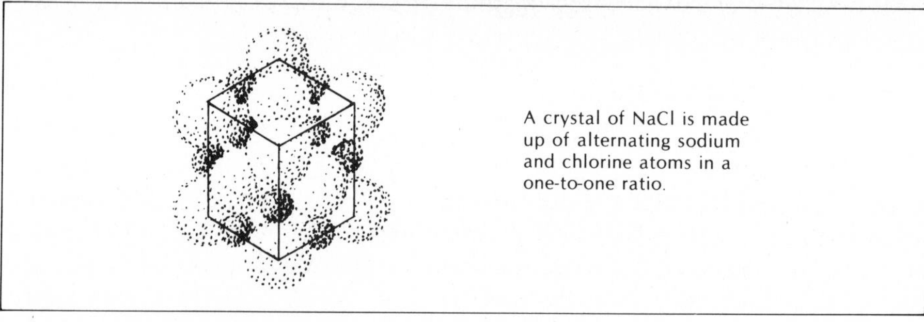

Figure 6.5
A model showing the arrangement of sodium and chlorine in sodium chloride.

Strictly speaking, NaCl can be called the formula for salt, but it does *not* represent a *molecule*. You will sometimes find that NaCl is referred to as a "formula unit" and the mass of one formula unit referred to as the "formula mass" rather than the molecular mass. This terminology has much to recommend it. But it means that, when you see a formula, you must know whether that material forms molecules or not—something the beginning student does not know. Therefore, we will normally make no distinction between molecules and formula units. We will use molecular mass to describe the relative mass of one molecule or one formula unit. When we talk about 1 mol of a compound, it will mean the molecular mass expressed in grams and will contain 6.02×10^{23} molecules if the substance forms molecules, or 6.02×10^{23} formula units if the substance does not form molecules. In other words, we will assume that all compounds are made up of molecules with a composition described by the formula.

MOLE AND MOLECULE

It is perhaps unfortunate that the amount of substance that contains 6.02×10^{23} elemental particles is called a mole, because it looks very much like a shortened form of the word "molecule." It is not. Actually, "mole" comes from the Latin *moles,* which means a "mass, hump, or pile." "Molecule" is a diminutive form of *moles* that means "the smallest piece" of that hump or pile. You must keep in mind that mole refers to a large amount that can easily be seen. Molecule refers to an invisible particle too small to imagine.

I have frequently asked the following questions on a test.

1. How much does one *mole* of oxygen gas weigh?
2. How much does one *molecule* of oxygen gas weigh?
3. How much does one *mole* of oxygen *atoms* weigh?
4. How much does one *atom* of oxygen weigh?

I always get gray hairs when students give the same answers to the first two questions. (Some students give the same answer to all four!) To say that a molecule of oxygen gas weighs the same as a *mole* of oxygen gas is equivalent to saying that an apple weighs the same as 6.02×10^{23} apples. We can expose the foolishness of such a statement by doing some simple calculations. The medium apple I had for lunch weighed 115 g. Then the mass of a mole of apples would be:

$$\frac{115\text{ g}}{\text{apple}} \times \frac{6.02 \times 10^{23}}{1\text{ mol}} = 6.92 \times 10^{25}\text{ g/mol apples}$$

By comparison, the mass of the earth is 5.97×10^{27} g, not quite 100 times as heavy as 1 mol of apples! Please be sure that you can answer the four questions just listed and that you *understand* why you get the answers you do.

6.10 EMPIRICAL FORMULA

Finally, we have enough information to determine the formula for a compound. Before we do so, recall once more what a formula represents. Water will do. The formula is H_2O. The formula indicates that

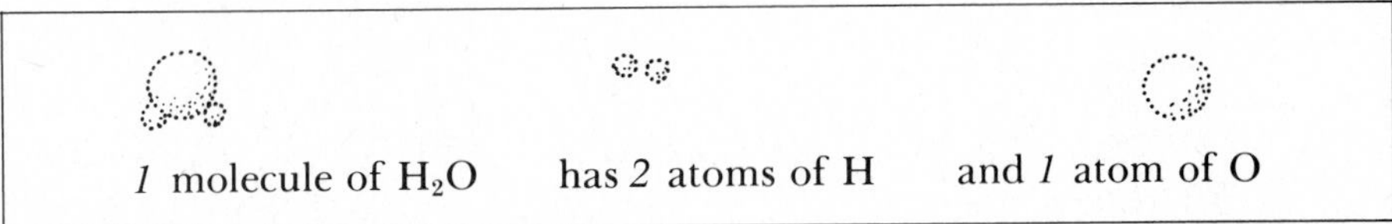

But if this is true, we could also say:

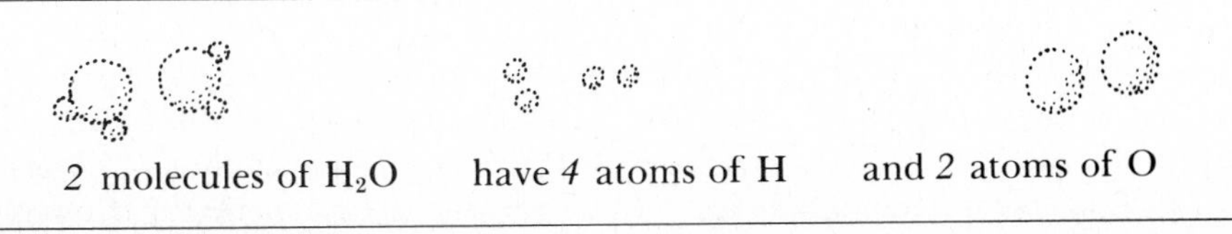

Increasing the amount of water even more, we could say:

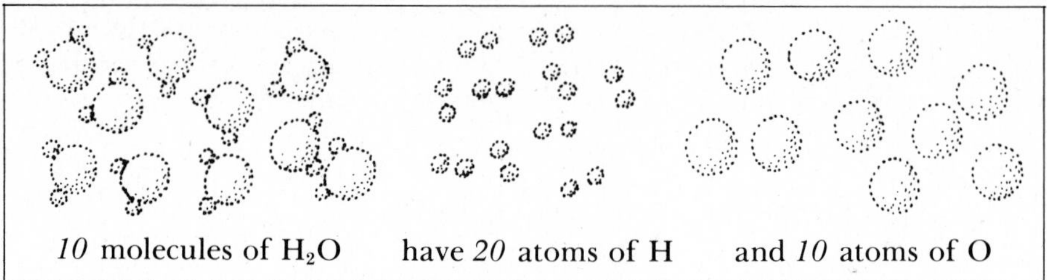

If we had an entire mole of water we could say:

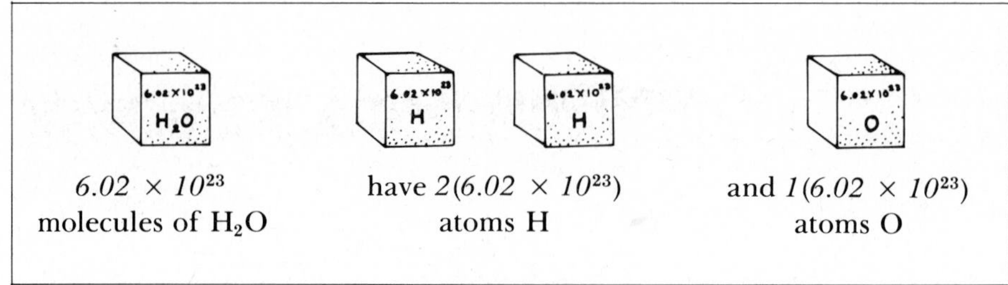

This last sentence could have been written

1 mol H_2O contains *2 mol* H and *1 mol* O

Even though you are accustomed to thinking about a formula as representing the number of *atoms* of each element in one *molecule* of a compound, it is just as correct to think of the formula as indicating the *moles* of each element in one *mole* of the compound. The two ways of reading the formula for water are illustrated in Figure 6.6.

Figure 6.6

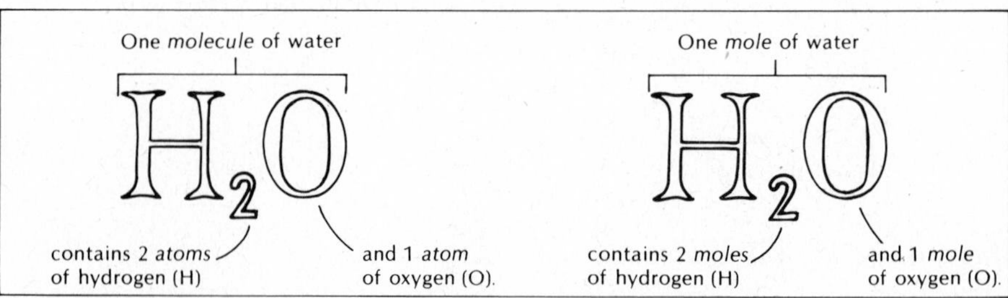

To determine the formula for a compound, it isn't necessary to count the atoms in a single molecule. The same information is obtained by finding the moles of each element in the compound. If we know the mass of the element, we can find the number of moles. This is the way we determine formulas.

Get a sheet of paper and work through the following example with me to see how the formula is found. Again, with a sheet of paper, cover each step of the solution until you have worked it yourself.

Example 6.5

How could we find the formula for water if we didn't know it?

Water decomposes when an electric current passes through it. The hydrogen and oxygen gas can be collected and weighed. An apparatus for the decomposition of water by electrolysis is shown in Figure 6.7

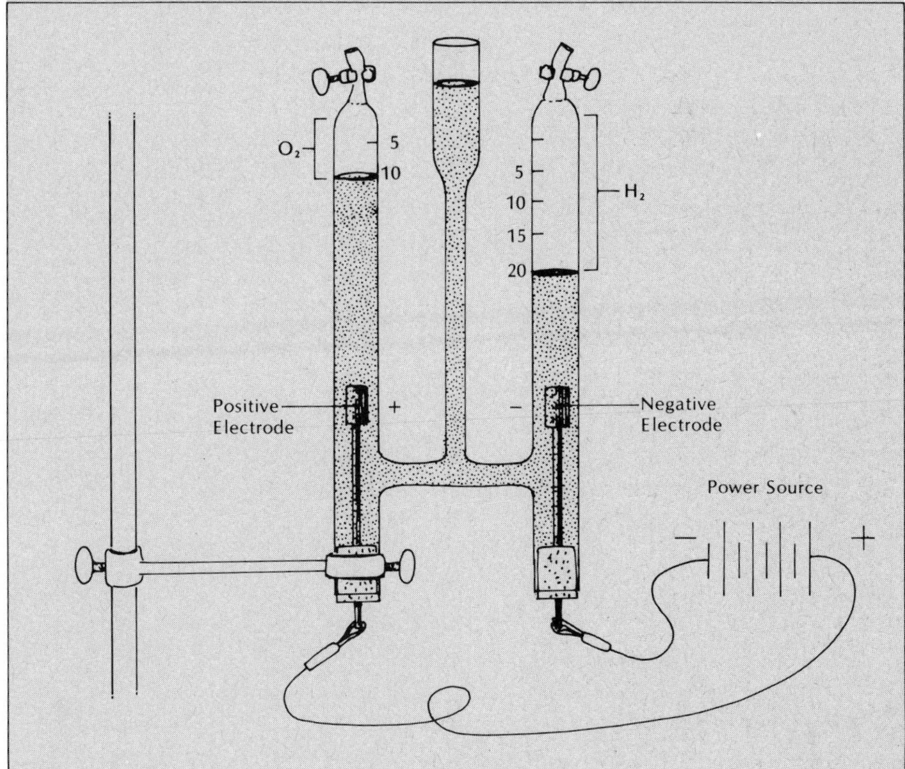

Figure 6.7 *Water can be decomposed to form hydrogen and oxygen.*

A cup of water weighs about 250 g. When it decomposes, 27.8 g of hydrogen and 222.2 g of oxygen gas are produced.

How many moles of H atoms are produced?

How many moles of O atoms are produced?

The 27.8 g of hydrogen gas must contain 27.5 mol H atoms and the 222.2 g of oxygen gas must contain 13.9 mol O atoms. The calculations are as follows:

$$? \text{ mol H} = 27.8 \text{ g H} \times \frac{1 \text{ mol H}}{1.01 \text{ g H}} = 27.5 \text{ mol H}$$

$$? \text{ mol O} = 222.2 \text{ g O} \times \frac{1 \text{ mol O}}{16.0 \text{ g O}} = 13.9 \text{ mol O}$$

The cup of water contains 27.5 mol of hydrogen atoms and 13.9 mol of oxygen atoms. How many moles of H atoms are there for each mole of O atoms?

There are 1.98 mol H for 1.00 mol O. Do you see why this is so? If not, the following reasoning may help.

We have:	13.9 mol O	and	27.5 mol H
We could divide each pile in half to get:	6.95 mol O	and	13.75 mol H
Dividing in half again, we get:	3.48 mol O	and	6.88 mol H
Repeating, we get:	1.74 mol O	and	3.44 mol H
Finally:	0.87 mol O	and	1.72 mol H

Working this way we never get the amount of hydrogen corresponding to exactly 1.00 mol O, but we can see that it would be a little more than 1.72. Let's start again.

We have: 13.9 mol O and 27.5 mol H

What can we divide by to get exactly 1.00 mol O?

Divide by 13.9.

We get: 1.00 mol O and 1.98 mol H

The general rule is to *divide the number of moles of each element by the smaller of the two numbers.* In this case, the smaller number was 13.9.

We now know that there are approximately 2 mol of hydrogen for every 1 mol of oxygen in water. It seems safe to assume that the formula for water is H_2O.

Before working through another example, let us summarize what we did to find the formula for water.

1. We experimentally determined the mass of each element in some amount of the compound.
2. We divided the mass of each element by the mole mass of that element to determine the number of moles of that element in the compound.
3. We divided the number of moles of each element by the smallest number of moles to give the ratio of atoms in the compound.

These steps are enough to find the formula for any molecule that contains no more than one atom of some element. An additional step is required for compounds like P_2O_5. The following example illustrates that additional step and suggests a neater way to organize your work. Work through it with me, covering each step of the solution, as done previously.

Example 6.6

In the laboratory, 27.28 g of rust (an oxide of iron) was mixed with charcoal and heated in a dish until all of the oxygen in the rust combined with the carbon and left behind a blob of pure iron. The blob of iron weighed 19.10 g. What is the formula for the rust?

To begin solving the problem, organize the experimental data in a table. The labels are listed. Fill in the values.

Mass of rust analyzed = ______ g

Mass of iron in the rust = ______ g

Mass of oxygen in the rust = ______ g

Mass of rust analyzed = 27.28 g (stated in the problem)

Mass of iron in the rust = 19.10 g (the mass of the iron blob)

Mass of oxygen in the rust = 8.18 g (not given, but the rust is iron and oxygen, so it can be found by subtraction)

As experimental data are obtained in the laboratory, it is common practice to determine the mass of all elements but one and find the mass of that final element by subtracting the mass of the known elements from the total mass of the compound. That is how the mass of oxygen was found here.

What is the next step in finding the formula?

Finding the moles of each element in the compound. Using factor-label, the problem can be set up like this:

$$? \text{ mol Fe} = 19.10 \text{ g Fe}$$

and

$$? \text{ mol O} = 8.18 \text{ g O}$$

Complete the problems to find the number of moles.

$$? \text{ mol Fe} = 19.10 \text{ g } \cancel{\text{Fe}} \times \frac{1 \text{ mol Fe}}{55.8 \text{ g } \cancel{\text{Fe}}} = 0.342 \text{ mol Fe}$$

$$? \text{ mol O} = 8.18 \text{ g } \cancel{\text{O}} \times \frac{1 \text{ mol O}}{16.0 \text{ g } \cancel{\text{O}}} = 0.511 \text{ mol O}$$

The sample of rust that was analyzed contained 0.342 mol Fe and 0.511 mol O. What is the next step in finding the formula?

Divide both numbers by the smaller number of the two.

$$\frac{0.342 \text{ mol Fe}}{0.342 \text{ mol Fe}} = 1 \qquad \frac{0.511 \text{ mol O}}{0.342 \text{ mol Fe}} = 1.49 \text{ mol O/mol Fe}$$

There are 1.49 mol O atoms for every 1 mol Fe atoms in the compound. What is the formula for the compound?

Fe_2O_3

The formula *could* be written $Fe_1O_{1.49}$, but this would be awkward and misleading. We could not have a single molecule with $1\frac{1}{2}$ oxygen atoms. In order to get a formula with whole-number subscripts, we must take an additional step.

When the subscripts calculated are not whole numbers, they must all be multiplied by some number to get whole-number subscripts.

We have: 1.00 mol Fe and 1.49 mol O

What could we multiply these numbers by to get whole numbers?

Multiplying by 2 we get: 2.00 mol Fe and 2.98 mol O

Because of uncertainty in any data obtained from an experiment, 2.98 can be rounded to 3 mol O. The formula is Fe_2O_3.

6.11 WHEN IS A NUMBER WHOLE?

In the last example, 2.98 was called a whole number and rounded to 3; but earlier in the example, 1.49 was not considered a whole number. Neither 2.98 nor 1.49 is *exactly* a whole number. When can you assume that the number *should* be whole?

We talked a lot about uncertainty in Chapter 2. Here you can see why. Unless we know how much uncertainty there is in a number, we have no way of knowing when we are justified in calling it a whole number. We must keep track of significant digits in calculations, and we must estimate the errors that could have occurred during an experiment. This includes such errors as spilling some of the compounds used, contaminating the compound with material left on dirty glassware, losing some compound during filtering or distillation, absorption of water from the atmosphere, and many other errors that cannot be accounted for exactly.

In the example, 2.98 was called a whole number (3) but 1.49 was not, because the data were good enough to produce uncertainty only in the third digit. The calculations should be certain when rounded to only two digits. When 2.98 is rounded to two places, we get 3.0. When 1.49 is rounded to two places, we get 1.5. I feel confident that uncertainty in the experiment could produce an answer of 2.98 when the true value is 3, but I am not confident that the uncertainty would be great enough to produce an answer of 1.5 when the true value is either 1 or 2. We must have some knowledge of the uncertainty in an experiment in order to get correct answers to important questions in science.

6.12 EMPIRICAL FORMULA VS. MOLECULAR FORMULA

There is one other point to be made about calculating formulas. If you had multiplied 1 and 1.49 by 4 or 6, you would have gotten a whole-number ratio within the uncertainty of the experiment. Doing so would produce formulas for rust of Fe_4O_6 and Fe_6O_9. There is really no way to be sure that the correct formula is Fe_2O_3 rather than one of these other formulas. All of the formulas indicate 1 mol of iron combined with 1.5 mol of oxygen, and that is all we were able to determine empirically. (Empirically means "by experiment.")

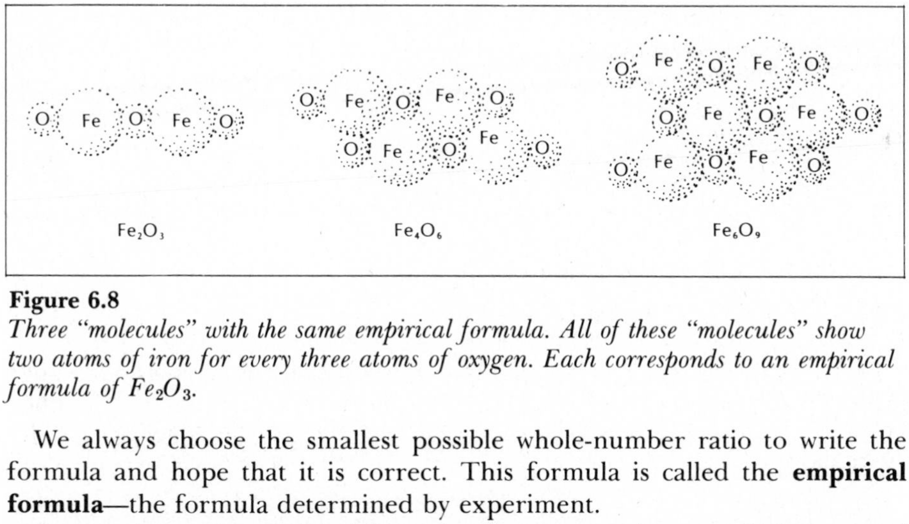

Figure 6.8
Three "molecules" with the same empirical formula. All of these "molecules" show two atoms of iron for every three atoms of oxygen. Each corresponds to an empirical formula of Fe_2O_3.

We always choose the smallest possible whole-number ratio to write the formula and hope that it is correct. This formula is called the **empirical formula**—the formula determined by experiment.

Once we know the empirical formula of a compound, we *can* find the actual formula if we know the molecular mass of the compound. Later in this course, you will learn one way to determine the molecular mass of a compound by experiment, but that procedure doesn't work for all compounds.

It sometimes takes all the ingenuity chemists can muster to be *certain* of the compound that they have prepared. The many techniques that are used to identify compounds constitute a whole branch of chemistry called analytical chemistry. In recent years, chemists have developed sophisticated instruments that make the job much easier than it was a century ago. Still, chemical detective work requires a lot of effort and a great deal of specialized knowledge. Of course, even the specialists started with the basic ideas that you are learning in this course. In order to get those ideas straight, they needed to practice—and so do you. Before going on, work the following problems.

6-59. 15.28 g of an oxide of manganese contain 9.66 g of Mn. What is the empirical formula of the compound?

6-60. 29.95 g of antimony react with iodine to produce 123.6 g of compound. What is the empirical formula of the compound?

6-61. 92.19 g of chlorine combine with 7.81 g of carbon to form a compound. What is the empirical formula of the compound?

↑ Read the example again if you don't see how to work the problem, but then try to do the first problem without going back to the example. After you finish, compare your work with the solution given at the end of the chapter. After you understand how the problem was solved, work the next problem *without* looking at your previous work or the solution at the end of the chapter. Check the answer only when you think you have the problem solved. Repeat this procedure for each problem. You *can* figure out the reasoning if you work at it!

6.13 PERCENT COMPOSITION

When a compound is analyzed in the laboratory, a particular sample of the compound is torn apart and the mass of each element is found, as illustrated in Example 6.6. However, it is easier to compare the composition of one compound with the composition of another one having the same elements in terms of **percent composition.**

The reasoning is the same as that used in assigning grades. If you score 20 out of a possible 30 on one test, and 35 out of a possible 50 on a second test, which score is better? Without some common basis for comparison, it is difficult to know.

In similar fashion, if a 36-g sample of iron oxide produces 28 g of iron and 8 g of oxygen, while a 160-g sample of iron oxide produces 112 g of iron and 48 g of oxygen, it is difficult to know whether the two samples are the same compound or different compounds. Once again, we need a common basis for comparison.

Percentages are frequently used to compare both grades and the composition of compounds. You have undoubtedly used percentages before. Review by working through the following example with me.

Example 6.7

If you score 20 on a 30-point test, what is your grade in percent?

- -

67 percent (%). Percent means "per 100 parts" and indicates the score that you would have gotten if the test had been graded on a 100-point basis. To find percentage, divide the portion of interest by the total. In this case, we would have

$$\frac{\text{The portion of interest} \longrightarrow 20}{\text{The total} \longrightarrow 30} \cong 0.67 \quad \text{or} \quad \frac{67}{100}$$

One way to interpret this result is to say that you got right sixty-seven hundredths of what was on the test. If that is true, you would have had a score of 67 on a test with 100 possible points. That's all that percent means.

$$\frac{20}{30} \cong 0.67 = \frac{67}{100} = 67\%$$

Those three statements are three ways of saying exactly the same thing.

If 36 g of an iron oxide contains 28 g of Fe and 8 g of O, what is the percentage of iron in the compound?

78%.

$$\frac{28 \text{ g}}{36 \text{ g}} = 0.78 = \frac{78}{100} = 78\%$$

Is this the same compound that has 112 g of Fe and 48 g of O in 160 g of the compound?

No, it isn't. There are several ways to prove this. What we had in mind was that you calculate the percentage of iron in this second compound: 112 g/160 g = 70% The other compound was 78 percent iron.

Since these are different compounds, we might find the formula for each one. We could find the formula from the actual mass in the sample of iron oxide, but we will find the formula from the percent composition.

The first compound is 78 percent iron. What is the percentage of oxygen in the compound?

---

22 percent. Since the compound contains only iron and oxygen, the easiest way to find the percentage of oxygen is to subtract 78 (the percentage of iron) from 100 (the percentage corresponding to everything).

When finding the formula of a compound from percentage data, we simply assume that we have 100 g of the compound. If the iron oxide is 78 percent iron, how many grams of iron are there in 100 g of the compound?

---

78 g. Remember, percent means "parts out of 100." If we have 100 g that are 78 percent iron, there must be 78 g of iron.

How many grams of oxygen are in 100 g of the compound?

---

22 g. The compound is 22 percent oxygen. Using 78 g for the mass of iron and 22 g for the mass of oxygen, find the formula for this compound.

FeO. The calculations are the same as before:

$$? \text{ mol Fe} = 78 \cancel{g} \text{ Fe} \times \frac{1 \text{ mol}}{55.8 \cancel{g}} \text{ Fe} = 1.4 \text{ mol Fe}$$

$$? \text{ mol O} = 22 \cancel{g} \text{ O} \times \frac{1 \text{ mol}}{16.0 \cancel{g}} \text{ O} = 1.4 \text{ mol O}$$

Dividing both numbers by the smaller number produces the one-to-one ratio shown in the formula.

Now work the following problems to find the empirical formula of a compound from the percent composition given.

6-62. The other iron oxide discussed in Example 6.7 was 70 percent iron. What is the formula of that oxide?

6-63. A compound containing oxygen, silver, and nitrogen is 63.5 percent Ag, 8.25 percent N, and 28.25 percent O. What is its empirical formula?

6-64. An acid is 2.055 percent hydrogen, 32.69 percent sulfur, and 65.255 percent oxygen. What is its empirical formula?

6.14 SUMMARY

We began this chapter by asking how to determine the formula of a substance like salt or water. Finally we answered the question, but not before we had covered a lot of ground.

The notion of *relative mass* was introduced using models. Then we talked about *atomic mass*—the mass of an element relative to that of hydrogen. Today, a particular carbon atom is used as the standard, but the idea that the relative mass of an atom indicates the number of hydrogen atoms that would weigh the same is approximately correct.

Atomic mass is important because the atomic mass of any element, when expressed in grams, contains 6.02×10^{23} atoms. This number of particles is called a *mole,* and the mass of that number of particles is called the *mole mass.* This relationship allows us to "count" atoms by weighing.

The sum of the atomic masses of the atoms in a molecule is the *molecular mass* of the compound. Like atomic mass, this is a relative mass and represents the number of hydrogen atoms that would weigh the same as one molecule of the compound. When the molecular mass is expressed in grams, it is the mole mass—the mass of the compound that contains 6.02×10^{23} molecules or 1 mol.

The mole is at the heart of most calculations in chemistry. The relationships that involve the mole must be understood. Knowing the mass of an element, you must be able to find the number of atoms or the number of moles of atoms. From the mass of a compound and the formula, you must be able to count molecules, talk about the mass of each element in the compound, or count the atoms of each kind in the compound. None of this is difficult once you have the various relationships clearly in mind, but it takes practice and thought to sort everything out.

In the end, we did what we set out to do. We showed how to find the formula of a compound from the mass of each element in the compound. This *empirical formula* tells the proportion of each element in the compound, but in some instances it isn't the true molecular formula.

Some compounds (such as salt and baking soda) do not form molecules, but you can't tell whether the compound forms molecules just by looking at a formula. We have not worried much about this distinction yet. In the next chapter, however, you will begin to learn the difference between compounds that form molecules and those that don't.

Questions and Problems

6-65. If models such as those described in Table 6.1 were made for all elements, those whose names begin with *a* would have the following masses. Use the mass for the hydrogen atom model given in Table 6.1 to calculate the relative mass of each of these elements. Check your answers with values given in the periodic table.

a. actinium	390 g
b. aluminum	46.4 g
c. americium	418 g
d. antimony	209 g
e. argon	68.7 g
f. arsenic	129 g
g. astatine	361 g

6-66. Using the information given in Table 6.1, show that 16 kg of oxygen atom models contain the same number of models as 35.5 kg of chlorine atom models. How many models would be in each pile of models?

6-67. Calculate the number of moles of the following particles that are contained in the given mass.

a. grains of sand in 1 million tonns of sand, if each grain of sand weighs 5.5×10^{-4} g (a tonn is 10^6 g)

b. grains of salt in 1.97×10^{17} kg of salt, if each grain weighs 3.27×10^{-4} g

c. drops of water in 2.87×10^{35} kg of water, if each drop weighs 52 mg

6-68. Calculate the molecular mass of each of the following compounds.

a. K_2SO_4
b. H_3PO_4
c. NH_4Cl
d. Li_3PO_4
e. $Ni(CO)_4$
f. KH_2PO_4
g. $SOFN_3$
h. CH_3MgI
i. $Fe(CO)_4H_2$
j. copper(II) chloride
k. iron(III) sulfide
l. nickel(II) iodide
m. diphosphorus pentoxide
n. aluminum oxide
o. disilicon hexabromide

6-69. How much does 1 mol of each of the following compounds weigh?

a. $AgC_2H_3O_2$
b. $MgCr_2O_4$

c. $(KSO_3)_2NO$
d. $Cu(SCN)_2$
e. FeI_2
f. zinc bromide
g. chromium(III) oxide
h. calcium fluoride
i. potassium sulfide
j. mercury(II) chloride

6-70. From the information given, calculate the molecular mass of each compound.
a. 1 mol of compound A weighs 40 g
b. 6.02×10^{23} molecules of compound B weigh 18 g
c. 8.5 g of compound C contain 3.01×10^{23} molecules
d. 1.62×10^{22} molecules of compound D weigh 6.18 g
e. 0.0021 g of compound E contain 8.02×10^{18} molecules

More Challenging Problems

6-71. a. How many moles of Al are there in 62 g of aluminum?
b. How many atoms of O are there in 32 g of oxygen?
c. How many molecules of Cl_2 are there in 71 kg of chlorine?
d. How many atoms of iron are there in 2.3×10^{-8} g of Fe?
e. How many moles of Mg weigh 28 g?
f. How many atoms of magnesium weigh 2.8×10^{-10} g?
g. If a piece of copper weighs 37.28 g, how many moles of Cu does it contain?
h. How many atoms are there in 2.13 g of H_2 gas?
i. How many molecules of Br_2 gas weigh 1.00 g?
j. If you have 23 g of sodium, how many atoms do you have?

6-72. How many moles of compound are contained in the following samples?
a. 36 g of $CuCl_2$
b. 128 g of FeI_2
c. 2.5 kg of Li_3PO_4
d. 527 g of NH_4Cl
e. 1.2 kg of NaCl
f. 21 g of zinc bromide
g. 1.5 kg of calcium chloride
h. 321 g of silver oxide
i. 5.3 mg of sulfur dioxide
j. 111 g of iron(III) oxide

6-73. How many molecules of the compound are contained in the samples described in Question 6-72?

6-74. Calculate the number of each kind of atom contained in the samples described in Question 6-72. (For example, how many atoms of Cu are there in 36 g of $CuCl_2$? How many atoms of Cl?)

6-75. Calculate the empirical formula for each of the following compounds.
a. What is the formula of a compound that is 79.9 percent copper and 20.1 percent oxygen?
b. When decomposed, a 29.5-g sample of uranium oxide produces 25.0 g of uranium. What is the formula of the oxide?
c. 27.5 g of cobalt combine with 22.5 g of sulfur to produce a pure compound. What is the formula of the compound?
d. 30.0 g of mercury react with sulfur to produce 34.8 g of compound. What is the formula of the compound?
e. Teflon® is 24 percent carbon and 76 percent fluorine. What is the empirical formula of Teflon®?
f. Glycerine is 39.1 percent carbon, 52.2 percent oxygen, and the rest hydrogen. What is the empirical formula for glycerine?

6-76. Find the percent composition of each compound listed in Question 6-68.

Answers to Questions in Chapter 6

6-1. 1.5 (The roll of cello tape is taken as the standard. We need to find how many times as heavy the typewriter ribbon is. The ribbon weighs 42 g and the tape weighs 28 g. Dividing 42 by 28, we find that the typewriter ribbon is 1.5 times as heavy.)

6-2. 4.1 (115 ÷ 28)

6-3. 8.5

6-4. 0.97 (Here the stapler is taken as the standard, so we must divide the mass of the tape by the mass of the stapler: $28\,\cancel{g}/115\,\cancel{g} = 0.243$. This says that the tape is 0.243 times as heavy as the stapler. However, the stapler does not have a relative mass of 1.00, so the answer is *not* 0.243. We gave 4.00 as the relative mass of the stapler; $0.243 \times 4.00 = 0.97$.)

6-5. 1.5

6-6. 8.28

6-7. 1.00

6-8. 5.81×10^5. This answer is obtained as follows:

$$? \text{ H models} = 1.00 \times 10^6\,\cancel{\text{g H models}} \times \frac{1 \text{ H model}}{1.72\,\cancel{\text{g H models}}} = 5.81 \times 10^5 \text{ H models}$$

6-9. 12.0

6-10. 5.81×10^5

$$\left(? \text{ C} = 12.0 \times 10^6\,\cancel{\text{g C}} \times \frac{1 \text{ C model}}{20.64\,\cancel{\text{g C}}} = 5.81 \times 10^5 \text{ C models}\right)$$

6-11. 16.0

6-12. 5.81×10^5

6-13. 35.5

6-14. 5.81×10^5

6-15. 5.81×10^5

6-16. 581

6-17. 7.52×10^{23} molecules

$$\left(1.25\,\cancel{\text{mol}} \times \frac{6.02 \times 10^{23} \text{ molecules}}{1\,\cancel{\text{mol}}}\right)$$

6-18. 20 mol CO

6-19. 1.50×10^{24} O atoms

$$\left(1.25\,\cancel{\text{mol } SO_2} \times \frac{2 \text{ O atom}}{1\,\cancel{SO_2}} \times \frac{6.02 \times 10^{23}}{1\,\cancel{\text{mol}}}\right)$$

6-20. 100 mol O

$$\left(1.2 \times 10^{25}\,\cancel{N_2O_5} \times \frac{5\,\cancel{\text{O atom}}}{1\,\cancel{N_2O_5}} \times \frac{1 \text{ mol O}}{6.02 \times 10^{23}\,\cancel{\text{atoms}}}\right)$$

6-21. 3.21×10^{24} molecules

$$\left(5.33\,\cancel{\text{mol}} \times \frac{6.02 \times 10^{23} \text{ molecules}}{1\,\cancel{\text{mol}}}\right)$$

6-22. 3.21×10^{24} Cu atoms (There is one atom in each molecule.)

6-23. 6.42×10^{24} Cl atoms (There are 2 atoms in each molecule.)

6-24. 0.10 mol $CuCl_2$

6-25. 1.90×10^{24} molecules SnO_2

6-26. 3.79×10^{24} atoms O (There are 2 atoms in each molecule of SnO_2.)

6-27. $$28\,\cancel{\text{g Na}} \times \frac{1\,\cancel{\text{mol Na}}}{23\,\cancel{\text{g Na}}} \times \frac{6.02 \times 10^{23} \text{ atoms Na}}{1\,\cancel{\text{mol Na}}} = 7.3 \times 10^{23} \text{ atoms Na}$$

6-28. $$122\,\cancel{\text{g Fe}} \times \frac{1\,\cancel{\text{mol Fe}}}{55.8\,\cancel{\text{g Fe}}} \times \frac{6.02 \times 10^{23} \text{ atoms Fe}}{1\,\cancel{\text{mol Fe}}} = 1.32 \times 10^{24} \text{ atoms Fe}$$

6-29. 2.54×10^{24} atoms Cl

6-30. 6.7×10^{23} atoms fluorine

6-31. 4.6×10^{22} atoms Zn

6-32. 3.6×10^{22} atoms calcium

6-33. $$42\,\cancel{\text{g Na}}\,\frac{1 \text{ mol Na}}{23\,\cancel{\text{g Na}}} = 1.8 \text{ mol Na}$$

6-34. 0.516 mol O

6-35. $$2.18\,\cancel{\text{mol Cu}} \times \frac{63.54 \text{ g Cu}}{1\,\cancel{\text{mol Cu}}} = 139 \text{ g Cu}$$

6-36. 16 g

6-37. $$7.2\,\cancel{\text{mol Cl}} \times \frac{6.02 \times 10^{23} \text{ atoms Cl}}{1\,\cancel{\text{mol Cl}}} = 4.3 \times 10^{24} \text{ atoms Cl}$$

6-38. $$1.0 \times 10^9\,\cancel{\text{atoms}} \times \frac{1 \text{ mol}}{6.02 \times 10^{23}\,\cancel{\text{atoms}}} = 1.7 \times 10^{-15} \text{ mol}$$

6-39. $$36\,\cancel{\text{g Br}} \times \frac{1\,\cancel{\text{mol Br}}}{79.9\,\cancel{\text{g Br}}} \times \frac{6.02 \times 10^{23} \text{ atoms Br}}{1\,\cancel{\text{mol Br}}} = 2.7 \times 10^{23} \text{ atoms Br}$$

6-40. $$1.2 \times 10^{25}\,\cancel{\text{atoms S}} \times \frac{1\,\cancel{\text{mol S}}}{6.02 \times 10^{23}\,\cancel{\text{atoms S}}} \times \frac{32.1 \text{ g}}{1\,\cancel{\text{mol S}}} = 640 \text{ g}$$

6-41. 12.0 + 16.0 = 28.0 (CO). Note that there are no units for molecular mass. They are relative masses. If you were asked to find the weight of a *mole* of the compound, there would be units. The mole mass is the weight of 1 mol and has units of grams. For example, 98.1 is the molecular mass of H_2SO_4; 98.1 g is the mass of 1 mol of H_2SO_4.

6-42. $(2 \times 12.0) + (6 \times 1.01) = 30.1$

6-43. 98.1

6-44. 100.1

6-45. 158.0

6-46. 153.8 (CCl_4)

6-47. 150.7 (SnO_2)

6-48. 159.7 (Fe_2O_3)

6-49. 1.9 mol
6-50. 4.12 mol
6-51. 0.36 mol
6-52. 245 g
6-53. 40 g
6-54. 1.1×10^{24} molecules
6-55. 8.95×10^{22} molecules
6-56. 4.54×10^{23} molecules
6-57. $3.01 \times 10^{23}\ \cancel{\text{molecules } Fe_2O_3}$

$$\times \frac{1\ \cancel{\text{mol}}\ Fe_2O_3}{6.02 \times 10^{23}\ \cancel{\text{molecules } Fe_2O_3}} \times \frac{159.7\ \text{g}}{1\ \cancel{\text{mol}}}$$
$$= 79.8\ \text{g } Fe_2O_3$$

6-58. 5.6×10^2 g
6-59. MnO_2 (We are told that the compound weighs 15.28 g and that the manganese in it weighs 9.66 g. By subtraction, we can find the mass of the oxygen in the compound:

$$15.28\ \text{g} - 9.66\ \text{g} = 5.62\ \text{g of oxygen}$$

Next we need to find the number of moles of Mn and O in the compound:

$$9.66\ \cancel{\text{g Mn}} \times \frac{1\ \text{mol Mn}}{54.938\ \cancel{\text{g Mn}}} = 0.176\ \text{mol Mn}$$

$$5.62\ \cancel{\text{g O}} \times \frac{1\ \text{mol O}}{15.9994\ \cancel{\text{g O}}} = 0.351\ \text{mol O}$$

In order to get a whole-number ratio for the formula, we now divide the number of moles of each element by the smaller number:

$$0.176 \div 0.176 = 1$$
$$0.351 \div 0.176 = 2$$

This tells us that we have 2 mol of oxygen for every 1 mol of manganese.)

6-60. SbI_3 Mass of Sb = 29.95 g
Mass of I = 123.6 g − 29.95 g = 93.65 g

$$29.95\ \cancel{\text{g Sb}} \times \frac{1\ \text{mol Sb}}{121.75\ \cancel{\text{g Sb}}} = 0.2460\ \text{mol Sb}$$

$$93.65\ \cancel{\text{g I}} \times \frac{1\ \text{mol I}}{126.904\ \cancel{\text{g I}}} = 0.7380\ \text{mol I}$$

$$0.2460 \div 0.2460 = 1$$
$$0.7380 \div 0.2460 = 3$$

6-61. CCl_4

$$92.19\ \cancel{\text{g Cl}} \times \frac{1\ \text{mol Cl}}{35.453\ \cancel{\text{g Cl}}} = 2.600\ \text{mol Cl}$$

$$7.81\ \cancel{\text{g C}} \times \frac{1\ \text{mol C}}{12.0111\ \cancel{\text{g C}}} = 0.650\ \text{mol C}$$

$$2.600 \div 0.650 = 4$$
$$0.650 \div 0.650 = 1$$

6-62. Fe_2O_3
6-63. $AgNO_3$ The number of moles of each element in a 100-g sample of the compound is as follows:

$$63.5\ \cancel{\text{g Ag}} \times \frac{1\ \text{mol Ag}}{107.87\ \cancel{\text{g Ag}}} = 0.589\ \text{mol Ag}$$

$$8.25\ \cancel{\text{g N}} \times \frac{1\ \text{mol N}}{14.0067\ \cancel{\text{g N}}} = 0.589\ \text{mol N}$$

$$28.25\ \cancel{\text{g O}} \times \frac{1\ \text{mol O}}{15.999\ \cancel{\text{g O}}} = 1.766\ \text{mol O}$$

If you now divide all of these numbers by the smallest (0.589), you find that the formula is $AgNO_3$.
6-64. By now you should have the general idea. If you do, you found that the formula for this compound is H_2SO_4.

READING The number of particles in 1 mol is called Avogadro's number. It is determined experimentally and, like all measurements, has uncertainty. Because of its fundamental importance in science, efforts continue to reduce the uncertainty in Avogadro's number. This short account tells of one recent effort and the value obtained.

NEW AVOGADRO'S NUMBER

Avogadro's number has been changed, but chemistry students needn't worry. The new figure, whose uncertainty is one-thirtieth previous direct measurements, is so close to the old that only extremely fine measurements will be affected. In fact, most significant impact will not be in use of Avogadro's number at all, but in use of the techniques developed by scientists at the National Bureau of Standards (NBS)

Source: *Chemistry*, Vol. 48, No. 3 (March 1975), pp. 24–25. Reprinted with the permission of the copyright owner, THE AMERICAN CHEMICAL SOCIETY.

to determine it (*Dimensions NBS*, October 1974, page 219).

Avogadro's number is the mass of one mole (atomic weight in grams) of an element divided by mass of one of its atoms. NBS scientists, under direction of R. D. Deslattes, chose to do the calculation for silicon because the mass of one atom of cube-type crystals such as silicon is readily found by multiplying density (ρ) by the cube of the lattice spacing (a_0) and dividing by the number of atoms in the crystal's unit cell. The scientists' achievement was to improve techniques for measuring ρ and a_0. They also devised more accurate methods for determining an element's atomic mass.

		Uncertainty, ppm
Old	$6.022\,045\,3 \times 10^{23}$ mol^{-1}	5
New	$6.022\,094\,3 \times 10^{23}$ mol^{-1}	1

Sophisticated refinements of Archimedes' principle were used to measure the density of silicon. Volume of a crystal was found by measuring the amount of liquid it displaced when submerged. One difficulty with this method is that density of the liquid changes continuously as it absorbs gas from the air and dissolves slightly the object being measured. To combat this, scientists did a series of measurements on steel balls of known densities and objects of unknown densities. By treating these data mathematically in such a way that density of the liquid was eliminated as a variable, scientists compared density of the silicon to density of the balls, not the liquid (density of the balls was found by determining mass on an air balance and volume using a laser interferometer accurate to 1 part in 10^7). Inert fluorocarbon liquid was used instead of water because its density, being twice that of water, allows weight to be measured with twice the precision. Also, fluorocarbons have less surface tension and so less attraction for the wire used to suspend the objects; thus the reading is more accurate.

Length of the lattice spacing was found using X-rays and optical interferometry. In the past, such measurements were made with X-ray diffraction but this method was rejected because X-ray wavelengths are known less precisely than Avogadro's number. In the new method, X-rays are used as a marker only, and their wavelength need not be known. Two pieces of silicon were arranged so that one could be moved with respect to the other. X-rays, passing through the pieces, registered maximum intensity when lattices in both crystals lined up. Thus, each time the piece of silicon was moved one lattice spacing, X-ray intensity peaked. Distance between peaks was measured with an optical interferometer.

To determine the atomic weight, data from a mass spectrometer were analyzed by techniques accurate to 1 part in 10,000. The spectrometer was used to determine abundances of each isotope of silicon-28, -29, or -30 in the sample prepared for measurement. Problem was to calibrate the spectrometer to provide necessary correction factors. To do this, scientists at NBS prepared a series of synthetic mixtures of silicon isotopes obtained from Oak Ridge National Laboratory, Oak Ridge, Tenn., and put these mixtures through the mass spectrometer. Such mixtures are available to the public as NBS Standard Reference Material 990 at about $50 per wafer.

NBS achievements have made possible other advances in measurement. Perhaps the most intriguing is the eventual possibility of eliminating the last standard of measurement based on a manufactured object. This is the kilogram, defined as the mass of cylinder made from platinum and iridium and kept by the International Bureau for Weights and Measures in Sevres, France. The new standard would be the product of Avogadro's number and one-twelfth the mass of a carbon-12 atom. Refinements of some other measurements have already been completed. For example, Deslattes and A. Henins, also of NBS, have determined the correction factor for measuring X-ray wavelengths by using the lattice spacing in silicon crystals as the diffraction medium in a reversal of the usual X-ray diffraction technique. Others are using the methods developed for silicon to determine atomic weights of several elements and to determine trace contaminants.

The Electrical Nature of Matter

You know that elements are made up of atoms and that compounds are made up of molecules. You will now learn what atoms are made of and how the structure of atoms can account for things like household electricity and static cling.

OBJECTIVES

Atomic Structure

1. Name the three major particles that make up an atom, tell where each is located within the atom, and indicate the electrical charge on each particle.
2. Given the atomic number of an element, identify the element, state the number of protons in an atom of the element, and state the number of electrons in a neutral atom. (See Problem 7-46.)

Ions

3. Describe experimental observations that are explained by assuming that some compounds dissociate to form ions.
4. Given the charge on an ion and its chemical symbol, state the number of protons and electrons in the ion. (See Problem 7-46.)

Polyatomic Ions

5. Distinguish between a group of symbols that represents a molecule and a group of symbols that represents a piece of a molecule. (See Problem 7-43.)
6. Given the formulas, name common compounds containing more than two elements. (See Problem 7-44.)

7. Given the name, write the formula of common compounds containing more than two elements. (See Problem 7-45.)
8. Given the formula, calculate the molecular mass or mole mass of a compound containing polyatomic ions. (See Problem 7-38.)

Words You Should Know

electron	quantized	insulator
positive charge	atomic number	electropositive
negative charge	ion	electronegative
nucleus	static electricity	ionic compound
proton	current electricity	polyatomic ion
neutron	conductor	

It is common knowledge that water is an excellent conductor of electricity. That common knowledge is *wrong*. *Pure* water is a very *poor* conductor of electricity. Gasoline, alcohol, and vegetable oil are also poor conductors. Table salt is a poor conductor of electricity when it is a dry solid, but it is an excellent conductor when it melts or when it is dissolved in water. Sugar, on the other hand, is a poor conductor under any conditions. Why does some matter conduct electricity while other matter doesn't? Why does salt conduct as a liquid but not as a solid? To find out, we must review some basic ideas about electricity.

7.1 ELECTRICAL CHARGE

Everyone knows that there are two kinds of people, male and female. In a similar way, there are two kinds of electricity.

How do we *know* there are two kinds of electricity? By observing differences in properties, of course. We can make electricity by rubbing two kinds of matter together. Rub your leather shoes on a nylon rug, or scoot across plastic seat covers in a wool suit. I'm sure you have had such shocking experiences.

What you may not have realized in your informal experience with electricity is that sometimes you obtain one kind of electrical charge and at other times you obtain a different kind of charge. You can observe the different kinds of charge by rubbing different kinds of plastic.

Two common types of plastic are polyvinyl (commonly called vinyl) and cellulose triacetate (commonly called acetate). When rubbed with a piece of wool (or almost any other fabric, for that matter), both types of plastic take on an electrical charge. After both plastics are rubbed, bits of paper or lint are attracted to them.

When a piece of vinyl and a piece of acetate are brought together after being rubbed, they are attracted, as shown in Figure 7.1. When two pieces of vinyl are rubbed and brought near one another, they are repelled, as shown in Figure 7.2. When two pieces of acetate are rubbed and brought near one another, they also repel each other, as shown in Figure 7.3.

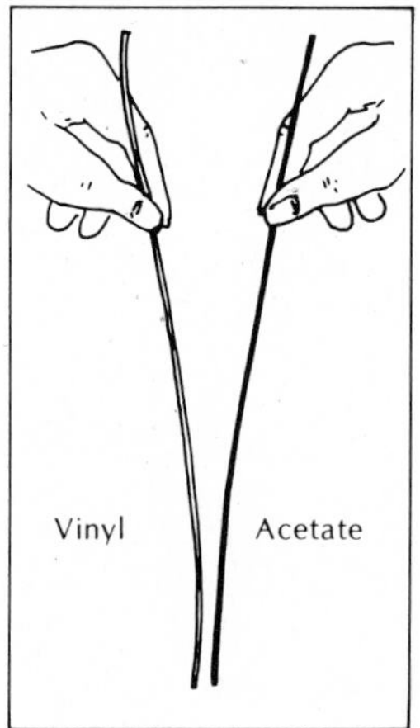

Figure 7.1
After being rubbed with wool, a piece of vinyl plastic and a piece of acetate plastic are attracted to each other.

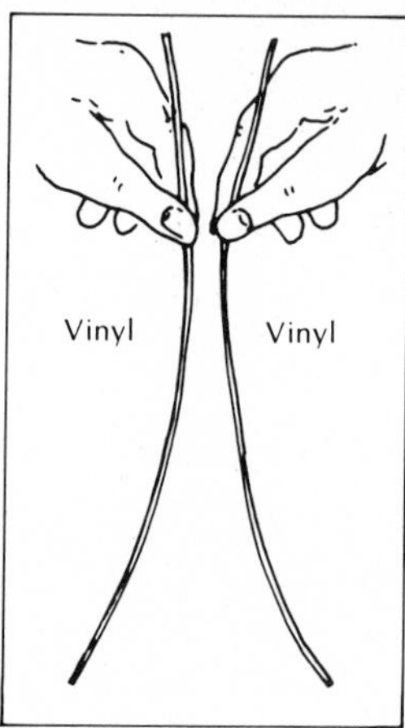

Figure 7.2
Two pieces of vinyl plastic that have been rubbed with wool repel each other.

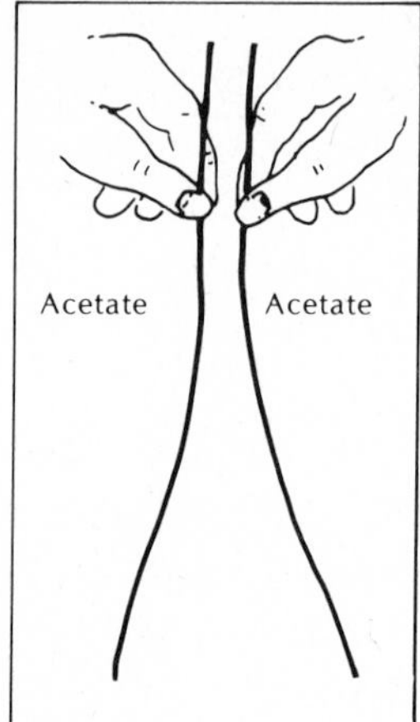

Figure 7.3
Two pieces of acetate plastic that have been rubbed with wool repel each other.

The fact that like pieces of plastic repel, while unlike pieces of plastic attract, leads us to believe that there are two kinds of electricity. Benjamin Franklin called the two kinds of electricity positive (+) and negative (−). Franklin thought that he could explain the behavior of electricity by assuming that some kind of "electrical fluid" exists. When objects are rubbed together, he explained, some of this fluid is transferred from one object to the other. The object with the excess of electrical fluid was called positive, and the object that had the deficiency was called negative.

Franklin's notion was a good one, because it *could* explain most observations of electricity. It is not far from what we believe today. Instead of an electrical fluid, we talk about particles called **electrons.** When objects are rubbed together to produce an electrical charge, we believe that electrons are rubbed off one object onto the other. However, the electron has the kind of charge that Franklin named negative. The object that has an *excess* of electrons has a **negative charge.** The object that has a *deficiency* of electrons has a **positive charge.**

When Franklin named the two kinds of electricity, he had no way of knowing which way his "electrical fluid" moved. If he had known about moving electrons, there would be virtually no difference between his theory of electricity and the ideas that we accept today.

7.2 ELECTRONS AND ATOMS

The story of how we got from Franklin's idea of an electrical fluid to our present notion of electrons and atoms is an interesting one. It is also long and filled with subtleties that are difficult to appreciate before one knows certain chemical facts. In this book, we will concentrate on the conclusions and leave the details for later.

The fundamental conclusion is that all matter contains electrons that can be made to move. Since matter is made of atoms, we believe that atoms contain electrons that can be made to move.

What *are* atoms anyway? We have said that they are the smallest pieces of an element. We have imagined them as tiny spheres that may bounce about in the gas phase and, under the proper circumstances, stick together to form larger particles called molecules. Now we need to look more closely at the parts of an atom.

Atoms of different elements are different. If they were not, we would not have different elements. However, we believe that all atoms are built in much the same way. We need to emphasize that we *believe* atoms are built in a certain way. Nobody knows *for sure* what an atom is like. Nobody has seen an atom.* Still, bits of information derived from a number of observations have led to a picture of atoms that has proven useful even though it may not be entirely correct.

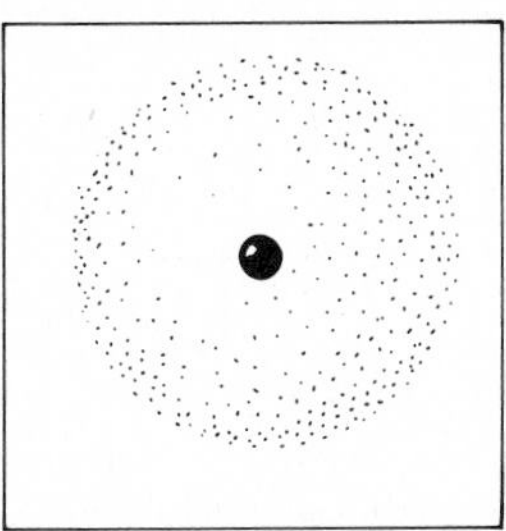

Figure 7.4
If you could see a single atom, it would probably look like a BB in a fluffy cotton cloud.

Instead of an image of hard, solid spheres, the picture that has emerged of the atom is more like a fluffy ball of cotton with a very dense BB (the small pellet, used as ammunition in a BB gun) at its center. This dense center is called the **nucleus** of the atom. The nucleus itself is composed of still smaller particles.

The exact composition of the nucleus is of only passing interest to chemists. However, two facts are very important.

1. The nucleus has a positive electrical charge.
2. Virtually all of the mass of the atom resides in the nucleus.

* You may have seen news reports indicating that atoms have been photographed. This is true and it is a great accomplishment, but these pictures show only dots where atoms are located. They don't provide any information about the structure of individual atoms.

You may wonder why we believe that nearly all the mass of the atom is in the center and that the center has a positive charge. Several experimental facts are explained by these properties of the nucleus. The experiment that contributed most grew out of the study of radioactivity in the early 1900s.

7.3 THE RUTHERFORD EXPERIMENT

Ernest Rutherford, at that time in Canada, noticed that alpha particles from a radioactive source form a hazy pattern on a photographic film if they pass through a thin sheet of metal before striking the photographic plate. Alpha particles are like bullets that are smaller than atoms. They have a positive charge and move very fast. Rutherford expected them to go straight through the thin metal and form sharp dots on the photographic plate.

Two years later, in Manchester, England, Rutherford continued his study of alpha particles with the help of his assistant, Hans Geiger. Later, Geiger and Marsden conducted a detailed study of the scattering of fast-moving alpha particles. They found that, when the alpha particles were passed through a thin sheet of platinum,* about 1 particle in 8,000 was scattered at an angle of 90°, as shown in Figure 7.5. Describing their work, Geiger and Marsden commented, "If the high velocity [about 1.8×10^9 cm/sec or almost one-tenth of the speed of light] and mass of the alpha particle be taken into account, it seems surprising that some of the alpha particles . . . can be turned within a layer of 6×10^{-5} cm of gold* through an angle of 90°, and even more." In describing these results several years later, Rutherford said, "It was

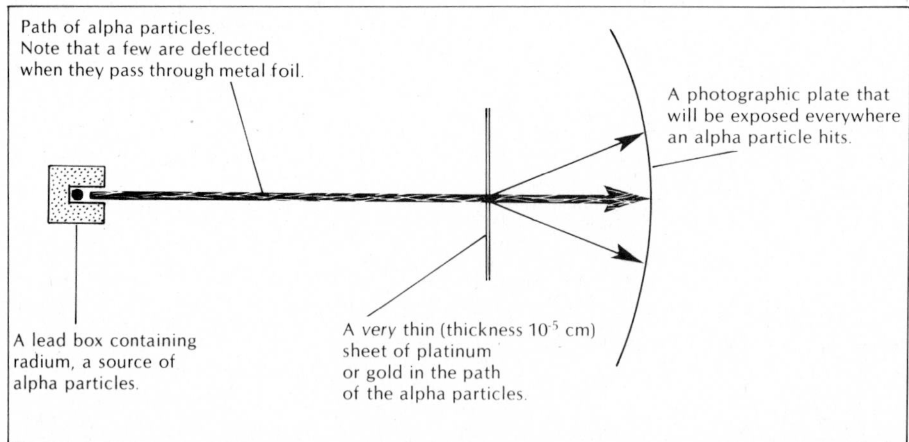

Figure 7.5
The Rutherford experiment. Alpha particles from radium leave the lead box through a small hole and pass through a thin metal sheet. Rutherford expected all alpha particles to pass through the metal undisturbed. Instead, some alpha particles were deflected.

* The metal used is not entirely clear. Reports of the work mention both platinum and gold.

about as credible as if you had fired a 15-inch shell at a piece of tissue paper and it came back and hit you."

After doing a number of careful calculations, Rutherford concluded that the most likely explanation for this result was that all of the positive charge and mass of an atom is concentrated in a very small region at the center of the atom, as shown in Figure 7.6. Further work by Geiger and Marsden in Rutherford's laboratory tended to confirm Rutherford's theory, and the idea of a nuclear atom has been generally accepted since that time.

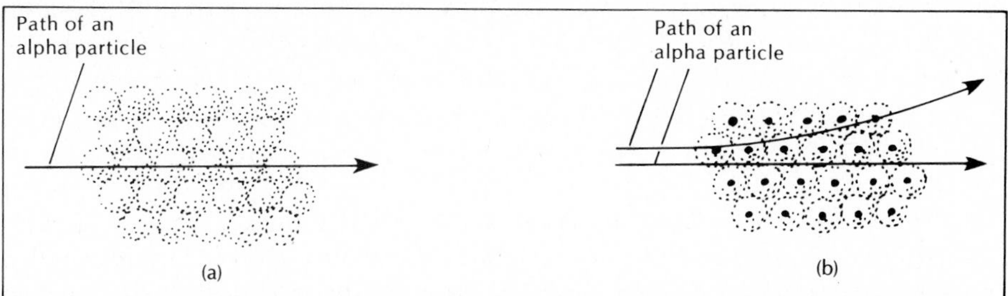

Figure 7.6
On the basis of the deflection of alpha particles passing through a thin metal foil, Rutherford proposed that atoms have a dense center that is positively charged. (a) The atoms in the thin metal foil would look like this if atoms were uniform throughout. (b) The atoms in the metal foil look like this, according to Rutherford's explanation. When the positively charged alpha particles approach the dense, positively charged nucleus of an atom in the foil, they are repelled and deflected slightly off course. Alpha particles whose course is equidistant between nuclei are not deflected.

Other research by a number of scientists showed that the positive charge on the nucleus of an atom is **quantized.** By "quantized" we simply mean that the charge comes in "packages" of some given size and that you can get one package or two or three or some other number, but nothing in between. In an analogous way, you can buy one egg or two or a dozen, but you can't buy 1.5 eggs or 1.38 eggs. The "packages" of positive charge on the nucleus of an atom were found to be the same size as the negative charge on an electron, a particle discovered earlier by J. J. Thomson.

Later in the development of atomic theory, the positive charge on the nucleus was associated with a particle called a **proton.** Each proton has a positive charge equal in magnitude to the negative charge on the electron. Unlike the electron, the proton is very heavy. The mass of the protons in the nucleus of the atom accounts for about half of the mass of most atoms. Most of the rest of the mass of an atom is due to other particles in the nucleus that have no electrical charge. They are called **neutrons.** Neutrons and protons have almost the same mass, and the two together account for virtually all of the mass of an atom. By comparison, the electrons in an atom are so light that their mass can usually be ignored.

The picture of the atom that began to emerge in the years following Rutherford's work is one of a dense nucleus made up of protons and neutrons with very light electrons surrounding the nucleus in some fashion. Since atoms of an element are neutral (that is, they have no overall electrical charge), the number of protons in the nucleus must be equal to the number of electrons that surround the nucleus.

7.4 THE DIFFERENCE BETWEEN ATOMS: ATOMIC NUMBER

We have said that the atoms of one element differ from the atoms of a second element. We can now explain *how* the atoms of different elements differ. All atoms of one element have a certain number of protons; atoms of a second element have a different number of protons. Hydrogen has only one proton in its nucleus. Helium has two protons and lithium has three protons. The number of protons in the nucleus of an atom is indicated by the **atomic number.** This is the whole number shown in the periodic table along with the atomic mass. Since an atom has the same number of electrons as it has protons, *the atomic number also indicates the number of electrons in an atom.*

From the atomic numbers found in the periodic table, fill in the missing information in Table 7.1.

Table 7.1. Protons and Electrons in Selected Atoms

Symbol for the element	*Number of protons in one atom*	*Number of electrons in one atom*
Cu	7-1. ______	7-2. ______
7-3. ______	17	7-4. ______
7-5. ______	7-6. ______	13
P	7-7. ______	7-8. ______
7-9. ______	20	7-10. ______
7-11. ______	7-12. ______	53

As you can see, the chemical symbol for an atom, or the number of protons in the atom, or the number of electrons in the atom is all we need to know to identify the element represented by that atom. At least this is true so long as we are talking about *neutral* or uncharged atoms. As indicated in the following section, atoms do not always remain neutral.

7.5 GIVING ATOMS A CHARGE: IONS

Even though atoms are electrically neutral, it is fairly easy to give them a charge. When you drag your feet on a nylon rug or comb your hair with a hard rubber or plastic comb, you produce an electrical charge on both articles that rub together. The charge on each article is the same in size but opposite in sign. One has a positive charge and the other a negative charge. These opposite charges attract, producing all of the phenomena that we associate with static electricity.

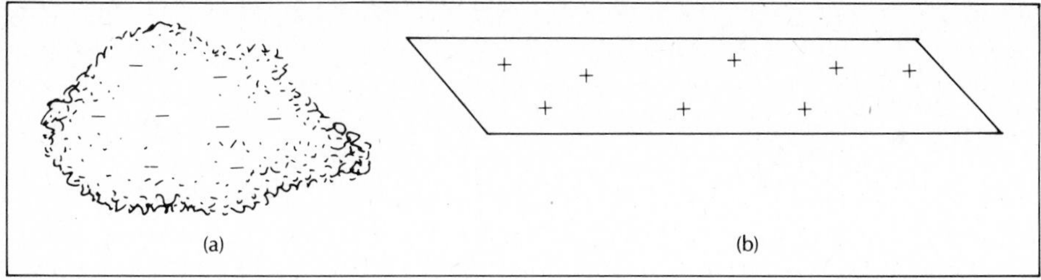

Figure 7.7
(a) When acetate plastic is rubbed with wool, electrons are transferred from the plastic to the wool, giving it an excess of electrons and a negative charge. (b) The acetate strip is left with more protons than electrons, giving it a positive charge.

Since it is the electrons and protons in the atom that have electrical charge, and since the electrons are on the surface of the atom, it seems certain that charges result from electrons being rubbed off the atoms of one object and sticking to the other.

When acetate plastic is rubbed with wool, some of the electrons on atoms near the surface of the plastic are rubbed off and onto the wool (Figure 7.7). As a result, the acetate has more protons than it has electrons, so the net charge on the acetate is positive. Meanwhile, the wool cloth has extra electrons from the plastic, so it has acquired a negative charge.

Since acetate and vinyl strips behave differently, we assume that the opposite happens when we rub vinyl plastic with wool. In this case, some of the electrons are rubbed off the atoms making up the wool and stick to the vinyl. In this way, the vinyl acquires a negative charge and the wool takes on a positive charge (Figure 7.8).

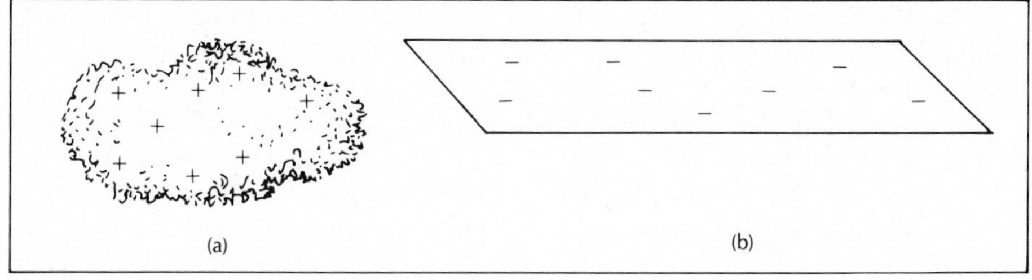

Figure 7.8
(a) When vinyl plastic is rubbed with wool, electrons are transferred from the cloth to the vinyl, leaving the wool with more protons than electrons and a positive charge. (b) The excess electrons on the surface of the vinyl give it a negative charge.

Somewhere in the molecules that make up the wool and the plastic, there are atoms that have lost or gained electrons. If electrons are lost, the atom has more protons than electrons and thus has a positive charge. If electrons are gained, the atom has more electrons than protons and thus has a negative charge. These charged atoms are called ions. Any atom or group of atoms having an electrical charge is called an **ion.**

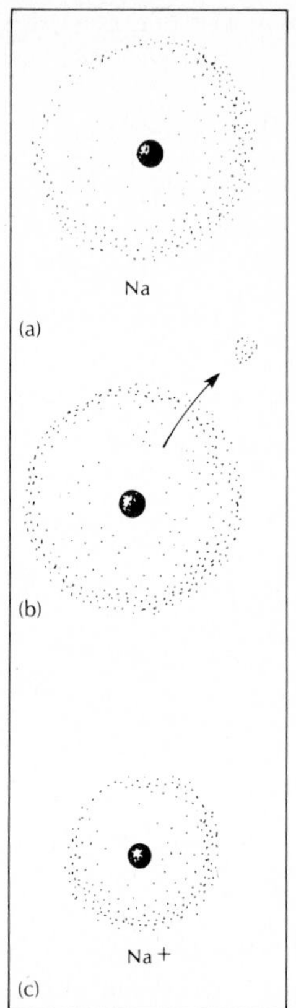

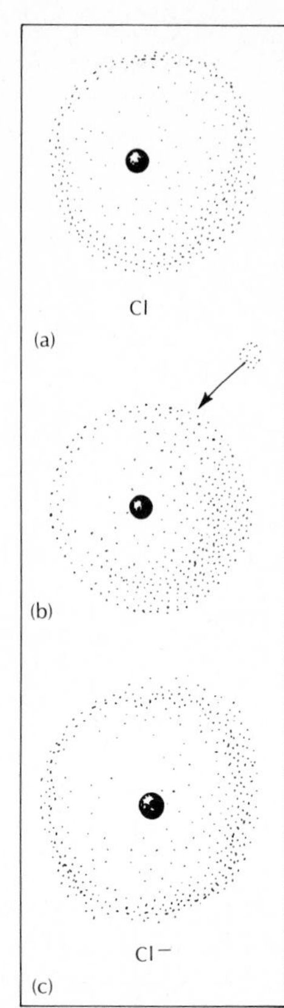

Figure 7.9 (left)
A sodium atom gives up an electron to form a sodium ion. (a) A neutral sodium atom (Na). The eleven positively charged protons are in the center, and the eleven negatively charged electrons are in the fuzzy region. (b) To form a positive ion, one or more electrons must be removed. (c) After an electron is removed, the nucleus is unchanged, but the fuzzy area around the nucleus is smaller. The sodium ion (Na^+) is smaller than the sodium atom.

Figure 7.10 (right)
A chlorine atom accepts an electron to form a chloride ion. (a) A neutral chlorine atom (Cl). It has seventeen protons in the nucleus and seventeen electrons surrounding it. (b) To form a negative ion, one or more electrons must be added to the atom. (c) The nucleus is unchanged in the chloride ion (Cl^-), but the fuzzy area contains one more electron.

What happens when an atom forms a positive ion is illustrated in Figure 7.9, and what happens when an atom forms a negative ion is illustrated in Figure 7.10.

Note the difference between the symbol for an atom and the symbol for an ion. Sodium atoms normally form ions by losing a single electron. This leaves the ion with one more proton than it has electrons. It would have one more positive charge than negative charge. Hence the symbol is Na^+, as shown in Figure 7.9.

In Figure 7.10, note the way the chloride ion (Cl^-) and the chlorine atom (Cl) are represented. The ion has one more electron than it has protons. The negative charge that results is indicated by the minus sign that appears with the symbol.

Sodium and chlorine were used to illustrate the formation of positive and negative ions because you are familiar with a substance that is composed of these ions. It is ordinary salt. We will explain why we think that salt is composed of sodium ions (Na^+) and chloride ions (Cl^-) rather than sodium chloride molecules (NaCl) after describing the difference between static and current electricity. Before we do, make sure you understand what an ion is by completing Table 7.2. The first row is completed for you as an example.

Table 7.2. Protons and Electrons in Selected Ions

Symbol for the ion	*Number of protons in the ion*	*Number of electrons in the ion*
Na^+	11	10
7-13. ______	17	18
Mg^{2+}	12	7-14. ______
7-15. ______	30	28
S^{2-}	7-16. ______	7-17. ______
7-18. ______	26	23
Ca^{2+}	7-19. ______	7-20. ______

7.6 STATIC AND CURRENT ELECTRICITY

What we have been discussing is **static electricity.** Static means "still, not moving." Once we get an electrical charge on a piece of plastic, it just sits there. In very dry air, the plastic remains charged for a long time if it doesn't touch anything but the air.

If you bring a charged piece of acetate very close to a charged piece of vinyl, a spark jumps from one piece of plastic to the other. The electrical charge moves. We say there is an **electric current.** An electric current is the movement of electrically charged particles.

CONDUCTORS AND INSULATORS

Normally an electric current is due to the movement of electrons. In metals, some electrons near the surface of the atoms move very easily from one atom to another. Electrons move easily through a metal wire. Anything through which electrons move easily is called a **conductor.** In nonmetals such as sulfur and phosphorus, the electrons are arranged in such a way that they do not move freely from one atom to another. An electric current does not move easily through these elements, and they are called nonconductors or **insulators.** Just as some elements are electrical conductors and some are not, some compounds are electrical conductors and some are not. Those compounds that conduct an electric current as solids have electrons that are free to move. They conduct like metals. Other compounds conduct an electric current only when they are melted or dissolved in water or some other liquid. Salt is one of those compounds.

7.7 CONDUCTIVITY BY IONS

Figure 7.11 (p. 194) is a diagram of a simple apparatus to test for conductivity. When the wires in the apparatus are placed in dry salt, as shown in Figure 7.12 (p. 194), the bulb does not light. Dry salt is not an electrical conductor. However, when the salt is melted or dissolved in water, it does conduct an electric current, as shown in Figure 7.13 (p. 194).

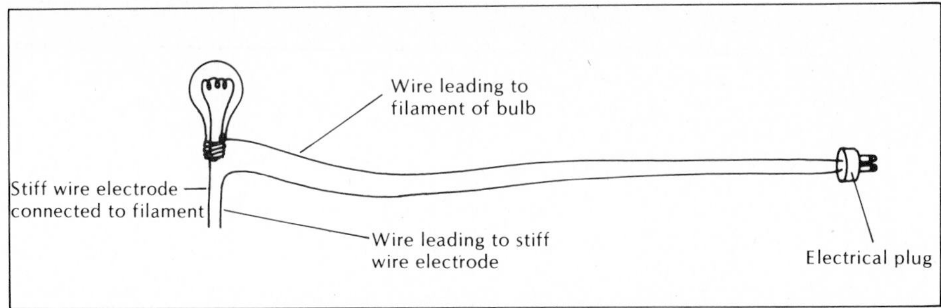

Figure 7.11 *Conductivity apparatus. The bulb will light only when an electrical conductor connects the two stiff wires.*

Figure 7.12 *Dry salt does not conduct electricity.* (Photo by Tom Greenbowe and Keith Herron.)

Figure 7.13 *Dissolved or melted salt does conduct electricity.* (Photo by Tom Greenbowe and Keith Herron.)

It would appear that solid salt does not have electrons that are free to conduct an electric current but that melted salt does. However, other experiments suggest that it is not electrons that carry the current in compounds like salt.

If carbon electrodes are connected to a car battery and placed in melted salt, a chemical reaction occurs. Sodium metal forms at the carbon rod connected to the negative terminal of the battery, and chlorine gas forms at the carbon rod connected to the positive terminal of the battery. If the connections on the battery are reversed, the reactions occurring at the carbon rods reverse. Sodium always appears at the rod connected to the negative terminal of the battery, and chlorine appears at the rod connected to the positive terminal of the battery. It appears that the sodium has a positive charge and is attracted to the negatively charged rod. The chlorine in the salt seems to have a negative charge and is attracted to the positively charged rod. Figure 7.14 illustrates what appears to take place. Many other compounds behave in a similar way when they are melted or dissolved in water.

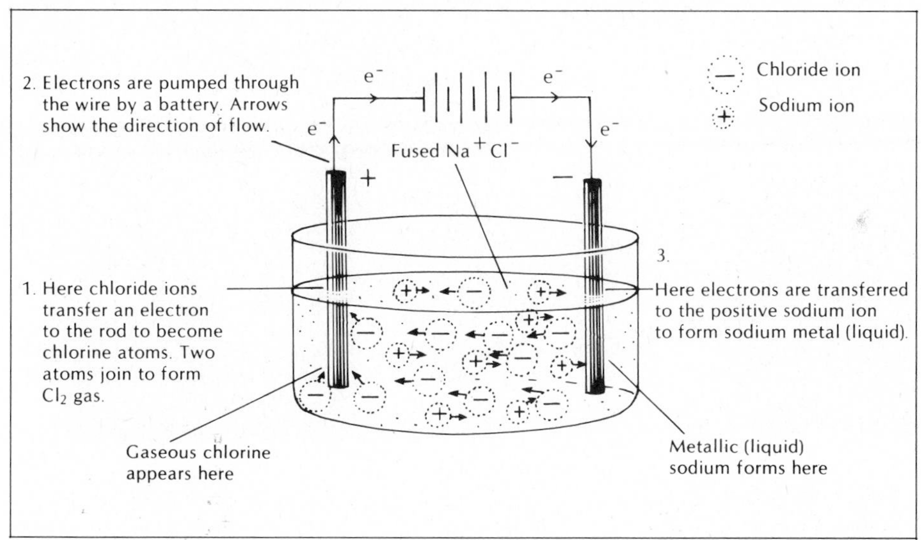

Figure 7.14
Electrolysis of molten salt.

Copper(II) sulfate, copper(II) chloride, copper(II) nitrate, and many other copper compounds dissolve in water to form blue solutions. We are fairly sure that the blue color is due to the presence of copper, because copper is common to all of these compounds.

If any one of these solutions is placed in a beaker, and two carbon rods are placed in the solution and then connected to a car battery, a chemical reaction occurs (Figure 7.15, p. 196). We will discuss only the part of the reaction that occurs at the carbon rod connected to the negative terminal of the battery. There, copper metal plates out on the rod. You can see the copper coating the rod connected to the negative side of the battery in Figure 7.16, p. 196. If the battery connections are switched, the copper disappears from the rod and begins to coat the other rod—the one now connected to the negative side of the battery.

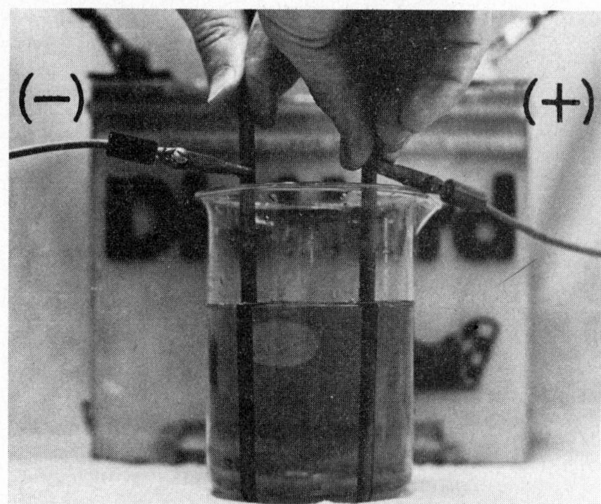

Figure 7.15
Carbon rods in a solution of copper(II) sulfate and connected to a battery. (Photo by Dudley Herron.)

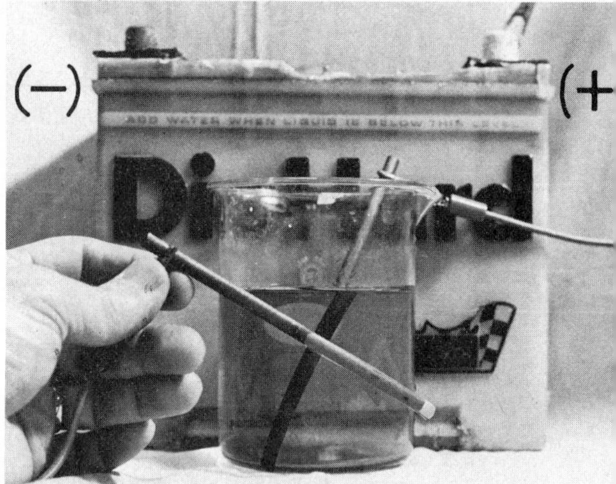

Figure 7.16
Copper metal appears on the rod connected to the negativ electrode. (Photo by Dudley Herron.)

Figure 7.17
The negatively charged carbon rod is in a colorless solution of NH_4NO_3. The positively charged rod is in a blue solution of $CuSO_4$. The solutions are separated by a porous clay wall that liquid can pass through very slowly. (Photo by Dudley Herron.)

Figure 7.18
After the rod in the colorless solution has been connected to the negative terminal of a battery for several minutes, both sides appear to be light blue. (Photo by Dudley Herron.)

We can make another observation with these solutions of copper compounds. If we separate two sides of a beaker with a piece of unglazed pottery, we can pour a blue solution of a copper compound in one side and a colorless solution of ammonium nitrate in the other side (Figure 7.17). The solutions will remain separated for days. However, if we now place a carbon rod in each side and connect them to the battery, the blue color moves to the side that is colorless, *when* the rod in that side is connected to the negative side of the battery (Figure 7.18). This does not happen when the connections on the carbon rods are switched.

These observations all suggest that, when copper compounds are dissolved in water, the copper part of the compound has a positive charge and moves toward a negatively charged electrode.

As you might have guessed, there are also negatively charged pieces of the compounds that move toward the positively charged electrode. This can be observed better when electrodes are placed in a pure substance such as the salt discussed earlier. When potassium iodide is melted and electricity is passed through the melt, potassium metal forms on the negative electrode and iodine forms at the positive electrode.

Similar results are obtained for a large number of compounds, but certainly not all compounds. *Pure* water doesn't seem to be affected when the electrodes from a battery are placed in it. Neither is ordinary alcohol, antifreeze, liquid carbon dioxide, cooking oil, gasoline, and a large number of other compounds. You can easily test liquids for electrical conductivity by using the apparatus shown in Figure 7.11 (p. 194). Some conduct an electric current and some do not.

In those compounds that conduct electricity, the metal part of the compound moves toward the negative electrode. The metal appears to exist in these compounds as a positively charged ion. Hence metals are said to be **electropositive** elements. The nonmetal part of compounds that conduct an electric current moves toward the positive electrode. Nonmetals seem to exist in such compounds as negatively charged ions. Hence nonmetals are said to be **electronegative** elements. When such compounds melt or are dissolved in water, they appear to come apart as separate ions rather than as molecules. What happens when salt dissolves in water is illustrated in Figure 7.19.

We emphasize that only *some* compounds behave like salt. These compounds are called **ionic compounds,** because they appear to be made up of ions. Other compounds do not appear to be made up of ions. They do not appear to separate into ions when dissolved in water. They do not conduct an electric current. They do not undergo electrolysis. One such compound is ordinary sugar.

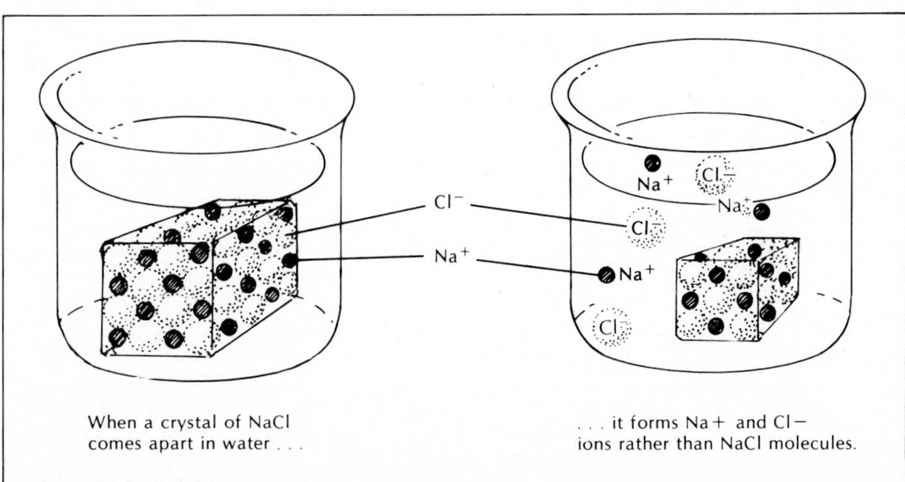

Figure 7.19
When ionic compounds melt or dissolve in water, they come apart into charged ions rather than neutral atoms or molecules.

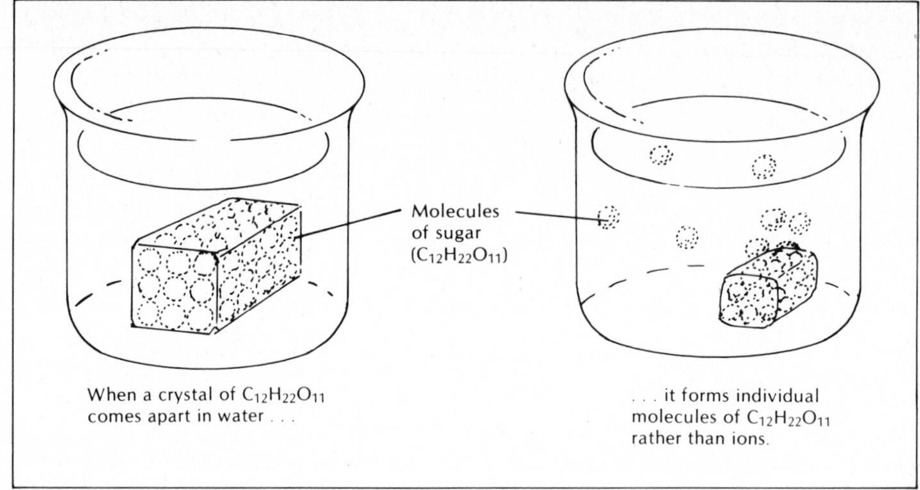

Figure 7.20 *When nonionic compounds melt or dissolve in water, they come apart into neutral molecules.*

Sugar is a relatively large molecule with the formula $C_{12}H_{22}O_{11}$. It is difficult to show the molecule in as much detail as we showed for NaCl in Figure 7.19. However, Figure 7.20 illustrates the important difference between dissolving salt and sugar in water. The salt separates into ions. The sugar remains hooked together as a molecule.

7.8 POLYATOMIC IONS

When NaCl, KI, or $CuBr_2$ comes apart as ions, each atom in the compound forms a separate ion. For NaCl, we get Na^+ and Cl^-; KI forms K^+ and I^-; $CuBr_2$ forms Cu^{2+} and $2Br^-$. At the other extreme, as we have said, molecules such as sugar ($C_{12}H_{22}O_{11}$) don't seem to come apart at all. *And* there is an "in-between."

Vinegar is a water solution of a compound with the formula $C_2H_4O_2$. Vinegar conducts an electric current. It contains ions. One of the ions is H^+; the other is $C_2H_3O_2^-$, an ion composed of several atoms joined together with one extra electron among them. Such ions are often called **polyatomic ions** (*poly* means "many"). Sulfuric acid, the substance added to water to increase its conductance in an electrolysis experiment, has the formula H_2SO_4. It breaks apart in water to give H^+ and HSO_4^-, another polyatomic ion with a single extra electron. Under certain conditions, this HSO_4^- ion breaks apart to form two ions, H^+ and SO_4^{2-}.

It is difficult to predict what molecules will break apart into ions or what ions they will form until you have learned more facts about chemistry. In the meantime, it is necessary to talk about these ionic compounds containing three or more elements. This is not difficult to do if you remember a few of the common polyatomic ions.

Table 7.3. Common Polyatomic Ions and Representative Compounds

Ion name	*Formula*	*Combining number*	*Compound name*	*Formula*
acetate*	$C_2H_3O_2^-$	1	hydrogen acetate (acetic acid, vinegar)	$HC_2H_3O_2$
ammonium*	NH_4^+	1	ammonium nitrate (fertilizer)	NH_4NO_3
carbonate*	CO_3^{2-}	2	calcium carbonate (limestone)	$CaCO_3$
hydrogen carbonate* (bicarbonate)	HCO_3^-	1	sodium bicarbonate (baking soda)	$NaHCO_3$
hypochlorite	ClO^-	1	sodium hypochlorite (bleach; Clorox®)	$NaClO$
chlorate	ClO_3^-	1	potassium chlorate	$KClO_3$
chromate	CrO_4^{2-}	2	potassium chromate	K_2CrO_4
dichromate	$Cr_2O_7^{2-}$	2	iron(III) dichromate	$Fe_2(Cr_2O_7)_3$
hydroxide*	OH^-	1	sodium hydroxide (lye; Drāno®)	$NaOH$
nitrate*	NO_3^-	1	hydrogen nitrate (nitric acid)	HNO_3
nitrite	NO_2^-	1	calcium nitrite	$Ca(NO_2)_2$
oxalate	$C_2O_4^{2-}$	2	hydrogen oxalate (oxalic acid)	$H_2C_2O_4$
permanganate	MnO_4^-	1	potassium permanganate	$KMnO_4$
phosphate*	PO_4^{3-}	3	sodium phosphate	Na_3PO_4
sulfate*	SO_4^{2-}	2	calcium sulfate (plaster of Paris)	$CaSO_4$
sulfite	SO_3^{2-}	2	sodium sulfite	Na_2SO_3

*Common ions that should be memorized.

Table 7.3 lists the names, formulas, and combining numbers of these common ions. (Those marked with an asterisk are frequently used in this book and should be memorized. Your instructor may want you to memorize others.) Table 7.3 also names a common compound containing each ion and shows the formula for that compound.

There are several things about Table 7.3 that you should notice. First, note that the charge on each polyatomic ion is the same as the combining number for that ion. This is true of all ionic compounds. You will recall that sodium always has a combining number of 1. You were just told that sodium forms Na^+ ions. The charge on the sodium ion is $+1$; the combining number for sodium is 1. The same is true of chlorine. When combined with a metal in binary compounds, chlorine has a combining number of 1; the Cl^- ion has a charge of -1.

The second thing to note in Table 7.3 is that all but one of the ions have a negative charge. When these negatively charged ions combine with metals, they take the place of nonmetals and are written last in the formula. Like the nonmetals that they replace, they are electronegative. In similar fashion, the one positively charged ion in the list, (NH_4^+) takes the place of a metal in compounds, and it is written first in the formula. Like metals, it is electropositive.

When formulas were first discussed, you were told that the more metallic element appears first in the formula. A more general rule would be to *write the more electropositive element or ion first.* In most cases, the more electropositive element is a metal, but hydrogen and the ammonium ion are not metals even though they are electropositive.

A third point about the ions shown in Table 7.3 is that they are *not* molecules. These ions never occur alone. They are *pieces* of a compound and are always accompanied by ions of opposite charge to make a neutral compound. For example, you will encounter both SO_3 and SO_3^{2-}. These look very much alike but they are not. The first is the formula for the compound called sulfur trioxide; the second is the formula for the sulfite ion. SO_3 represents a compound that exists alone; SO_3^{2-} represents a part of a compound that always has some positive ion with it.

7.9 NAMES AND FORMULAS OF COMPOUNDS CONTAINING POLYATOMIC IONS

There is a systematic way to name polyatomic ions and the compounds containing these ions. However, most compounds that you will encounter are known by common names. Chemists refer to these common, unsystematic names of compounds as *trivial* names. Since these are the names that you will see on reagent bottles and hear in class, these are the names given here. You are not likely to encounter the systematic names unless you study a great deal of chemistry.

The trivial names for compounds containing polyatomic ions are easy to learn once you memorize the names and charges of the various ions. They are named just like binary compounds, but the ending for most polyatomic ions is not *ide.* (The most common exception is hydrox*ide.*)

To name a compound containing one of the electronegative ions, name the metal and then name the polyatomic ion. For example, Na_2SO_4 is sodium sulfate. (The systematic name for this compound is sodium tetraoxosulfate.) In the name of a compound containing the electropositive ion NH_4^+, ammonium is mentioned first and followed by the name of the electronegative part of the compound. NH_4Br is ammonium bromide. Note that the name ends in *ide,* just like a binary compound.

FORMULAS

There is nothing new about writing the formula for these compounds except remembering that the *combining number of the complex ion refers to the entire ion.*

To write the formula for calcium hydroxide, begin just as though you were writing the formula for a binary compound. The symbol for the more electropositive (or more metallic) ion is written first, followed by the symbol for the more electronegative ion:

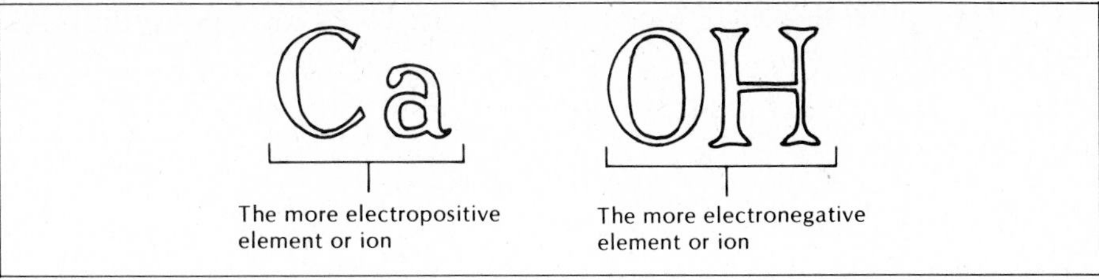

As a reminder, you may temporarily write the combining number of each element or ion above it, like this:

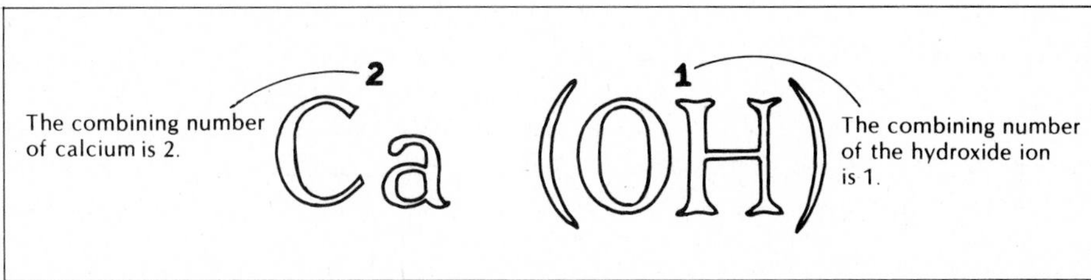

Keep in mind that the 1 over the (OH) is the combining number of the *hydroxide ion.* It does not refer to the hydrogen atom or the oxygen atom. Parentheses have been drawn around the ion to emphasize that it acts as a single unit.

Finally, subscripts are placed after each atom or ion so that the subscript for the electropositive ion times the combining number of the electropositive ion is the same as the subscript for the electronegative ion times the combining number of the electronegative ion:

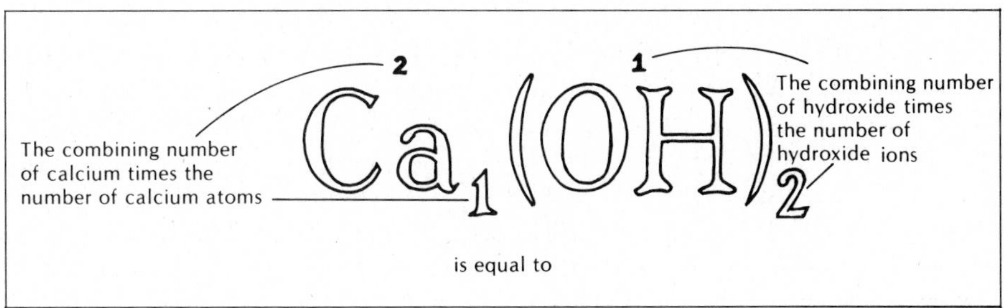

Note that the 2 after the (OH) means "two hydroxide ions." It refers to the entire ion, not just to the hydrogen. To make this clear, the parentheses are retained in the formula:

Parentheses are not used if there is only one of the ions in the formula. The formula for ammonium sulfate is:

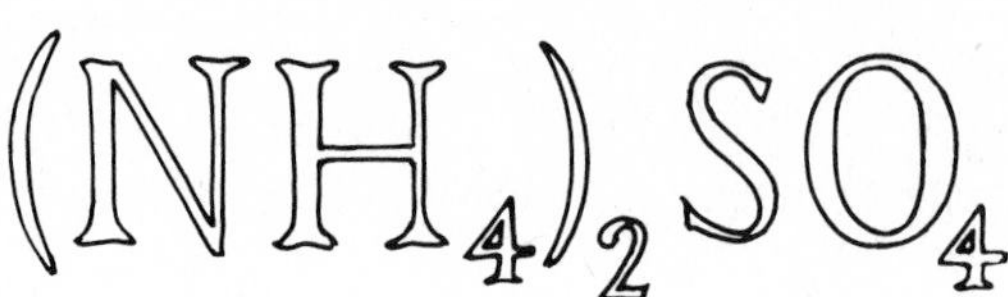

There are two ammonium ions in this formula, which is indicated by the 2 after the parentheses surrounding NH_4^+. There are no parentheses drawn around SO_4^{2-}, because there is only one of these ions in the formula and the parentheses are not needed.

One final point. In writing the formula of a compound containing ions, we do not show the charge on the ion. The molecule as a whole is neutral, and no charges are indicated. When ions are discussed separately, however, the charge is always included. This is to distinguish between molecules such as SO_3 and pieces of molecules such as SO_3^{2-}. The charge is also necessary to indicate the difference between sodium atoms (Na) and sodium ions (Na^+).

Distinguishing between atoms and ions is not trivial. Atoms and ions have different properties. If you ate a chunk of sodium atoms or chlorine atoms, you would surely die. However, you eat sodium ions and chloride ions every time you salt your food, with no harmful effects. In fact, these ions are a necessary part of your diet.

Before you practice naming compounds and writing formulas, draw pictures to represent the following compounds.

7-21. Draw a picture to represent a molecule of $Ca(OH)_2$.
7-22. Draw a picture to represent a molecule of ammonium sulfate.

When you drew the pictures, I hope that you showed the atoms in each polyatomic ion stuck together. These atoms generally act as a single unit and would be bonded together.

Now practice what you have learned by writing the formulas for the compounds in the following list and naming the compounds for which formulas are given. Try to do this without referring to Table 7.3, but use the table when you forget the name or combining number of an ion.

7-23. $Al(NO_3)_3$
7-24. H_2CO_3
7-25. $Sn(NO_3)_2$
7-26. $K_2Cr_2O_7$
7-27. Cu_2CO_3
7-28. NH_4Cl
7-29. NiC_2O_4
7-30. calcium acetate
7-31. sodium phosphate
7-32. copper(II) sulfate
7-33. iron(II) phosphate
7-34. aluminum hydroxide
7-35. calcium hydrogen carbonate
7-36. ammonium phosphate
7-37. lithium dichromate

MOLECULAR MASS

Some students make errors when they use the formula of a compound containing a polyatomic ion to find the molecular mass of the compound. The error usually occurs because the student forgets that the subscript outside of a parentheses refers to *everything* inside the parentheses. Try not to make that mistake as you answer the following questions. Do the calculations to three digits.

7-38. What is the molecular mass of $Ca(NO_3)_2$?
7-39. What is the mass of 1 mol of ammonium sulfate?
7-40. How many moles of copper(II) sulfate are there in 126 g of the compound?
7-41. What is the molecular mass of $Al(C_2H_3O_2)_3$?
7-42. How many moles of aluminum acetate are there in 1 kg of the compound?

7.10 SUMMARY

Since ancient times, human beings have used mental pictures to describe things that they could not see directly. The Hebrews spoke of "the *King* of Kings" and the psalmist wrote, "The Lord is my *shepherd.*" Christians pray to "Our *Father* who art in Heaven." King, shepherd, and father are images used to describe something of the nature of God. Other religions have used similar images to describe what was believed to exist but impossible to observe directly.

In a similar way, scientists have developed mental pictures to describe the miscroscopic nature of matter. The images may not be entirely correct, but they help us understand something about matter.

In this book, we began by describing atoms as solid spheres that join together in some way to form molecules. Soon we came to a point where this picture was no longer adequate. To explain the electrical nature of matter, atoms must have an electrical charge.

Evidence from many sources has led us to picture atoms with a dense, positively charged *nucleus* surrounded by *electrons* arranged in a hazy cloud of negative charge. The positive charge on the nucleus is attributed to particles called *protons.* The number of protons in an atom is called the *atomic number.* Each element has a different number of protons and a different atomic number.

In neutral atoms, the number of protons is equal to the number of electrons, so the overall charge on the atom is zero. However, it is possible to remove electrons from an atom and deposit them on other atoms. This process results in positively or negatively charged atoms known as ions.

Compounds that come apart into ions when they melt are known as *ionic compounds.* We can distinguish ionic compounds from nonionic compounds by checking the electrical conductivity of the compound when it melts or is dissolved in water.

Many compounds contain groups of atoms joined together in a single, polyatomic ion. The simplest way to name such compounds is to memorize the names of the polyatomic ions along with the charge on each ion. The compound is then named like a binary compound.

Now that the idea of electrical charge has been introduced, *electropositive* has been given as a more general term for "more metallic" and *electronegative* has been substitued for the less descriptive "more nonmetallic." In writing formulas and naming compounds, we write the more electropositive part first.

We discussed two kinds of electricity: static and current electricity. Static electricity refers to electrical phenomena that result from fixed charges on objects. However, when fixed charges become large or are brought near oppositely charged objects, they don't stay fixed, as evidenced by the lightning that accompanies every summer thunderstorm. Moving electrical charges are referred to as *electric current.*

The electric current in your house is due to electrons moving through metal wires, but the electric current in most solutions is due to the movement of oppositely charged ions. The behavior of ions in solution, chemical reactions that produce electrons in a battery, and chemical reactions that occur when a battery is connected to electrodes in a solution constitute an interesting and important branch of chemistry called electrochemistry. Some electrochemistry will be discussed in a later chapter of this book and in other courses that you are likely to study.

Questions and Problems

7-43. For each of the following formulas, decide whether a molecule or a piece of a molecule is represented. If a molecule is represented, name the compound. If a piece of a molecule is represented, name the polyatomic ion.

a. SO_3	f. PO_4^{3-}	k. NO_2
b. NO_2^-	g. CO	l. SO_4^{2-}
c. PO_5	h. HCO_3^-	m. CS_2
d. NH_4^+	i. HNO_3	n. $Fe_2(SO_4)_3$
e. CO_3^{2-}	j. OH^-	o. $C_2H_3O_2^-$

7-44. For each of the following formulas, give the name of the compound.

a. $Cu_3(PO_4)_2$ d. $FeSO_3$ g. $Ca(NO_3)_2$
b. H_2SO_3 e. $KHCO_3$ h. FeS
c. SnC_2O_4 f. $MgCO_3$

7.45. For each of the following compounds, give the formula.

a. lead(IV) oxide e. magnesium sulfite
b. lead(II) chromate f. manganese(II) sulfide
c. sodium hydroxide g. manganese(II) sulfate
d. ammonium carbonate h. calcium acetate

7-46. Complete the following table.

Chemical symbol	*Atomic number*	*Number of protons*	*Number of electrons*
Fe			
	12		12
Al^{3+}			
		33	33
Au^{+}			
	53		54
Pb			
		36	36
C			
	9		10

More Challenging Problems

7-47. What is the percent composition of the following compounds?

a. magnesium acetate c. tin(IV) chromate
b. ammonium sulfite d. iron(III) hydroxide

7-48. A compound known to be either hydrogen oxalate or hydrogen acetate contains 26.7 percent carbon. Which compound could it be?

7-49. Liquid hydrogen chloride does not conduct an electric current. Liquid water does not conduct an electric current. However, when hydrogen chloride is dissolved in water, the solution is an excellent conductor of electricity. What could explain this experimental observation?

7-50. Glass is an excellent electrical insulator. However, when glass is heated until it gets soft, it is an electrical conductor. Give a possible explanation for this conductivity in glass.

Answers to Questions in Chapter 7

7-1. 29
7-2. 29
7-3. Cl
7-4. 17
7-5. Al
7-6. 13
7-7. 15
7-8. 15
7-9. Ca
7-10. 20
7-11. I
7-12. 53
7-13. Cl^-
7-14. 10
7-15. Zn^{2+}
7-16. 16
7-17. 18
7-18. Fe^{3+}

7-19. 20

7-20. 18

7-21.

H O Ca O H

Other arrangements are possible, but you should show two OH^- groups and one Ca.

7-22.

H N H O S O H N H (with H above and below each N; O above and below S)

Other arrangements are possible, but you should show two NH_4^+ groups and one SO_4^{2-} group.

7-23. aluminum nitrate

7-24. hydrogen carbonate (carbonic acid)

7-25. tin(II) nitrate

7-26. potassium dichromate

7-27. copper(I) carbonate

7-28. ammonium chloride

7-29. nickel(II) oxalate

7-30. $Ca(C_2H_3O_2)_2$

7-31. Na_3PO_4

7-32. $CuSO_4$

7-33. $Fe_3(PO_4)_2$

7-34. $Al(OH)_3$

7-35. $Ca(HCO_3)_2$

7-36. $(NH_4)_3PO_4$

7-37. $Li_2Cr_2O_7$

7-38. 164 (Note that there are no units. Molecular masses are relative masses. The mole mass is 164 g.)

7-39. 132 g (Note that there *are* units. The mole mass has units of grams.)

7-40. 0.789 mol

7-41. 204

7-42. 4.90 mol

READING Chemistry is put to work in many ways. One of the most common is the storage of electricity in batteries. The following article, written in June 1977, describes some of the technology involved in making better batteries for your car.

THE INSIDE STORY ON THOSE NO-FILL, NO-FUSS BATTERIES

Robert Gorman

By now, you've heard of the automobile batteries that never need maintenance—not even an occasional check of the water level. Some don't even have filler caps, so you can't check the electrolyte if you want to. The new no-maintenance batteries—with or without filler caps—are called different names, and may be made differently inside. But they all share one thing: a claim by the seller that once you've installed one, you never have to do anything to maintain it.

Are such claims justified? And how do these batteries differ from the old ones that need periodic attention? There are no simple answers to these questions, because some very different types of batteries are being sold under the no-maintenance label. One type is completely new—at least to the automotive field. Another is just an improvement in the same old technology that's riding under the hood of your car now.

Source: *Popular Science,* Vol. 210, No. 6 (June 1977), pp. 141–185. Reprinted from *Popular Science* with permission.

Needless to say, each manufacturer claims that his approach is best. Which point of view will eventually prevail is, at this time, anybody's guess. But to understand what the controversy is about and what you can and can't expect from the new batteries, take a look at what goes on inside the molded case.

As explained in the box (pp. 207–208), a battery—either lead-antimony or lead-calcium—generates electricity by an electrochemical reaction between its plates and the electrolyte. Batteries have traditionally needed maintenance because some of the water in the electrolyte is broken down by electrolysis during recharging and converted into hydrogen and oxygen, which escape through the battery's vent. If the level falls enough so that part of the plate is exposed to air, the plate can be damaged. If the level falls even lower, the battery can be destroyed.

Making the New Batteries

This involves a single simple principle: Some way must be found to stop the loss of water caused by

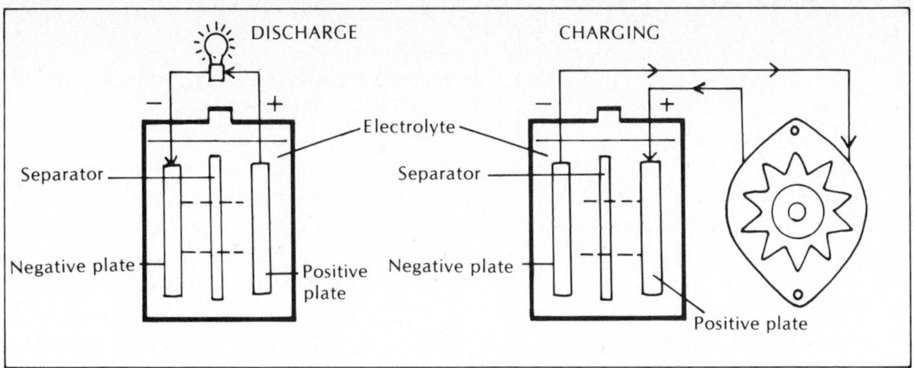

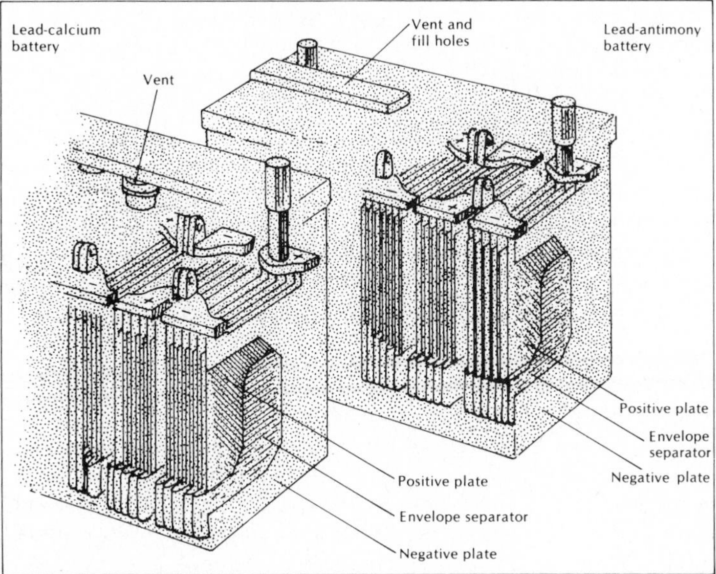

Lead-Calcium and Conventional Lead-Antimony—How Batteries Work

Both types of batteries use lead plates and electrolyte (sulfuric acid and water) to make electricity. Battery plates start life as lead grids. To give the grids rigidity, a hardener is added. Traditionally that hardener has been antimony. But the new generation of batteries uses calcium to harden the grids. After the grids are formed, a lead paste is applied; and the grids are called plates.

To make positive plates, the paste added is in the form of lead peroxide. For negative plates, the paste is in the form of spongy lead. When a cell is discharged by the closing of an external circuit, as in switching on the lights, the sulfuric acid in the electrolyte solution acts on both the positive and negative plates, forming a new chemical compound, lead sulfate. As the sulfate forms, the chemical reaction releases electrons, which flow in the external circuit from the negative to the positive plates. As the discharge continues, the sulfuric acid concentration of electrolyte becomes weaker. The amount of sulfuric acid consumed is in direct proportion to the quantity of electricity generated by the cell.

When the acid in the electrolyte is partially used up, the battery can no longer deliver electricity at a useful voltage, and the battery is said to be discharged. To recharge it, current is passed through the battery in a direction opposite to that of discharge—from positive to negative. For the current to pass in reverse through the battery, it must be at a higher voltage than the normal battery voltage. (That's why the charging system of an automobile with a 12-volt battery is set to deliver in the neighborhood of 13.8–15 volts to the battery.) As the current passes through the battery in the direction opposite to that of the discharge, the lead sulfate on the plates is decomposed. Expelled from the plates, it returns to the electrolyte, reforming sulfuric acid, and gradually restoring the electrolyte to original strength. The plates are restored to their original condition, ready to deliver electricity again. Hydrogen and oxygen gases are given off at the negative and positive plates during recharging—the result of the decomposition of water by an excess of charging current not used by the plates. This highly explosive combination of gases is vented to the outside and through the top of the battery. Construction of lead-calcium (LC) and lead-antimony (LA) batteries is similar. Each battery has six cells, with negative and positive plates in each cell. The number of plates and amount of exposed surface area determine how much energy can be stored. Internal differences between the two batteries are shown by the cutaway of two Gould batteries. . . . Structural details of other LC and LA batteries may vary somewhat, but all are similar. The two-way charge-discharge process can be repeated some large—but finite—number of times before the active materials in the plates wear out and lose their ability to deliver or hold an electrical charge.

Both batteries use separators between plates to keep plates from touching and short-circuiting. Separators in an LC battery, however, are envelopes that cover positive plates. With an LA battery, separators are just flat sheets. Both separators are microporous, allowing electrolyte to flow through. Flaking lead in an LA battery accumulates in the bottom of the battery case. With LC, it accumulates in envelopes. Plates in LC batteries rest on the bottom of the case, and the electrolyte level can be higher above the plates than with an LA battery, so it can lose more water before battery operation and life are affected. This factor, plus electrochemical differences that make an LC battery more resistant to overcharge (explained in text), slows water loss sufficiently so that the battery can be made without filler caps.

electrolysis, or slow it sufficiently so the battery can go through its entire normal life span without needing a refill. The new batteries do this in two ways. First, they use the envelopes (described in the box) to keep the plates electrically separated. This means that the flaking of the electrodes that normally occurs is contained within the envelopes. Traditionally, manufacturers have put the plates well above the bottom of the case so the flakes would fall to the bottom, well away from the plates and thus unable to short them out.

With the plates lower in the battery, the water level can be considerably above the top of the plates. Thus far more water can be lost without exposing the plates.

Second, the new batteries are designed to reduce the amount of water lost by electrolysis to a small fraction of what it has traditionally been. This improvement involves the metal alloyed with the lead plates.

Battery engineers have always wished that they didn't have to use *any* other metal to give the plates the necessary mechanical stiffness, because any alloying of the lead grid can have undesirable effects. Chemical, physical, and electrical alterations can interfere with adhesion between grids and reactive coatings; they can lead to faster oxidation (which would shorten effective life); and they can increase electrical resistance through the conducting pathways. Perhaps most important, higher resistance increases heating and gassing of the electrolyte and causes more power to be dissipated inside a battery instead of getting outside where it's wanted.

Why, then, are the lead grids alloyed with other materials? Because otherwise, says Robert W. Stoll, Gould Battery's director of product development, "plates would have some of the characteristics of a wet noodle."

So for workability and survivability, lead grids must be hardened. Virtually from the beginning of car-battery manufacture, antimony—a brittle white metal—has been the hardener of choice for reasons of economy and production convenience.

But also from the beginning it has been known that antimony is an important cause of a battery's perishability. As soon as a battery is activated, antimony begins to leach out of the grids and deposit on negative plates. This changes their chemical and electrical characteristics and lowers the voltage, which increases the charge current. This causes disassociation of the electrolyte solution into hydrogen and oxygen (gassing). In other words, antimony

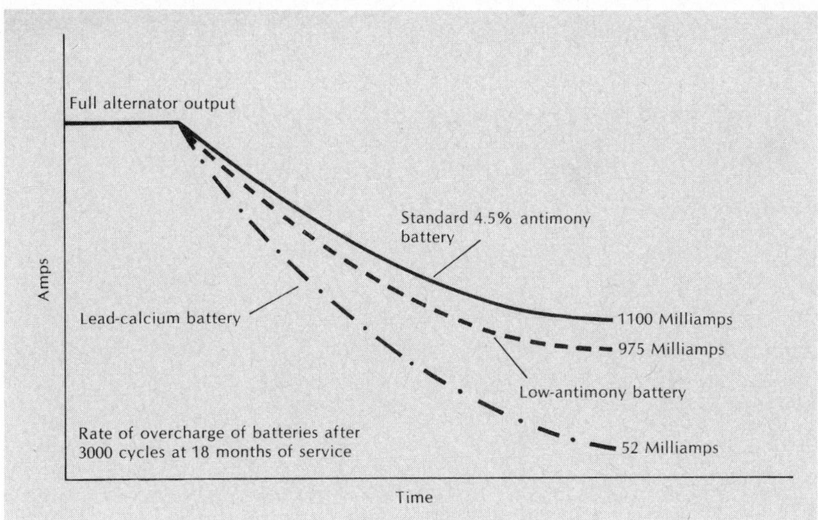

promotes gassing and speeds water loss. As a side effect, its gaseous discharge causes terminal corrosion.

Antimony also decreases a battery's resistance to overcharge. It does this by reducing the countervoltage inside a battery that serves to limit the amount of current a fully charged battery will accept. As you can see in the accompanying graphs, this is really the heart of the controversy.

At a given time in the recharge cycle, the countervoltage in the lead-calcium battery has reduced the charging current to 52 milliamperes—52/1000ths of an amp. The conventional battery made with 4.5-percent antimony, on the other hand, has a charging current of 1100 milliamperes—over an amp—which means it would lose water from the electrolyte many times faster than the lead-calcium battery.

You'll notice that the above comparisons were made between a lead-calcium battery and one made with 4.5-percent antimony. That was standard in the 1960's. Actually, the percentage of antimony had been coming down—from about 11 percent in the 1930's to seven percent right after World War II, and then later to 4.5 percent.

But That Wasn't the End

Meeting recent environmental and energy concerns has put a lot of new strains on battery performance. Pollution-control hardware, for one thing, has been steadily raising underhood temperatures. It's harder to keep batteries alive in this environment, and the job has often added to car weight in the form of additional heat-baffling and cooling equipment.

For some years now, battery designers have been aware that they had to do something to improve their product's resistance to heat, overcharge, water loss, and neglect. And, if possible, to move a battery to a more hospitable location that would probably be less accessible than its present up-front spot.

So the engineers developed both better alloys and improved mechanical design of grids, eventually developing thinner grids able to pack more power into a case of given size and weight. At the same time, they brought the level of antimony down to about 2.5 percent, which meant that batteries lost water at a much slower rate. These batteries were a big improvement, able to give four or five years of relatively trouble-free life.

Designers Weren't Satisfied

They wanted to get rid of *all* the antimony. There was nothing new about this idea. Many companies have been building batteries with antimony-free grids for the last 30 or 40 years. The grids, hardened with calcium, are widely used in batteries made for telephone exchanges and other stationary or standby applications. They're designed for—and reliably deliver—more than 20 years of service.

Adapting lead-calcium technology to high-volume car-battery production, however, has not been easy. Car-battery grids are much thinner (commonly about 0.07″ compared to 0.25″ in a stationary battery), and

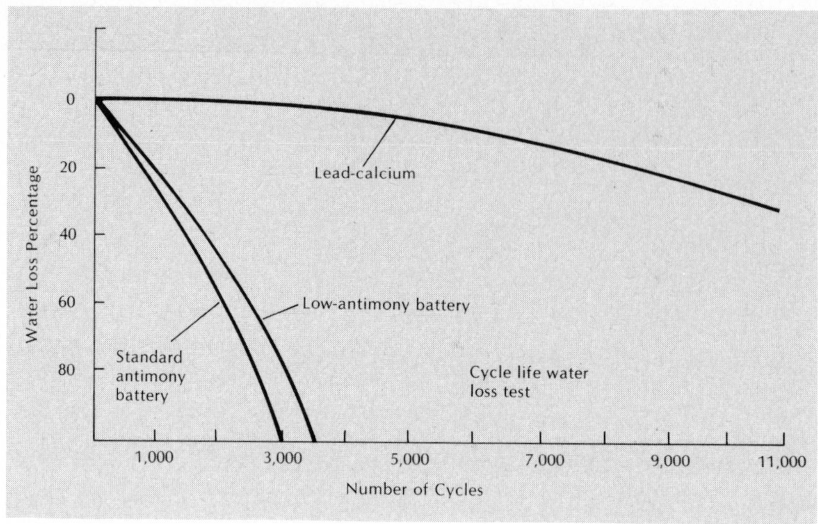

they have to be made much more resistant to vibration. Back in the middle 1960's, when Delco-Remy first cast a few experimental calcium-hardened grids, they looked, as one competitor remembers, "like they were put together by a drunken plumber."

Delco must have thought so, too, because the company finally decided to abandon—rather than adapt—traditional casting technology. It turned to a system of stamping and working expanded metal. The present production line, says Everett Foust, Delco's director of manufacturing engineering, "is more like a sheet-metal factory than an ordinary battery plant."

Like any alloying material, calcium also changes lead's electrical and chemical characteristics. But one thing it has going for it is that it can be used more sparingly for equal hardening effect. A typical survivable grid has about 0.07-percent calcium (plus traces of tin); this leaves the lead much nearer to its pure state than does the $2\frac{1}{2}$-percent alloy of a low-antimony grid.

Critics point out that calcium is not without problems of its own. It reacts with oxygen to form calcium oxide, which could interfere with adhesion of active plate materials. Lead-calcium also exhibits a tendency to flake, causing particles to separate from positive plates and grow or "tree" on negatives. This is one of the reasons lead-calcium plates have to be packaged in microporous envelope separators. Since these loosened particles are contained in their envelopes, they can't form a short circuit between positive and negative plates. But if enough of the live material becomes inert it could shorten battery life for reasons other than water loss.

Characteristics like these are cited by some battery specialists as a reason why lead-calcium batteries are somewhat less likely than other types to recover fully from deep discharge. Auto batteries, to be sure, aren't intended for deep-discharge service as in boats or rec vehicles. But it can happen if you leave your lights on overnight or wear down a well-charged battery cranking an engine that refuses to start.

Makers of lead-calcium batteries, such as Delco, say this type is the only one that's truly maintenance-free. But manufacturers who make the new low-antimony batteries say that they too have a maintenance-free battery.

Who's right? That's hard to say. There are no industry standards governing the use of the term "maintenance-free." And millions of the new batteries that contain no calcium are being marketed as maintenance-free. Some have fill holes. Some do not.

Like the lead-calcium types—most of which do not have fill holes—many of these lead-antimony batteries incorporate a range of significant design improvements. For example, grid profiles have been improved. Internal circuitry has been shortened. Better separator materials are in use. Modern space-age plastics give the molded containers stronger physical characteristics.

And the extremely low levels of antimony slow water loss significantly. In fact, say makers, water loss will not normally be the first cause of failure. And if it isn't, then further improvement in this direction is really of no importance.

Less Gas, Less Water Loss

To sum up, lead-calcium batteries clearly produce less gas than lead-antimony batteries. But those who have stuck with lead antimony say their new batteries have cut water losses enough, and that further cutting isn't necessary. In addition they say, production difficulties make it difficult, if not impossible, to mass-produce lead-calcium batteries so that most of them perform adequately. Naturally, makers of lead-calcium batteries say they have solved production problems.

Meanwhile, Another Trend

Makers of both kinds of batteries are working on still other kinds of batteries with grid hardeners such as strontium. This and other metals may give many of the advantages of calcium without its problems.

In seeking to get a fix on this growing argument, I talked to engineers and officials of the industry's "Big Seven"—the companies that account for about 90 percent of America's total battery production—and to some major private-label distributors.

The manufacturers, more or less in size order, are Globe-Union, Delco, ESB, General Battery, Gould, Prestolite, and Chloride USA.

Gould and Delco (General Motors' largest division) have led the way in calcium construction and are firmly committed to this and similar systems as the wave of the future. Gould, for example, is planning a new type of battery with cadmium in the positive grid and calcium in the negative.

Globe-Union (largest supplier to Sears and Ford) and ESB (Willard, Exide, and prime supplier to Chrysler and Montgomery Ward) are the main doubters. But Globe, it should be noted, has recently gone into production with a lead-strontium battery that's being introduced as standard equipment on the mid-1977 Lincoln Versailles; it's probably very similar to the lead-calcium type, though reportedly easier to make.

The other three companies have moved in varying—sometimes reluctant—steps into lead-calcium construction. They don't dispute its advantages but are less sure whether these benefits really justify higher cost.

The Winner?

So what's going to come out on top? Despite the fact that lead-calcium may have slightly less overcharge tolerance, such disputed drawbacks don't loom very large. You might well conclude, therefore, that lead-calcium will soon take the fore in the maintenance-free battle. But if you did so, you'd be wrong. The reason: those production difficulties. In

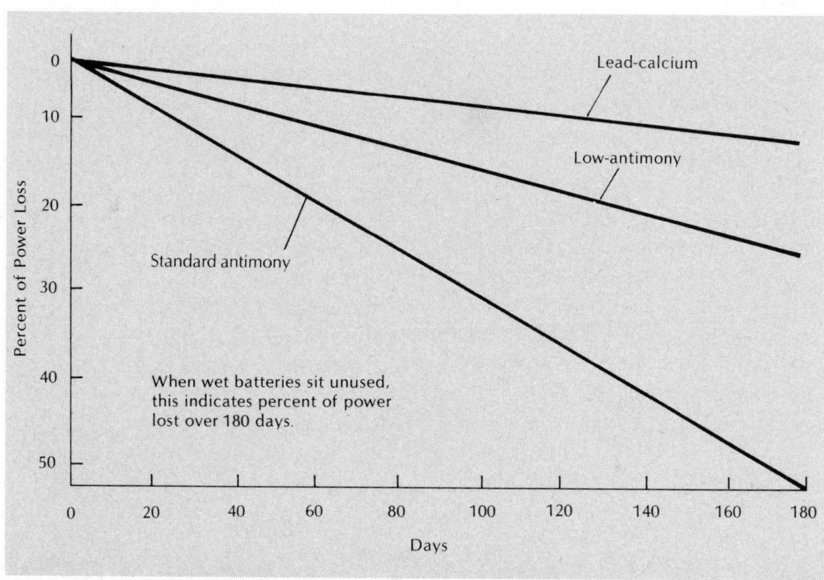

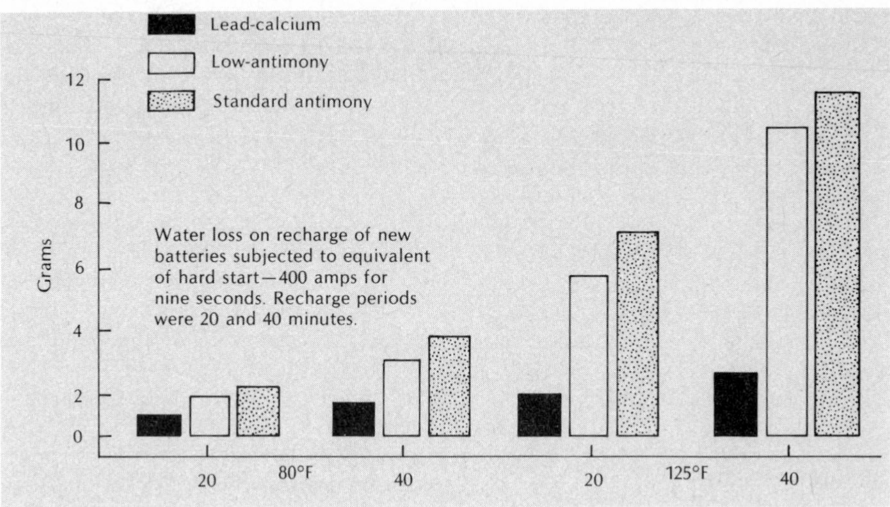

addition, calcium alloys are highly susceptible to contamination; they have to be physically separated in manufacture from operations in which antimony is involved.

Scores of new manufacturing facilities might be needed to effect a full switch. Since they are not likely to materialize soon, most observers expect low-antimony batteries to dominate the so-called "maintenance-free" market for some time to come. They will probably account for nearly half of total production [in] 1980, roughly twice anticipated lead-calcium volume.

Whatever the outcome, you stand to gain from the accelerating ferment in the battery industry, for many improvements are common to low-antimony as well as to antimony-free designs. The low antimonies, moreover, are built to fill a much greater market need—including bottom-of-the-line units with prices that start at about $20.

How much longer? The point is still in dispute and field evidence may not settle the issue for another couple of years. Anthony Sabatino, Gould Battery's vice president for engineering, feels very strongly, he told me, that at least half of these low-antimony batteries will fail within 24 months unless water is added. "If they are maintained," he adds, they should average about 42 months—"just like conventional batteries."

Other companies, however, and major marketers such as Sears and Wards, put the figures much higher and back them up with warranties.

Which Battery Should You Buy?

On present evidence, lead-calcium probably should be your first choice as a replacement battery for a car you expect to keep for more than a couple of years. But that brings up a problem. How do you tell which battery type you're getting for how much? The truth is that you may not be able to tell.

If the literature on the label isn't specific or comprehensive, your best clue is to look for filler caps. It's not a certain guide (some lead-calcium batteries can be uncapped for testing and the addition of water) but a sealed top does give a clue that the battery is lead-calcium.

Don't get confused, in this examination, between filler caps and vent plugs. Antimony-free batteries all have little openings through which gases can escape. In other respects, however, you shouldn't see much that's different. Lead-calcium batteries all have standard top or side terminals by which they can be connected—and charged or jumped—just like the conventional ones.

So you don't have to go back to school to learn how to love or live with a maintenance-free battery. Aside from the fact that you can't—or shouldn't have to—add water, you select and use it just like any other battery.

Chemical Equations

You have already learned some chemical "words"—the formulas of chemical compounds. In this chapter you will learn to put those words into sentences, or chemical equations. You will get more practice in describing chemical reactions by equations in later chapters. The primary purpose of this chapter is to teach you to read chemical equations and to balance them. In the process, we discuss some common chemical reactions that we hope you will find interesting.

OBJECTIVES

1. Given a chemical equation, express what it says in ordinary language. (See Problem 8-41.)
2. Given a description of a reaction, write a chemical equation that says the same thing. (See Problem 8-42.)
3. Given the names or formulas of the reactants and products of a reaction, write a balanced equation for the reaction. (See Problem 8-43.)
4. Given the names of two elements that react, predict the product of the reaction and write a balanced equation for the reaction.
5. Given the name or formula of a compound that decomposes, predict the products of the decomposition and write a balanced equation for the reaction.

Words You Should Know

chemical equation
law of conservation of mass
reactant
product
balanced
coefficient
refluxing
corrosion
combustion

The more we observe in nature, the more we need a special language to express what we see. Grass grows; logs rot; trees change color in the fall; people grow old and die; an embryo develops in a mother's womb. All of these processes involve chemical change. For some changes, we do not know enough to describe what happens, but those that we do understand deserve careful discussion. We need a language in which to express all we know about all that is happening.

In an earlier chapter, you learned about chemical symbols and formulas. With that shorthand notation, you are able to say a great deal about chemical compounds. You can indicate what elements go into the compound and how many atoms of each kind make up a molecule. Although we have not developed the ideas here, you saw that formulas can be elaborated to indicate how atoms are connected in a molecule as well.

We now turn from compounds and elements to reactions involving them. We want to talk about **chemical equations.**

8.1 CHEMICAL EQUATIONS AND CONSERVATION OF MASS

A chemical equation is a shorthand sentence explaining what takes place in a reaction.

Earlier in this book, it was argued that copper must be an element because all of the reactions involving copper produce some substance that weighs more than the original copper. We discussed one experiment in which a copper penny was heated in a flame to produce a black solid, copper(II) oxide.

If you have not done the experiment or don't remember it, perhaps you can repeat the experiment now. Hold a clean penny with pliers or tongs and place it in a hot flame from a gas burner or lay it on the burner of an electric stove. As the penny gets hot, you will see a black film form on the surface. When the hot penny is quickly placed in cold water, the black material usually pops off the surface. (Take care: A hot penny looks like a cold penny, but it doesn't feel the same!)

If you heat the penny in a gas flame, you can observe an interesting effect. When you move the penny back and forth through the flame, the penny turns black and then back to copper color again. (What happens is described later.)

When this experiment is done quantitatively, it is found that 100 g of copper produce 125 g of the black powder. The increase in mass convinces us that the copper must be combining with something. Without saying so, we indicate that we believe in the law of conservation of mass. We believe that matter doesn't just appear out of nothing or disappear into nothing. If we end up with 125 g of black powder, we conclude that the 100 g of copper must have combined with something that weighs 25 g. *The total mass before a chemical reaction must be the same as the total mass after a chemical change.* This is the **law of conservation of mass.** A chemical equation is based on this law.

Since matter is made up of atoms, the law of conservation of mass might be thought of as the law of conservation of atoms. The total mass does not change in a chemical reaction, because the atoms that take part in the reaction do not change. We end up with the same atoms that we started with. These atoms are rearranged to form new molecules, but the same atoms are there. Consequently we have the same total mass.

8.2 HOW CHEMICAL EQUATIONS ARE WRITTEN

We will begin writing equations by describing a reaction that occurs when charcoal burns. Charcoal is mostly carbon, and it produces carbon dioxide gas when it burns. What appears to happen is illustrated in Figure 8.1 and described by the equation shown in Figure 8.2.

The English "translation" of the equation shown in Figure 8.2 is only one of many possible translations. The translation given in Figure 8.2 states a

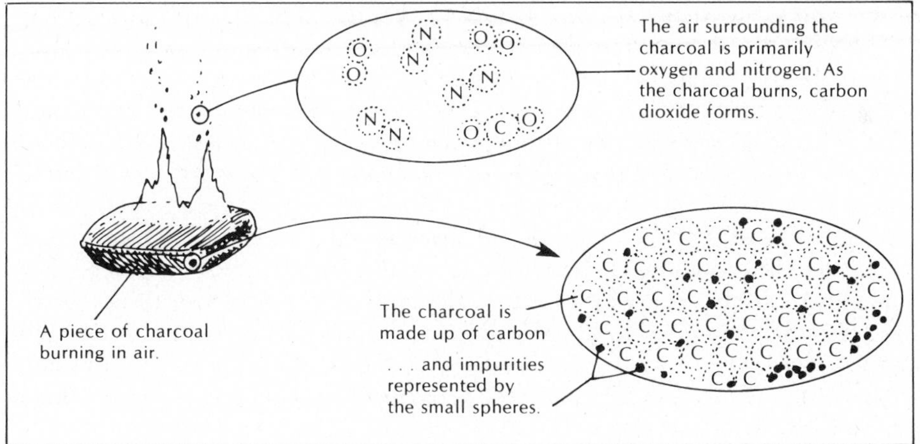

Figure 8.1
A microscopic view of charcoal burning in air.

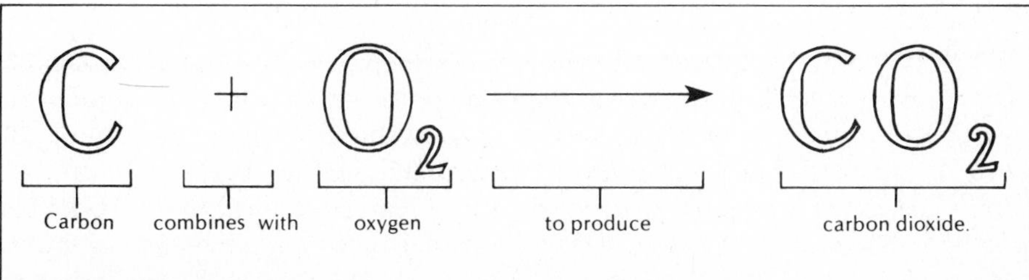

Figure 8.2
An equation describing the reaction shown in Figure 8.1.

simple fact about the reaction taking place when charcoal burns: "Carbon combines with oxygen to produce carbon dioxide."

Before we consider some other ways to read this equation, look at how it is written. First, note that carbon is represented by the symbol C, but oxygen is represented by the formula for an oxygen molecule, O_2. Certainly the solid charcoal does not exist as individual atoms of carbon, so why is it written this way?

Carbon and most other elements exist in the solid as many atoms packed together. There is little evidence of individual molecules, unless one considers the entire piece of solid to be a giant molecule consisting of some unknown number of atoms. *Elements in the solid or liquid phase are normally represented by the symbol for the element with no subscript.*

Avogadro's hypothesis that elemental gases consist of diatomic molecules has generally been accepted. Experimental evidence indicates that oxygen gas consists of molecules with the formula O_2 rather than individual atoms with the symbol O. Therefore, when oxygen gas is shown in an equation, it is represented the way it exists, as diatomic molecules. *In any equation, the symbols and formulas shown must represent substances as we think they exist.* The only exception is the case of solid elements such as the carbon in our charcoal. In these cases, we write only the symbol for the element, because any number of atoms may be stuck together in the solid.

Only those things that are known to be involved in a reaction are shown in the equation. When charcoal burns in air, the nitrogen gas in the air is not involved, so it is not shown in the equation. The impurities in the charcoal are also not involved in the reaction between carbon and oxygen, so they are not shown in the equation. Only those things that change are written on the left side of the arrow. These are the **reactants** in the chemical change.

If there is a chemical change, there must be some product. **Products** of the chemical change are written on the right side of the arrow in the equation. In this case, carbon dioxide gas is produced. It is written to the right of the arrow.

To indicate our faith that matter is conserved, the equation is always written to show the same atoms before reaction (in the reactants shown on the left) and after the reaction (in the products shown on the right). Figure 8.2 shows one atom of carbon before reaction and one atom of carbon after reaction. It shows two atoms of oxygen before reaction and two atoms of oxygen after reaction. The equation is said to be **balanced.**

Balancing simple equations is easy if you remember a few points. First, you *must* know the correct formula for all reactants and products. Second, you must add numbers in front of the formulas to show the same atoms on both sides of the equation. These numbers are called **coefficients.**

The equation shown in Figure 8.2 describes only one of two possible reactions between carbon and oxygen. At high temperatures, carbon reacts with oxygen to produce carbon monoxide gas. Work through the following example with me to see how an equation is balanced.

Example 8.1

You can do the first part. Write carbon plus oxygen to the left of an arrow and write the product, carbon monoxide, to the right of the arrow. This step indicates the reactants and the product of this chemical change.

$$C + O_2 \rightarrow CO$$

I trust that you wrote CO instead of CO_2. If the product is carbon monoxide, the formula *must* be CO.

This equation tells what we started with and what we end up with, but it violates our faith that atoms must be conserved in any chemical change. The following illustration shows what has been written. What is wrong with it?

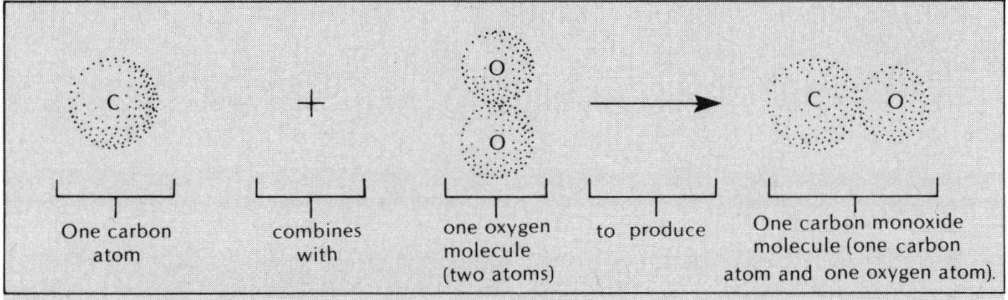

An oxygen atom has disappeared. We don't think this can happen, and we modify the equation to show the same atoms to the left of the arrow (reactants) as to the right of the arrow (products).

$2C$	$+$	O_2	$\rightarrow$	$2CO$
Two carbon atoms	and	one oxygen molecule (two atoms)	react to give	two carbon monoxide molecules (two carbon atoms and two oxygen atoms)

Draw pictures to illustrate the equation as written.

There are two atoms of carbon on the left and two on the right; there are two atoms of oxygen on the left and two on the right. The equation shows that atoms are conserved; it is balanced.

Beginning students often want to balance the equation for the reaction of carbon and oxygen to give carbon monoxide like this:

$$C + O_2 \rightarrow CO_2$$

This equation *is* balanced. Why is it wrong?

It does not describe what was produced. This equation says that the product is carbon *dioxide.* The reaction that we want to talk about produces carbon *monoxide.*

In balancing equations, we write the correct formula for reactants and products first, and work forward from formula to equation. An equation can't be written or balanced until you know the formula. *The formula must never be changed in order to balance the equation.* Doing so would be lying about what takes place.

8.3 HOW CHEMICAL EQUATIONS ARE READ

We have said that equations can be read in several ways. Let's illustrate this, using the equation that you just balanced:

$$2C + O_2 \rightarrow 2CO$$

The normal way to read this equation would be:

"*Two atoms* of carbon combine with *one molecule* (which, incidentally, contains two atoms) of oxygen to produce *two molecules* of carbon monoxide."

Like most of the things we talk about in chemistry, the equation is just indicating proportions. When we burn a piece of charcoal, there are certainly more than two atoms of carbon involved. We might emphasize this proportional relationship by reading the equation like this:

"For every *two atoms* of carbon that burn, *one molecule* of oxygen is needed and *two molecules* of carbon monoxide are produced."

This being the case, we know that:

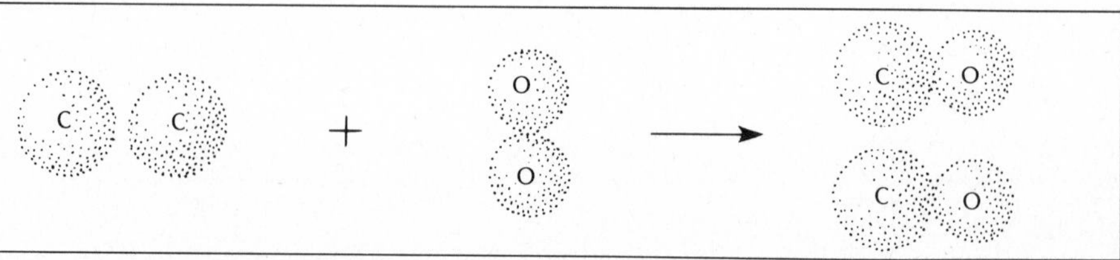

2 atoms C plus 1 molecule O_2 give 2 molecules CO

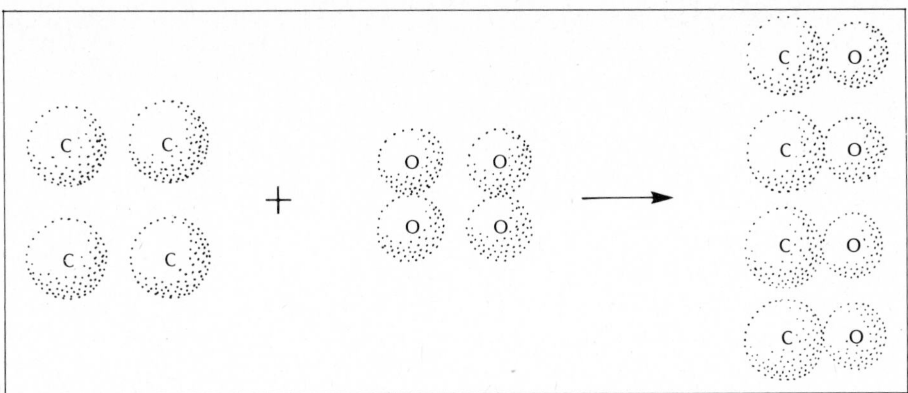

4 atoms C plus 2 molecules O_2 give 4 molecules CO

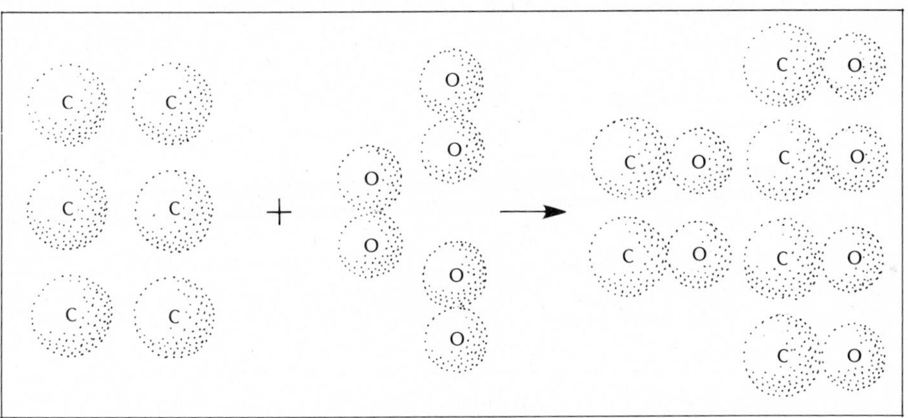

6 atoms C plus 3 molecules O_2 give 6 molecules CO

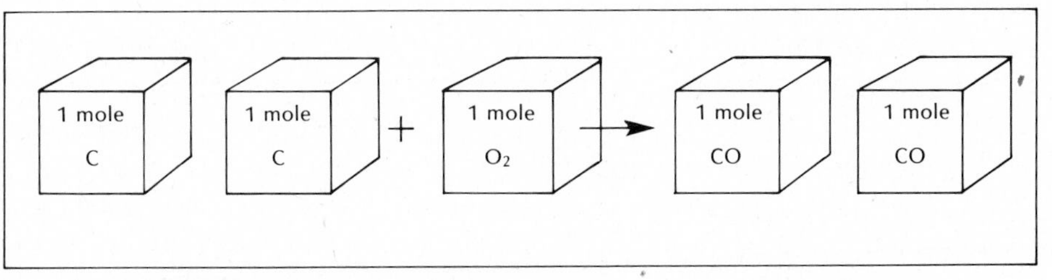

$2(6.02 \times 10^{23})$ atoms C plus $1(6.02 \times 10^{23})$ molecules O_2 give $2(6.02 \times 10^{23})$ molecules CO

The equation can be read:

"*Two moles* of carbon (atoms) combine with *one mole* of oxygen (molecules) to produce *two moles* of carbon monoxide (molecules)."

The words in parentheses are normally omitted, but they are understood. When an equation is read this last way, the balanced equation may be written a different way and still make sense:

$$C + \tfrac{1}{2}O_2 \rightarrow CO$$

This says:

"One mole of carbon combines with *one-half mole* of oxygen to produce *one mole* of carbon monoxide."

When balanced this way, the equation also shows that:

"One atom of carbon combines with *one-half molecule* (one atom) of oxygen to produce *one molecule* of carbon monoxide."

The only objection to this latter reading is that one-half molecule of oxygen is a single atom. Someone could get the idea that oxygen gas exists as individual atoms, and it does not. For this reason, it is generally preferable to use whole numbers in balancing an equation. This can be done by multiplying the last equation by the number in the denominator of the fraction:

$$2(C + \tfrac{1}{2}O_2 \rightarrow CO) = 2C + O_2 \rightarrow 2CO$$

Before going on to balance more complicated equations, do the following exercises to be sure that you know how to translate from English sentences to chemical sentences, and vice versa. For the equations, write an English sentence that says the same thing. For the sentences, write an equation that conveys the same meaning.

8-1. When one mole of water decomposes, it produces one mole of diatomic hydrogen gas and one-half mole of diatomic oxygen gas. (Rewrite the equation using whole numbers.)

8-2. Vinegar (hydrogen acetate) reacts with baking soda (sodium hydrogen carbonate) to produce carbon dioxide gas, water, and sodium acetate.

8-3. $NaOH + HCl \rightarrow H_2O + NaCl$

8-4. $4CuO + CH_4 \rightarrow 4Cu + 2H_2O + CO_2$

The last equation is the chemical description of what you saw happening if you heated copper in a gas flame. The black copper(II) oxide that is originally formed reacts with the methane in natural gas to reproduce the bright copper color of the penny. Water and carbon dioxide are also produced in the reaction. Thus, when you heat a copper penny in a gas flame, you can observe two reactions. In one, the bright copper combines with oxygen from the air to form the black oxide of copper. However, in regions of the flame where there is an excess of methane gas, the black copper oxide reacts with the methane to produce the bright copper once more.

8.4 CHEMICAL EQUATIONS AND CHEMICAL REACTIONS

You must keep in mind that a chemical equation is nothing more than a shorthand description of a chemical reaction. It tells us a lot about a reaction but certainly not everything. Let us talk about a chemical reaction and use it to illustrate the relationship between a balanced chemical equation and the reaction it describes. The one we have selected is a reaction between antimony and iodine. It is a simple reaction that you can do in the laboratory.

Powdered antimony is mixed with iodine in a boiling flask. A liquid, toluene, is added to dissolve the iodine and give better contact between the iodine and pieces of antimony metal. The mixture is heated to boiling and cooked for several minutes. Since toluene is a flammable liquid, a "cooling tower" (condenser) is added to change the toluene back to a liquid and let it run back down into the boiling flask. In that way the mixture can boil indefinitely without losing any of the liquid solvent. This procedure, commonly used in organic chemistry, is called **refluxing.** The apparatus is shown in Figure 8.3.

After cooking for several minutes, the hot liquid is poured off, leaving the excess antimony metal in the bottom of the flask. The single compound produced is dissolved in the hot toluene, but as it cools, beautiful orange

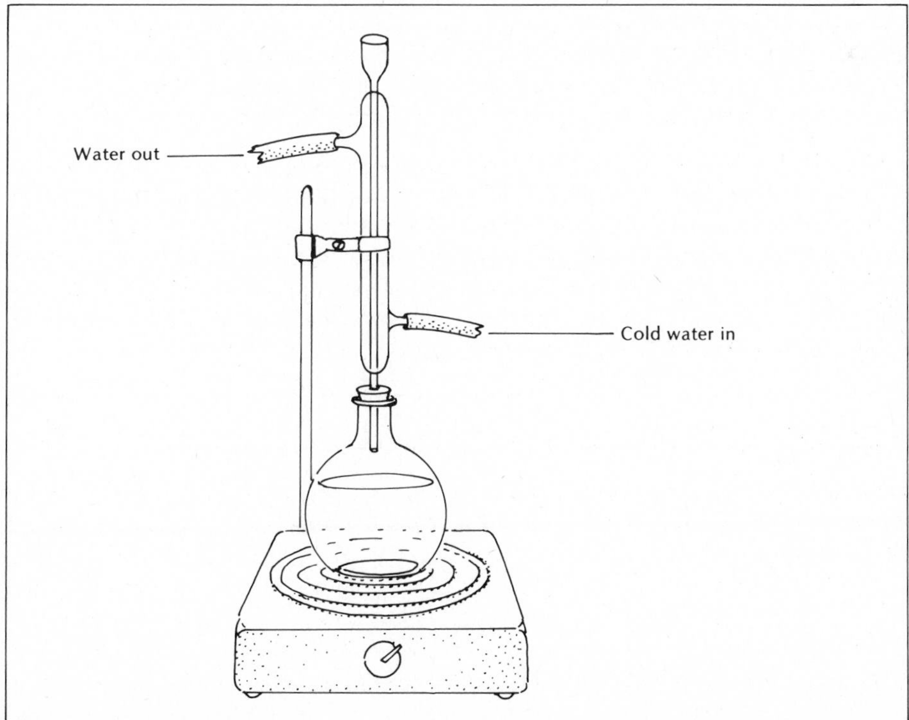

Figure 8.3
Apparatus for refluxing. Vapor from the boiling liquid is cooled to form a liquid once more. The liquid then runs back into the flask to boil again.

crystals settle to the bottom. Filtration of the cooled liquid separates the solid crystals of product from the liquid toluene. Any toluene wetting the crystals quickly evaporates to leave the pure compound.

If we want to write a balanced equation to describe what happened in this chemical change, we need more information. Three things were placed in the boiling flask: antimony, iodine, and toluene. Were all three substances involved in this chemical change?

When a little of the compound formed is heated, purple vapors of iodine are given off. The compound contains iodine, so we can be sure that iodine was a reactant. Careful analysis of the product indicates that it also contains antimony, but nothing else. Furthermore, the liquid remaining after the reaction stops has the properties of the original toluene. All of the evidence suggests that antimony and iodine were reactants in this chemical change but that toluene was not. We can write the first half of the equation like this:

$$\mathrm{Sb} + \mathrm{I} \rightarrow$$

To complete the equation, we must know the formula of the product. There is only one way to know that, if this compound has never been prepared before. The compound must be analyzed to find its percent composition and empirical formula. You know how to do that.

8-5. The compound formed is 24.3 percent antimony. What is the formula of the compound?
8-6. What is the balanced equation for the reaction?

As the reactants were written, the equation was very easy to balance, once the formula for the product was known. It is a good equation and it makes sense, but it is not the way most chemists would write the equation. Iodine is one of those elements that exist as molecules, even as a solid or in solution. Knowing this, the chemist would want to write the equation to represent each reactant as it actually exists. The reactants and products would be written as shown in Question 8-7. Balance the equation.

8-7. $\mathrm{Sb} + \mathrm{I_2} \rightarrow \mathrm{SbI_3}$

To balance the equation for the reaction between antimony and iodine, we need an additional atom of iodine on the left. One way to show this would be as follows:

$$\mathrm{Sb} + \tfrac{3}{2}\mathrm{I_2} \rightarrow \mathrm{SbI_3}$$

Since whole numbers are preferred, the equation can now be multiplied by 2:

$$2\,\mathrm{Sb} + 3\,\mathrm{I_2} \rightarrow 2\,\mathrm{SbI_3}$$

This is the way the equation would normally be written. However, there are times when we need to include more information in an equation. Two facts that we might include in describing this reaction are that the reaction was carried out in toluene and that it was necessary to heat the mixture to get the reaction to occur. This kind of information is sometimes included by writing the name of the solvent over the arrow and by writing "heat" under the arrow:

$$2\,Sb + 3\,I_2 \xrightarrow[\text{heat}]{\text{toluene}} 2\,SbI_3$$

The equation now says:

"Two moles of antimony react with three moles of iodine (molecules) to form two moles of antimony triiodide. The reaction takes place in the presence of toluene and must be heated."

As usual, the equation can also be read in terms of atoms and molecules instead of moles. Either reading is acceptable, even though you could never detect a reaction involving only a few atoms and molecules.

8.5 THINGS EQUATIONS DON'T SAY

It is important to remember that the balanced equation tells you what *might* happen; it does *not* tell you what *will* happen. We can write many balanced equations for reactions that could never occur. For example, we could write the following equation for the production of ordinary sugar from graphite and water:

$$12\,C + 11\,H_2O \rightarrow C_{12}H_{22}O_{11}$$

At the present time, nobody knows how to get this reaction to occur.

Another point you should remember is that a balanced equation describes the amounts of reactants that would *completely* react to give the amount of product indicated in the equation. (In practice, many reactions do not go to completion, and some of one or more reactants remains.) Furthermore, it represents the *proportions* and not the actual amounts that we must use in the reaction.

The balanced equation for the reaction between antimony and iodine told us that 2 mol of Sb react with 3 mol of I_2 to give 2 mol of SbI_3. This doesn't mean that we *must* use 2 mol of Sb and 3 mol of I_2 to get a reaction. We could use much smaller or much larger amounts of reactants to produce smaller or larger amounts of the product. The important point is that the *proportions that react* will always be 2 atoms of Sb for every 3 molecules of I_2.

Reactions are seldom carried out in the laboratory with the amounts of reactants suggested by the balanced equation. The balanced equation for the reaction between antimony and iodine says that 2 mol of antimony (243.52

g) are needed to react with 3 mol of I_2 (761.52 g). These amounts of reactants would produce far more antimony triiodide than we are likely to want.

Calculations using the mole concept show that the proportions are the same if 0.320 g of Sb are used for each 1.00 g of I_2. According to the equation, these proportions will completely react, leaving nothing but 1.32 g of SbI_3 dissolved in the toluene.

The instructions given in a laboratory manual for this experiment call for 2 g of Sb to be used with 1 g of I_2. This is over 6 times the amount of Sb that can react with 1 g of I_2! Why the excess?

It is virtually impossible to have every atom or molecule of reactants come together so that they combine. Consequently, it is common practice to add an excess of one reactant—usually the cheaper one or the one that can be easily separated from the product. In this way, we usually use up all of one reactant and can recover what is left of the other reactant to use again.

The point is that an equation tells us something about the amounts of reactants that are actually *transformed* in the reaction. It does not tell us much about what is done in the laboratory in carrying out the experiment. As normally written, the equation for this reaction did *not* tell us that toluene was used as a solvent. It did *not* tell us that an excess of antimony was used in the experiment. It did *not* tell us that the mixture had to be heated. As you have seen, some of this information can be added to an equation. It *is* added when that information is considered important.

8.6 EVERYDAY CHEMICAL CHANGES

Do you have the feeling that chemistry has little to do with what happens in everyday life? I can understand why you might. We haven't talked much about chemical reactions that you may have observed outside the chemical laboratory. That often happens in beginning courses. There are some basic ideas that must be developed first. Developing those ideas takes time, and we sometimes forget to show you how they relate to things that you have seen. Now is a good time to talk about some everyday chemistry. And it will give you an opportunity to practice using equations, the sentences in the language of chemistry.

Let's start by talking about some ordinary chemical reactions that involve elements. About the only substances you observe as pure elements are the nitrogen and oxygen that make up the major part of air and metals such as iron, copper, aluminum, zinc, and tin. Most of the metals that you see are not entirely pure, but we can talk about the reactions they take part in as though the elements were pure.

RUSTING

The steel that is used to make your car is mostly iron, and the reaction that turns your car body to rust is primarily a reaction between iron and oxygen gas to produce iron oxide. As you already know, there are two oxides of iron, iron(II) oxide and iron(III) oxide. I have written a balanced chemical sentence saying "iron combines with oxygen gas to produce *iron(II) oxide*": $2\,Fe(s) + O_2(g) \rightarrow 2\,FeO(s)$

8-8. Write a chemical sentence that says, "iron combines with oxygen to produce *iron(III) oxide.*" Be sure that your equation is balanced too.

In the sentence above, you see something new. I have added letters in parentheses after each formula. You will often see these letters in chemical sentences. They indicate the physical state of the materials. Just about any substance can exist as a solid, liquid, or gas. The (s) after Fe means that the iron is solid in this reaction, and the (g) after O_2 indicates that the oxygen is a gas. As you might guess, (ℓ), is used to indicate a liquid. One other symbol is frequently seen. It is (aq) and it stands for aqueous. You probably recognize that "aqueous" comes from the Latin word *aqua,* which means "water." $O_2(aq)$ means oxygen gas dissolved in water.

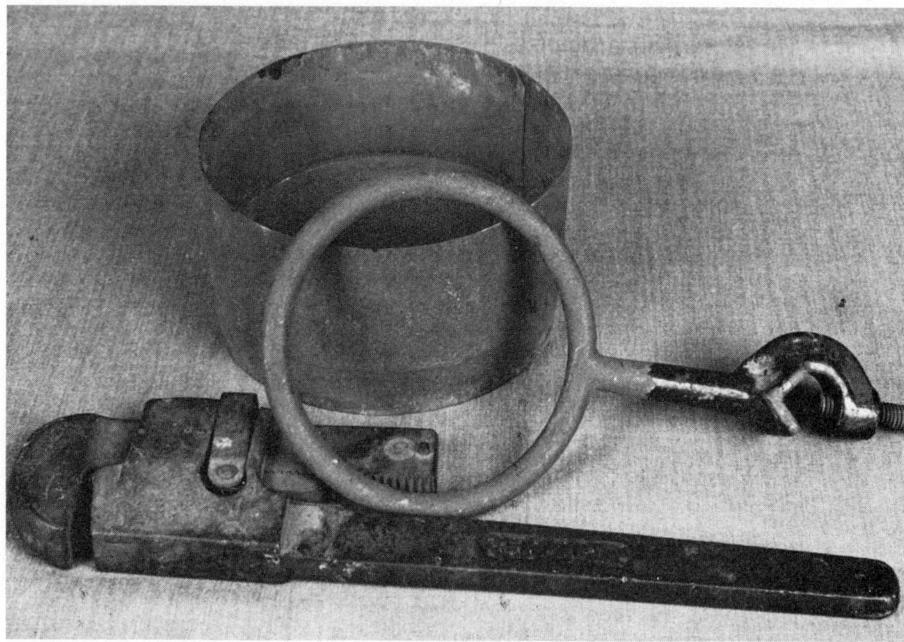

Figure 8.4 *Iron reacts with oxygen in the air to form rust.* (*Photo by Tom Greenbowe.*)

You are going to need the (aq), because the equation that you have just written to describe the rusting of iron doesn't tell the whole story. Iron rusts much faster when there is some water around. Water seems to play a role in the reaction. In fact, the first step in the reaction appears to be

$$2\,Fe(s) + 2\,H_2O(\ell) + O_2(aq) \rightarrow 2\,Fe(OH)_2(s)$$

8-9. Translate the equation just given (bottom of p. 225) into an ordinary English sentence.

After the iron(II) hydroxide forms, it reacts further with water and oxygen to produce iron(III) hydroxide. Complete and balance the following equation to indicate this.

8-10. $4\,Fe(OH)_2$ + ______ H_2O + ______ $O_2 \rightarrow$

Once the sun comes out and the iron(III) hydroxide heats up, it decomposes to form iron(III) oxide and water.

8-11. Write an equation to indicate this decomposition.

Both iron(III) hydroxide and iron(III) oxide have the color that is characteristic of rust. What you see when your car body starts to fall apart is probably a mixture of these two compounds.

Rusting iron is just one example of a class of chemical reactions called **corrosion.** Often the metals that we want for their strength and beauty combine with oxygen or some other element to form compounds that do not have the desired properties.

SILVER TARNISH

Take the case of your favorite silverware. When it is new, it has the shiny luster of untarnished metal, but after a while it turns black—provided that it is real silver rather than stainless steel. This is another example of corrosion. The black deposit that forms on silver may be silver oxide or silver sulfide.

Figure 8.5 *Silver tarnishes by combining with sulfur or oxygen in the air.* (*Photo by Tom Greenbowe and Keith Herron.*)

Egg yolk contains a fair amount of sulfur, which can combine with the silver in the fork that you use to eat the egg. The silver sulfide that results can be formed by rubbing silver briskly with powdered sulfur. The equation looks like this:

$$16\,Ag(s) + S_8(s) \rightarrow 8\,Ag_2S(s)$$

Note that sulfur is written as S_8, whereas we usually just write the symbol for solid elements without any subscript. It is written this way because sulfur exists as molecules containing eight atoms of sulfur. This is not obvious from looking at solid sulfur and was not known until fairly recently. It need not concern you. You can write a balanced equation if you assume that the sulfur exists as individual atoms.

8-12. Write the equation.
8-13. Translate the equation into an English sentence.

PROTECTING ALUMINUM

Not all corrosion is bad. All of the aluminum that you see is slightly oxidized (combined with oxygen). Aluminum reacts *very* rapidly with oxygen in the air to form aluminum oxide.

Figure 8.6
Aluminum products such as these resist corrosion because aluminum oxide forms a tight coating on the surface of the metal. (*Photo Courtesy of Aluminum Company of America.*)

8-14. Write the equation for this reaction.

Fortunately, aluminum oxide forms a very tight coating on the surface of aluminum metal. Once the surface is coated with aluminum oxide, air can no longer get to the aluminum and the reaction stops. If the surface of the aluminum gets scratched, aluminum oxide rapidly forms in the scratch and oxidation stops again. The aluminum oxide is a kind of self-healing paint that protects the metal from further corrosion. It's a shame that the oxide of iron doesn't stick tightly. If it did, we wouldn't need to paint cars to protect them.

Don't get the idea that the oxide on aluminum protects it from everything. It doesn't. The aluminum oxide seems to be easily attacked by chlorides. Since sea water has a lot of sodium chloride in it, aluminum doesn't hold up very well in the ocean—or even nearby where the salty mist blows in from the sea. Household ammonia and sodium hydroxide also attack the aluminum oxide and then react with the aluminum. It isn't a good idea to clean aluminum doors with solutions containing ammonia or other compounds classified as bases. If your great-grandmother had tried to make lye soap in an aluminum pan, the reaction described by the following equation would have taken place. (The same reaction occurs when you use Drāno®.) See whether you can finish balancing the equation and translate the equation into English.

8-15. $\_\_\_\_\_ \text{Al} + \_\_\_\_\_ \text{NaOH} + \_\_\_\_\_ H_2O \rightarrow 2\,NaAlO_2 + 3\,H_2$

8-16. Express this equation in an English sentence.

Figure 8.7 *The bubbles formed when Drāno® dissolves in water are hydrogen gas produced from the reaction between aluminum and sodium hydroxide.* (Photo by Tom Greenbowe and Keith Herron.)

Now would be a good time to reemphasize an important point. Even though you could write a balanced equation for the reaction between aluminum and sodium hydroxide, it doesn't necessarily mean that the equation is a correct description of what takes place. Some books give $NaAl(OH)_4$ as a product of this reaction in place of $NaAlO_2$, and one can easily write a balanced equation with this as the product. Do it and you will see.

8-17. ______ Al + ______ NaOH + ______ $H_2O \rightarrow 2\,NaAl(OH)_4 + 3\,H_2$

The point is that writing a balanced equation doesn't mean that what the equation says is true. To correctly describe what happens in a chemical reaction, you must *know* what happens. And knowing what happens may require years of careful investigation. Many of the most important chemical reactions that occur in nature are still not understood. We generally try to write equations only for those that we *do* understand—or at least *think* that we understand.

BURNING

Corrosion is not the only kind of chemical reaction you have encountered. Whenever something burns, there is a chemical reaction. Burning describes a chemical reaction that produces a visible flame and heat.

Burning, or **combustion,** usually involves a reaction with oxygen from the air. This is why one of the most common procedures for putting out a fire is to smother it. If you can cut off the supply of oxygen, the fire will go out. When water is thrown on a fire, the water absorbs heat produced by the fire and changes to gaseous water—steam. The steam forces air away from the fire. The fire is extinguished by a combination of the cooling effect and the removal of oxygen.

Not all fires can be extinguished by water. Some metals react with water to produce great quantities of heat. The heat produced can be so great that it causes water and other gases to expand rapidly in an explosion. Sodium metal can react with water in this way. The products of the reaction are sodium hydroxide and hydrogen gas.

8-18. Write a balanced equation for the reaction between sodium metal and water. 

Sodium is in Group IA of the periodic table. Elements in the same group often have similar chemical properties; that is, they undergo similar reactions.

8-19. Write an equation for the reaction of lithium and water.
8-20. Write an equation for the reaction of potassium and water.

The only observable difference between these reactions is the speed of the reaction. Lithium reacts at a moderate speed, while sodium reacts rapidly, causing the metal to melt and dance about rapidly on the surface of the water. If the piece of sodium is very large, an explosion occurs. Even a small piece of potassium reacts explosively.

Metals in Group IIA of the periodic table also react with water, though more slowly than those in Group IA. However, the trend of decreasing speed of the reaction as you go from bottom to top holds true.

8-21. Write an equation for the reaction of barium and water.
8-22. Write an equation for the reaction of calcium and water.

The reaction between calcium and water is slower than the reaction between barium and water. It is much slower than the reaction between potassium and water.

BURNING IN WATER

Magnesium metal does not react rapidly with water the way barium and calcium do, but burning magnesium continues to react when placed in steam. The reaction is

$$Mg(s) + H_2O(g) \rightarrow MgO(s) + H_2(g)$$

Figure 8.8
Magnesium burning in steam. (*Photo by Tom Greenbowe.*)

8-23. Translate $Mg(s) + H_2O(g) \rightarrow MgO(s) + H_2(g)$ into an English sentence.

Writing an equation to describe burning is difficult, because the products that form often depend on the conditions under which the burning takes place. Most combustion is incomplete. The smoke that you see in a fire is a mixture of many compounds that result from incomplete combustion. The equations that we will write here are for complete burning.

Paper is mostly cellulose. Cellulose does not have a simple formula. It is a giant molecule made up of units that have the empirical formula $C_6H_{10}O_5$. If combustion of cellulose is complete, the only products are water and carbon dioxide.

8-24. Write the equation to describe this reaction.

METABOLISM

Using food in your body is essentially a slow combustion process. The reaction occurs through a number of steps that involve enzymes. Enzymes allow the process to occur at a lower temperature than is possible when food reacts directly with oxygen. Still, the amount of energy produced in the body is the same as the amount of energy produced by burning sugar in air. Once again, the only products are carbon dioxide and water. Complete the following equation for the combustion of sugar, and balance the equation.

8-25. $C_{12}H_{22}O_{11} + O_2 \rightarrow$

BAKING

The mention of sugar brings us to the kitchen. Most of you know that baking soda and baking powder are used in cakes and breads to make them rise. The reaction involves the production of carbon dioxide gas, which forms tiny bubbles in the dough.

Baking soda forms carbon dioxide through the reaction

$$2\,NaHCO_3(s) \underset{\Delta}{\rightarrow} NaCO_3(s) + CO_2(g) + H_2O(g)$$

The Δ under the arrow indicates that heat must be applied to get the reaction to occur.

8-26. Translate the above equation into an English sentence.

Although some carbon dioxide is probably formed by this thermal decomposition when you are baking, the process is very slow except at high temperatures. Generally an acid is added to the baking soda to get the job done.

Baking powder is a mixture of soda and a dry acid, and the reaction begins as soon as the two compounds are dissolved in water. Recipes that include baking *powder* usually call for sweet milk or water as the liquid. When baking *soda* is specified, buttermilk or some other liquid that contains an acid is used. The reaction that occurs can be illustrated nicely by the reaction between soda and acetic acid, the active ingredient in ordinary vinegar. (By the way, if a recipe calls for buttermilk and you don't have any, sweet milk with a little vinegar added is usually a suitable substitute.) You know the formula for soda (sodium hydrogen carbonate) and acetic acid (hydrogen acetate). The products for the reaction are sodium acetate, water, and carbon dioxide.

8-27. Write the equation and balance it.

Well, there wasn't much balancing required for that one!

Figure 8.9 *Bread rises because of CO_2 bubbles formed in the dough through some chemical reaction.* (*Photo by Tom Greenbowe.*)

CARBONATED DRINKS

While we are on the subject of carbon dioxide, do you know that carbonated drinks are simply drinks that have CO_2 dissolved in them? Carbon dioxide dissolves in water, and some of it reacts to form dihydrogen carbonate (carbonic acid).

8-28. Write the equation for the reaction.

It is this acid that gives the fizzy quality to soft drinks. As you know, the drinks lose their fizz after standing open for some time. CO_2 isn't very soluble in water unless it is kept under pressure. As soon as you open a bottle of pop, you begin to lose some of the CO_2, and after a few hours it is just about gone.

CAVE FORMATION

Interestingly enough CO_2 dissolves in water at ordinary temperature and pressure to cause caves to form. How? Well, there is always some CO_2 in the air, so when it rains, the rain water contains a small amount of H_2CO_3. This compound reacts with limestone very slowly to dissolve the rock, leaving behind a hole in the ground—a cave. Limestone is primarily calcium carbonate, which is insoluble in water. Calcium hydrogen carbonate, which is formed from the reaction between calcium carbonate and dihydrogen carbonate, *is* soluble in water.

8-29. Write the equation for the reaction. Be sure to write (aq) after the calcium hydrogen carbonate to indicate that it is in water solution.

The calcium hydrogen carbonate that is produced in this reaction does not exist except in solution. When enough heat is applied to drive away the water, carbon dioxide is driven away as well, and the solid product is calcium carbonate again. The interesting rock formations found in caves are formed by the slow deposition of calcium carbonate as water that contains calcium hydrogen carbonate evaporates. It takes thousands of years for these formations to grow just a few inches.

Before leaving the subject, we might mention two other reactions that involve limestone. You may have heard of quicklime and slaked lime. Quicklime is calcium oxide, which is formed by heating limestone (calcium carbonate) to decompose it.

8-30. Predict the other product (a common gas), and write the equation.

Quicklime reacts with water to form calcium hydroxide, which is commonly called slaked lime.

8-31. Write the equation for the reaction.

One might think that quicklime got its name from the speed with which it reacts with water to produce slaked lime, but this probably isn't the case. Today "quick" has come to mean "fast," but the Middle English word *quik,* from which it comes, meant "alive." Calcium oxide (quicklime) bubbles and produces a great deal of heat when it is added to water. It will also "eat" flesh if left on it for very long. To the ancients, this compound must have seemed to have some of the qualities of living things, and quicklime seemed to be a good name.

The fact that "quick" originally meant "alive" also seems to explain the origin of "slaked lime" to describe the calcium hydroxide produced when calcium oxide reacts with water. "Slake" comes from a Middle English word that means "to die down." The behavior of calcium hydroxide in water is very calm compared to the "live" action of calcium oxide. It is indeed slaked!

Lime is applied to fields that are too acid, and in the past it was frequently used as a whitewash in place of paint. Remember Tom Sawyer's job of whitewashing the fence—a job he skillfully managed to get his friends to do for him? I use lime to make lime pickles. They are delicious, but I don't know the equation for the reaction that takes place.

CAR BATTERIES

The battery is made of two kinds of plates, one lead and the other lead (IV) oxide. The plates are covered with a solution of dihydrogen sulfate (sulfuric acid). When the battery discharges, the lead, lead (IV) oxide, and sulfuric acid react to form lead(II) sulfate and water.

8-32. Write the equation and try to balance it.

You were probably able to write the equation but not able to balance it. Equations like this are difficult to balance by inspection. Later in your study of chemistry, you will learn a systematic procedure for balancing equations like this one. For now, let me point out just one or two facts about this reaction.

First, note that before the battery discharges, the liquid in the battery is a solution of sulfuric acid. Sulfuric acid is a rather dense liquid. One cubic centimeter of the liquid weighs 1.8 g. When the reaction you have just described is complete, the liquid in the battery is water. Water has a density of 1 g/cm^3. The density of the liquid in a battery decreases as the battery discharges. Because of this, it is possible to see how well the battery is charged by measuring the density of the fluid. This is done routinely when you have your battery checked at a service station.

When the battery is charged, the reaction described in Question 8-32 is reversed so that the $PbSO_4$ reacts with the water to form lead and lead(IV) oxide once more. It is because the reaction is easily reversed by passing an electric current through the battery that a car battery can be recharged. The reactions that occur in ordinary dry cells are not easily reversed, and such batteries cannot be recharged.

Making a good battery involves more than knowing the chemistry of the reaction. The battery must withstand stresses and strains during use and must be capable of being recharged many times. Some of the technology that has gone into making modern car batteries is described in the Reading at the end of Chapter 7.

8.7 PREDICTING PRODUCTS OF REACTIONS

In order to balance the equation for any reaction, you must know what the reaction is. That is, you must know the products. There is only one way to know that for sure. The products must be collected and analyzed to determine the formula. Nevertheless, chemists write equations for many reactions that they have never observed. They are fairly confident that the equations are correct. They are using their knowledge of the properties of elements and compounds to predict what will happen in a reaction.

You can predict reactions too. You haven't learned enough chemical facts to predict many reactions at this point, but you can predict some.

COMBINATION REACTIONS

Several elements have the same combining number in all compounds. You can predict the formula for a compound formed when two such elements react. Knowing the formula for the product, you can write an equation for the reaction. Write the formula for the product and then write a balanced equation for the reaction that occurs when the following elements combine.

8-33. Potassium metal reacts with bromine gas.
8-34. Zinc metal reacts with oxygen gas.
8-35. Sodium reacts with sulfur.
8-36. Aluminum reacts with chlorine.

DECOMPOSITION

The reverse of a combination reaction is called decomposition. Some compounds can be decomposed into the elements they are made of by heating. Many others can be decomposed by electrolysis.

Many compounds decompose to form other compounds. You do not know enough chemical facts to make accurate predictions about *all* decompositions, but you can write balanced equations for the reactions that occur when the following compounds decompose into elements.

8-37. HI
8-38. HgO
8-39. KI (by electrolysis)
8-40. NH_3

In the next two chapters, you will study other types of reactions and learn to predict the products and the equations for many more chemical changes.

8.8 SUMMARY

This chapter has been devoted to chemical equations—the shorthand sentences used to describe chemical reactions. Mastering it is primarily a matter of learning the symbols used to express ideas in an equation. Table 8.1 lists symbols that are used in equations and explains their meaning.

Table 8.1. Symbols Used in Equations

Symbols used	*Meaning*
+	"plus" Used between two formulas to indicate reactants combined or products formed.
$\rightarrow$	Used to separate reactants (on the left) from products (on the right). The arrow points in the direction of the change.
Δ or heat	"heat" Written under the arrow to indicate that heat must be added for the reaction to occur.
(s)	"solid" Written after a symbol or formula to indicate that the substance is solid.
(ℓ)	"liquid" Written after a symbol or formula to indicate that the substance is liquid.
(g)	"gas" Written after a symbol or formula to indicate that the substance is gaseous.
(aq)	"aqueous" Written after a symbol or formula to indicate the substance is dissolved in water.
$\xrightarrow{\text{toluene}}$	Some substance that must be present for the reaction to occur but does not change during the reaction.
=	Used in place of an arrow and has the same meaning.
$\rightleftharpoons$	Used in place of a single arrow to show that the reaction can occur in both directions.
$\uparrow$	Written after the formula of a product that occurs as a gas. The arrow suggests that the substance bubbles out of the reaction mixture. This symbol is usually replaced by (g) in more recent books.
$\downarrow$	Written after the formula of a product that occurs as a solid and settles to the bottom of the reaction mixture. This symbol is replaced by (s) in more recent books.

Figure 8.10 shows how some of these symbols are used in a chemical equation.

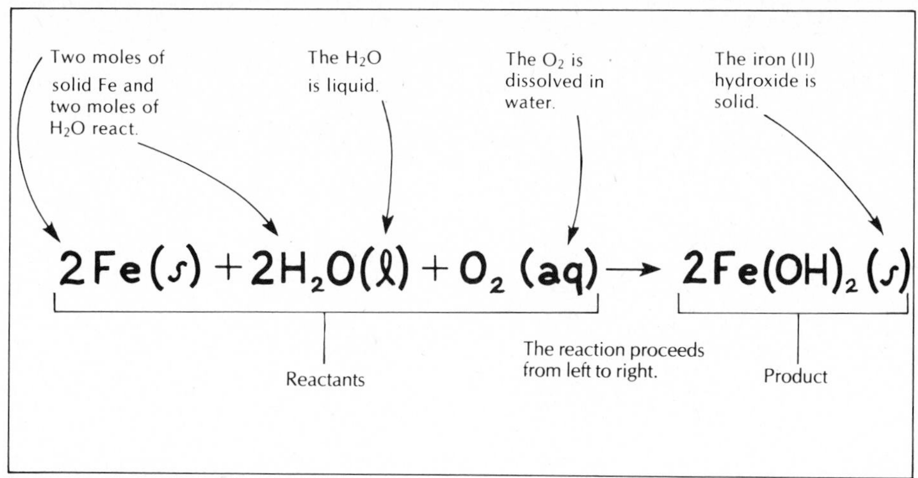

Figure 8.10

Questions and Problems

8-41. Translate the following equations into English sentences.

a. $2\,AgNO_3 + CuCl_2 \rightarrow 2\,AgCl + Cu(NO_3)_2$
b. $BaCl_2 + (NH_4)_2CO_3 \rightarrow BaCO_3 + 2\,NH_4Cl$
c. $3\,Mg + N_2 \rightarrow Mg_3N_2$
d. $2\,KClO_3 \rightarrow 2\,KCl + 3O_2$

8-42. Write chemical equations for the following processes.

a. Barium oxide combines with sulfur trioxide to form barium sulfate.
b. When carbon dioxide gas passes through pellets of solid lithium hydroxide, lithium carbonate and water are formed.
c. Aqueous silver nitrate reacts with aqueous sodium chloride to form solid silver chloride and aqueous sodium nitrate.
d. Hydrochloric acid (hydrogen chloride) reacts with active metals such as zinc to form hydrogen gas and a chloride for the metal.
e. Magnesium metal reacts with ammonia gas (NH_3) to produce hydrogen gas and magnesium nitride.
f. Hydrogen gas can be produced by passing steam over metallic zinc at high temperatures. In addition to hydrogen, zinc oxide is a product of the reaction.
g. When left in air, calcium reacts with oxygen to form calcium oxide.
h. Calcium reacts directly with bromine to produce calcium bromide.
i. When silver oxide is heated to a high temperature, it decomposes into the elements.
j. Antimony triiodide can be decomposed by heating.

8-43. Balance the following equations.

a. $KNO_3 \rightarrow KNO_2 + O_2$
b. $Fe + H_2O \rightarrow Fe_3O_4 + H_2$
c. $CO_2 + NaOH \rightarrow NaHCO_3$
d. $Zn + H_2SO_4 \rightarrow ZnSO_4 + H_2$

e. $FeCl_2 + Na_3PO_4 \rightarrow Fe_3(PO_4)_2 + NaCl$
f. $ZnO + HCl \rightarrow ZnCl_2 + H_2O$
g. $CuSO_4 + Fe \rightarrow FeSO_4 + Cu$
h. $Br_2 + KI \rightarrow KBr + I_2$
i. $HgO \rightarrow Hg + O_2$
j. $C_6H_6 + O_2 \rightarrow CO_2 + H_2O$
k. $FeS + HBr \rightarrow FeBr_2 + H_2S$
l. $KClO_3 \xrightarrow[\Delta]{MnO_2} KCl + O_2$
m. $NaOH + Bi(NO_3)_3 \rightarrow NaNO_3 + Bi(OH)_3$
n. $Pb(NO_3)_2 + KCl \rightarrow PbCl_2 + KNO_3$
o. $H_2 + Cl_2 \rightarrow HCl$
p. $KBr + Cl_2 \rightarrow Br_2 + KCl$
q. $Sb_2S_3 + HCl \rightarrow SbCl_3 + H_2S$
r. $Mg(ClO_3)_2 \rightarrow MgCl_2 + O_2$
s. $NH_4NO_2 \rightarrow N_2 + H_2O$
t. $MgCO_3 + HCl \rightarrow MgCl_2 + CO_2 + H_2O$
u. $CaI_2 + H_2SO_4 \rightarrow HI + CaSO_4$
v. $Mg(CN)_2 + HCl \rightarrow HCN + MgCl_2$
w. $(NH_4)_2S + HgBr_2 \rightarrow NH_4Br + HgS$
x. $Fe_2(SO_4)_3 + Ba(OH)_2 \rightarrow BaSO_4 + Fe(OH)_3$
y. $Al + CuSO_4 \rightarrow Al_2(SO_4)_3 + Cu$

More Challenging Problems

8-44. Cesium is a metal in Group IA of the periodic table. Write the equation for the reaction of cesium with water.

8-45. Magnesium is in the same group of the periodic table as calcium. On the basis of that fact, predict the reaction that occurs when magnesium carbonate is heated. Write an equation for the reaction.

8-46. Again, on the basis of the chemical similarity between magnesium and calcium, can you write an equation to suggest how $Mg(OH)_2$ could be made? (Magnesium hydroxide is a common ingredient in medicines taken for acid indigestion.)

8-47. Any of the metals in Groups IA or IIA of the periodic table will combine directly with any of the elements in Group VIIA. Write equations for all possible reactions of elements in Groups IA and IIA with elements in Group VIIA.

8-48. When iron rusts, any one of several oxides can be formed. Write the equation for the reaction between iron and oxygen when the product is 72.3 percent iron.

8-49. What is the equation for the reaction between iron and chlorine when 54.1 g of the product contain 18.6 g of iron?

8-50. When a compound of nickel and bromine is decomposed by electrolysis, 150 g of the compound produce 109.7 g of Br_2 gas. What is the equation for this decomposition reaction?

Answers to Questions in Chapter 8

8-1. $H_2O \rightarrow H_2 + \frac{1}{2}O_2$ or $2H_2O \rightarrow 2H_2 + O_2$

8-2. $HC_2H_3O_2 + NaHCO_3 \rightarrow CO_2 + H_2O + NaC_2H_3O_2$

8-3. One mole of sodium hydroxide reacts with one mole of hydrogen chloride to produce one mole of water and one mole of sodium chloride. (There are other equivalent wordings that are acceptable.)

8-4. Four moles of copper(II) oxide react with one mole of methane to produce four moles of copper, two moles of water, and one mole of carbon dioxide.

8-5. SbI_3. Assume that we have 100 g of the compound. Then there would be 24.3 g of Sb and 75.7 g of I. Next we find the moles of each element in 100 g of the compound:

$$? \text{ mol Sb} = 24.3 \cancel{\text{g Sb}} \times \frac{1 \text{ mol Sb}}{122 \cancel{\text{g Sb}}} = 0.199 \text{ mol Sb}$$

$$? \text{ mol I} = 75.7 \cancel{\text{g I}} \times \frac{1 \text{ mol I}}{127 \cancel{\text{g I}}} = 0.596 \text{ mol I}$$

Dividing the moles of each element by the smaller of the two, we have

$$\frac{0.596 \text{ mol I}}{0.199 \text{ mol Sb}} = \frac{2.99 \text{ mol I}}{1 \text{ mol Sb}} = \frac{3 \text{ mol I}}{1 \text{ mol Sb}}$$

There are 3 mol of iodine for each mole of antimony. The formula is SbI_3.

8-6. $Sb + 3I \rightarrow SbI_3$

8-7. $2Sb + 3I_2 \rightarrow 2SbI_3$

8-8. $4Fe(s) + 3O_2(g) \rightarrow 2Fe_2O_3(s)$

8-9. Two moles of solid iron react with two moles of water and one mole of oxygen gas dissolved in water to produce two moles of iron(II) hydroxide that is solid. (There are other ways of saying this, but the message should be the same.)

8-10. $4Fe(OH)_2 + 2H_2O + O_2 \rightarrow 4Fe(OH)_3$

8-11. $2Fe(OH)_3 \rightarrow Fe_2O_3 + 3H_2O$

8-12. $2Ag(s) + S(s) \rightarrow Ag_2S(s)$

8-13. Two moles of solid silver combine with one mole of sulfur to form one mole of silver sulfide.

8-14. $4Al + 3O_2 \rightarrow 2Al_2O_3$

8-15. $2Al + 2NaOH + 2H_2O \rightarrow 2NaAlO_2 + 3H_2$

8-16. Two moles of aluminum, two moles of sodium hydroxide, and two moles of water combine to form two moles of sodium aluminum oxide and three moles of hydrogen gas. (Note: The name for $NaAlO_2$ is not one that you are likely to know. However, you should be able to do the rest of the translation without difficulty.)

8-17. $2Al + 2NaOH + 6H_2O \rightarrow 2NaAl(OH)_4 + 3H_2$

8-18. $2Na + 2H_2O \rightarrow 2NaOH + H_2$

8-19. $2Li(s) + 2H_2O(\ell) \rightarrow 2LiOH(aq) + H_2(g)$

8-20. $2K(s) + 2H_2O(\ell) \rightarrow 2KOH(aq) + H_2(g)$

8-21. $Ba(s) + 2H_2O(\ell) \rightarrow Ba(OH)_2(aq) + H_2(g)$

8-22. $Ca(s) + 2H_2O(\ell) \rightarrow Ca(OH)_2(s) + H_2(g)$ (Note: $Ca(OH)_2$ is not very soluble in water. If more than a pinch of calcium is added to water, solid is likely to form.)

8-23. One mole of magnesium metal reacts with one mole of gaseous water (steam) to produce one mole of magnesium oxide and one mole of hydrogen gas.

8-24. $C_6H_{10}O_5 + 6O_2 \rightarrow 6CO_2 + 5H_2O$

8-25. $C_{12}H_{22}O_{11} + 12O_2 \rightarrow 12CO_2 + 11H_2O$

8-26. When sodium bicarbonate is heated, two moles of it decompose to form one mole of sodium carbonate, one mole of carbon dioxide gas, and one mole of water.

8-27. $NaHCO_3 + HC_2H_3O_2 \rightarrow NaC_2H_3O_2 + H_2O + CO_2$

8-28. $CO_2 + H_2O \rightarrow H_2CO_3$

8-29. $CaCO_3(s) + H_2CO_3(aq) \rightarrow Ca(HCO_3)_2(aq)$

8-30. $CaCO_3 \rightarrow CaO + CO_2$

8-31. $CaO + H_2O \rightarrow Ca(OH)_2$

8-32. $Pb(s) + PbO_2(s) + 2H_2SO_4(aq) \rightarrow 2PbSO_4(s) + 2H_2O(\ell)$

8-33. $2K + Br_2 \rightarrow 2KBr$

8-34. $2Zn + O_2 \rightarrow 2ZnO$

8-35. $2Na + S \rightarrow Na_2S$ or $16Na + S_8 \rightarrow 8Na_2S$

8-36. $2Al + 3Cl_2 \rightarrow 2AlCl_3$

8-37. $2HI \rightarrow H_2 + I_2$

8-38. $2HgO \rightarrow 2Hg + O_2$

8-39. $2KI \rightarrow 2K + I_2$

8-40. $2NH_3 \rightarrow N_2 + 3H_2$

READING

The same kind of reaction that makes batteries possible causes corrosion. Corrosion costs us billions of dollars each year and depletes our natural resources. This article gives some of the facts about corrosion and what can be done to fight it.

CORROSION
PART II. PASSIVITY AND MECHANISM

Wayne L. Smith

Many metals when exposed to the environment develop an outer passive coat which inhibits further corrosion. In other cases the passive coat can be formed by treatment with certain chemicals. In either case the passive film renders the metal nearly noble in character.

From practical experience we know that aluminum airplanes, cooking pots, and beer cans last a long time despite the large positive oxidation potential of aluminum. When aluminum is exposed to air, a tightly adhering layer of aluminum oxide rapidly forms. Because of this oxide layer the oxidation potential of passivated aluminum is about −0.6 volt (between copper and silver in the metal activity series).

If the metal is scratched, the protective oxide layer rapidly reforms. However, aluminum is subject to atmospheric corrosion, particularly in the presence of chloride ions. An occasional chloride ion disrupts the crystalline continuity between aluminum and its oxide layer, and the underlying metal is oxidized, leading to pitting.

Zinc also might be expected to oxidize rapidly but, in the presence of air, it forms a passive coating of basic zinc carbonate, $ZnCO_3 \cdot nZn(OH)_2$, rather than oxide. Galvanized ware consists of steel with a thin coating of zinc on each side, usually applied by dipping, spraying, or electrodeposition. Galvanized objects are more durable in rural regions than in cities where the rain is more acidic.

Tin also forms a passive coat in air, which explains its use in tin foil (now largely replaced by aluminum foil and plastics) and in "tin cans" which are really steel cans coated with a thin layer of tin. Tin is generally electrodeposited because this produces a thin coating relatively free from pores or defects. The nontoxic nature of tin salts makes tin plate ideal for handling beverages and foods.

Although copper in the atmosphere would not be expected to oxidize to any great extent because it is quite electropositive, corrosion is further inhibited by formation of a basic copper carbonate $CuCO_3 \cdot Cu(OH)_2$ layer. This layer, sometimes called patina, creates the pleasing green color of copper roofs and bronze statues. Bronze is an alloy of copper and tin.

On exposure to air, chromium forms a passive oxide coating that has the additional advantage of taking a high polish. The chrome decorations on cars are steel which has been electroplated with chromium, generally with an intervening layer of nickel to improve adhesion properties.

Stainless steel contains chromium and nickel as the major alloying elements with iron, often about 18 and 8%, respectively. These alloys also form passive coatings, mainly Cr_2O_3 which is extremely durable and readily repairs itself if oxygen is available.

Unlike the passive coatings already mentioned, that of iron does not normally adhere well to the metal surface. The reddish oxide formed on untreated iron slowly flakes off, exposing additional metal for corrosion. In addition the oxide layer retains moisture, which also speeds rusting.

However iron can be passivated by suitable chemical treatment. For example, concentrated nitric acid oxidizes the surface of iron to a layer of ferric oxide, and the metal assumes noble properties. This can be illustrated by the failure of passive iron to plate out copper from a solution of copper sulfate; in contrast, copper is readily plated out by untreated iron. The passive coating is easily damaged, however, and scratching the coating leads to rapid attack on the underlying metal.

A more permanent "gun blue" protective layer on iron can be obtained by dipping the iron in a strong oxidant such as molten potassium nitrate. These thicker passive coatings are composed of adherent hard layers of oxides such as Fe_3O_4.

Another means of protecting iron is through use of dilute solutions of oxidizing agents such as potassium dichromate, $K_2Cr_2O_7$, sodium chromate, Na_2CrO_4, or sodium nitrite, $NaNO_2$. Iron becomes several tenths of a volt more noble when dipped in 10^{-3} to 10^{-4} M solutions of these agents; presumably this is caused by formation of an oxide layer. Chromates are commonly used as rust inhibitors for circulating cooling

Source: *Chemistry,* Vol. 49, No. 5 (June 1976), pp. 7–9.

waters—for example, in internal combustion engines. Unlike chromates, nitrites have little tendency to react with alcohol or ethylene glycol and are therefore frequently used for rust inhibition in antifreeze cooling waters.

Mechanism of corrosion

Although studied extensively, the corrosion process is not well understood. But it seems clear that electrochemical principles are involved, that the first step must involve oxidation of the metal, and that a corresponding reduction must occur somewhere. Although the theory may not be complete, means of corrosion prevention are fairly obvious. Because of the importance of iron in our society and extensive information available, this discussion is limited to the mechanism of iron rusting.

Clearly both oxygen and water must be present for rapid rusting at ordinary temperatures. In dry climates (relative humidity less than 50%) rusting of exposed iron is almost negligible. Another example is found in home heating systems that involve circulating hot water. Copper pipes are commonly connected to cast iron boilers, but rusting of the boiler is slow because the hot water is deaerated. Perhaps the most common rust-preventive measure involves coating the object with paint, which effectively excludes oxygen if the barrier is complete. Electrolytes expedite rusting, which supports the idea that rusting involves electrochemistry.

When iron corrodes, the rate is usually controlled by the cathodic reaction, which usually is much slower than the anodic reaction. A mechanism that is consistent with the observed facts is:

$$4\,Fe(c) \rightarrow 4\,Fe^{2+}(aq) + 8e^- \qquad (1)$$

$$8e^- + 8\,H^+(aq) \rightarrow 4\,H_2(g) \qquad (2)$$

$$4\,H_2(g) + 2O_2(g) \rightarrow 4\,H_2O(\ell) \qquad (3)$$

$$4\,Fe^{2+}(aq) + O_2(g) + (4 + 2x)\,H_2O(\ell) \rightarrow 2[Fe_2O_3 \cdot xH_2O(s)] + 8\,H^+(aq) \qquad (4)$$

In Step 1, ferrous ions are produced, but this step cannot proceed far without a means of removing electrons. Even a neutral solution contains a small concentration of aqueous protons that might react with the hydrated electrons as in Step 2. The nascent hydrogen atoms might be expected to form H_2 molecules, but hydrogen gas is not observed in rust formation in neutral solutions. Iron is a good catalyst for hydrogenation reactions, however, and hydrogen atoms may react with oxygen as in Step 3.

Meanwhile the ferrous ions react with oxygen to form rust and regenerate aqueous protons; in this sense the reaction is self-catalytic. In acidic solutions (pH less than 4), the ferrous oxide film dissolves, and rusting proceeds rapidly with hydrogen evolution. Note that the sum of equations 1 through 4 is

$$4\,Fe(c) + 3O_2(g) + (2x)H_2O(\ell) \rightarrow 2[Fe_2O_3 \cdot xH_2O(s)]$$

In any mechanism, the sum of the steps must give the overall reaction equation.

This mechanism also accounts for other observations in connection with rusting. For example, when

Figure 1. Galvanized iron–copper joint. The layer of zinc inside the pipe is rapidly destroyed followed by rusting of the steel, with copper providing an "electron sink." The cathodic reaction is reduction of molecular oxygen.

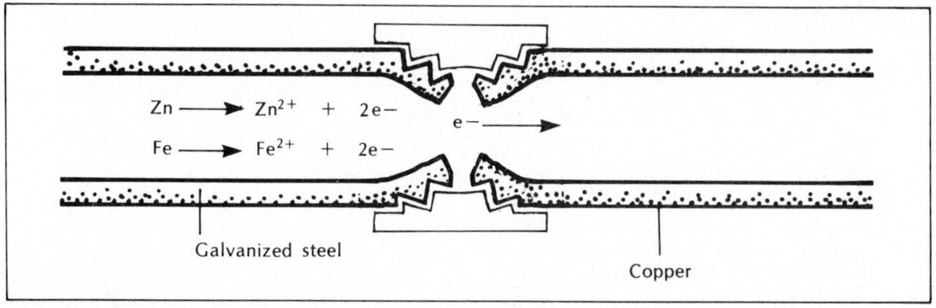

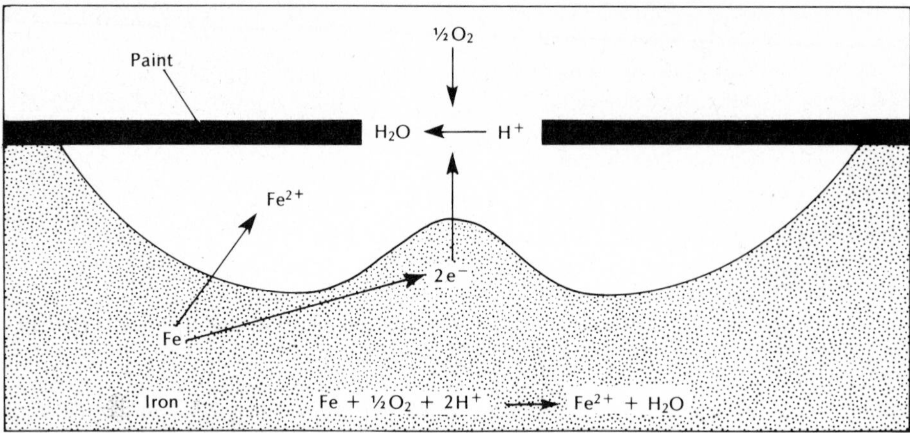

Figure 2. Corrosion of painted metal. Generally, oxidation occurs under the paint where the oxygen supply is limited.

iron and copper pipes are connected together, the iron rusts more rapidly than usual, even though it may be galvanized. A possible household example of such a cell is shown in Figure 1. The copper provides a convenient means by which electrons can be removed and Step 1 is expedited. In addition, the $H^+(aq)$ ions, attracted to the copper and neutralized, are less tightly bound to the copper than they are to the iron, thus accelerating Step 3.

Rusting is more rapid in the presence of electrolytes, which can remove electrons and accelerate Step 1. Thus steel ships rust much more rapidly in seawater than in fresh water. In addition, chloride ions prevent formation of a passive coat. Cars also rust more rapidly when salt is present, either from the winter salting of roads or from sea spray. If the salt is hygroscopic, such as calcium chloride sometimes used for salting roads, the moisture retained expedites Step 4.

Iron rusts rapidly in acidic solutions, and hydrogen gas is generally evolved. Air pollutants, such as sulfur and nitrogen oxides, form acidic salts in water and thus accelerate rusting. Rainwater having a pH as low as 4 [0.0001 M $H^+(aq)$ solution] has been reported. Such acidic mists are responsible for nylon stockings disintegrating and cars sizzling.

The weathering of sulfur compounds in coal ashes also produces acidic solutions, presumably sulfuric acid. Iron pipes in contact with cinders, for example near railroad tracks, rust much more rapidly than normally.

That oxidation of metals occurs in areas where oxygen is denied supports the stepwise mechanism. Thus steel generally corrodes near a defect in the finish underneath the paint where the oxygen supply is restricted, as indicated in Figure 2. Where the oxygen supply is limited, ferrous ions may diffuse before encountering enough oxygen to be oxidized to iron(III) and eventually form the hydrated ferric oxide. Thus rust forms away from the spot where the metal is oxidized. A small spot of rust on a fender is usually a sign of a more serious problem underneath the paint.

Other examples of rust formation away from the site of oxidation include rivets and iron posts partially immersed in water. The rivet shank, although protected from air, actually deteriorates while only rust forms on the rivet head where oxygen is plentiful. In the case of iron posts immersed in water, rust forms at the waterline, and pitting occurs underwater where the oxygen supply is limited.

The electrochemical corrosion of metals, as opposed to film formation of oxides, sulfides, and similar tarnishes, requires water. Electrolytes expedite rusting by functioning as charge carriers, but this mechanism is also valid for atmospheric rusting processes. Iron generally begins to rust at relative humidities above 60% with the rate increasing as the relative humidity increases. In the presence of sodium chloride in marine atmospheres, iron rusts at relative humidities as low as 40%.

Differential aeration or differences in supply of oxygen to different parts of a metal surface is responsible for corrosion of underground iron pipes in

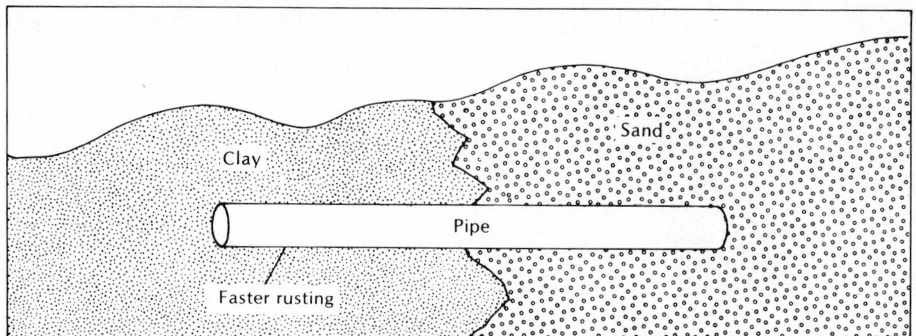

Figure 3. Corrosion of underground pipes. Oxygen has better access through the sand and is reduced at the partially passivated surface. The anodic reaction (corrosion) is more rapid in the wetter, compacted clay.

various kinds of soil—for example, in dense clay and porous sand as shown in Figure 3. The more rapid corrosion of a pipe in clay is related not only to the greater availability of oxygen in the sand but also to the slower diffusion of oxidation products in the sandy soil. In relatively dry sand, the oxidation products can form a partially passive coat on the iron surface.

Despite its agreement with observed facts, the stated mechanism should not be considered the final word on the rusting process. Mechanisms are theories and can never really be proved or disproved. We accept them as long as they can explain observations, but new observations often require formulation of new mechanisms.

9

What Do We Get and How Much? Predicting Products and Stoichiometry

Chemistry wouldn't be so bad if you could forget an idea after you have studied it and gone on to the next. Unfortunately, each new idea builds on the ideas that you have already learned. You can't forget and expect to understand what is coming up next. This chapter illustrates what we mean.

Do you remember how to write formulas? Good! You will use that skill in this chapter.

Do you remember how to write equations and balance them? Good! You will need to do that in this chapter.

Do you remember how to find the molecular mass of a compound? Good! You will need to do that, too.

Do you remember all that you learned about moles? No? Then go back and review. We *told* you you had to remember that!

In this chapter you will see how all of these ideas are put together in order to answer questions such as "How much lye do I need to make a pound of soap?" and "How much alcohol can I get by fermenting a bag of sugar?" You won't answer these particular questions, but you will answer a number of questions like them, and the procedure is essentially the same.

OBJECTIVES

Stoichiometry

1. Given a balanced equation for a reaction and the number of moles of a

reactant that undergoes reaction, find the number of moles of any product that is produced. (See Problems 9-53a and 9-54a.)

2. Given the balanced equation for a reaction and the mass of any reactant, find the number of moles of any product produced. (See Problems 9-53c and 9-54b.)

Limiting Reagents

3. Given the balanced equation for a reaction and the mass of two reactants that are mixed, determine which reactant is completely consumed and the mass of the reactant that remains unreacted. (See Problem 9-59.)
4. Given the mass of two reactants that are mixed, determine the mass of product formed and the mass of any reactant that remains unreacted. (See Problem 9-59.)

Predicting Products

5. Predict the products and write a balanced equation for decomposition, combination, and replacement reactions. Formulas for unusual products will be supplied. (See Problems 9-59 and 9-60.)
6. When told whether a particular replacement reaction occurs, indicate the relative activity of the two elements involved in the replacement. (See Problem 9-61.)

Words You Should Know

combination reaction
combustion
stoichiometry
decomposition reaction
replacement reaction
limiting reagent
active
activity series
halogen
halide

We would do well to return to some of the questions that interest people. We began this course by pointing out one question of interest: "What kind of material do we have?" You saw that we can answer this question by comparing the properties of the unknown material to the properties of matter that is familiar to us. If the properties are the same, the matter is the same. From this beginning, we asked such questions as "Is it pure?" Once again, we saw that a careful examination of such properties as melting point and boiling point usually provides an answer. We continued to talk about elements and compounds as different kinds of pure substances and showed how compounds can be made from elements.

In many cases it is not enough to ask what we have. We also need to know how much. We want to know how much ammonia can be produced from a ton of hydrogen gas, how much cement can be made from a ton of limestone, or how much carbon monoxide is produced when a gallon of gasoline burns in an automobile. Often we can answer such questions by examining the equation that describes the reaction that occurs.

In this chapter we will examine some simple chemical reactions and discuss two things. First, we will try to make predictions about the products that are formed. Second, we will determine how much product is formed from a given amount of reactants.

Your instructor will probably demonstrate many of the reactions that are discussed in this chapter. If so, try to recall what the materials look like when the reactions take place. Always keep in mind that equations and formulas are merely shorthand methods of talking and writing about chemistry. Real chemistry is what happens in a growing plant, in the fireplace, and in your stomach.

You will be happy to know that there is little new material in this chapter. In the last several chapters, you were introduced to a number of important skills: calculating moles, writing formulas, and balancing equations. In this chapter you will practice using these skills to answer quantitative questions about chemistry. Along the way, we will discuss some simple chemical reactions.

9.1 CALCULATIONS INVOLVING COMBINATION REACTIONS

Perhaps the simplest chemical reaction that we could imagine is the combination of two elements to form a compound. If a reaction takes place when two elements are mixed, there is little chance that anything other than a **combination reaction** has occurred. Consequently, it is often possible to predict the product of such a reaction, as you saw in Chapter 8.

Combustion can produce some spectacular reactions. When iron wool is heated in a flame and placed in a bottle of pure oxygen, the result looks like a sparkler on the Fourth of July (see Figure 9.1). Combustion reactions need not involve oxygen. Similar results are obtained when reactive metals are placed in chlorine. The first reaction that we will consider occurs when sodium metal is heated and then placed in a bottle of chlorine gas (see Figure 9.2).

The only possible product for this reaction is NaCl, ordinary table salt. Only sodium and chlorine atoms are present, and only a compound of Na and Cl can form. Since the combining number of both Na and Cl is 1, the formula must be NaCl.

Although we would never make salt in this way, we can use the reaction to begin asking the "how much" questions that we want to consider in chemistry. Get a piece of paper and work Example 9.1 with me.

Figure 9.1
Steel wool burning in pure oxygen. (Photo by Tom Greenbowe.)

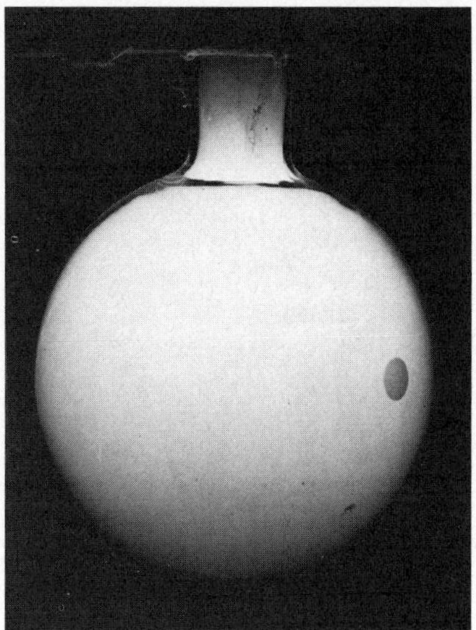

Figure 9.2
Sodium metal burns in chlorine gas to form clouds of solid sodium chloride. (Photo by Tom Greenbowe.)

Example 9.1

How many moles of chlorine gas would be required to produce 1 mol of salt?

This question asks about a relationship between reactants and products in a chemical reaction. What kind of statement describes such relationships?

- -

The equation for the reaction.

The first thing that we need to answer questions such as this one is a balanced equation for the reaction. Write a balanced equation for the reaction between sodium and chlorine gas.

- -

$$2\,Na + Cl_2 \rightarrow 2\,NaCl$$

Now translate the equation into ordinary English.

- -

There are two translations:

1. Two *atoms* of sodium combine with one *molecule* of chlorine to give two *molecules* of sodium chloride.

2. Two *moles* of sodium combine with one *mole* of chlorine (molecules) to give two *moles* of sodium chloride.

The translation in terms of moles is more useful when calculations are to be made. It practically answers the question that was asked in the beginning. The equation indicates that in this reaction 1 mol of chlorine gas produces 2 mol of salt. Then how many moles of chlorine gas are required to produce just 1 mol of salt?

0.5 mol Cl_2

You probably got the correct answer without doing any written calculations. You may not even be aware of the mathematics needed to get the answer. That's O.K., but when the numbers are not so easy, you need to know what to do mathematically. The key to the mathematical solution is the relationship between moles of Cl_2 and moles of NaCl given in the equation. What is that relationship?

One mole of Cl_2 for every *two moles* of NaCl.

This represents a proportional relationship and can be used to construct unit-factors. What are the two unit-factors that show this relationship?

$$\frac{1 \text{ mol } Cl_2}{2 \text{ mol NaCl}} \quad \text{and} \quad \frac{2 \text{ mol NaCl}}{1 \text{ mol } Cl_2}$$

Translate each of these unit-factors into ordinary English.

The exact wording will vary, but the idea is expressed as "one mole of chlorine for every two moles of sodium chloride" and "two moles of sodium chloride for every one mole of chlorine."

Use the appropriate unit-factor to finish setting up the following problem.

$$? \text{ mol } Cl_2 = 1 \text{ mol NaCl} \times \underline{\qquad\quad} = ?$$

$$? \text{ mol } Cl_2 = 1 \cancel{\text{mol NaCl}} \times \frac{1 \text{ mol } Cl_2}{2 \cancel{\text{mol NaCl}}} = 0.5 \text{ mol } Cl_2$$

Unit-factors derived from this equation can be used to answer all of the following questions. Work them before going on.

9-1. How many moles of chlorine would be needed to produce 0.13 mol of NaCl?
9-2. How many moles of sodium would react with 5.2 mol of chlorine gas?
9-3. How many moles of salt could be made from 1.26 mol of Cl_2?
9-4. How many moles of sodium metal would be needed to make the 737 g of salt in the box commonly sold in the grocery store?

Did you have difficulty with the last question? It requires an additional step but no new kind of calculation. Work through another example and you will see how the two steps are put together.

Example 9.2

How many moles of sodium metal would be needed to make 737 g of salt?

The question is still asking about a relationship between reactants and products in a chemical reaction, and the relationship is shown by the chemical equation. Write the equation once more.

$$2\,Na + Cl_2 \rightarrow 2\,NaCl$$

The equation is always written in terms of *moles* of reactants and products. In this example, you are asked about *moles* of reactant but are given *grams* of product.

Many students find that a kind of "road map" of such problems helps them organize their thinking. The "road map" shows what you start with and what you want to end up with, like this:

$$\text{g NaCl} \longrightarrow \text{mol Na}$$

The problem gives information about *grams of NaCl.*

The problem asks for information about *moles of Na.*

This road map reminds us that we know grams of NaCl from the problem and that we must find the moles of sodium required to produce it.

After diagramming the road map, look for relationships that allow you to calculate what you are after. Sometimes this can be done directly, as in Example 9.1. In that problem we were asked about *moles* of chlorine and *moles* of salt. The road map is

$$\text{mol NaCl} \rightarrow \text{mol } Cl_2$$

Since the equation gives a relationship between moles of NaCl and moles of Cl_2, the problem can be worked with a single unit-factor. This can be represented on the road map.

$$\text{mol NaCl} \xrightarrow{\frac{1\text{ mol }Cl_2}{2\text{ mol NaCl}}} \text{mol } Cl_2$$

The relationship found in the equation that allows one to go from moles of NaCl to moles of Cl_2 is written over the arrow in the road map.

At times there is not a single relationship that will allow the "trip" you want to take from the known to the unknown, so you must look for alternative routes. This must be done in the problem we are working now. The road map can be revised as follows:

$$\text{g NaCl} \xrightarrow{?} \text{mol Na}$$
$$\text{g NaCl} \rightarrow \text{mol NaCl} \rightarrow \text{mol Na}$$

I know of no relationship that will allow calculation of moles of Na from grams of NaCl. But I *do* know a relationship that allows calculation of *moles of NaCl from grams of NaCl* and another relationship that allows me to go from *moles of NaCl to moles of Na.* What are the two relationships?

$$\frac{1 \text{ mol NaCl}}{58.5 \text{ g NaCl}} \quad \text{and} \quad \frac{2 \text{ mol Na}}{2 \text{ mol NaCl}}$$

Note that the first relationship requires information that is not in the problem. This common situation may cause difficulty at first. You must recall that you can add the atomic masses of elements in a compound to find the molecular mass, and you must recall that the molecular mass in grams is the mass of 1 mol. This relationship between mass and moles is frequently used in chemical calculations.

Now that you have the unit-factors, show them on the road map and use them to answer the question posed in the beginning.

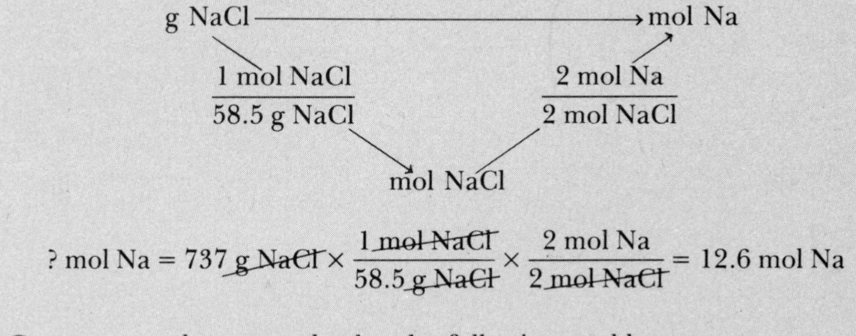

$$? \text{ mol Na} = 737 \cancel{\text{g NaCl}} \times \frac{1 \cancel{\text{mol NaCl}}}{58.5 \cancel{\text{g NaCl}}} \times \frac{2 \text{ mol Na}}{2 \cancel{\text{mol NaCl}}} = 12.6 \text{ mol Na}$$

Construct road maps and solve the following problems.

9-5. How many grams of NaCl can be made from 2.5 mol of Cl_2?

9-6. How many grams of sodium will react with 1.15 mol of Cl_2?

9-7. How many grams of sodium are needed to produce 52.0 g of sodium chloride?

9-8. How many grams of chlorine will react with 20.4 g of sodium metal?

9-9. When 20.4 g of sodium metal are mixed with 0.05 g of chlorine gas, are 52.0 g of NaCl produced? Explain your answer.

9-10. In view of the answer to Problem 9-9, what assumption have we been making in solving the problems in this section?

Later we will deal with problems in which we do not assume that there is enough of both reactants to allow completion of the reaction. Right now, let's discuss some simple combination reactions and practice using the skills that you have learned.

9.2 BURNING MAGNESIUM

Burning a piece of magnesium ribbon in air is almost as interesting as burning sodium in a bottle of chlorine (see Figure 9.3). The magnesium can be lit in a burner. The flame is bright enough to burn the retina of your eye if you look at it very long, so don't!

Figure 9.3 *Magnesium ribbon burns to give a brilliant light.* (*Photo by Tom Greenbowe and Keith Herron.*)

It is not so easy to predict the product of this reaction. Air isn't pure. It contains oxygen and nitrogen as well as small amounts of water, carbon dioxide, sulfur dioxide, nitrogen dioxide, and other "pollutants."

9-11. What is the *only* way to know the product of this (or any other) reaction?

The material produced when magnesium ribbon burns in air has been analyzed many times. The major product is an oxide of magnesium.

9-12. Write an equation for the reaction that produces this product.

Most combustion in air involves only a reaction with oxygen. At the high temperature of this reaction, however, nitrogen also reacts with magnesium.

9-13. Assuming that the nitrogen has a combining number of 3, write the equation for the reaction between magnesium and nitrogen gas.
9-14. How many moles of nitrogen gas would react with 25 g of magnesium ribbon?

As we have seen, predictions must be made with caution. When several substances are present, several reactions are likely to be possible. We are on safer ground to say, "I *think* the reaction will be . . ." rather than "I *know* the reaction will be . . ." Only a careful examination of the products can tell us what actually happens.

There are cases in which accurate prediction of products is impossible even when only two elements are present. We have already mentioned the reaction between iron and oxygen (Figure 9.1).

9-15. Write an equation to describe the reaction that you think would occur between iron and oxygen.

You probably wrote FeO or Fe_2O_3 as the product, depending on whether you assumed that iron has a combining number of 2 or 3. Under some reaction conditions, the product formed when iron combines with oxygen has an empirical formula of Fe_3O_4.

9-16. What is the combining number of iron in Fe_3O_4?

It probably seems strange to you to find that the combining number of iron in this compound is not a whole number. Nevertheless, this is the empirical formula for iron oxide in such substances as the natural ore, magnetite.

It has been suggested that the fractional combining number results from half of the iron atoms assuming a combining number of 2 while the other half assume the combining number of 3. This might be represented by the formula $FeO \cdot Fe_2O_3$.

Fractional combining numbers such as that found in Fe_3O_4 present no real problem in writing equations or doing chemical calculations.

9-17. Write an equation for the production of Fe_3O_4 from iron and oxygen gas.
9-18. How many grams of Fe_3O_4 would be produced from 1.5 kg of iron?
9-19. How many moles of O_2 would be needed to produce 3.6 kg of Fe_3O_4?

Many substances that can assume more than one combining number combine with other elements to form a mixture of reaction products. Which product predominates depends on the conditions under which the reaction takes place.

When any carbon-containing fuel is burned, either CO or CO_2 may be formed. When burning takes place with a large amount of oxygen, CO_2 is the more likely product. When the amount of oxygen is limited or the temperature of combustion is high, CO is favored.

Charcoal fires burn at a high temperature, and some CO is always produced. This is why you should *never* cook with charcoal in an enclosed area. The CO produced can kill you.

9-20. Charcoal is mainly carbon. Write an equation to represent a possible reaction between carbon and oxygen gas.

9-21. Write an equation to represent the other possible reaction.

9-22. How many moles of oxygen gas are needed to burn 1 kg of charcoal (assume that it is pure carbon) when CO_2 is the product? How many moles are needed when CO is the product?

9.3 STOICHIOMETRY

The study of chemical calculations such as those you have been doing is called **stoichiometry.** The word is formed from a Greek word, *stoicheion,* which means "element," and the *metry* suffix, which means "to measure." In stoichiometry we measure (or calculate) the amounts of elements or compounds involved in a chemical change.

The calculations that you have done so far have been relatively simple. More complex problems involve nothing more, but they may require several steps before you find the answer you want. Furthermore, you have already seen that a problem may not give all the information that you need. In order to solve the problem, you must realize what necessary information is missing and either recall that information from memory or look it up somewhere.

Few people have trouble with the calculations in chemical problems. The difficulty is in recognizing what you need to know in order to do the calculation and then recalling the relationships that will get you from what is given to what you want to find. At times the task seems impossible, but patient analysis and a little trial-and-error usually disclose some relationship that will yield the answer.

As problems become more complex, road maps like those described in Example 9.2 can be particularly helpful. Keep in mind that there is seldom only one way to solve a problem. It is not a good idea to memorize mathematical formulas or patterns for each kind of chemical problem. The more efficient strategy is to thoroughly understand basic relationships such as the relationship between molecular mass and number of moles, the relationships described by a chemical equation, the relationship defined as density, and the relationships between various units of measurement. When you are given a problem, read it carefully to identify the relationship asked for. Then try to construct a logical road map connecting the information given to the information requested. Keep in mind that information such as

atomic mass and molecular mass is generally not stated in a problem—even though it may be needed. You have to recall such information or look it up. Having a periodic table or table of atomic masses at hand is a must when you are doing stoichiometric calculations.

In this chapter you can learn more if you study with a partner. Work through the problems with someone else and think aloud. Your partner should listen carefully and point out errors that you make. A good procedure for this kind of study is described in Appendix F.

9.4 PREDICTING DECOMPOSITION PRODUCTS

In Chapter 8 you learned that some compounds decompose into their elements when they are heated or when an electric current is passed through the melted compound. **Decomposition reactions** do not always produce elements, however. Predicting the products of such reactions is difficult and requires a great deal of factual information about the behavior of matter. However, knowing a few generalizations will enable you to make reasonable predictions about several decomposition reactions.

1. Experience with a number of compounds containing either CO_3^{2-} or HCO_3^- ions has shown that they decompose to form CO_2 gas. The other product is usually an oxide of the metal in the compound.
2. Water is a very stable compound. Compounds that contain both hydrogen and oxygen frequently decompose to produce water. The HCO_3^- ion contains both hydrogen and oxygen. Both water and CO_2 are formed when compounds containing this ion decompose.
3. Oxygen gas can be produced from the decomposition of several compounds, but it is difficult to predict the formula for the other product. In some cases, another element is the product. However, it is more common to find that only a part of the oxygen in the compound is removed and that a new compound containing oxygen is the other product.

Note that all three of these generalizations involve compounds that contain oxygen. How can we know what product will form? "With difficulty!" is a cavalier answer, but true. No one is good at such predictions until she or he has learned a large number of chemical facts. However, you can be fairly confident that heating a carbonate will produce CO_2 rather than O_2 gas. Heating a bicarbonate will almost certainly produce both CO_2 and H_2O, but not O_2. Other compounds containing both hydrogen and oxygen are *likely* to produce water as a product, but other possibilities exist. Compounds that contain oxygen but are not carbonates or bicarbonates frequently decompose to produce O_2, but the formula of the other product often depends on temperature and other reaction conditions.

To help you make predictions for the following decomposition reactions, one of the products is indicated. Write balanced equations for the reactions.

9-23. $H_2O_2 \rightarrow H_2O +$
9-24. $HgO \rightarrow Hg +$
9-25. $KClO_3 \rightarrow KCl +$
9-26. $KNO_3 \rightarrow KNO_2 +$
9-27. $CaCO_3 \rightarrow CaO +$
9-28. $NaHCO_3 \rightarrow Na_2CO_3 +$
9-29. $C_{12}H_{22}O_{11} \rightarrow C +$

9.5 STOICHIOMETRY INVOLVING DECOMPOSITION REACTIONS

A number of practical questions can be asked about decomposition reactions. Work through the following questions with a partner. If you have difficulty, check the solutions given at the end of the chapter.

9-30. How many grams of oxygen gas can be formed from 1.0 mol of mercury(II) oxide?
9-31. How many grams of mercury form when mercury(II) oxide decomposes to form 1.5 mol of oxygen gas?
9-32. How many grams of water are produced when exactly 100 g of baking soda (sodium bicarbonate) decompose?
9-33. How much limestone (calcium carbonate) is required to produce 10 kg of lime (calcium oxide)?

9-34. Both the decomposition of $KClO_3$ and the decomposition of H_2O_2 are used to produce small quantities of oxygen gas for laboratory use. Reagent-grade $KClO_3$ sells for about \$8.62 per kilogram, and H_2O_2 sells for about \$25.67 per kilogram. Which is the least expensive source of oxygen gas?
9-35. If the density of CO_2 is 0.0019 g/cm³, what volume of CO_2 can be produced by heating 1.0 kg of limestone?

9.6 REPLACEMENT REACTIONS

Replacement reactions occur between an element and a compound in solution. Usually the element is a metal. When the metal is placed in a solution of some other metallic compound, a reaction may occur in which the metal added to the solution replaces the electropositive element in the compound to form a new compound in solution. The metal that was originally in the combined form then precipitates as a free element. The following equation describes such a reaction.

$$Cu(s) + 2AgNO_3(aq) \rightarrow 2Ag(s) + Cu(NO_3)_2(aq)$$

Sometimes the metal produced by the reaction forms a smooth coating on the surface of the piece of metal added to the solution, but more often it forms a soft, fluffy solid that falls to the bottom of the beaker. This fluffy solid may not look like a metal at all, because it is so finely divided. It is seldom shiny like metals. It is often a dull black or muddy brown. However,

Figure 9.4
A sheet of copper coated with silver metal. (Photo by Dudley Herron.)

Figure 9.5
A copper wire coated with silver metal. (Photo by Tom Greenbowe and Keith Herron.)

when it is collected and beaten with a hammer or melted, it looks like other samples of the same metal. Figure 9.4 shows silver formed as a smooth coating, and Figure 9.5 shows it as a fluffy solid.

Write balanced equations for the following replacement reactions.

9-36. $Zn(s) + CuSO_4(aq) \rightarrow \quad + $
9-37. $Al(s) + Pb(NO_3)_2(aq) \rightarrow \quad + $
9-38. $Na(s) + H_2O(\ell) \rightarrow NaOH(aq) + $

In the last reaction, hydrogen was the electropositive element in the compound, and when sodium replaces the hydrogen, a gas rather than a solid is produced. Otherwise the reaction is quite similar to the others shown.

All of the metals in these reactions have a single combining number, so you could predict the formula of the new compound produced in the reaction. When metals such as tin, copper, and iron are reactants in replacement reactions, you may not be able to predict the formula of the new compound. These metals may have more than one combining number, so more than one formula is possible. Experienced chemists know which combining number is most common for these elements and can usually predict the formula of the product by taking the conditions of the reaction into account. As you gain experience and learn more chemical facts, the number of predictions that you can make correctly will increase. For the time being, see whether you can answer the following questions on the basis of the equations that you just completed. These questions are like those you did before, but they require information not given in the problem.

9-39. When 10.0 g of sodium metal are thrown into a liter of water, how many grams of water remain when the reaction is over?

9-40. How many moles of aluminum nitrate can be produced when 15 g of aluminum react with lead(II) nitrate?

9-41. A chunk of aluminum with a volume of 10.0 cm^3 reacts with a solution of lead(II) nitrate to produce 310.5 g of lead metal. What is the density of the aluminum metal?

9-42. The density of hydrogen gas is about 0.082g/L at normal temperature and pressure. What volume of H_2 is produced when 46.0 g of sodium metal react with water?

9-43. How many grams of zinc sulfate are formed when 50.0 g of zinc metal are added to a solution containing 108 g of copper(II) sulfate?

9.7 LIMITING REAGENTS

The last question raises an issue that you have not faced before. You missed it, if you worked it as follows:

$$? \text{ g } ZnSO_4 = 50.0 \cancel{\text{ g Zn}} \times \frac{1 \cancel{\text{ mol Zn}}}{65.4 \cancel{\text{ g Zn}}} \times \frac{1 \cancel{\text{ mol } ZnSO_4}}{1 \cancel{\text{ mol Zn}}} \times \frac{161.5 \text{ g } ZnSO_4}{1 \cancel{\text{ mol } ZnSO_4}} = 123 \text{ g } ZnSO_4$$

What is wrong with this solution? It is worked in exactly the same way that you have been working problems. You will see what is wrong when you answer the following question.

9-44. How many grams of copper(II) sulfate are needed to react with 50.0 g of zinc?

You should have found that 122 g of $CuSO_4$ are needed to react with 50.0 g of zinc. If you reread Question 9-43, you will see that 50.0 g of Zn are added to *108 g of copper(II) sulfate.* That isn't enough $CuSO_4$ to react with all of the zinc metal. The reaction will stop with some zinc metal left over.

In the situation described in Question 9-43, the $CuSO_4$ would be described as a **limiting reagent.** The amount of $ZnSO_4$ produced in the reaction is limited by the amount of $CuSO_4$ that is available to react.

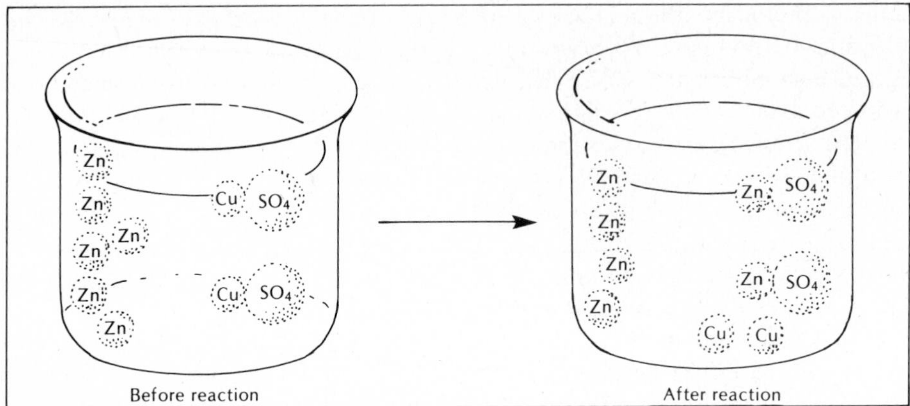

Figure 9.6
The extent of reaction is limited by the amount of each reactant present. In this illustration, $CuSO_4$ is the limiting reagent. When the reaction stops, some unreacted zinc will remain. ($ZnSO_4$ and $CuSO_4$ exist as separate ions in solution. They are represented as molecules to make it easier to visualize the effect of the limiting reagent.)

Either reactant can be the limiting reagent. If 0.5 g of Zn is added to a solution containing 108 g of $CuSO_4$, the zinc is the limiting reagent. The reaction proceeds only until the 0.5 g of Zn is gone. With no more zinc available to react, no more $ZnSO_4$ can form.

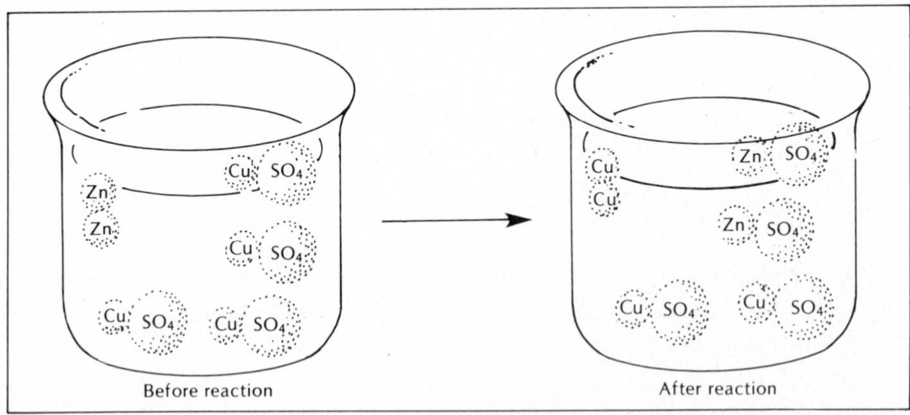

Figure 9.7
In this illustration, zinc is the limiting reagent. When the reaction stops, some unreacted $CuSO_4$ remains. ($ZnSO_4$ and $CuSO_4$ exist as separate ions in solution. They are represented as molecules to make it easier to visualize the effect of the limiting reagent.)

Problems involving a limiting reagent are solved in the same way as other stoichiometric problems, but they generally require twice as much work. Work through the following example with me and you will see why.

Example 9.3

How much SbI_3 can be formed when 1 g of I_2 and 2 g of Sb are allowed to react?

Since we do not know whether antimony or iodine will be totally consumed in this reaction, we have to ask two questions: How much SbI_3 can be formed from 1.0 g of I_2, and how much SbI_3 can be formed from 2.0 g of Sb?

What is the first thing that we need in order to answer either of these questions?

The balanced equation for the reaction.

Write the equation and balance it.

$$2Sb + 3I_2 \rightarrow 2SbI_3$$

From the information in the equation, calculate the following:

$$? \text{ mol } SbI_3 = 1 \text{ g } I_2$$
$$? \text{ mol } SbI_3 = 2 \text{ g Sb}$$

$$? \text{ mol } SbI_3 = 1.0 \cancel{\text{g } I_2} \times \frac{1 \cancel{\text{mol } I_2}}{254 \cancel{\text{g } I_2}} \times \frac{2 \text{ mol } SbI_3}{3 \cancel{\text{mol } I_2}} = 2.6 \times 10^{-3} \text{ mol } SbI_3$$

$$? \text{ mol } SbI_3 = 2.0 \cancel{\text{g Sb}} \times \frac{1 \cancel{\text{mol Sb}}}{122 \cancel{\text{g Sb}}} \times \frac{2 \text{ mol } SbI_3}{2 \cancel{\text{mol Sb}}} = 1.6 \times 10^{-2} \text{ mol } SbI_3$$

Which reactant will produce the *smaller* amount of product?

2.6×10^{-3} (0.0026) is smaller than 1.6×10^{-2} (0.016). The 1.0 g of I_2 will produce the smaller amount of product.

Which reactant is the limiting reagent?

The I_2 is the limiting reagent. It determines the amount of product formed.

How much SbI_3 is formed when 1.0 g of I_2 and 2.0 g of Sb are allowed to react?

$$2.6 \times 10^{-3} \text{ mol } SbI_3\text{, or } 1.3 \text{ g } SbI_3$$

The questions that follow require that you determine which reactant is the limiting reagent.

9-45. When 4 mol of copper are heated in 3 mol of oxygen gas, will there be enough oxygen for all of the copper to form CuO?

9-46. When 20.0 g of NaOH and 20.0 g of HCl are mixed, will there be any NaOH left when the reaction to form water and sodium chloride stops?

9-47. When CuO and CH_4 are heated together, Cu, water, and CO_2 are the products. Which reactant will be left when 6 mol of CuO are heated with 2 mol of CH_4?

9.8 ACTIVITY SERIES

You have been cautioned that equations can be written to describe many reactions that do not occur. The following is such an equation.

$$\text{(No reaction)} \qquad 2Ag(s) + CuSO_4(aq) \not\rightarrow Cu(s) + Ag_2SO_4(aq)$$

Metals do not react to replace all other metals from compounds in solution. If a metal *does* replace another metal in a compound, the metal is said to be more **active** than the metal it replaces. Since zinc reacts with copper(II) sulfate as described in Question 9-36, we say that zinc is more active than copper.

When silver metal is placed in a solution of copper(II) sulfate, there is no reaction. We say that silver is less active than copper. We could just as easily say that copper is more active than silver.

Note that we get the same information about the relative activity of silver and copper metal from either of the equations that we have written:

$$Cu(s) + 2AgNO_3(aq) \rightarrow 2Ag(s) + Cu(NO_3)_2(aq)$$
$$2Ag(s) + CuSO_4(aq) \rightarrow \text{No reaction}$$

Because a reaction *does* occur when copper is added to silver nitrate, we know that *copper is more active than silver.* Because a reaction *does not* occur when silver is added to copper(II) sulfate, we know that *silver is less active than copper.* The conclusions are equivalent.

By observing whether a reaction occurs when metals are placed in solutions of metallic compounds, we can place metals in a list from most active to least active. Such a list is called an **activity series.** Once such a list has been compiled, it can be used to predict other replacement reactions. You can construct an activity series for several metals by observing replacement reactions in the laboratory. More complete lists are found in many reference books.

Replacement reactions are not confined to reactions involving metals. Chlorine gas reacts with potassium iodide by replacing the iodine in the compound.

9-48. Write an equation for this reaction.

Bromine, chlorine, and iodine can be placed in an activity series similar to the one just described for metals. From the equation that you just wrote, you can conclude that chlorine is more active than iodine.

9-49. $Br_2(aq) + KCl(aq) \rightarrow$ No reaction. What information does this equation convey about the relative activity of bromine and chlorine?

9-50. If you wanted to list bromine, chlorine, and iodine from the most active to the least active, what additional information would you need?

Following are several equations for replacement reactions involving the halogens.*

1. $Cl_2(aq) + 2NaBr(aq) \rightarrow Br_2(aq) + 2NaCl(aq)$
2. $Br_2(aq) + 2KI(aq) \rightarrow I_2(aq) + 2KBr(aq)$
3. $I_2(aq) + LiCl(aq) \rightarrow$ No reaction
4. $I_2(aq) + KBr(aq) \rightarrow$ No reaction

9-51. Which of these four equations provides information that you could use with the equations given in Questions 9-48 and 9-49 to list bromine, chlorine, and iodine in an activity series?

9-52. Which of these four equations provides the same information about activity that is provided in Questions 9-48 and 9-49?

9.9 SUMMARY

This chapter has been an exercise in recall and thinking. Little new material has been introduced, but you have been asked to apply a number of ideas that you have already learned. Here are some of the ideas that you have used.

1. You were asked to predict the products of several chemical reactions. You were told what kind of reaction would take place, but you had to write the formula for the compound formed. To do that, you had to remember the combining numbers of several elements and polyatomic ions (or look them up), and you had to recall how to use combining numbers to write a formula.

2. Once you predicted the products of a reaction and wrote the correct formula, you had to balance the chemical equation. This meant that you had to recall the rules you learned in Chapter 8.

3. Knowing just what an equation says proved to be important. An equation tells you the moles of each reactant and product involved in the reaction.

* **Halogen** is a name given to the elements in Group VIIA of the periodic table. Binary compounds of these elements are called **halides.**

This is *not* the actual amount of reactants and products that might be present in a given reaction. It represents the proportions involved when all reactants are consumed and only products remain after the reaction stops.

4. From the proportions of reactants and products described in the equation, you were able to calculate the amount of product that could be formed from a given amount of reactant or the amount of a reactant needed to produce a given amount of product. If the amounts of reactants and products were expressed in moles, the calculation was not very difficult. But this wasn't always the case.

5. At times it was necessary for you to express moles of a compound in terms of grams or to express grams of a compound in terms of moles. You were able to do this by recalling that the molecular mass of a compound in grams is the molar mass.

6. You sometimes needed to recall other information. For example, density was sometimes needed to convert from the mass of a substance to the volume.

7. The biggest task seemed to be keeping the logic straight. A "road map" showing a logical path from what is given in a problem to what the problem asks for was introduced as one way to keep track of the logic. You were encouraged to work complex problems with a partner, alternating so that you worked one problem while your partner checked your reasoning, and then switching so that your partner worked a problem while you checked your partner's reasoning.

8. You were given descriptions of experimental situations in which reactants were mixed, and you were asked to decide which reactant was left after the reaction stopped. You found that you usually had to perform two stoichiometric calculations in order to decide which reactant was the limiting reagent that determined the amount of product that could be made.

9. Finally, you studied some replacement reactions that do take place and some that do not take place. You were told how the results of these reactions (or nonreactions) could be used to determine the activity of metals and halogens. You were encouraged to use your logical powers to construct some simple activity series from experimental results.

The work you did in this chapter required you to put together many ideas. This kind of synthesis is not easy. You should practice these skills by working several of the supplementary problems.

Questions and Problems

9-53. Ammonia is made commercially from the direct combination of hydrogen gas with nitrogen gas. The equation is

$$3\,H_2(g) + N_2(g) \rightarrow 2\,NH_3(g)$$

a. How many moles of ammonia can be produced from 2.5 mol of hydrogen gas?

b. How many moles of nitrogen gas would react with 2.5 mol of hydrogen gas?
c. How many moles of ammonia would be produced from 56 g of nitrogen gas?

9-54. The equation for the complete combustion of gasoline (assuming that it is pure isooctane) can be represented by the equation

$$2C_8H_{18} + 25O_2 \rightarrow 16CO_2 + 18H_2O$$

a. How many moles of CO_2 are produced when 4 mol of O_2 react?
b. How many moles of water are produced when 1 gal of gasoline burns? (A gallon of gasoline weighs about 2,500 g.)
c. How many grams of O_2 are needed to completely burn 1 gal of gasoline?
d. Air is about 21 percent O_2 by volume. What volume of air is needed to burn 1 gal of gasoline if the density of air is 0.0012 g/cm^3?

9-55. When gasoline burns in your automobile, combustion is incomplete. The exact nature of the reaction varies a lot, depending on a number of conditions. One possible reaction is described by the equation

$$2C_8H_{18} + 7O_2 \rightarrow 3C_3H_8 + 6CO + 6H_2O + CO_2$$

a. How many moles of carbon monoxide are produced when 3 mol of oxygen gas react?
b. How many grams of CO are produced from 1 gal of gasoline? (See Question 9-54b.)
c. If gasoline burns as indicated in this equation, will 6 mol of O_2 be enough to burn 200 g of the gasoline?

9-56. Welders produce the high temperatures needed to weld by burning acetylene gas in pure oxygen. The equation for the reaction is

$$2C_2H_2(g) + 5O_2(g) \rightarrow 4CO_2(g) + 2H_2O(g)$$

a. How many kilograms of acetylene would burn before a tank of oxygen containing 500 mol of O_2 is consumed?
b. How many grams of carbon dioxide would be produced when the 500 mol of oxygen gas are consumed?
c. What volume of water would be produced when 1 kg of oxygen reacts? Assume that the water is liquid with a density of 1 g/cm^3.
d. A tank of acetylene and a tank of oxygen gas both contain 50 kg of gas. Which tank will empty first?

9-57. Acetylene can be produced by dropping water on calcium carbide—a reaction that I often saw in my grandfather's mining lamp, which burned acetylene to produce light. The equation for this reaction is

$$CaC_2(s) + 2H_2O(\ell) \rightarrow C_2H_2(g) + Ca(OH)_2(s)$$

a. How many milliliters of water must react to produce 1 mol of acetylene gas?
b. Calcium carbide is a solid with a density of 2.22 g/cm^3. How many grams

of acetylene gas can be made from a cube of calcium carbide measuring 3 cm on a side?

c. How many grams of water must react to produce 1.2 mol of calcium hydroxide?

9-58. The "hypo" used in developing pictures is sodium thiosulfate with the formula $Na_2S_2O_3$. It reacts with silver bromide in the film to give two soluble products that are washed away. The equation for the reaction is

$$2\,Na_2S_2O_3(aq) + AgBr(s) \rightarrow Na_3Ag(S_2O_3)_2(aq) + NaBr(aq)$$

a. How many grams of silver bromide can be dissolved by a 1-kg package of "hypo"?

b. How many moles of sodium bromide are formed when 1 kg of hypo reacts?

c. What mass of sodium thiosulfate is needed to produce 1.2 mol of $Na_3Ag(S_2O_3)_2$?

More Challenging Problems

9-59. Write a balanced equation for the reaction that occurs when the following reagents are mixed, and determine the amount of the excess reagent that remains when the reaction is complete.

a. Three grams of hydrogen gas are mixed with 30 g of oxygen gas and lit with an electric spark.

b. A 5.5-g piece of zinc metal is placed in a solution containing 0.1 mol of $CuCl_2$.

c. A 2.5-g chunk of potassium is placed in 25 mL of water.

d. 0.15 mol of K is heated and placed in 4.5 g of $Cl_2(g)$.

e. 1.62 g of magnesium are placed in 150 mL of solution containing 0.018 mol HCl.

f. 1.2 g of KI are dissolved in water containing 0.02 mol of $Cl_2(aq)$.

9-60. When heated, each of the following substances decomposes to produce a gas. Write a balanced equation for each reaction, and calculate the moles of gas produced from the amount of solid given.

a. 125 g of $PbCO_3$

b. 125 g of $Pb(NO_3)_2$

c. 125 g of $KHCO_3$

d. 125 g of $NaClO_3$

e. 125 g of $MgCO_3$

9-61. On the basis of the information provided by the following equations, list Cu, Fe, and H from most active to least active.

$$Fe(s) + 2\,HCl(aq) \rightarrow H_2(g) + FeCl_2$$
$$Fe(s) + CuSO_4(aq) \rightarrow Cu(s) + FeSO_4(aq)$$
$$Cu(s) + HCl(aq) \rightarrow \text{No reaction}$$

9-62. Which of the three equations given in Question 9-61 do you *not* need in order to answer the question asked?

9-63. If you are given the task of arranging lead, iron, copper, aluminum, and zinc in an activity series, what is the smallest number of reactions that you would need to try before you would have the necessary information?

9-64. If you have solutions of lead(II) nitrate, iron(II) nitrate, copper(II) nitrate, aluminum nitrate, and zinc nitrate, what is the least number of metals that you would need to work with to determine an activity series for the five metals?

Answers to Questions in Chapter 9

9-1. 0.065 mol Cl_2

9-2. 10.4 mol Na

9-3. 2.52 mol NaCl

9-4. 12.6 mol Na

9-5. mol Cl_2 ⟶ g NaCl (via mol NaCl)

$$? \text{ g NaCl} = 2.5 \text{ mol } Cl_2 \times \frac{2 \text{ mol NaCl}}{1 \text{ mol } Cl_2} \times \frac{58.5 \text{ g NaCl}}{1 \text{ mol NaCl}} = 290 \text{ g NaCl}$$

9-6. mol Cl_2 ⟶ g Na (via mol Na)

$$? \text{ g Na} = 1.15 \text{ mol } Cl_2 \times \frac{2 \text{ mol Na}}{1 \text{ mol } Cl_2} \times \frac{23.0 \text{ g Na}}{1 \text{ mol Na}} = 52.9 \text{ g Na}$$

9-7. g NaCl ⟶ g Na (via mol NaCl ⟶ mol Na)

$$? \text{ g Na} = 52.0 \text{ g NaCl} \times \frac{1 \text{ mol NaCl}}{58.5 \text{ g NaCl}} \times \frac{2 \text{ mol Na}}{2 \text{ mol NaCl}} \times \frac{23.0 \text{ g Na}}{1 \text{ mol Na}} = 20.4 \text{ g Na}$$

9-8. g Na ⟶ g Cl_2 (via mol Na ⟶ mol Cl_2)

$$? \text{ g } Cl_2 = 20.4 \text{ g Na} \times \frac{1 \text{ mol Na}}{23.0 \text{ g Na}} \times \frac{1 \text{ mol } Cl_2}{2 \text{ mol Na}} \times \frac{71.0 \text{ g } Cl_2}{1 \text{ mol } Cl_2} = 31.5 \text{ g } Cl_2$$

9-9. No. Question 9-8 shows that 31.5 g of Cl_2 are needed to react with 20.4 g of Na, and Question 9-7 shows that 20.4 g of Na must react to produce 52 g of NaCl. If there is only 0.05 g of Cl_2 around, all of the 20.4 g of Na can't react. If 20.4 g of Na can't react, 52 g of NaCl can't be formed.

9-10. We have assumed that there is always enough of each reactant available for all reactants mentioned to be consumed. In other words, we have assumed that reactants are available in the proportions decribed by the equation. This isn't always the case.

9-11. Laboratory analysis of the products

9-12. $2\,Mg(s) + O_2(g) \rightarrow 2\,MgO(s)$

9-13. $3\,Mg(s) + N_2(g) \rightarrow Mg_3N_2(s)$

9-14. $$?\text{mol } N_2 = 25 \text{ g Mg} \times \frac{1 \text{ mol Mg}}{24.31 \text{ g Mg}} \times \frac{1 \text{ mol } N_2}{3 \text{ mol Mg}} = 0.34 \text{ mol } N_2$$

9-15. Possible answers are

$$2\,Fe(s) + O_2(g) \rightarrow 2\,FeO(s)$$
$$4\,Fe(s) + 3\,O_2(g) \rightarrow 2\,Fe_2O_3(s)$$

9-16. 8/3, or about 2.7

9-17. $3\,Fe(s) + 2\,O_2(g) \rightarrow Fe_3O_4(s)$

9-18. $$? \text{ g } Fe_3O_4 = 1.5 \text{ kg Fe} \times \frac{1 \text{ mol Fe}}{55.85 \text{ g Fe}} \times \frac{1 \text{ mol } Fe_3O_4}{3 \text{ mol Fe}} \times \frac{231.5 \text{ g } Fe_3O_4}{1 \text{ mol } Fe_3O_4} \times \frac{1000 \text{ g}}{1 \text{ kg}} = 2{,}100 \text{ g } Fe_3O_4$$

9-19. $$? \text{ mol } O_2 = 3.6 \text{ kg } Fe_3O_4 \times \frac{1 \text{ mol } Fe_3O_4}{231.5 \text{ g } Fe_3O_4} \times \frac{2 \text{ mol } O_2}{1 \text{ mol } Fe_3O_4} \times \frac{1000 \text{ g}}{1 \text{ kg}} = 31 \text{ mol } O_2$$

9-20. $2\,C(s) + O_2(g) \rightarrow 2\,CO(g)$

9-21. $C(s) + O_2(g) \rightarrow CO_2(g)$

9-22. $$? \text{ mol } O_2 = 1 \text{ kg C} \times \frac{1 \text{ mol C}}{12 \text{ g C}} \times \frac{1 \text{ mol } O_2}{1 \text{ mol C}} \times \frac{1000 \text{ g}}{1 \text{ kg}}$$

= 80 mol O_2 (one significant figure)
(40 mol O_2 when CO is the product)

9-23. $2\,H_2O_2 \rightarrow 2\,H_2O + O_2$

9-24. $2\,HgO \rightarrow 2\,Hg + O_2$

9-25. $2KClO_3 \rightarrow 2KCl + 3O_2$

9-26. $2KNO_3 \rightarrow 2KNO_2 + O_2$

9-27. $CaCO_3 \rightarrow CaO + CO_2$

9-28. $2NaHCO_3 \rightarrow Na_2CO_3 + H_2O + CO_2$

9-29. $C_{12}H_{22}O_{11} \rightarrow 12C + 11H_2O$

9-30. $? \text{ g } O_2 = 1.0 \cancel{\text{mol HgO}} \times \frac{1 \cancel{\text{mol } O_2}}{2 \cancel{\text{mol HgO}}} \times \frac{32.00 \text{ g } O_2}{1 \cancel{\text{mol } O_2}} = 16 \text{ g } O_2$

9-31. $? \text{ g Hg} = 1.5 \cancel{\text{mol } O_2} \times \frac{2 \cancel{\text{mol Hg}}}{1 \cancel{\text{mol } O_2}} \times \frac{200.59 \text{ g Hg}}{1 \cancel{\text{mol Hg}}} = 6.0 \times 10^2 \text{ g Hg}$

9-32. $? \text{ g } H_2O = 100 \cancel{\text{g NaHCO}_3} \times \frac{1 \cancel{\text{mol NaHCO}_3}}{84.00 \cancel{\text{g NaHCO}_3}} \times \frac{1 \cancel{\text{mol } H_2O}}{2 \cancel{\text{mol NaHCO}_3}} \times \frac{18.00 \text{ g } H_2O}{1 \cancel{\text{mol } H_2O}} = 10.71 \text{ g } H_2O$

9-33. $? \text{ g } CaCO_3 = 10 \cancel{\text{kg CaO}} \times \frac{1 \cancel{\text{mol CaO}}}{56.08 \cancel{\text{g CaO}}} \times \frac{1 \cancel{\text{mol CaCO}_3}}{1 \cancel{\text{mol CaO}}} \times \frac{100.09 \text{ g } CaCO_3}{1 \cancel{\text{mol CaCO}_3}} \times \frac{1000 \cancel{\text{g}}}{1 \cancel{\text{kg}}}$

$= 20{,}000 \text{ g } CaCO_3$ (17,848 ignoring significant figures)

9-34. This problem is rather involved, and it would be a good idea to outline how we will work it. Several approaches are possible. I did it like this.

1. Calculate the moles of O_2 that I can get from 1 kg of $KClO_3$.
2. Use the price of 1 kg of $KClO_3$ to calculate the cost of 1 mol of O_2 produced.
3. Calculate the moles of O_2 that I can get from 1 kg of H_2O_2.
4. Use the price of 1 kg of H_2O_2 to calculate the cost of 1 mol of O_2.
5. Compare the prices of 1 mol of O_2 from the two reactions.

SOLUTION:

1. $? \text{ mol } O_2 = 1.000 \cancel{\text{kg KClO}_3} \times \frac{1 \cancel{\text{mol KClO}_3}}{122.55 \cancel{\text{g KClO}_3}} \times \frac{3 \text{ mol } O_2}{2 \cancel{\text{mol KClO}_3}} \times \frac{1000 \cancel{\text{g}}}{1 \cancel{\text{kg}}} = 12.24 \text{ mol } O_2$

2. $? \text{ \$/mol } O_2 = \frac{\$8.62}{1 \cancel{\text{kg KClO}_3}} \times \frac{1 \cancel{\text{kg KClO}_3}}{12.24 \text{ mol } O_2} = \$0.7042/\text{mol } O_2$

3. $? \text{ mol } O_2 = 1.000 \cancel{\text{kg } H_2O_2} \times \frac{1 \cancel{\text{mol } H_2O_2}}{34.01 \cancel{\text{g } H_2O_2}} \times \frac{1 \text{ mol } O_2}{2 \cancel{\text{mol } H_2O_2}} \times \frac{1000 \cancel{\text{g}}}{1 \cancel{\text{kg}}} = 14.70 \text{ mol } O_2$

4. $? \text{ \$/mol } O_2 = \frac{\$25.67}{1 \cancel{\text{kg } H_2O_2}} \times \frac{1 \cancel{\text{kg } H_2O_2}}{14.70 \text{ mol } O_2} = \$1.746/\text{mol } O_2$

5. A mole of O_2 can be made from reagent-grade $KClO_3$ for about \$0.70. It costs over twice as much (\$1.75) to make a mole of O_2 from H_2O_2. In spite of this, the reaction with H_2O_2 is *definitely* the one that you should use if you want to make O_2 in the laboratory. Why? Because it is *much* safer!

9-35. We can use the density to find the volume of CO_2 if we know the mass of CO_2. We can use the equation and molecular weights to find the mass as follows:

$? \text{ g } CO_2 = 1.0 \cancel{\text{kg CaCO}_3} \times \frac{1 \cancel{\text{mol CaCO}_3}}{100.09 \cancel{\text{g CaCO}_3}} \times \frac{1 \cancel{\text{mol } CO_2}}{1 \cancel{\text{mol CaCO}_3}} \times \frac{44.01 \text{ g } CO_2}{1 \cancel{\text{mol } CO_2}} \times \frac{1000 \cancel{\text{g}}}{1 \cancel{\text{kg}}}$

$= 439.7 \text{ g } CO_2$ (440 to correct number of significant figures)

$? \text{ cm}^3 \text{ } CO_2 = 439.7 \cancel{\text{g } CO_2} \times \frac{1 \text{ cm}^3 \text{ } CO_2}{0.0019 \cancel{\text{g } CO_2}} = 2.3 \times 10^5 \text{ cm}^3 \text{ } CO_2$

9-36. $Zn(s) + CuSO_4(aq) \rightarrow Cu(s) + ZnSO_4(aq)$

9-37. $2Al(s) + 3Pb(NO_3)_2(aq) \rightarrow 3Pb(s) + 2Al(NO_3)_3(aq)$

9-38. $2Na(s) + 2H_2O(\ell) \rightarrow 2NaOH(aq) + H_2(g)$

9-39. First we need to know how much of the water reacted with the sodium. The rest will be left over.

$? \text{ g } H_2O = 10.0 \text{ g Na} \times \frac{1 \cancel{\text{mol Na}}}{23.0 \cancel{\text{g Na}}} \times \frac{1 \cancel{\text{mol } H_2O}}{1 \cancel{\text{mol Na}}} \times \frac{18.00 \text{ g } H_2O}{1 \cancel{\text{mol } H_2O}}$

$= 7.83 \text{ g } H_2O$ reacted

We started with 1 liter of water. Since the density of water is 1.00 g/cm³ and 1 liter is 1,000 cm³, we also started with 1,000 g of water. We are left with 1,000 g − 7.83 g = 992 g of water. (In leaving the answer to the nearest gram, I assume that we knew the volume of the water to the nearest cm³.)

9-40. $? \text{ mol } Al(NO_3)_3 = 15 \cancel{\text{g Al}} \times \frac{1 \cancel{\text{mol Al}}}{27.0 \cancel{\text{g Al}}} \times \frac{2 \text{ mol } Al(NO_3)_3}{2 \cancel{\text{mol Al}}} = 0.56 \text{ mol } Al(NO_3)_2$

9-41. Since the volume of the aluminum is known, we can calculate the density if we know the mass of the aluminum.

$? \text{ g Al} = 310.5 \cancel{\text{g Pb}} \times \frac{1 \cancel{\text{mol Pb}}}{207 \cancel{\text{g Pb}}} \times \frac{2 \cancel{\text{mol Al}}}{3 \cancel{\text{mol Pb}}}$

$$\times \frac{27.0 \text{ g Al}}{1 \cancel{\text{mol Al}}} = 27.0 \text{ g Al}$$

Clearly a 10-cm^3 piece of aluminum contains 1 mol of the metal and weighs 27.0 g. The density would be

$$\frac{27.0 \text{ g Al}}{10.0 \text{ cm}^3 \text{ Al}} = 2.7 \text{ g/cm}^3$$

9-42. I first find the mass of hydrogen gas produced.

$$? \text{ g } H_2 = 46.0 \cancel{\text{g Na}} \times \frac{1 \cancel{\text{mol Na}}}{23.0 \cancel{\text{g Na}}} \times \frac{1 \cancel{\text{mol } H_2}}{2 \cancel{\text{mol Na}}}$$

$$\times \frac{2.02 \text{ g } H_2}{1 \cancel{\text{mol } H_2}} = 2.02 \text{ g } H_2$$

One mol H_2 would be produced. It weighs 2.02 g. The volume would be

$$? \text{ liters } H_2 = 2.02 \cancel{\text{g } H_2} \times \frac{1 \text{ liter}}{0.082 \cancel{\text{g } H_2}} = 25 \text{ liters } H_2$$

(Note: The volume of a gas changes a great deal with a change in temperature or pressure. A mole of hydrogen gas would have a different volume at other temperatures and pressures. This relationship will be discussed in Chapter 12.)

9-43. $$? \text{ g } ZnSO_4 = 108 \cancel{\text{g } CuSO_4} \times \frac{1 \cancel{\text{mol } CuSO_4}}{159.6 \cancel{\text{g } CuSO_4}}$$

$$\times \frac{1 \cancel{\text{mol } ZnSO_4}}{1 \cancel{\text{mol } CuSO_4}} \times \frac{161.5 \text{ g } ZnSO_4}{1 \cancel{\text{mol } ZnSO_4}} = 109 \text{ g } ZnSO_4$$

9-44. $$? \text{ g } CuSO_4 = 50.0 \cancel{\text{g Zn}} \times \frac{1 \cancel{\text{mol Zn}}}{65.4 \cancel{\text{g Zn}}}$$

$$\times \frac{1 \cancel{\text{mol } CuSO_4}}{1 \cancel{\text{mol Zn}}} \times \frac{159.6 \text{ g } CuSO_4}{1 \cancel{\text{mol } CuSO_4}} = 122 \text{ g } CuSO_4$$

9-45. You first need a balanced equation for the reaction. It is

$$2Cu + O_2 \rightarrow 2CuO$$

The equation shows that 2 mol of Cu react for every 1 mol of O_2. Then 4 mol of Cu would require just 2 mol of O_2. The problem states that there are 3 mol of O_2 available, so there will be enough oxygen. The problem can be solved more systematically by either of the calculations shown.

$$? \text{ mol } O_2 = 4 \cancel{\text{mol Cu}} \times \frac{1 \text{ mol } O_2}{2 \cancel{\text{mol Cu}}} = 2 \text{ mol } O_2$$

This says that only 2 mol of O_2 are needed to react with 4 mol of Cu. The other solution is

$$? \text{ mol Cu} = 3 \cancel{\text{mol } O_2} \times \frac{2 \text{ mol Cu}}{1 \cancel{\text{mol } O_2}} = 6 \text{ mol Cu}$$

This indicates that 3 mol of O_2 are enough to react with 6 mol of Cu. Since you only had 4 mol of Cu, there is plenty of O_2 around.

9-46. The equation for the reaction is

$$NaOH + HCl \rightarrow H_2O + NaCl$$

The equation indicates that 1 mol of NaOH is required for each mole of HCl. One approach to the problem is first to calculate the moles of each reactant:

$$? \text{ mol NaOH} = 20.0 \cancel{\text{g NaOH}} \times \frac{1 \text{ mol NaOH}}{39.997 \cancel{\text{g NaOH}}}$$

$$= 0.500 \text{ mol NaOH}$$

$$? \text{ mol HCl} = 20.0 \cancel{\text{g HCl}} \times \frac{1 \text{ mol HCl}}{36.46 \cancel{\text{g HCl}}}$$

$$= 0.549 \text{ mol HCl}$$

We see that we have more moles of HCl than moles of NaOH. Therefore all of the NaOH will react, leaving 0.049 mol of HCl unreacted.

9-47. The equation is

$$4CuO + CH_4 \rightarrow 4Cu + 2H_2O + CO_2$$

$$? \text{ mol } CH_4 = 6 \cancel{\text{mol CuO}}$$

$$\times \frac{1 \text{ mol } CH_4}{4 \cancel{\text{mol CuO}}} = 1.5 \text{ mol } CH_4$$

This tells us that we need 1.5 mol of CH_4 to react with 6 mol of CuO. We were told that we have 2 mol of CH_4, which is more than enough. But 0.5 mol of CH_4 will remain unreacted when the CuO has been converted to copper.

9-48. $Cl_2 + 2KI \rightarrow I_2 + 2KCl$

9-49. The equation indicates that bromine is less active than chlorine. (You could also say that it indicates that chlorine is more active than bromine.)

9-50. From Questions 9-48 and 9-49, we know that chlorine is more active than bromine and iodine, but we do not know whether bromine is more or less active than iodine.

9-51. Either equation 2 or equation 4 provides the necessary information.

9-52. Equation 3 gives the same information as Question 9-48, and equation 1 gives the same information as Question 9-49.

READING Not everything dissolves in water. Alumina, the ore from which aluminum is obtained, doesn't dissolve in water or most other solvents. The key to the production of inexpensive aluminum was finding a suitable solvent for alumina. You may already know the story of how Charles Martin Hall, the son of an Ohio minister, developed the process for the electrolysis of alumina dissolved in cryolite. However, you probably don't know about the important role played by his sister Julia. This short article tells the tale.

JULIA B. HALL AND ALUMINUM

Martha M. Trescott

The establishment of the Pittsburgh Reduction Company, the forerunner of ALCOA (the Aluminum Company of America), has long been attributed to the inventive efforts of Charles Martin Hall. Hall developed a process for the electrolytic reduction of alumina in a molten bath of cryolite to yield aluminum metal, thus solving the decades-old search for a way to produce the metal cheaply. The process was patented in 1889, three years after he first produced the metal by this method (*1*). What is not always known is that much of the invention and innovation which culminated in the formation of Pittsburgh Reduction in 1888 owed a great deal to the activities of Julia Brainerd Hall, an older sister of Charles.

Like Charles, Julia had graduated from Oberlin, with the date of her matriculation, 1881 and his, 1885. Like Charles, Julia took chemistry at Oberlin, also in her junior year and also under Professor Franklin F. Jewett. In fact, overall at Oberlin, Julia completed slightly more credits in science than Charles, even though she was officially enrolled in what was called the "Literary Course," an outgrowth of the earlier "Ladies' Course." While Charles received a degree for his four year course of study, Julia received a diploma (*2*). In childhood, Julia was Charles's closest companion and confidante, a role which she continued to play in the adulthood of this shy, seclusive man. Neither Charles nor Julia ever married. Yet, upon graduation, Julia assumed the responsibilities of raising her two younger sisters and of directing other household tasks, as her mother was ailing and in 1885 died.

Charles used the Oberlin home as his base of operations, primarily before 1887, and set up his laboratory in the woodshed, next to the kitchen, Julia's headquarters. Being in the home, Julia was often present in Charles's lab, helping out with the experiments and consulting with him on scientific and technical matters. She served also as a scientifically astute, well-educated, and competent eye witness for Charles's experiments and for the letters and papers he wrote concerning the aluminum invention. She also acted in this capacity for certain of his other inventive ideas (*3*). These and other such activities resulted in the issuance of a family of patents to Charles Hall on April 2, 1889 for the production of electrolytic aluminum. This inventive activity since 1882 was non-random and planned by Julia and Charles, with an eye toward potential markets from the beginning of serious work on the invention. Just as today, with R & D teams, invention in the nineteenth century was also a team effort.

Besides assisting in the lab and offering her technical advice and expertise, Julia faithfully and minutely recorded the steps in the invention processs—that is, the results of a given day's work, along with technical details and the date and evidence which could substantiate the date. She also contacted family and friends who might have leads on financial backing or finances themselves; she advised Charles on difficult questions such as to whom he might wish to sell his process. She acted as an information center, relaying information to and from Charles about people potentially interested in the process. On occasion, she advised him what he should write to the *Scientific American* for their "free" advice on patents, a service apparently offered to inventors of the times by the journal. She dated his papers and letters pertaining to the invention and made sure that the copies he

Source: *Journal of Chemical Education,* Vol. 55, No. 1 (January 1977), pp. 24–25.

sent were clearly legible, and she acted as a censor of important names, dates and other facts in the letters Charles wrote her, in case the letters fell into the wrong hands, including relatives. She also advised Charles not to leave her letters lying around for the same reasons.

Finally, she composed a "History of C. M. Hall's Aluminum Invention," a six-page document, as her 1887 eye-witness account for the important Hall-Hèroult patent interference case. And her testimony served to clinch Hall's victory in this case. She was the necessary, and perhaps would have been sufficient, witness to win this case for Hall. Since Hall had filed his patent application on July 9, 1886, and Hèroult in May, Hall had to establish without a doubt that he had reduced his invention to practice before May and in particular before April 23, 1886, the date on which the Frenchman Hèroult's French patent had been granted. The witnesses were Charles himself, Julia, Charles's father and two of Charles's professors, one of whom was Jewett. Only Julia could positively identify as an eye witness the production of aluminum, verifying the identity of the product, produced on February 23, 1886. Julia was the first human being to whom Hall had fully disclosed his ideas about the invention on Ferbuary 10, 1886, a date which Julia fully documented in court (*4*).

Yet Julia received little real recognition for her informational, managerial, and entrepreneurial contributions here. Charles went on to become extremely wealthy, with $170,000 annual income from ALCOA stock alone at the time of his death in 1914. Julia at that time until her death in 1925 averaged about $8,000 income from her stock (*5*). And not even Charles credited her in his 1911 acceptance speech for the cherished Perkin Medal, when he told of family involvements in the invention and innovations leading to the establishment of Pittsburgh Reduction Company (*6*). Not even Julia's 1887 account was referenced as such by Hall's biographer Junius D. Edwards nor by Charles C. Carr, ALCOA company historian, although both authors evidently used it and both knew about it (*7*). It is apparent that the team of Charles and Julia Hall brought to the market cheap aluminum.

Notes

1. (a) Junius D. Edwards, "The Immortal Woodshed," New York, Dodd, Mead and Company, **1955** and also by Edwards, "A Captain in Industry," **1957;** (b) Charles C. Carr, "Alcoa, An American Enterprise," **1952.** (c) This essay is abstracted from a longer forthcoming paper with complete bibliography. (d) For an abstract of the oral presentation before the Society for the History of Technology, October 19, 1975, cf. Deborah Shapley, "History of American Technology—A Fresh Bicentennial Look," *Science,* **190,** 763 (November 21, 1975).
2. [The author is] especially indebted to Mrs. Gertrude Jacobs of the Oberlin College Alumni Office for providing [her] with alumni necrology on the Hall family and also with transcripts of college courses for both Charles and Julia.
3. This is not only borne out in the letters but also in the documents used in court in the interference case. ALCOA furnished us with both Julia Hall's written statement of 1887, "History of C. M. Hall's Aluminum Invention" and with the actual testimonies before the patent examiner.
4. Cf. Charles M. Hall vs. P. L. V. Hèroult, In Interference in the United States Patent Office, October 24, 1887, pp. 5–8 for Julia's testimony and pp. 5–6 for commentary on February 10, 1886.
5. Edwards, Ref. 1a, p. 226. From ALCOA [the author] obtained ALCOA stock ledgers for Julia B. Hall, 1909–25, and her sisters Edie and Louie, 1909–1919 and 1909–25, respectively, along with a table of "Dividends Paid on Common Stock," 1895–1943, all supplied by Anna G. Lydon.
6. "The Perkin Medal, Remarks in Acknowledgment by Mr. Hall," *Industrial and Engineering Chemistry,* **III,** 146–148 (1911).
7. Edwards had Julia Hall's account typed, along with the letters from Charles to Julia and others, and transmitted to company historian Charles Carr in 1936, according to the file from ALCOA's archives.

10

Reactions in Solution

Mixtures are far more common than pure substances. Homogeneous mixtures containing particles the size of molecules are called solutions, and they are common scenes for chemical reactions. That is one of the reasons why chemists are interested in solutions.

In this chapter you will learn to describe solutions, make solutions, and predict what happens when solutions are mixed.

OBJECTIVES

Solutions

There are certain terms that we use to describe solutions and the strength of a solution. You must know these terms and be able to use them in solving chemical problems. You can if you can do the following:

1. Define solute, solvent, solution, solubility, saturated, concentration, and molarity, and use these terms correctly.
2. Given the mass of solute or the moles of solute dissolved in a given volume of solution, calculate the molarity of the solution. (See Problem 10-41.)
3. Calculate the mass of solute required to prepare a specified volume of solution of a specified concentration. (See Problem 10-42.)
4. Given the volume and molarity of a solution, find the number of moles of solute or the mass of solute in a given volume of solution. (See Problem 10-43.)
5. Explain how a mixture, a solution, and a pure substance differ.

Metathesis Reactions

Many reactions that take place in solution do so because an insoluble product is formed. The products of these reactions can be predicted from information about solubilities of compounds. You should be able to do the following:

6. Given the formulas of two compounds in solution, predict whether a metathesis reaction will occur when the solutions are mixed. (See Problem 10-44.)
7. Predict the formulas of the products of a metathesis reaction. (See Problem 10-44.)
8. State general rules for solubility, and, when given the name or formula of a compound, use these rules to predict the solubility of the compound. (See Problem 10-44.)

Net Ionic Equations

Metathesis reactions involve ions in solution. Reactions between ions can be described by equations that ignore other ions that are present but do not react. These equations are net ionic equations. You will understand them when you are able to do the following:

9. Write the net ionic equation for any given metathesis reaction. (See Problem 10-45.)
10. Write the net ionic equation for a replacement reaction.
11. State the limitations of net ionic equations.

Words You Should Know

solution	solvent	saturated
suspension	concentration	dissociate
opaque	dilute	precipitate
clear	concentrated	metathesis
colorless	molarity	ionic equation
solute	solubility	net ionic equation

10.1 SOLUTIONS AND SUSPENSIONS

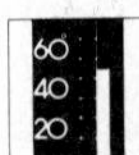

Make a cup of instant coffee or brew some tea. If you prefer, mix a batch of Kool-Aid® or instant lemonade. You have made a **solution.** If your taste buds yearn for hot chocolate or frozen orange juice, the mixture you prepare will *not* be a solution. It will be a **suspension** instead.

"What," you may ask, "is the difference?" Not much, really. One drink tastes as good as the other, provided that you like the respective flavors. You may notice that the chocolate and orange juice tend to settle out when the

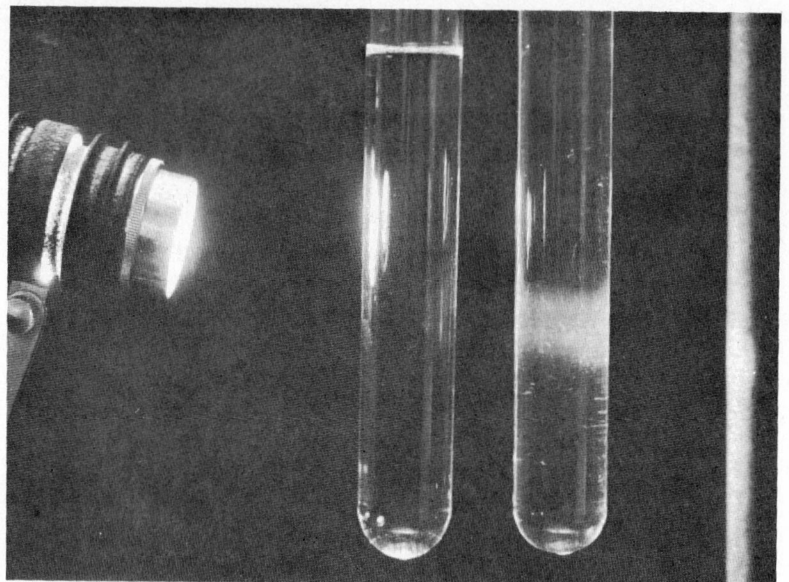

Figure 10.1
When a beam of light passes through a solution, the light is not scattered and is not visible when viewed from the side. When light passes through a suspension, the light is scattered by the suspended particles and is visible from the side. (*From* Introduction to Chemical Principles, *Second Edition, by Edward I. Peters. Copyright © 1978 by W. B. Saunders Company. Reprinted by permission of Holt, Rinehart and Winston.*)

drink is left standing for several minutes. By contrast, instant coffee and filtered tea do not settle, even when they are left standing for several hours. Another difference is that light passes through a solution of coffee or tea without being scattered, whereas hot chocolate and orange juice look cloudy, as shown in Figure 10.1. These are the major differences that we can observe at the macroscopic level.

At the microscopic level, the difference between solutions and suspensions is in the size of the particles, as shown in Figure 10.2. In solutions, the particles are so small that light passes through the solution without being deflected. That's why solutions look clear.* In a suspension, the particles are large enough to deflect light, so the suspension looks cloudy or **opaque.**

The size of the particles explains why suspensions tend to settle out while solutions do not. If the dissolved particles are much larger than the molecules of the liquid they are mixed with, the force of gravity pulls them toward the bottom. They move slowly, because the water molecules keep bumping into them and bouncing them back toward the top, but eventually many of the larger particles make it to the bottom of the container.

In solutions, the dissolved particles are closer to the size of the water molecules, and gravity has little more effect on the particles than it has on the molecules of water. Even when it isn't stirred, the mixture remains uniform throughout almost indefinitely.

*"Clear" and "colorless" are sometimes confused. **Clear** means "free from cloudiness." A clear solution may be blue, yellow, green, or any other color, but it cannot be cloudy. **Colorless** means "without color." Pure water is colorless. Tea is not colorless. This paper and other white, opaque materials can also be described as colorless, but they are usually described as white.

Solutions exist in which the dissolved particles are very small clumps of molecules, but normally the dissolved particles are either single molecules or ions (Figure 10.3).

TYPES OF SOLUTIONS

The solutions we have mentioned are made by dissolving some solid or liquid in water. Water solutions are by far the most common, but solids can dissolve in other liquids. In Chapter 8 an experiment was described in which solid iodine was dissolved in toluene, a liquid similar to gasoline or fuel oil. It is possible to make solutions of gases in liquids (carbonated water), solids in solids (a gold ring), gases in gases (any gas mixture), liquids in liquids (alcohol in water), and other combinations. We will limit our discussion in this book to liquid solutions that you will encounter most often.

Suspensions—especially colloids that don't settle out but do deflect light—can also be interesting, but we must save some interesting material for courses that follow this one. We will not discuss suspensions here. Rather, we will

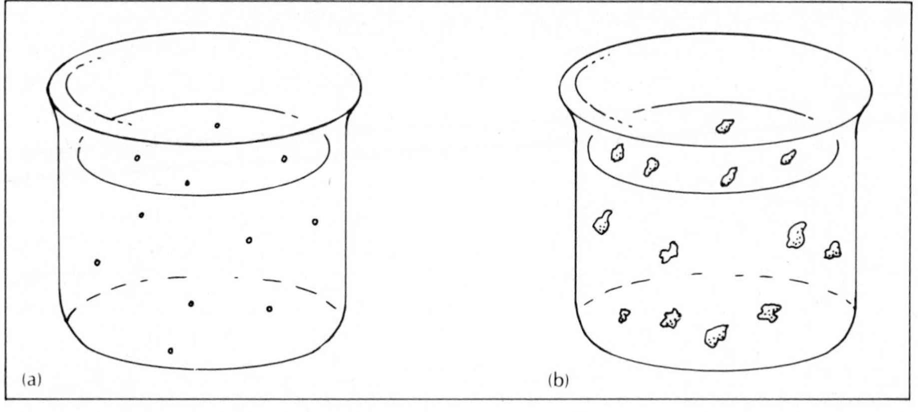

Figure 10.2 *The major difference between a solution and a suspension is the size of the particles in the mixture. (a) A solution has very small particles that stay evenly distributed in the mixture. (b) A suspension has larger particles that settle to the bottom after standing for a long time.*

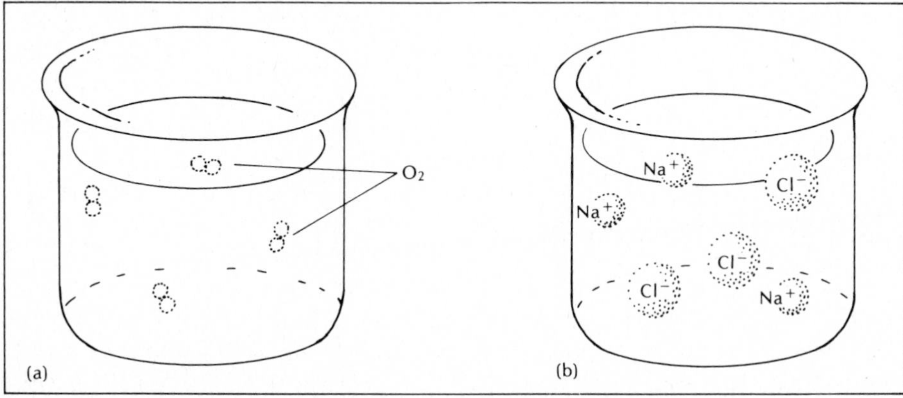

Figure 10.3 *In most solutions, the dissolved particles are separate molecules or ions. (a) A solution of oxygen gas dissolved in water. The O_2 molecules are greatly magnified. (b) A solution of salt (NaCl) dissolved in water. The Na^+ and Cl^- ions are greatly magnified.*

confine our discussion to why chemists are interested in solutions and how they describe solutions quantitatively.

10.2 WHY SOLUTIONS ARE IMPORTANT

By now you have seen several chemical reactions and have described them with chemical equations. You have observed that these equations describe what *might* happen when atoms and molecules come together. We emphasize "might," because atoms and molecules must come into contact under favorable conditions for a chemical change to occur. When you turn on a gas stove that has no pilot light, and you do not place a match to the flow of gas, no reaction occurs. When you just mix antimony and iodine, you don't get antimony triiodide. When you place sodium metal in contact with chlorine gas without warming the sodium, there isn't much visible change.

Two conditions must be satisfied in order for a chemical change to occur. First, the atoms that combine to form a new substance must come in contact with one another. Second, they must come in contact with enough energy to make them "stick together." The things that we do to speed up chemical reactions either bring atoms into better contact, increase the energy associated with this contact, or reduce the energy needed to bring about the change.

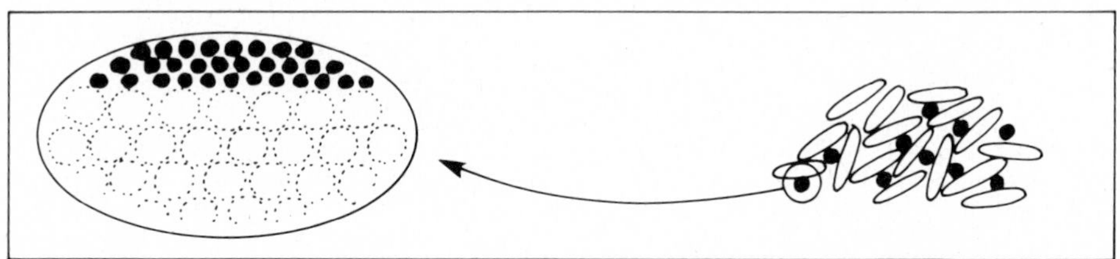

Figure 10.4
A mixture of solids allows relatively little contact between the different kinds of atoms present. In this mixture of powdered antimony and iodine crystals, only a few atoms of iodine and antimony on the surface of the solids come into contact.

Reactions between two solids are usually slow, because the different kinds of atoms are in contact only at the surface, as shown in Figure 10.4. When a reaction occurs at the surface, the product usually coats the surface, preventing other atoms from coming into contact, and the reaction stops. Reactions between gases or well-mixed liquids are generally much faster, because the different atoms and molecules that undergo change are able to come into contact and the product usually does not coat the surface, as illustrated in Figure 10.5. Obviously, getting substances into a form in which atoms can bump into other atoms can be important in causing a reaction to occur. One of the most common ways to do this is to make a solution.

Liquid solutions are easy to handle and provide good contact between molecules of different kinds. Many chemical reactions are carried out in

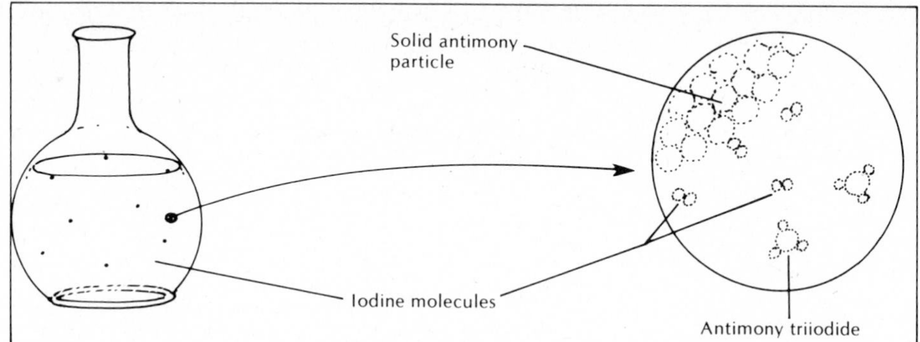

Figure 10.5
Iodine molecules are dissolved in the toluene. They can move throughout the solution. Any antimony triiodide formed dissolves in the toluene, leaving the surface of antimony free to react with more iodine.

solution. In your body, complex solutions make possible the thousands of chemical reactions that maintain life. Both blood and lymph are solutions that carry molecules from one part of the body to another to take part in chemical change. (There are also particles suspended in blood and lymph.)

In most solutions, it is convenient to speak of one part of the mixture as the **solute** and the other part as the **solvent.** *The solvent is normally the part of the mixture that is present in the larger amount.* It is usually of little interest when we are discussing chemical reactions that occur in solution. *The solute is the part of the mixture that is normally present in the lesser amount.* The solute is usually the "active ingredient" in chemical reactions and the component of the solution we are interested in. Consequently, when we discuss solutions, we usually focus on the solute rather than the solvent.

10.3 CHARACTERISTICS OF SOLUTIONS

1. Solutions are homogeneous mixtures. They appear to be uniform throughout.

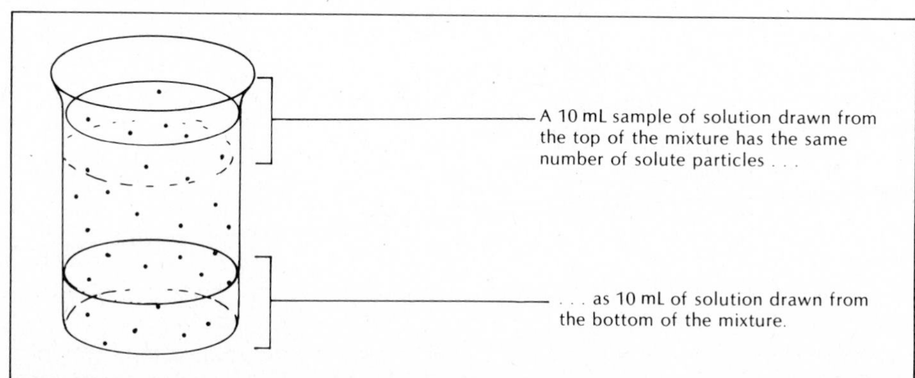

Figure 10.6
Solutions are homogeneous.

2. Solute particles are small. They are usually individual molecules or ions moving about at random.

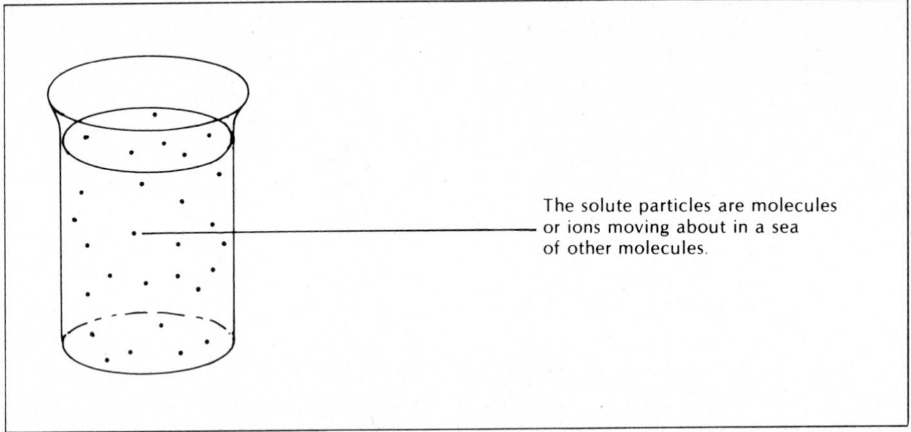

Figure 10.7
Solute particles are small.

3. A solution is a mixture rather than a compound. The proportions of solute and solvent can vary over a wide range of values.

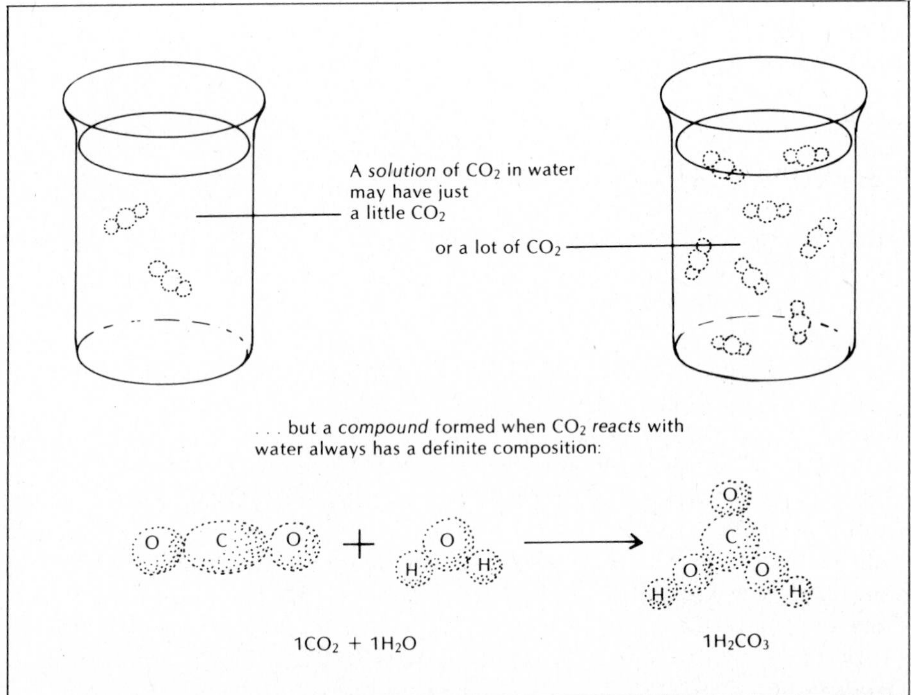

Figure 10.8
A solution is a mixture rather than a compound.

4. The particles in solution neither settle out nor deflect a beam of light.

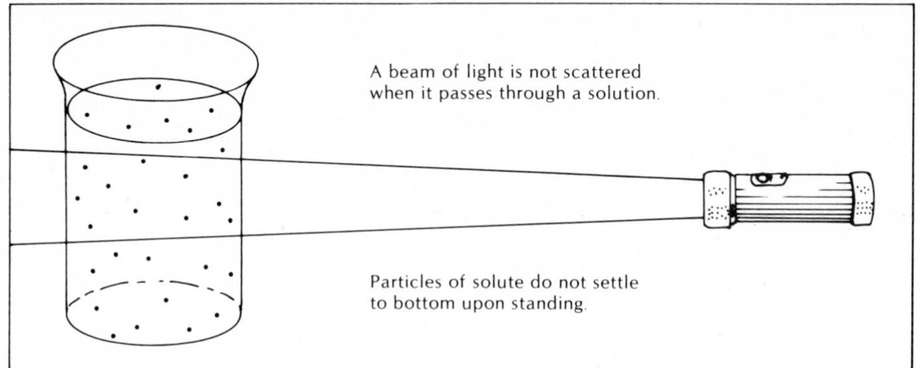

Figure 10.9
A solution stays mixed indefinitely.

5. Although many kinds of solutions are possible, the most common solutions are liquids. The most common solvent is water.

10.4 CONCENTRATIONS OF SOLUTIONS

Some people like their coffee "strong" and some like it "weak." Instant coffee drinkers can vary this property readily by changing the amount of powder dissolved in the water. The chemist would describe the strong coffee as **concentrated** and the weak coffee as **dilute.**

"Concentrated" and "dilute" are relative terms like "tall" and "short." They are fine for describing which of two cups of coffee is stronger, but they aren't satisfactory when you want to describe *how much* coffee powder is in a cup. **Concentration** *describes how much solute is in a given amount of solution.*

There are many ways to describe concentration. Directions on a jar of instant coffee suggest you "use one rounded teaspoon per cup." Although not very explicit, these instructions mean you are to use one teaspoon of the solute (the powdered coffee) to make one cup of solution (the coffee you drink).

An examination of several bottles around the house will turn up another method of expressing concentration. Vinegar is "5 percent acetic acid," household bleach is "5.25 percent sodium hypochlorite," and rubbing alcohol is "70 percent isopropyl alcohol *by volume.*"

"By volume" was emphasized in the excerpt from the last label to call attention to the fact that percentage may be expressed in terms of mass or volume. The 5 percent acetic acid in vinegar means 100 *grams* of vinegar contain 5 *grams* of acetic acid. The 70 percent isopropyl alcohol in rubbing alcohol indicates 100 *liters* of rubbing alcohol contain 70 *liters* of isopropyl alcohol.

Although percentages are used to express concentrations of commercial products, they are seldom used in chemistry and are not discussed here. The only concentration term used in this book is molarity. Definitions of other concentration terms are given in Appendix D.

MOLARITY

Hydrochloric acid is a solution of hydrogen chloride gas dissolved in water. When this solution is poured on zinc metal, the reaction described by the following equation occurs.

$$2HCl(aq) + Zn(s) \rightarrow ZnCl_2(aq) + H_2(g)$$

The equation indicates that 2 mol of HCl are needed for every mole of Zn that reacts. It is the HCl in the solution that reacts. When a solution of HCl is used, it is important to know how many *moles of solute* are present.

The normal way to measure liquids is by volume. It is much easier to measure 100 mL of liquid in a graduated cylinder than to weigh 100 g of liquid on a balance. For convenience, chemists would like to express the concentration of a solution in such a way that they can pour out a certain *volume of solution* and know the number of *moles of solute* measured out. The concentration term that provides this relationship is called molarity.

One mole of HCl weighs 36 g. When we dissolve 36 g of HCl gas in enough water to form 1 liter of solution, we have *one mole of HCl per liter of solution.* This English phrase can be translated into a mathematical expression like this:

$$\frac{1 \text{ mol HCl}}{1 \text{ L solution}}$$

Such a solution would be described as a one molar solution. This is abbreviated "1 *M*" and it means that there is 1 mol of HCl for every liter of solution.

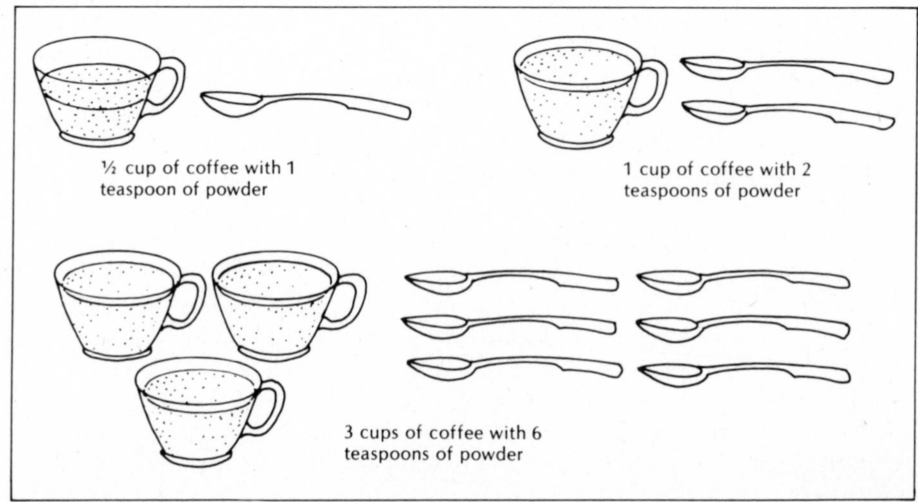

Figure 10.10 *Solutions of the same concentration can be prepared in many different ways.*

You need not make 1 liter of solution to have a concentration of 1 M any more than you need to make one cup of coffee to have strong coffee. You could make $\frac{1}{2}$ cup of coffee with a teaspoon of powder, 1 cup of coffee with 2 teaspoons of powder, or 3 cups of coffee with 6 teaspoons of powder and still have coffee of the same concentration. In each case there would be 2 teaspoons of powder per cup of coffee.

In a similar fashion, whether you make 500 mL of solution with 18 g of HCl, 1 liter of solution with 36 g of HCl, or 2 liters of solution with 72 g of HCl, the amount of HCl per liter of solution is the same. The concentration is 1 M (one molar).

Work through the following example and you will see how to calculate the molarity of a solution.

Example 10.1

What is the concentration (molarity) of a solution when 250 mL of the solution contain 1.5 mol of HCl?

We want to find the molarity of the solution, so the first thing we need is the definition of molarity. Write the mathematical statement that describes molarity.

- -

$$M = \frac{\text{moles solute}}{\text{1 liter solution}}$$

The problem states that we have 1.5 mol of the solute, HCl. However, that is *1.5 mol of HCl per 250 mL of solution.* Write the mathematical equivalent of the italicized statement.

- -

$$\frac{\text{1.5 mol HCl}}{\text{250 mL solution}}$$

This statement is similar to the definition of molarity. The only difference is that the denominator is expressed in terms of milliliters instead of liters. Rewrite the statement in such a way that the denominator is expressed in liters.

- -

$$\frac{\text{1.5 mol HCl}}{\text{250 mL solution}} \times \frac{\text{1,000 mL}}{\text{1 L}} = \frac{\text{1.5 mol HCl}}{\text{0.25 L solution}}$$

Do the arithmetic indicated in the expression to find the moles of HCl per liter of solution.

- -

$$\frac{\text{1.5 mol HCl}}{\text{0.25 L solution}} = \text{6 mol HCl/L solution}$$

What is the molarity of this HCl solution?

- -

$6\ M$

By definition, **molarity** is the moles of solute per liter of solution. You just found that this solution has the equivalent of 6 mol of HCl per liter of solution. Its concentration is 6 *M*.

Now see whether you can summarize what we did to find the molarity of this solution.

1. We recalled that molarity means

$$\frac{\text{moles solute}}{\text{1 liter solution}}$$

2. We used what was stated in the problem to write an expression showing moles of solute in some volume of solution.
3. We expressed the volume of solution in liters.
4. We did the indicated division to find the number of moles of solute in a liter of solution. This is the molarity.

Work the following problems to reinforce the idea.

10-1. What is the molarity of a sugar solution that contains 2.2 mol of sugar dissolved in 3 liters of solution?

10-2. A solution of salt was made by dissolving 0.5 mol of NaCl in water to make 750 mL of solution. What is the molarity of the solution?

10-3. 125 mL of tincture of iodine (a solution made by dissolving iodine in alcohol) contain 0.025 mol of I_2. What is the molarity of the solution?

10-4. A 1,000-liter tank of gasoline has 0.5 mol of lead tetraethyl (an antiknock ingredient) dissolved in it. What is the concentration of the lead tetraethyl?

The only convenient way to measure moles of a compound is to calculate the mole mass and then weigh out the desired amount on a balance. The following problems are more realistic in that they indicate the *mass* of solute dissolved in a given volume of solution. See whether you can find the molarity of the solutions described by adding one more step. If you get stuck, check the solution to the first question (the solution appears at the end of the chapter), but please don't look until you have tried to do it alone.

10-5. What is the molarity of a salt solution that contains 58.4 g of NaCl in 2 liters of solution?

10-6. What is the concentration of an iodine solution that contains 1.27 g of I_2 in 10 mL of solution? (Note that iodine exists as I_2 molecules. Then the molarity is the "moles of I_2 *molecules* per liter of solution.")

10-7. What is the concentration of household ammonia (NH_3 in water) when there are 4.25 g of ammonia in 500 mL of solution?

10-8. How much ammonia would you have to dissolve in water to form 250 mL of solution with a concentration of 4 *M*?

The last problem is a little different. If you got it right, it is a good sign that you understand the notion of molarity. If you had trouble with this problem, work through it with me in the following example.

Example 10.2

How much ammonia would you have to dissolve in water to form 250 mL of solution with a concentration of 4 *M*?

In this problem, you are given the concentration of the solution and asked to find how much ammonia is in the solution. "How much" can be expressed in several ways. Give two interpretations of "how much ammonia."

It could mean "how many *moles* of ammonia."

It could mean "how many *grams* of ammonia."

We will determine both. First, write down what we mean when we say, "The ammonia solution has a concentration of 4 *M*."

$$\frac{4 \text{ mol NH}_3}{1 \text{ L solution}}$$

Molarity means moles of solute per liter of solution. This tells us the moles of solute in 1 liter of solution. If you multiply the liters of solution that we have by this factor, you will find the moles of ammonia in that volume of solution.

How many liters of solution are described in the problem?

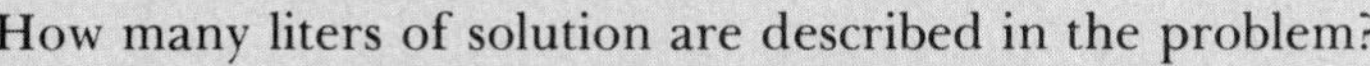

0.25 L solution (250 mL)

Now use the proper factor to do the indicated calculation.

$$? \text{ mol NH}_3 = 0.25 \text{ L solution} \times \underline{\qquad\qquad} =$$

$$? \text{ mol NH}_3 = 0.25 \cancel{\text{L solution}} \times \frac{4 \text{ mol NH}_3}{1 \cancel{\text{L solution}}} = 1 \text{ mol NH}_3$$

How many grams of ammonia would you need to form 250 mL of a 4 *M* solution?

---

17 g, the mass of 1 mol NH_3

The problem could be worked in one step:

$$? \text{ g NH}_3 = 0.25 \cancel{\text{L solution}} \times \frac{4 \cancel{\text{mol NH}_3}}{1 \cancel{\text{L solution}}} \times \frac{17 \text{ g NH}_3}{1 \cancel{\text{mol NH}_3}} = 17 \text{ g NH}_3$$

Now work the following problems.

10-9. How many moles of HCl are there in 150 mL of a 4 *M* HCl solution?

10-10. How many moles of NaCl are there in 2.5 liters of a 0.5 *M* salt solution?

10-11. How many moles of I_2 are there in 0.25 mL of a 0.25 *M* solution of iodine?

10-12. Which contains more H_2SO_4, 250 mL of a 1.4 *M* solution or 300 mL of a 1.3 *M* solution?

10-13. How many grams of H_2SO_4 are there in 250 mL of a 1.4 *M* solution?

10-14. How many grams of HCl are there in 150 mL of 4 *M* HCl?

10-15. How many grams of I_2 are there in 0.25 mL of a 0.25 *M* solution of iodine?

10-16. How many grams of NaCl will it take to make 2.5 liters of a 2 *M* salt solution?

MAKING A SOLUTION OF KNOWN CONCENTRATION

Logically, it is very simple to make a solution of known concentration. To make a 1 *M* solution of a solid compound such as NaOH, you simply weigh out exactly 1 mol of NaOH on a balance, place it in a 1-liter volumetric flask, dissolve the NaOH in water, and then add water until the total volume of solution is exactly 1 liter—the volume contained when the liquid reaches the scratch on the neck of the 1-liter volumetric flask. This process is illustrated in Figure 10.11. (Why would it be wrong to measure 1 liter of water and add it to the solid NaOH?)

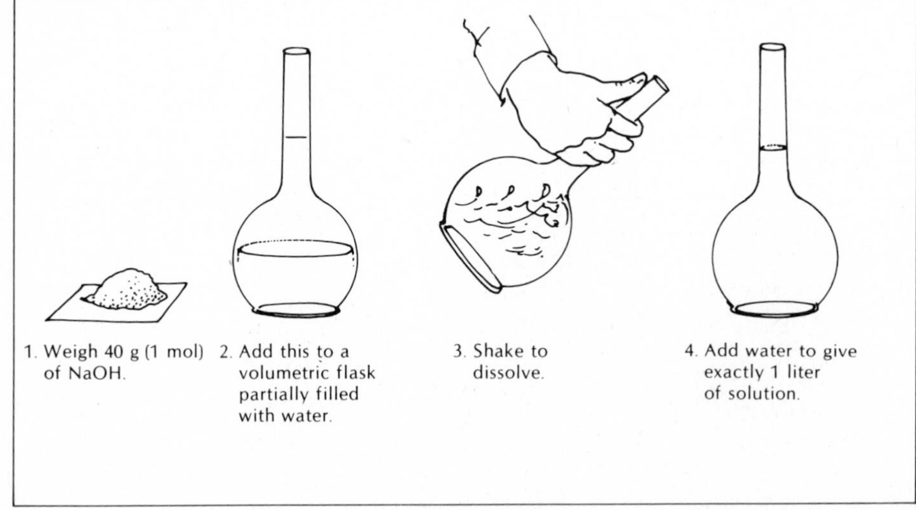

Figure 10.11 *Making a solution of known concentration.*

10.5 SOLUBILITY AND TEMPERATURE

Some things are soluble in water and others are not. This is fortunate. If everything dissolved in water, we could find nothing to put it in.

Even those things that do dissolve in water have limits of **solubility**. If you have a sweet tooth, you have probably left solid sugar in the bottom of your iced tea glass because you added more sugar than could be dissolved. A solution that has all of the dissolved solute that can possibly dissolve is called a **saturated** solution.

The amount of solute that will dissolve in a solvent depends on temperature. You can dissolve far more sugar in hot tea than in cold tea. Clothes generally get cleaner in hot water than in cold water, because the solubility of soap and dirt increases with temperature. As a matter of fact, most (but not all) substances are more soluble at high temperature than at low temperature.

A major exception to the generalization that solubility increases with temperature occurs with gases. The solubility of all gases *decreases* as temperature increases. This is why hot carbonated drinks foam more than cold ones when they are opened. It is also why lakes and streams contain less dissolved oxygen at higher temperatures, making it difficult for fish to survive in hot weather.

The solubility of a substance can be used to identify the substance, and we can often take advantage of this property to separate one substance from another.

When we described the synthesis of SbI_3 in Chapter 8, we pointed out that solubility was used to separate antimony that did not react from the antimony triiodide that was produced. Antimony is not soluble in toluene, the solvent used in the reaction. It was removed from the reaction mixture by pouring off the solution. The antimony triiodide that was produced is soluble in *hot* toluene, but it is not very soluble in *cold* toluene. When the solution was cooled, the antimony triiodide formed solid crystals and was easily separated from the solvent. Similar separations based on solubility are made routinely in industry and in the laboratory.

Solubilities for several compounds are given in Table 10.1, page 284. The table shows some interesting variations in solubility.

One of the first things that you notice in Table 10.1 is that solubility increases with temperature for most compounds. There are exceptions. For example, the solubility of SbF_3 and $Ca(C_2H_3O_2)_2$ decreases as the temperature increases.

The change in solubility as the temperature increases varies a great deal from one compound to another. The solubility of ordinary salt (NaCl) changes very little with temperature, whereas the solubility of NaOH and $AgNO_3$ increases about eightfold when the temperature is raised from 0°C to 100°C. The increase for $Ba(IO_3)_2$ is even more dramatic.

We have emphasized that formulas showing the same elements in different proportions represent different compounds with different properties. This fact is evident when we compare the solubility of $FeCl_2$ to that of $FeCl_3$ or the solubility of Hg_2Cl_2 to that of $HgCl_2$.

Formulas that look very similar can represent compounds that have very different properties. AgF and AgI look a lot alike when written on paper. (The compounds look a lot alike too!) However, Table 10.1 shows that the two compounds are drastically different in solubility.

Table 10.1. Solubility of Common Compounds

Formula	*Solubility in grams per 100 cm³ H₂O* Cold*	Hot*
$Al_2(SO_4)_3$	31.3	98.1
NH_4Cl	29.7	75.8
NH_4I	154.2	250.3
SbF_3	384.7	$5.636^{(30°)}$
$Ba(IO_3)_2$	0.008	197.
$Ca(C_2H_3O_2)_2$	37.4	29.7
$CaCl_2$	$74.5^{(20°)}$	159.
CaC_2O_4	$0.00067^{(13°)}$	$0.0014^{(95°)}$
$CaSO_4$	$0.209^{(30°)}$	0.1619
$CuCl_2$	70.6	107.9
$CuSO_4$	14.3	75.4
Cu_2S	1×10^{-14}	—
HCl	82.3	56.1
$FeCl_2$	$64.4^{(10°)}$	105.7
$Fe(OH)_2$	0.00015	—
$FeCl_3$	74.4	535.7
PbI_2	0.044	0.41
$Pb(NO_3)_2$	37.65	127.
$PbSO_4$	$0.00425^{(25°)}$	$0.0056^{(40°)}$
$MgCl_2$	$54.25^{(20°)}$	72.7
$Mg(OH)_2$	$0.0009^{(18°)}$	0.004
$MgSO_4$	26.	73.8
Hg_2Cl_2	$0.00020^{(25°)}$	$0.001^{(43°)}$
$HgCl_2$	$6.9^{(20°)}$	48.
KI	127.5	208.
KNO_3	13.3	247.
AgF	$182.^{(15.5°)}$	$205^{(108°)}$
AgI	$2.8 \times 10^{-7(25°)}$	$2.5 \times 10^{-6(60°)}$
$AgNO_3$	122.	$952.^{(190°)}$
$NaCl$	35.7	39.12
$NaOH$	42.	347.

* Unless noted, all solubilities for cold water are at 0°C and solubilities for hot water are at 100°C. Other temperatures are given in superscript following the solubility.

All data are from Weast, Robert C. (Ed.), *Handbook of Chemistry and Physics, 53rd Edition.* Cleveland, Ohio: The Chemical Rubber Company, 1972.

Table 10.1 gives solubilities for only a few compounds. The *Handbook of Chemistry and Physics* gives solubility data for thousands of compounds. A portion of the table of physical constants is shown in Figure 10.12. For most compounds the table shows i for insoluble, s for soluble, vs for very soluble, or d to indicate that the compound decomposes in water. This information is not very precise, but it can be helpful.

We might ask what "insoluble" means at this point. Does it mean that absolutely *none* of the substance dissolves? No, it doesn't. It simply means that

PHYSICAL CONSTANTS OF INORGANIC COMPOUNDS (Continued)

No.	Name	Synonyms and Formulae	Mol. wt.	Crystalline form, properties and index of refraction	Density or spec. gravity	Melting point, °C	Boiling point, °C	Solubility, in grams per 100 cc		
								Cold water	Hot water	Other solvents
	Siloxane									
s123	**Siloxane, (di-), oxide.**	$[H(O)Si]_2.O$	106.19	wh volum subst		expl *ca* 300		sl s		s, d HF; d al
s124	**Silver**	Ag	107.868	wh met, cub 0.54	10.5^{20}	961.93	2212	i	i	s HNO_3, h H_2SO_4, KCN; i alk
s125	acetate	$AgC_2H_3O_2$	166.92	wh pl	3.259^{15}	d		1.02^{20}	2.52^{80}	s dil HNO_3
s126	acetylide	Ag_2C_2	239.76	wh ppt		expl		i		s a; sl s al
s127	*ortho*arsenate	Ag_3AsO_4	462.53	dk red, cub	6.657^{25}	d		0.00085^{20}		s NH_4OH, ac a
s128	*ortho*arsenite	Ag_3AsO_3	446.53	yel, powd		d 150		0.00115^{20}	i	s ac a, NH_4OH, HNO_3; i al
s129	azide	AgN_3	149.89	wh rhomb pr, expl		252	297	i	0.01^{100}	s KCN, dil HNO_3; sl s NH_4OH
s130	benzoate	$AgC_7H_5O_2$	228.99	wh powd				0.262^{25}	s	0.017 al
s131	*tetra*borate	$Ag_2B_4O_7.2H_2O$	407.01	wh cr				sl s		s a
s132	bromate	$AgBrO_3$	235.78	col, tetr, 1.874, 1.920	5.206	d		0.196^{25}	1.33^{90}	s NH_4OH; sl s HNO_3
s133	bromide	Bromyrite: AgBr	187.78	pa yel, 2.253	6.473^{25}	432	d>1300	8.4×10^{-6}	0.00037^{100}	s KCN, $Na_2S_2O_3$, NaCl sol; sl s NH_4OH; i al
s134	carbonate	Ag_2CO_3	275.75	yel powd	6.077	d 218		0.0032^{20}	0.05^{100}	s NH_4OH, $Na_2S_2O_3$; i al
s135	chlorate	$AgClO_3$	191.32	wh, tetr	4.430^{20}_{4}	230	d 270	10^{15}	50^{80}	sl s al
s136	*per*chlorate	$AgClO_4$	207.32	wh, cr, deliq	2.806^{25}	d 486		557^{25}	s	s al; 101 tol; 5.28 bz
s137	chloride	Nat. cerargyrite. AgCl	143.32	wh, cub, 2.071	5.56	455	1550	0.000089^{10}	0.0021^{100}	s NH_4OH, $Na_2S_2O_3$, KCN
s138	chlorite	$AgClO_2$	175.32	yel cr		105 expl		0.45^{25}	2.13^{100}	
s139	chromate	Ag_2CrO_4	331.73	red, monocl	5.625			0.0014^{0}	0.008^{70}	s NH_4OH, KCN
s140	*di*chromate	$Ag_2Cr_2O_7$	431.72	red, tricl	4.770	d		0.0083^{15}	d	s a, NH_4OH, KCN
s141	citrate	$Ag_3C_6H_5O_7$	512.71	wh need		d		0.028^{18}	sl s	s a, NH_4OH, KCN, $Na_2S_2O_3$
s142	cyanate	AgOCN	149.89	col	4.00	d		sl s	s	s KCN, HNO_3, NH_4OH
s143	cyanide	AgCN	133.84	wh, hex	3.95	d 320		0.000023^{20}		s HNO_3, NH_4OH, KCN, $Na_2S_2O_3$
s144	ferricyanide	$Ag_3Fe(CN)_6$	535.56					0.000066^{20}		i a; s NH_4OH, h $(NH_4)_2CO_3$
s145	ferrocyanide	$Ag_4Fe(CN)_6.H_2O$	661.45	wh				i	i	s KCN; i a, NH_4 salts, NH_4OH
s146	fluogallate	$Ag_3[GaF_6].10H_2O$	687.47	col, orthorhomb cr, 1.493	2.90			v s		i al
s147	fluoride	AgF	126.87	yel, cub, deliq	$5.852^{15.5}$	435	*ca* 1159	$182^{15.5}$	205^{108}	sl s NH_4OH
s148	fluoride, di-	AgF_2	145.87	brn, rhomb	4.57–4.58	690	d 700	d	d	
s149	(di-)fluoride	Ag_2F	234.74	yel, hex	8.57	d 90		d		
s150	fluosilicate	$Ag_2SiF_6.4H_2O$	429.88	wh powd or col cr, deliq		>100	d	v s		
s151	fulminate	$Ag_2C_2N_2O_2$	299.77	need		expl		0.075^{13}	s	i HNO_3; s NH_4OH
s152	iodate	$AgIO_3$	282.77	col, rhomb	$5.525^{16.5}$	>200	d	0.003^{10}	0.019^{80}	s HNO_3, NH_4OH, KI
s153	*per*iodate	$AgIO_4$	298.77	or yel, tetrag	5.57	d 180		d		s HNO_3
s154	iodide(α)	Nat. iodyrite. AgI	234.77	yel, hex 2.21, 2.22	5.683^{30}_{4}	tr 146 to β		$2.8\times10^{-7.25}$	$2.5\times10^{-6.60}$	s KCN, $Na_2S_2O_3$, KI; sl s NH_4OH
s155	iodide (β)	AgI	234.77	or, cub	$6.010^{14.4}_{4}$	558	1506			
s156	iodomercurate (α)	Ag_2HgI_4	923.95	yel, tetrag	6.02	tr to β 50.7		i		s KI, KCN; i dil a
s157	iodomercurate (β)	Ag_2HgI_4	923.95	red, cub	5.90	d 158		i		s KI, KCN; i dil a
s158	hydrogen(tri-) *para*periodate	$Ag_2H_3IO_6$	441.69	yel, rhomb	5.68^{25}	60 d		1.68^{25}		s HNO_3
s159	hyponitrite	$Ag_2N_2O_2$	275.75	yel	5.75^{30}	d 110		v sl s		d HNO_3, H_2SO_4
s160	lactate	$AgC_3H_5O_3.H_2O$	214.96	wh or sl gray cr, powd				*ca* 7.7		
s161	laurate	$AgC_{12}H_{23}O_2$	307.19	wh, greasy powd		212.5				0.007^{25} al; 0.008^{15} eth
s162	levunilate	$AgC_5H_7O_3$	222.98	leaf				0.67^{17}	d	
s163	*per*manganate	$AgMnO_4$	226.81	dk vlt, monocl	4.27^{25}	d		0.55^{0}	$1.69^{28.5}$	d al
s164	mercury iodide (α)	Ag_2HgI_4	923.98	yel, tetrag	6.02	trst 50.7		i		
s165	mercury iodide (β)	Ag_2HgI_4	923.98	red, cub	5.90	158 d		i		
s166	myristate	$AgC_{14}H_{27}O_2$	335.24			211		0.007^{25}		0.006^{25} al; 0.007^{15} eth
s167	nitrate	$AgNO_3$	169.87	col, rhomb, 1.729, 1.744, 1.788	4.352^{19}	212	d 444	122^{0}	952^{190}	s eth, glyc; v sl s abs al

Figure 10.12
A sample of some of the data available in standard reference works.

the solubility is so low that it can normally be ignored. The AgI, Hg_2Cl_2, $Mg(OH)_2$, $PbSO_4$, Cu_2S, and CaC_2O_4 shown in Table 10.1 have very low solubilities and would be considered insoluble for most purposes. It does depend on the purpose, however.

Mercury and lead are poisons that we would like to avoid. The food industry would be *very* concerned about mercury and lead present in food even in the very small amounts suggested by the solubilities of Hg_2Cl_2 and $PbSO_4$.

10.6 SOLUBILITY AND CHEMICAL REACTIONS

Many budding chemists have been hooked for life when they poured two colorless solutions together and observed a beautiful red or yellow cloud swirl through the water and slowly settle to the bottom of the container. Most of us are still fascinated when we see two solutions that look like pure water produce brilliantly colored solids, even though we know that the chemistry of the reaction is rather simple.

The secret of reactions like these is that two soluble substances interact to form a compound that is insoluble. The reaction between lead nitrate and potassium iodide is an example.

As you can see in Table 10.1, $Pb(NO_3)_2$ is relatively soluble. Over 37 g of the solid dissolves in 100 mL of cold water, and almost three times as much dissolves in boiling water. When lead(II) nitrate dissolves, it **dissociates** (comes apart) into Pb^{2+} and 2 NO_3^- ions (Figure 10.13).

Figure 10.13

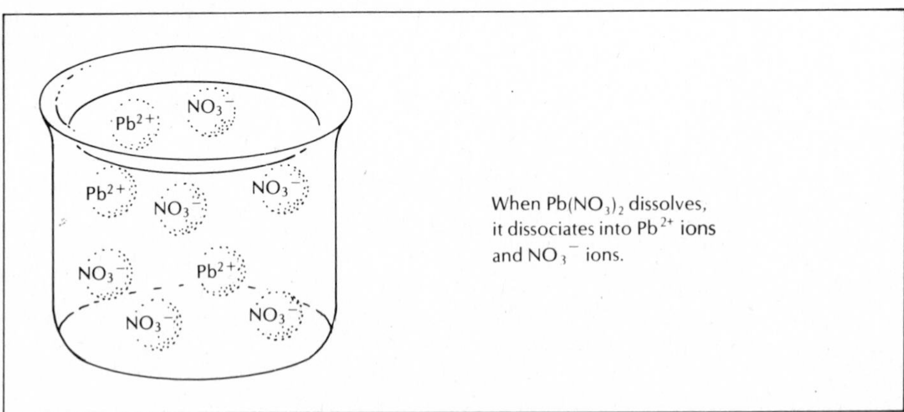

When $Pb(NO_3)_2$ dissolves, it dissociates into Pb^{2+} ions and NO_3^- ions.

Potassium iodide is even more soluble than lead(II) nitrate. Like lead(II) nitrate, it is an ionic compound and dissociates into K^+ ions and I^- ions when it dissolves in water (Figure 10.14).

When a solution of lead(II) nitrate and a solution of potassium iodide are

Figure 10.14

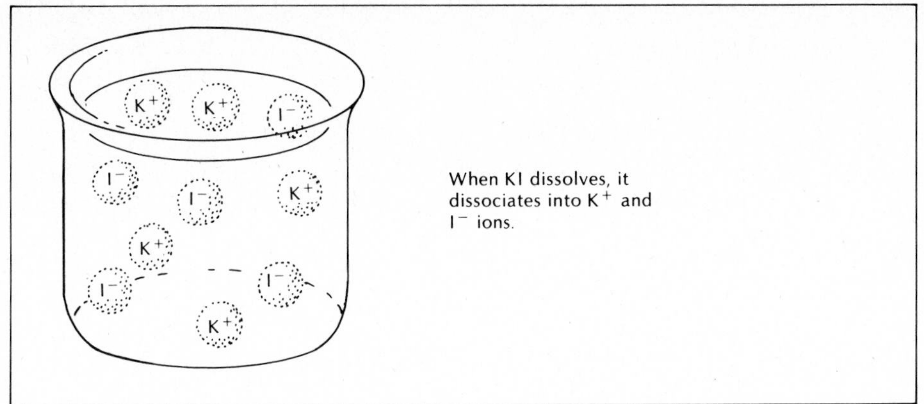

When KI dissolves, it dissociates into K^+ and I^- ions.

Figure 10.15

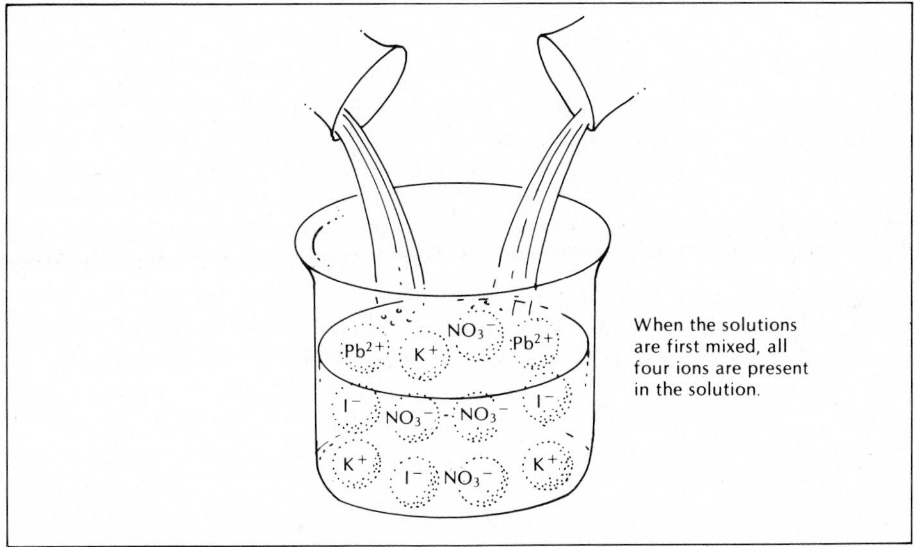

When the solutions are first mixed, all four ions are present in the solution.

poured together, the mixture contains four different ions moving about in a sea of water molecules, as suggested by Figure 10.15. This situation doesn't last long enough to see, however. As the lead ions and the iodide ions collide, they stick together to form beautiful yellow crystals of lead(II) iodide. The solid PbI_2 soon settles to the bottom of the container, as shown in Figure 10.16, page 288. The potassium ions and the nitrate ions stay separate in solution.

The reaction between lead(II) nitrate and potassium iodide is sometimes called a **metathesis** (me-tath′-i-sis) reaction. Metathesis comes from a Greek word that means "to transpose, or change positions." A glance at the equation for the reaction shows why this is a suitable name.

$$\mathbf{Pb}(NO_3)_2(aq) + 2K\mathbf{I}(aq) \rightarrow \mathbf{PbI}_2(s) + 2KNO_3(aq)$$

Figure 10.16

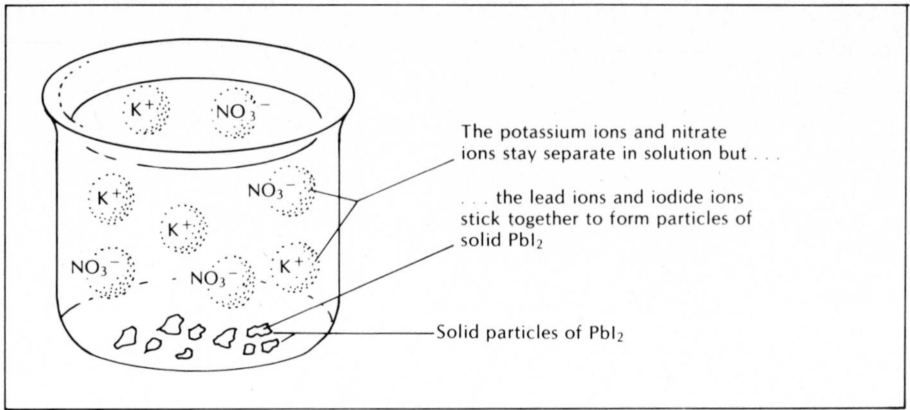

If you examine the formulas for the reactants and products in this equation, you will see that the lead was originally combined with the nitrate ion and that the potassium was combined with the iodide ion. In the products, the potassium and lead have exchanged partners. The negative ions have been transposed so that the nitrate is now combined with the potassium and the iodide is combined with the lead. Metathesis reactions get their name from this transposing.

Most ionic compounds can undergo metathesis reactions, provided that "exchanging partners" results in the formation of some compound that has a low solubility. In the example given, lead nitrate and potassium iodide are both soluble in water. However, when solutions of these compounds are mixed, lead iodide, which has a very low solubility, can be formed. This compound forms a solid material that precipitates out of the solution. (**Precipitate** means "to fall out." Rain and snow are forms of precipitation in which water falls out of the sky.)

Metathesis reactions are fun to do. I hope you will have an opportunity to do some in the laboratory.

It is less fun but just as instructive to predict whether a metathesis reaction will occur when two solutions are mixed. We make such predictions by referring to a table of solubilities. An example will show how it is done.

Example 10.3

Will a metathesis reaction occur when solutions of NaOH and $FeCl_2$ are mixed?

To answer the question, we must first predict the formulas of the products of the reaction. What ions will be combined in the products if a metathesis reaction takes place?

$$\mathbf{Fe}Cl_2(aq) + Na\mathbf{OH}(aq) \rightarrow ? + ?$$

Fe^{2+} with OH^- and Na^+ with Cl^-

Metathesis means "transpose," so the positive ion in the first compound will combine with the negative ion in the second compound, and vice versa.

Sodium and chlorine always have a combining number of 1, so the ions are Na^+ and Cl^-. However, iron can form Fe^{2+} ions and Fe^{3+} ions. How do you know that it is Fe^{2+} in $FeCl_2$?

Since the combining number of chlorine is always 1 in binary compounds with metals, the combining number of iron must be 2 in this compound.

In metathesis reactions, the combining numbers of the positive and negative ions do not change during the reaction. Write the correct formula for the products, and complete the equation for this reaction.

$$FeCl_2(aq) + 2NaOH(aq) \rightarrow Fe(OH)_2 + 2NaCl$$

What we have written is the balanced equation for the metathesis reaction *if it occurs.* It will occur if one of the products is insoluble in water. According to Table 10.1, is either product insoluble?

$Fe(OH)_2$ is insoluble.

Table 10.1 lists the solubility of $Fe(OH)_2$ as 0.00015 g in 100 g of water. This represents

$$\frac{0.00015 \text{ parts solute}}{100 \text{ parts solvent}} \quad \text{or} \quad \frac{1.5 \text{ parts solute}}{1{,}000{,}000 \text{ parts solvent}}$$

One part per million would be a very low solubility. When solutions normally encountered in the laboratory are mixed, a solid precipitate appears when the product has a solubility of less than 1 g/100 g of water. The solubility of $Fe(OH)_2$ is well below this value. However, the solubility of NaCl is about 36 g/100 g of water. We would not expect to see a precipitate of NaCl.

Now use Table 10.1 to decide whether a metathesis reaction occurs when solutions of the following compounds are mixed. If a reaction occurs, write the formulas of the products and balance the equation.

$$Pb(NO_3)_2(aq) + Al_2(SO_4)_3(aq) \rightarrow ? + ?$$

$$3\mathbf{Pb}(NO_3)_2(aq) + Al_2(\mathbf{SO_4})_3(aq) \rightarrow 3\mathbf{PbSO_4}(s) + 2Al(NO_3)_3(aq)$$

SOLUBILITY RULES

To be certain about the solubility of compounds, it is necessary to test the compound in the laboratory or look up the solubility determined by some other worker. However, several generalizations about solubility can be made.

If you memorize these generalizations, you will save a great deal of time spent looking up information in a handbook.

1. Compounds of Li^+, Na^+, K^+, Rb^+, Cs^+, Fr^+ (that is, all metals in Group IA of the periodic table), H^+, and the NH_4^+ ion are *soluble.*
2. Most chlorides (Cl^-), bromides (Br^-), and iodides (I^-) are *soluble.* Important exceptions are compounds formed between these ions and copper(I), silver, mercury(II), and lead(II).
3. Compounds containing the NO_3^-, $C_2H_3O_2^-$, and ClO_4^- ions are *soluble.*
4. Most sulfates (SO_4^{2-}) are *soluble.* Important exceptions are sulfates of Ca^{2+}, Sr^{2+}, Ba^{2+}, and Pb^{2+}.
5. Most sulfides (S^{2-}) are *insoluble.* Exceptions are the sulfides of metals found in Group IA and Group IIA of the periodic table.
6. Most compounds containing the OH^- ion are *insoluble.* Exceptions are the hydroxides of metals from Group IA, Sr^{2+}, and Ba^{2+}.
7. Most compounds containing the CO_3^{2-}, PO_4^{2-}, and SO_3^{2-} ions are *insoluble.* Exceptions are the compounds formed with metals from Group IA of the periodic table.

The following exercises will give you an opportunity to use these solubility rules to predict metathesis reactions. You are given the formulas for two compounds that are soluble. You should write the formula for the products of a metathesis reaction between these compounds (assuming that such a reaction will occur), balance the equation, and then check the solubility of the products to see whether the reaction could occur. If both products are soluble, draw a line through the products and write "no observable reaction". Write (s) after the formula of any product that is insoluble.

10-17. $PbCl_2(aq) + HCl(aq) \rightarrow$
10-18. $Bi(NO_3)_2(aq) + NaOH(aq) \rightarrow$
10-19. $Pb(C_2H_3O_2)_2(aq) + K_2SO_4(aq) \rightarrow$
10-20. $CuSO_4(aq) + FeCl_3(aq) \rightarrow$
10-21. $FeSO_4(aq) + (NH_4)_2S(aq) \rightarrow$
10-22. $K_2CO_3(aq) + Sr(NO_3)_2(aq) \rightarrow$
10-23. $NaCl(aq) + KOH \rightarrow$
10-24. $NaI(aq) + AgNO_3(aq) \rightarrow$
10-25. $Al_2(SO_4)_3(aq) + CaCl_2(aq) \rightarrow$
10-26. $Na_2SO_4(aq) + CaCl_2(aq) \rightarrow$
10-27. $K_2SO_4(aq) + Ca(C_2H_3O_2)_2(aq) \rightarrow$
10-28. $MgSO_4(aq) + CaBr_2(aq) \rightarrow$

10.7 IONIC AND NET IONIC EQUATIONS

The equation for the reaction between lead(II) nitrate and potassium iodide was written like this:

$$\mathbf{Pb}(NO_3)_2(aq) + 2K\mathbf{I}(aq) \rightarrow \mathbf{PbI_2}(s) + 2KNO_3(aq)$$

This equation shows the compounds in solutions used to produce the reaction, the compound that precipitates, and the compound present in solution after the reaction. It is a good equation, but it can be misleading. It suggests that these compounds are present in the form of molecules.

Both lead(II) nitrate and potassium iodide are ionic compounds. When they dissolve in water, they dissociate into ions, as shown in Figures 10.13 and 10.14. When the metathesis reaction occurs, only the lead(II) and iodide ions stick together to form a new compound. The potassium ions and nitrate ions are still moving about freely in the solution, and they stay there until the water is distilled away. What actually exists before and after the reaction would be described better by the following equation:

$$\mathbf{Pb^{2+}}(aq) + 2NO_3^-(aq) + 2K^+(aq) + \mathbf{2I^-}(aq) \rightarrow \mathbf{PbI_2}(s) + 2K^+(aq) + 2NO_3^-(aq)$$

This equation is an **ionic equation.** Rather than writing formulas for ionic compounds in solution, we write the ions making up the compound separately to emphasize the fact that these ions separate when the compound dissolves. If two ions combined to form an ionic solid like PbI_2, the formula for that product is written in the normal manner.

Note that, in the ionic equation, $2K^+$ and $2NO_3^-$ appear on both sides of the equation. This indicates that these ions are present before the reaction starts and after the reaction is over. In other words, they do not take part in the chemical change. What actually *changes* in this chemical process could be described by the following equation:

$$Pb^{2+}(aq) + 2I^-(aq) \rightarrow PbI_2(s)$$

This equation is called the **net ionic equation.** It ignores those ions that are present but do not take part in the reaction. It simplifies the ionic equation.

The metathesis reactions described in Questions 10-17 through 10-28 are all ionic reactions. The reactants exist as ions in solution, and the insoluble product represents the new compound that is formed. The other ions remain in solution.

10-29. through 10-40. Rewrite the equations for the reactions described in Questions 10-17 through 10-28 as net ionic equations.

Note that the last four net ionic equations are identical. Even though the reactants were different in each case, the only product was calcium sulfate, and the reaction could be described by the same net ionic equation. It doesn't seem to matter what compound furnishes the calcium ions or what compound furnishes the sulfate ions. When these ions are in solution together, they eventually come into contact and stick together to form the insoluble compound calcium sulfate. The net ionic equation emphasizes this fact.

OTHER IONIC REACTIONS

In all the metathesis reactions described so far, an insoluble solid was formed. This is not always the case. When a solution of sodium hydroxide and a solution of hydrogen chloride (hydrochloric acid) are mixed, the following metathesis reaction occurs:

$$\mathrm{Na^+(aq)} + \mathbf{OH^+}\mathrm{(aq)} + \mathbf{H^+}\mathrm{(aq)} + \mathrm{Cl^-(aq)} \rightarrow \mathrm{Na^+(aq)} + \mathrm{Cl^-(aq)} + \mathbf{HOH}(\ell)$$

The net ionic equation for this reaction is

$$\mathrm{OH^-(aq)} + \mathrm{H^+(aq)} \rightarrow \mathrm{HOH}(\ell)$$

You will recognize the product better when the formula is written in its ordinary form, H_2O. This reaction between ionic compounds containing H^+ and OH^- to form the nonionic compound water is a very common reaction. It will be discussed at length in the chapter on acids and bases.

In another reaction involving the H^+ ion, metals dissolve and hydrogen gas forms. One such reaction is described by the following molecular equation and net ionic equation.

$$\mathbf{Zn}\mathrm{(s)} + 2\mathbf{H}\mathrm{Cl(aq)} \rightarrow \mathbf{H_2}\mathrm{(g)} + \mathbf{Zn}\mathrm{Cl_2(aq)}$$

$$\mathrm{Zn(s)} + 2\mathrm{H^+(aq)} \rightarrow \mathrm{H_2(g)} + \mathrm{Zn^{2+}(aq)}$$

Perhaps you recognize this last equation as an example of a replacement reaction like those we discussed in Chapter 9. All of the replacement reactions described there could have been written as net ionic equations. Generally, in a replacement reaction, a metal is reacting with a metal ion to form a new metal and a new metal ion. The reaction between copper and silver nitrate could be described by the following net ionic equation:

$$\mathrm{Cu(s)} + 2\mathrm{Ag^+(aq)} \rightarrow \mathrm{Cu^{2+}(aq)} + 2\mathrm{Ag(s)}$$

You might ask yourself what could be taking place with the atoms in this reaction. How does the copper atom acquire that positive charge and become an ion? How does the positive silver ion become a neutral atom?

Still another ionic reaction is illustrated by the reaction between baking soda and vinegar. The first part of the reaction might be described as a metathesis reaction, but the reaction actually takes place because one of the products decomposes.

$$\mathrm{Na}\mathbf{HCO_3}\mathrm{(aq)} + \mathbf{H}\mathrm{C_2H_3O_2(aq)} \rightarrow \mathrm{NaC_2H_3O_2(aq)} + \mathbf{H_2CO_3}\mathrm{(aq)}$$

The H_2CO_3 that forms decomposes readily to form H_2O and CO_2:

$$\mathrm{H_2CO_3(aq)} \rightarrow \mathrm{H_2O}(\ell) + \mathrm{CO_2(g)}$$

The overall reaction can be described by the following net ionic equation:

$$\mathrm{HCO_3^-(aq)} + \mathrm{H^+(aq)} \rightarrow \mathrm{H_2O}(\ell) + \mathrm{CO_2(g)}$$

These are but a few examples of reactions that can occur when solutions containing ions are mixed.

LIMITATIONS OF NET IONIC EQUATIONS

Net ionic equations provide a simplified description of many reactions. They can often clarify what is taking place in a chemical reaction. However, like most good things, they have limitations.

It is impossible to write an equation if you do not know what the reactants and products are. To write an ionic equation, you need even more information. You must know which compounds dissociate into ions when they are in solution and which compounds do not dissociate into ions. Until you learn several chemical facts, this is not easily determined. Consequently, students in a beginning course may not be able to write ionic equations for many ionic reactions. This should not disturb you. As you learn more chemical facts, your ability to describe them with ionic equations will increase.

Another problem arises with net ionic equations. They provide a simplified description of a reaction, but it is possible to oversimplify. The net ionic equation implies that, regardless of the additional ions in solution, those ions mentioned in the net ionic equation combine as indicated. It just isn't so. Any time other chemical species are present, other chemical reactions are possible.

The reaction between zinc metal and a compound containing the H^+ ion was described in the last section by the following net ionic equation:

$$Zn(s) + 2H^+(aq) \rightarrow Zn^{2+}(aq) + H_2(g)$$

This reaction certainly does occur, but other reactions can occur, depending on the ions present in solution. When the H^+ ions come from HNO_3, the following reaction also occurs.

$$Zn(s) + 4H^+(aq) + 2NO_3^-(aq) \rightarrow N_2O_4(g) + 2H_2O\ (\ell) + Zn^{2+}(aq)$$

As you can see, net ionic equations that ignore other ions in solution can be misleading.

10.8 SUMMARY

Solutions are important to chemists, because they provide a good medium for atoms and molecules to come into contact so that a reaction can occur.

We are normally interested in the solute in a solution. We defined molarity as the number of moles of solute in 1 liter of the solution.

The amount of solute that can dissolve in a given amount of solvent varies from one compound to another, and differences in solubility can be used to identify compounds and to separate mixtures. In general, solubility of solids increases with temperature, but there are exceptions. The solubility of gases always decreases as temperature increases.

A number of reactions involving ionic compounds in solution were discussed, and net ionic equations were introduced as one way in which to describe these reactions.

To know whether a reaction has occurred when solutions are mixed, we look for some noticeable change.

1. Sometimes, when solutions containing ionic compounds are mixed, a solid material precipitates. The reaction usually involves "changing partners." The positive ion from one compound combines with the negative ion from the other compound, and vice versa. Such a reaction is known as a metathesis reaction. The following are the molecular and net ionic equations describing a metathesis reaction.

$$MgCl_2(aq) + 2AgNO_3(aq) \rightarrow 2AgCl(s) + Mg(NO_3)_2(aq)$$

$$Cl^-(aq) + Ag^+(aq) \rightarrow AgCl(s)$$

2. Some reactions take place because a nonionic compound is formed instead of an insoluble precipitate. Such reactions may show no visible change. The reaction to form water is the most common example. The net ionic equation is

$$H^+(aq) + OH^-(aq) \rightarrow H_2O(\ell)$$

3. Still other reactions take place because a gas is produced. Bubbles of gas in a solution signal that a reaction is taking place. The most common example of such a reaction is the production of CO_2 from a carbonate or bicarbonate. The following net ionic equation shows the reaction with a carbonate ion.

$$CO_3^{2-}(aq) + 2H^+(aq) \rightarrow H_2O(\ell) + CO_2(g)$$

Questions and Problems

10-41. Calculate the molarity of each of the following solutions.
 a. 250 mL of solution containing 36 g of NaCl
 b. 1.2 mol of NH_3 dissolved in 3.1 L of solution
 c. 110 mL of solution containing 0.062 mol of NaOH
 d. 1.25 L of solution containing 111 g of KOH
 e. 0.055 g of LiCl dissolved in 5.5 mL of solution

10-42. Calculate the mass of solute needed to prepare the following amounts of solution. In some of the examples, you have not been given enough information to solve the problem. In these cases, indicate what additional information is needed.
 a. 100 mL of a 0.5 *M* solution of LiCl
 b. 125 cm^3 of 2 *M* Na_2CO_3
 c. a 2.2 *M* solution of KOH
 d. 1 dm^3 of a 3 *M* solution

e. 10 dm^3 of HCl solution
f. 1.2 L of HCl with a concentration of 0.25 *M*
g. 126 L of acetic acid with a concentration of 1 *M*
h. 350 mL of a 0.5 *M* sugar solution

10-43. How much solute is contained in the following solutions? If insufficient information is given to solve the problem, indicate what additional information you need.

a. How many grams of LiCl are in 100 mL of a 0.25 *M* solution?
b. How many moles of HCl have you poured out when you pour 125 mL of a 1 *M* solution?
c. How many grams of NaCl are in a 2.5 *M* solution?
d. A 1-liter bottle of 6 *M* NaOH contains how many moles of NaOH?
e. How many moles of ammonia are in 1.5 L of household ammonia?
f. How many grams of HCl are required to make 2,000 L of Sno-Bol® when the HCl concentration is 4 *M*?

10-44. Use the general solubility rules given in this chapter to predict whether a reaction will occur when the following solutions are mixed. Write balanced equations for those reactions that will occur.

a. $NaCl(aq) + FeCl_2(aq) \rightarrow$
b. $NaCl(aq) + Pb(NO_3)_2(aq) \rightarrow$
c. $KClO_4(aq) + Mg(NO_3)_2(aq) \rightarrow$
d. $Ca(C_2H_3O_2)_2(aq) + Na_2SO_4(aq) \rightarrow$
e. $H_2S(aq) + CuSO_4(aq) \rightarrow$
f. $NaOH(aq) + NiCl_2(aq) \rightarrow$
g. $K_2CO_3(aq) + AgNO_3(aq) \rightarrow$
h. $Na_3PO_4(aq) + SnCl_4(aq) \rightarrow$

10-45. Write net ionic equations for each reaction that will occur in Question 10-44.

More Challenging Problems

10-46. How many milliliters of a 6 *M* HCl solution are needed to react with 10.5 g of Zn?

10-47. If 0.23 mol of zinc react with 125 mL of HCl solution, what is the concentration of the solution?

10-48. What volume of 1.0 *M* NaOH is required to react with 25 mL of 1.0 *M* H_2SO_4? The equation for the reaction is

$$2NaOH(aq) + H_2SO_4(aq) \rightarrow Na_2SO_4(aq) + 2H_2O(\ell)$$

10-49. What volume of 1.0 *M* NaOH is required to react with 25 mL of 0.25 *M* H_2SO_4?

10-50. What is the concentration of each ion in solution when 32 g of Na_2CO_3 are dissolved in 100 mL of solution?

10-51. What is the concentration of each ion in solution when 50 mL of 0.2 *M* KI are mixed with 50 mL of 0.1 *M* $Pb(NO_3)_2$?

10-52. What is the approximate concentration of the NaOH solution formed when 10.0 g of Na are placed in a liter of water?

10-53. When 0.2 mol of Zn is placed in 150 mL of 2.5 *M* $CuSO_4$, will all of the zinc react?

10-54. What is the concentration of the aluminum nitrate solution formed when an excess of aluminum reacts with 250 mL of 0.1 *M* $Pb(NO_3)_2$?

Answers to Questions in Chapter 10

10-1. 0.7 *M*

10-2. 0.7 *M*

10-3. 0.20 *M*

10-4. 5×10^{-4} *M*

10-5. Molarity is defined as the number of moles of solute in each liter of solution. Therefore, we must first calculate the moles of NaCl that we have.

$$? \text{ mol NaCl} = 58.4 \cancel{\text{g NaCl}} \times \frac{1 \text{ mol NaCl}}{58.4 \cancel{\text{g NaCl}}} = 1 \text{ mol NaCl}$$

This amount of NaCl is dissolved in 2 liters of solution. Then,

$$? \frac{\text{mol NaCl}}{1 \text{ L sol}} = \frac{1 \text{ mol NaCl}}{2 \text{ L sol}} = 0.5 \frac{\text{mol NaCl}}{\text{L sol}} = 0.5\,M$$

10-6. $? \text{ mol } I_2 = 1.27 \cancel{\text{g } I_2} \times \frac{1 \text{ mol } I_2}{254 \cancel{\text{g } I_2}} = 0.005 \text{ mol } I_2$

$$? \frac{\text{mol } I_2}{\text{L sol}} = \frac{0.005 \text{ mol } I_2}{10 \text{ mL sol}} \times \frac{1000 \text{ mL}}{1 \text{ L}} = \frac{0.5 \text{ mol } I_2}{\text{L sol}} = 0.5\,M\ I_2$$

10-7. 0.500 *M* NH_3

10-8. 17 g (solution appears in the text)

10-9. $? \text{ mol HCl} = 150 \cancel{\text{mL}} \times \frac{1 \cancel{\text{L}}}{1000 \cancel{\text{mL}}} \times \frac{4 \text{ mol HCl}}{\cancel{\text{L}}} = 0.6 \text{ mol HCl}$

10-10. 1 mol (1.25 mol ignoring significant digits)

10-11. 6.2×10^{-5} mol

10-12. 300 ml of 1.3 *M* solution (0.39 mol vs. 0.35 mol)

10-13. 34 g

10-14. 20 g (21.84 g ignoring significant digits)

10-15. 1.6×10^{-2} g

10-16. 300 g (292.2 g ignoring significant digits)

10-17. $PbCl_2 + HCl \rightarrow$ no observable reaction. (When both reactants have the same positive or negative ion, it is not possible to get a different substance.)

10-18. $Bi(NO_3)_2(aq) + 2NaOH(aq) \rightarrow 2NaNO_3(aq) + Bi(OH)_2(s)$

10-19. $Pb(C_2H_3O_2)_2(aq) + K_2SO_4(aq) \rightarrow 2KC_2H_3O_2(aq) + PbSO_4(s)$

10-20. $3CuSO_4(aq) + 2FeCl_3(aq) \rightarrow 3CuCl_2(aq) + Fe_2(SO_4)_3(s)$ Iron(III) sulfate is slightly soluble. Whether or not you observe a reaction depends on the concentration of the solutions used.

10-21. $FeSO_4(aq) + (NH_4)_2S(aq) \rightarrow (NH_4)_2SO_4(aq) + FeS(s)$

10-22. $K_2CO_3(aq) + Sr(NO_3)_2(aq) \rightarrow 2\ KNO_3(aq) + SrCO_3(s)$

10-23. $NaCl(aq) + KOH(aq) \rightarrow$ no observable reaction

10-24. $NaI(aq) + AgNO_3(aq) \rightarrow NaNO_3(aq) + AgI(s)$

10-25. $Al_2(SO_4)_3(aq) + 3CaCl_2(aq) \rightarrow 2AlCl_3(aq) + 3CaSO_4(s)$

10-26. $Na_2SO_4(aq) + CaCl_2(aq) \rightarrow 2NaCl(aq) + CaSO_4(s)$

10-27. $K_2SO_4(aq) + Ca(C_2H_3O_2)_2(aq) \rightarrow 2KC_2H_3O_2(aq) + CaSO_4(s)$

10-28. $MgSO_4(aq) + CaBr_2(aq) \rightarrow MgBr_2(aq) + CaSO_4(s)$

10-29. No observable reaction

10-30. $Bi^{2+}(aq) + 2OH^-(aq) \rightarrow Bi(OH)_2(s)$

10-31. $Pb^{2+}(aq) + SO_4^{2-}(aq) \rightarrow PbSO_4(aq)$

10-32. $3\ SO_4^{2-}(aq) + 2Fe^{3+}(aq) \rightarrow Fe_2(SO_4)_3(s)$

10-33. $Fe^{2+}(aq) + S^{2-}(aq) \rightarrow FeS(s)$

10-34. $CO_3^{2-}(aq) + Sr^{2+}(aq) \rightarrow SrCO_3(s)$

10-35. No observable reaction

10-36. $I^-(aq) + Ag^+(aq) \rightarrow AgI(s)$

10-37. $SO_4^{2-}(aq) + Ca^{2+}(aq) \rightarrow CaSO_4(s)$

10-38. $SO_4^{2-}(aq) + Ca^{2+}(aq) \rightarrow CaSO_4(s)$

10-39. $SO_4^{2-}(aq) + Ca^{2+}(aq) \rightarrow CaSO_4(s)$

10-40. $SO_4^{2-}(aq) + Ca^{2+}(aq) \rightarrow CaSO_4(s)$

READING

Many important mixtures of chemical substances are not solutions. A common example of such mixtures is ordinary toothpaste. Have you ever wondered what is in toothpaste or what the various ingredients are designed to do? The following article tells all.

CHEMISTRY IN ORAL HEALTH

What's in Toothpaste?

All dentifrices—toothpastes, gels, and powders—contain at least four types of ingredients: abrasives, sudsers, sweetening agents and flavoring agents. Many contain other ingredients—such as fluoride, whiteners, moisture-retainers, softeners and binding agents.

Among the abrasives used are

$CaHPO_4$	$NaPO_3$	$CaCO_3$
calcium hydrogen phosphate	sodium metaphosphate	calcium carbonate

Magnesium carbonate, hydrated aluminum oxides, silicates, and silica gel also are used. Abrasives remove debris and stains from teeth by polishing.

The American Dental Association and the American Pharmaceutical Association rate toothpastes on their average abrasivity. Table 1 gives the abrasiveness ratings for some common brands. Tooth enamel is the hardest substance produced by the human body. The abrasives used in dentifrices do not appear to wear away much of the tooth enamel.

Commonly used sudsers are the detergents sodium lauryl sulfate and sodium N-lauroyl sarcosinate. Sudsers also act as cleansers (detergents).

$$CH_3(CH_2)_{11}{-}O{-}SO_3^-Na^+$$

sodium lauryl sulfate

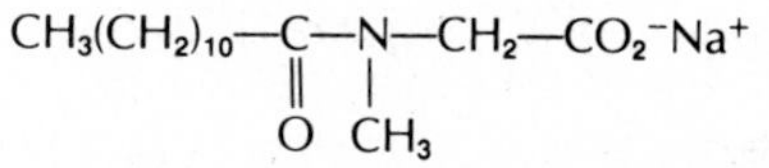

$$CH_3(CH_2)_{10}{-}\underset{\underset{O}{\|}}{C}{-}\underset{\underset{CH_3}{|}}{N}{-}CH_2{-}CO_2^-Na^+$$

sodium N-lauroyl sarcosinate

Saccharin is the most commonly used sweetening agent, though others may be substituted for it because of recent concern over its possible toxicity. Addition of sugar is discouraged by dental authorities because it is a key cause of cavities. Flavors such as spearmint, peppermint, wintergreen, and sassafras are choices of most manufacturers.

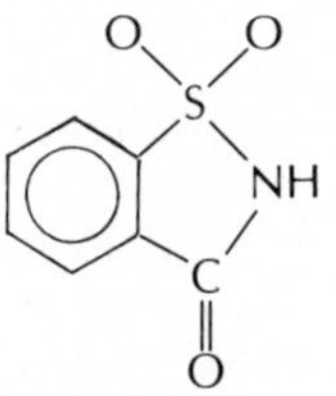

Saccharin

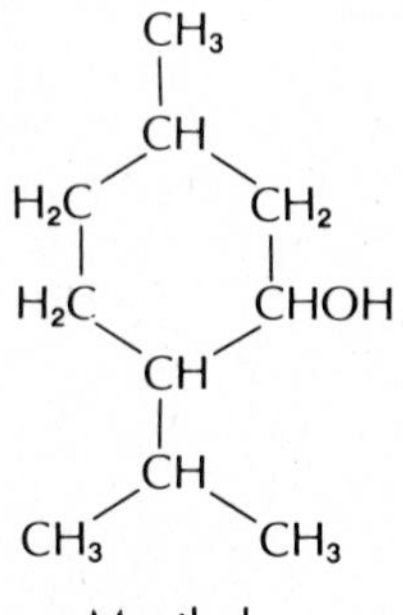

Menthol

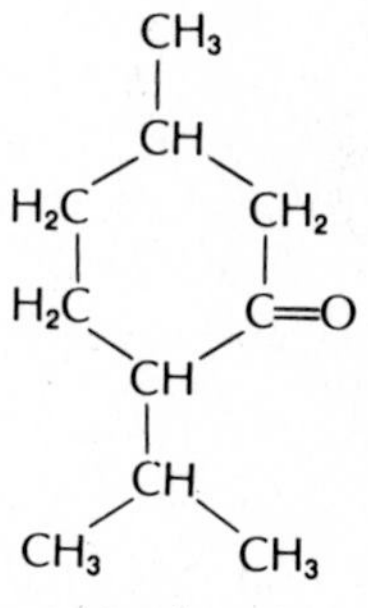

Menthone

major components of oil of peppermint

Table 1. Ratings of Toothpaste Abrasiveness[a] (Listed in Order of Increasing Abrasiveness)

Product	*Average abrasivity*	*Product*	*Average abrasivity*	*Product*	*Average abrasivity*
Listerine	26	Ultra Brite	64	Crest (mint)	81
Pepsodent (with zirconium silicate)	26	Macleans (spearmint)	66	Close-up	87
Colgate (with MFP)	51	Pearl Drops	72	Gleem II	106

[a] Taken from reference (1).

Source: *Journal of Chemical Education*, Vol. 55, No. 11 (November 1978), pp. 736–737.

$$\begin{array}{c} CH_2OH \\ | \\ CHOH \\ | \\ CH_2OH \end{array}$$

glycerin (glycerol)

NaH_2BO_4

sodium perborate

$CH_3—(CH_2)_{17}—CH_3$

nonadecane
component of mineral oil

Glycerin is added to keep toothpaste from drying out too rapidly when exposed to air. Mineral oil is added to keep it soft, and sodium perborate is added in certain products to enhance whitening and to assist in removing stains.

Advertisements making claims on "whitening" and "brightening" of teeth can be misleading. The teeth of many are simply not white and no commercial product can change this. Brushing with a good dentifrice can remove stains and polish the teeth, however.

Common Dental Problems

The two most prevalent dental diseases are tooth decay (dental caries) and gum inflammation (periodontal disease). Both can result in tooth loss—caries through disintegration or extraction, periodontal disease by destruction of gum tissue which supports the tooth in its socket.

Both diseases have been indisputably linked to dental plaque. Plaque is a soft film that forms readily and holds tenaciously on tooth surfaces. It consists chiefly of bacteria and bacterial products. Within minutes after a tooth is cleaned, a film composed of protein from saliva is deposited. Micro-organisms, mostly *streptococcus mutans* and *streptococcus sanguis* adhere to the film and begin to multiply. The bacteria and their products, carbohydrates and proteins, are the chief constituents of plaque. Under proper conditions, plaque can slowly absorb calcium salts, and become transformed into a hard material known as dental calculus.

Dental caries can form when sugars are absorbed into the plaque and are metabolized by the bacteria present producing acids which, in time, erode the tooth enamel. Lactic acid appears to cause the greatest damage, though pyruvic and acetic acids also are present. In effect, the acid creates chemical conditions that cause calcium ions in tooth enamel to slowly dissolve in saliva.

$$\begin{array}{l} CH_3—CHCO_2H \\ \qquad\quad | \\ \qquad\quad OH \end{array}$$

Periodontal disease appears to begin at the interface between plaque and the gums (gingiva). Products from bacteria in plaque cause inflammation of the gums. To rid themselves of the bacteria and their products, the gums marshall a host of chemicals to destroy the bacteria. If these chemicals are present in large enough quantities and over a long enough period, they also can destroy gum tissue and the fibers that hold teeth in place.

The Role of Fluoride

Research has shown convincingly that routine use of fluoride-containing dentifrices (in particular, those that contain stannous fluoride, SnF_2, or sodium monofluorophosphate, Na_2PO_3F) can reduce the rate of caries formation in young and old alike. The use of sodium fluoride, NaF, for this purpose has been less successful.

Evidence available suggests that, in the chemical environment of the teeth, fluoride ion can react with the key component of tooth enamel, hydroxyapatite, $Ca_{10}(PO_4)_6(OH)_2$, to form an extremely thin layer of densely packed particles of calcium fluoride, CaF_2, and stannous phosphate, $Sn_3(PO_4)_2$. This layer forms an effective protective surface film on dental enamel.

Fluoride-containing dentifrices approved by the American Dental Association or accepted by Consumer's Union are listed in Table 2.

Proper Oral Health Care

Although toothpastes help in scrubbing and cleaning, and are therefore an aid in plaque control, the toothbrush and the way it is used are far more important. The ADA recommends a five-part program of dental care:

1. regular and effective brushing with a suitable toothbrush

2. daily use of unwaxed dental floss
3. forced rinsing of teeth and mouth after eating
4. proper nutrition including sensible sugar intake (sugar is less damaging when taken in liquids, for example)
5. regular visits with dentist

Table 2. Fluoride Dentifrices Accepted by American Dental Association or Consumer's Union

ADA accepted	*CU accepted*
Colgate, with Gardol plus MFP fluoride	A&P Fluoride
Crest, with Fluoristan, mint flavor	Gleem II, New Special Formula, New Special Brightners
Crest, with Fluoristan, regular	Rexall Fluoride
	Safeway Fluoride
	Worthmore Stannous Fluoride

Mouthwashes

Mouthwashes sold without prescription are essentially dilute solutions of breath-perfuming substances that often are sweetened with saccharin and colored to sell. Many contain ethanol and antibacterial agents. The Council on Dental Therapeutics of the ADA has said it "finds no substantial contribution to oral health in the unsupervised use of medicated mouthwashes by the general public."

Adhesive Sealants

The painting of teeth with polymeric or plastic adhesive sealants has resulted in reports that these afford 80% or better protection against caries. However, the painting process is costly and must be repeated at six month intervals. Three types of sealants have been successfully used: cyanoacrylates, polyurethane materials and a polymer formed with bisphenol A, glycidyl methacrylate, and methylmethacrylate.

$$-CH_2-\underset{\displaystyle CN}{\overset{\displaystyle CO_2CH_3}{\overset{|}{\underset{|}{C}}}}-$$

portion of a cyanoacrylate adhesive sealant

$$-(CH_2)_n-O-\underset{\displaystyle O}{\underset{\|}{C}}-\underset{\displaystyle H}{\underset{|}{N}}(CH_2)_n-\underset{\displaystyle H}{\underset{|}{N}}-\underset{\displaystyle O}{\underset{\|}{C}}-O-$$

portion of a polyurethane sealant

Possibilities for the Future

Work now is underway to develop even more effective preventatives for dental caries and periodontal diseases. Some promising approaches are

1. serums for immunization against dental disease
2. dietary additives to reduce effect of bacteria
3. antimicrobial agents to reduce microbial plaque

References

(1) Handbook of Nonprescription Drugs, American Pharmaceutical Association, 5th ed., 1977.

(2) Cornacchia, Harold J., "Consumer Health," Mosby, St. Louis, 1976.

(3) Brown, W. E., "Oral Health, Dentistry and the American Public," University of Oklahoma Press, Norman, 1974.

11

Classifying Compounds as Acids or Bases

OBJECTIVES

Acids and Bases

The more facts you learn, the more important it is to group facts into categories. One of the most useful ways in which we group compounds is into the categories of acids and bases. You will know what acids and bases are when you can do the following:

1. Name three properties of acids.
2. Name three properties of bases.
3. Given a household product such as vinegar or soap, determine whether that product is an acid, a base, or neither.
4. Given a list of equations, identify those that describe acid-base reactions. (See Problem 11-48.)
5. Identify each acid and its conjugate base in an equation for an acid-base reaction. (See Problem 11-52.)

Strong and Weak

Some acids and bases are weak and some are strong. You should understand the difference when you can do the following:

6. When told that one acid is strong and another is weak, predict which will react more rapidly with limestone. (See Problem 11-26.)
7. When told that an acid is strong or weak, predict the degree of dissociation or conductivity of the acid in water. (See Problems 11-27 and 11-28.)

8. When given an equation for an acid-base reaction and told whether reactants or products are favored at equilibrium, identify the stronger of the two acids involved and the stronger of the two bases involved. (See Problems 11-30 and 11-32.)

Acid-Base Reactions

Classifying compounds into groups allows us to discuss reactions as a group. You will be able to describe many of the reactions involving acids and bases when you can:

9. Write a general equation for the reaction between an acid and an active metal.
10. Predict the products of a reaction between any acid and a carbonate.
11. Write an equation for the reaction between a metal oxide and an acid.
12. Write the net ionic equation for the reaction between a hydroxide and an acid.
13. Describe at least three practical uses of acids and bases found in the home, and write equations for the reactions.

Concentration and pH

Reactions of acids are reactions of the hydrogen ion. Hydrogen ion concentration varies so widely in water solution that a logarithmic scale called pH is used to describe it. You should understand pH when you can:

14. Indicate the hydrogen ion concentration when you are given the pH of a solution, or vice versa.
15. Identify a solution as acidic or basic when you know either the pH or the hydrogen ion concentration. (See Problem 11-53.)
16. Name a compound that is used in practical applications to raise (or lower) pH (for example, in a swimming pool or in soil).

Indicators and Titration

Compounds that change color when the pH of a solution changes are called indicators. They are especially useful in an analytical procedure called titration. This will be clearer to you when you can meet the following objectives:

17. Name at least two compounds that act as acid-base indicators, and give their color in acid and base forms.
18. Describe the relationship between the color change of an indicator during titration and the equation that describes the reaction.
19. Given the volume of acid (or base) of known concentration needed to react with a base (or acid) of unknown concentration, find the unknown concentration. (See Problem 11-55.)

Words You Should Know

indicator
acid
base
neutralization
proton donor
proton acceptor
hydronium ion
dissociate
strong
weak
reversible reaction
equilibrium
conjugate acid-base pair
pH
titration
equivalence point

11.1 WHY WE CLASSIFY

In the last three chapters you have been introduced to quite a few chemical reactions, but you have seen no more than the tip of the iceberg. Though there are less than 100 elements on the earth, there are millions of compounds, and each compound can react to form new substances in several ways. One begins to wonder how it is possible to make sense out of chemistry. There is just too much to know!

People deal with a large number of facts by lumping them together and talking about groups. That's what we do when we make a statement like "Birds lay eggs." We don't say that chickens lay eggs and ducks lay eggs and robins lay eggs and wrens lay eggs. We classify all of these animals as birds and economize with words by saying "*Birds* lay eggs."

Classifying in this way is so much a part of human behavior that we often do it unconsciously. We have certainly done a lot of classifying in this book. We have separated matter into the pure and the impure. We have divided pure substances into elements and compounds. We have classified elements as metals and nonmetals. We have referred to compounds as carbonates, sulfates, and hydroxides. We have also classified reactions. We have identified metathesis reactions, decomposition reactions, combination reactions, and replacement reactions.

All classification is done on the basis of some defining characteristic. As you study chemistry—or any subject, for that matter—it is usually a good idea to focus your attention on defining attributes. Sometimes there are only one or two, and sometimes there are many. How well have you identified the defining attributes of the various concepts we have discussed so far? Find out by answering the following review questions.

11-1. What do you look for in a chemical formula to decide whether a compound is a carbonate?

11-2. What characteristic distinguishes an element from a compound?

11-3. How would you identify an equation that describes a metathesis reaction?

11-4. How is a replacement reaction different from metathesis?

11-5. What observation would tell you whether a white rock is a carbonate?

11-6. What properties would you observe in order to classify a piece of black solid as an element or a compound?

11-7. When you pour two colorless solutions together and a yellow solid is produced, how can you decide whether the reaction you observe is metathesis?

11-8. What characteristics would you look for in a solid to decide whether that solid is a metal or a nonmetal?

11-9. Given the name of an element, where would you look to decide whether that element is a metal or a nonmetal?

11-10. What characteristics of a substance enable you to decide whether the substance is pure or impure?

11-11. Does the following equation represent a replacement reaction? a metathesis reaction?

$$Cu(s) + 4HNO_3(aq) \rightarrow Cu(NO_3)_2(aq) + 2H_2O(\ell) + N_2O_4(g)$$

Did you find the first four questions easier than the next three? Probably you did. The first four questions require you to apply a simple rule that you have learned. If you recall that a carbonate is any compound containing the CO_3^{2-} ion, you need only look at the formula to see whether it contains that group. Now contrast this with what you must do in Question 11-5.

Rocks don't come with formulas written on them. A rock with one chemical formula may look like hundreds of rocks with different chemical formulas. Unless you know a way to determine the formula or know some reaction that occurs only with carbonates, Question 11-5 is a tough one to answer.

Questions 11-2, 11-3, and 11-4 are much like Question 11-1. If you recall that an element has only one kind of atom, whereas compounds have at least two kinds of atoms, Question 11-2 is answered. The defining characteristic of metathesis reactions is that compounds "exchange partners," and replacement reactions are characterized by one element replacing another element in a compound. As long as these questions are discussed in terms of formulas and equations that represent atoms, there is no difficulty.

Questions 11-5, 11-6, and 11-7 are asking about macroscopic behavior of matter. Formulas and equations are unknown, and atoms and molecules cannot be seen. Identifying characteristics of the matter that will allow classification may be very difficult.

We have paused to discuss the classification of matter in order to emphasize several points. First, we think it is important for you to keep in mind that formulas and equations are not chemistry. They are merely ways of talking about chemistry. Chemistry is what happens to matter. It takes place in the air, in the earth, in plants, and in you. It is often complex. It is often difficult to understand. It is easier to talk about the chemistry we *do* understand by using formulas and equations and the other chemical language that you have been learning.

Figure 11.1 *These foods taste sour because they contain acids.* (*Photo by Tom Greenbowe and Keith Herron.*)

11.2 ACIDS

We want to talk about an important class of compounds called acids. We will begin with some of the observations that originally led to the classification of certain substances as acids and then move to a microscopic description that helps us explain these observations.

The English word "acid" comes from *acidus,* Latin for "sour." Sour taste is the most prominent characteristic of the compounds called acids. From taste alone, you can be quite confident that cherries, lemons, oranges, buttermilk, yogurt, rhubarb, strawberries, and tomatoes all contain acids. Hydrochloric acid (HCl), sulfuric acid (H_2SO_4), and nitric acid (HNO_3)—compounds that we have already discussed—also taste sour, but you should not taste them. Concentrated solutions of these acids can destroy flesh and cause death.

Many acids are toxic. The leaves of rhubarb contain high concentrations of oxalic acid ($H_2C_2O_4$), which is poisonous. Be sure you don't use the leaves when you make a rhubarb pie!

Vinegar, the end product of fermentation of the juice from apples, grapes, and other fruits, contains acetic acid. Vinegar made from grapes (wine vinegar) has been known for thousands of years, and the Latin name for vinegar, *acetum,* is closely related to the Latin name for acid. It is reported that Roman women checked the strength of the vinegar they bought by pouring a little on the ground to see whether bubbles formed. The dirt had carbonates in it. Carbonate reacts with an acid to produce CO_2 gas, and this reaction is often used to identify either a carbonate or an acid.

11.3 PROPERTIES OF ACIDS

Mention acid to most individuals and they think of a dangerous liquid that will "eat through" metal. Strong acids *do* dissolve many metals, but dilute solutions of weak acids react very slowly with most metals. Otherwise you would have trouble finding a pan to cook foods like tomatoes or rhubarb. Even so, the acids in sour fruits react fast enough with aluminum to make it unwise to store these foods in aluminum for very long. There is little danger that the pan will spring a leak, but enough aluminum may dissolve to affect the taste of the food.

One of the most interesting characteristics of acids is the reaction between acids and another class of compounds that have come to be called indicators. As you might guess, an **indicator** is a compound that indicates something. In this case, we refer to compounds that indicate whether an acid is present. The indicator sends its message by changing color.

You have probably observed an indicator changing color in the presence of acid, but may not have paid much attention. Ordinary tea contains a compound that acts as an acid indicator. When you add lemon juice to tea, the tea becomes lighter in color. The color change is due to a reaction between the acid in the lemon and the indicator compound in the tea. You can get the dark color back again by adding a little household ammonia or some baking soda, compounds that destroy acids (but don't drink it).

Figure 11.2
When lemon juice is added to tea, the tea becomes lighter in color. Tea acts as an acid indicator. (*Photo by Tom Greenbowe and Keith Herron.*)

There are many naturally occurring pigments that act as acid indicators. Extracts from many flowers work, and the juice of purple cabbage is effective as an indicator. The color change in many of these indicators is difficult to see, and they aren't used as indicators in practice. A few of the compounds that are commonly used as indicators are listed in Table 11.1 (p. 306).

Table 11.1. Common Acid Indicators

Indicator	*Color in acid*	*Color in base*
methyl orange	red	yellow
phenol red	yellow	red
bromothymol blue	yellow	blue
phenolphthalein	colorless	red
litmus	red	blue

11.4 DEFINING AN ACID

Listed below are equations for several reactions involving acids:

$$\mathbf{CaCO_3}(s) + \mathbf{2HCl}(aq) \rightarrow CaCl_2(aq) + \mathbf{CO_2}(g) + H_2O(l)$$
A carbonate plus an acid produces carbon dioxide gas.

$$\mathbf{CuO}(s) + \mathbf{H_2C_2O_4}(aq) \rightarrow CuC_2O_4(s) + \mathbf{H_2O}(l)$$
A metal oxide plus an acid produces water.

$$\mathbf{Zn}(s) + \mathbf{2HCl}(aq) \rightarrow ZnCl_2(aq) + \mathbf{H_2}(g)$$
An active metal plus an acid produces hydrogen gas.

$$\mathbf{Mg(OH)_2}(s) + \mathbf{2HNO_3}(aq) \rightarrow Mg(NO_3)_2(aq) + \mathbf{H_2O}(l)$$
A hydroxide plus an acid produces water.

Hydrochloric acid, oxalic acid, and nitric acid are shown in these reactions, but just about any acid could be substituted. Based on this information, answer Question 11-12.

11-12. You are handed a bottle containing a liquid solution. List observations that you could make in order to classify the liquid as an acid or a nonacid. Your list of observations may include properties such as color or smell as well as reactions that you might observe when the liquid is added to other chemicals.

In Question 11-12 you listed macroscopic observations that enable you to classify a substance as an acid. Knowing the microscopic nature of acids can lead to even greater understanding, just as the microscopic notion of atoms and molecules makes it easier to talk about elements and compounds.

You may have noticed that all of the compounds we have described as acids contain hydrogen. Furthermore, in the net ionic equations for all of the reactions involving acids, the acid is represented by the hydrogen ion, H^+. Compounds that behave as acids appear to provide hydrogen ions for reaction. Indeed, this is one way to define an acid. *An acid is any compound that provides hydrogen ions.*

Before going on, think about what a hydrogen ion is. In Chapter 7 you learned that all atoms are made up of subatomic particles called electrons, protons, and neutrons. Answer the following questions to see what you remember.

11-13. How many protons are there in the nucleus of a hydrogen atom? (Where do you get that information?)
11-14. How many electrons are there in a hydrogen atom? (Where do you get that information?)
11-15. What must be done to a hydrogen atom in order to produce a hydrogen ion?
11-16. There are no neutrons in the nucleus of a hydrogen atom. If this is so, what is another name for a hydrogen ion?

In view of the last question, I hope you see that *an acid is any compound that can donate a proton to some other particle.* Figure 11.3 illustrates the difference between a hydrogen atom and a hydrogen ion.

Figure 11.3
(a) A hydrogen atom consists of a single proton in the nucleus and a single electron surrounding it. (b) A hydrogen ion consists of a single proton.

11.5 BASES

A tiny proton with its positive charge does not remain unattached. If an acid molecule is to act as a proton donor, there must be something around to act as the proton acceptor. That something is called a **base.**

The net ionic equation for the reaction between an acid and a hydroxide is

$$OH^- + \mathbf{H^+} \rightarrow \mathbf{HOH}$$

This equation shows that the proton (hydrogen ion) becomes attached to the hydroxide ion to form the neutral molecule water (Figure 11.4). We might

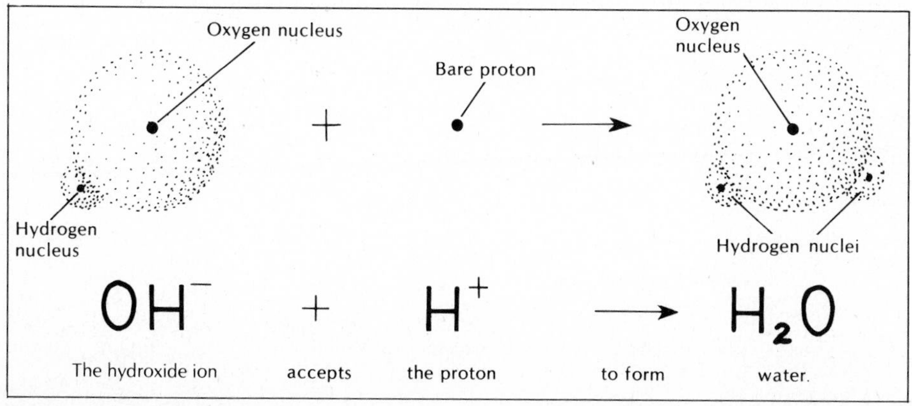

Figure 11.4
The hydroxide ion acts as a base, because it accepts a proton. A base is a proton acceptor.

describe what is happening by saying that *the hydroxide ion acts as a proton acceptor.* In other words, the hydroxide ion acts as a base.

The net ionic equation used to describe the reaction between an acid and a metallic oxide is

$$O^{2-} + 2H^+ \rightarrow H_2O$$

This equation indicates that two protons become attached to the oxide ion to form water. The oxide ion acts as a proton acceptor, as shown in Figure 11.5. It is a base.

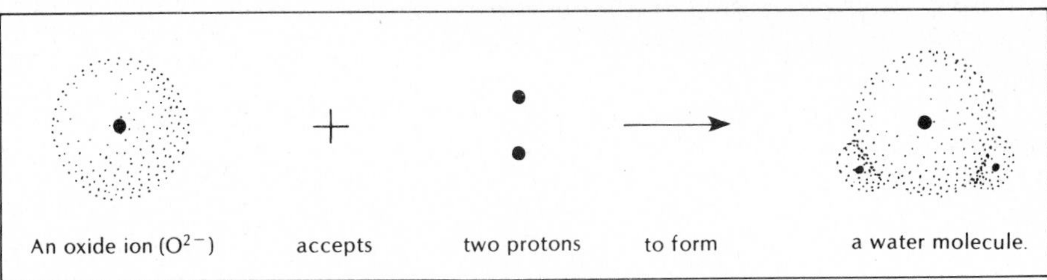

Figure 11.5
The oxide ion can also act as a base.

It is not apparent that CO_3^{2-} is a base from the net ionic equation for the reaction of an acid with a carbonate. The equation is

$$CO_3^{2-} + 2H^+ \rightarrow CO_2 + H_2O$$

The basic character of the carbonate ion is hidden by the fact that the equation describes the end result of three separate processes. In the first step, the carbonate ion accepts a proton to form the bicarbonate ion:

$$CO_3^{2-} + H^+ \rightarrow HCO_3^-$$

In this step the carbonate ion is clearly acting as a base. The bicarbonate ion that is formed can also act as a base, accepting an additional proton:

$$HCO_3^- + H^+ \rightarrow H_2CO_3$$

The carbonic acid shown as the final product is not a stable molecule. It easily dissociates into the two molecules shown as products in the original equation:

$$H_2CO_3 \rightarrow H_2O + CO_2$$

So far our examples of proton acceptors (bases) have all been ions. Although many bases are negative ions, neutral molecules can also act as bases.

Household ammonia is a base. It reacts with vinegar as indicated by the equation

$$NH_3 + \mathbf{HC_2H_3O_2} \rightarrow \mathbf{NH_4^+} + C_2H_3O_2^-$$

In this reaction, the proton from one of the hydrogen atoms in hydrogen acetate is transferred (leaving its electron behind) to the neutral ammonia molecule. The ammonia molecule accepts the proton and becomes an ammonium ion with a positive charge. The acetate ion has a negative charge, because only the hydrogen nucleus (proton) was transferred. The electron from the hydrogen atom is left behind. In this reaction, hydrogen acetate acts as a proton donor. It is an acid. The ammonia acts as a proton acceptor. It is a base.

Water molecules can also act as a base. When hydrogen chloride gas dissolves in water, the following proton transfer takes place, as shown in Figure 11.6.

$$H_2O + \mathbf{HCl} \rightarrow \mathbf{H_3O^+} + Cl^-$$

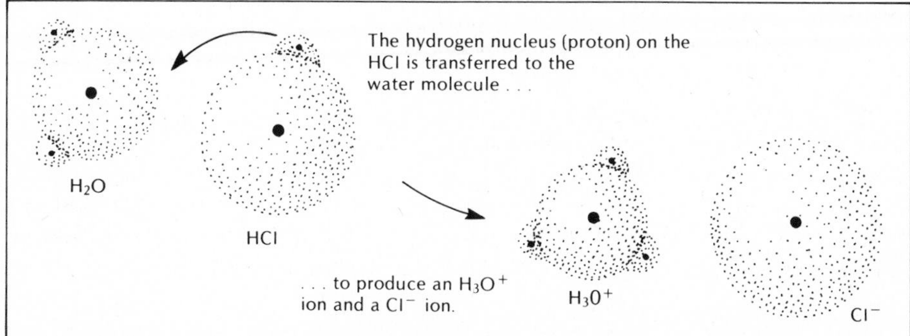

Figure 11.6 *Even water can act as a base.*

The H_3O^+ is the **hydronium ion,** and is present in water solutions of strong acids. Before going on, make sure that you have the idea by identifying the acid and base in the following equations.

In each equation, write "acid" under the formula of the reactant acting as an acid and "base" under the formula of the reactant acting as a base.

11-17. $HNO_3 + KOH \rightarrow KNO_3 + H_2O$

11-18. $NH_4^+ + NH_2^- \rightarrow 2NH_3$

11-19. $NH_3 + HCl \rightarrow NH_4Cl$

11-20. $H_2O + H_2SO_4 \rightarrow H_3O^+ + HSO_4^-$

11-21. $HCO_3^- + PO_4^{3-} \rightarrow CO_3^{2-} + HPO_4^{2-}$

11-22. $HCO_3^- + HC_2H_3O_2 \rightarrow H_2CO_3 + C_2H_3O_2^-$

Now that you know that a base is a proton acceptor and can identify one in an equation, could you identify one in real life? Probably not. A solution of ammonia doesn't look much different from a solution of hydrogen chloride, but ammonia acts as a base and HCl acts as an acid. Ordinary lye is a white crystalline solid that looks a lot like table salt. Lye is a very strong base; salt isn't.

11.6 PROPERTIES OF BASES

We began our discussion of acids with a description of their macroscopic properties. We will end our discussion of bases with the same kind of description.

BASES TASTE BITTER

When profanity was considered crude instead of chic, parents were known to wash their children's mouths out with soap. It was not a delightful taste sensation. Soap is basic and, like all bases, it tastes bitter. Don't go around tasting things to see whether they are bases. Many bases are toxic. Common bases such as lye and aqueous ammonia can do serious damage to your insides.

BASES NEUTRALIZE ACIDS

When a base acts as a proton acceptor, it neutralizes the acid. After ammonia has been added to vinegar, the vinegar no longer tastes sour, reacts with carbonates to give carbon dioxide, or combines with hydroxides to give water. The properties that made vinegar an acid are no longer present. Neither, we might add, are the properties present that made ammonia a base. To neutralize means "to make something ineffective; counteract; nullify." This is exactly what happens when bases and acids combine. The properties of both are nullified.

BASES AFFECT INDICATORS

The acid indicators listed in Table 11.1 (p. 306) have a different color in the presence of bases than they have in the presence of acids. They are called acid indicators in the table, but they are normally referred to as acid-base indicators, because they indicate both classes of compounds equally well. As you will see in a later section, indicators are actually weak acids or weak bases themselves.

OH^-, THE MOST COMMON BASE

The first example we gave of a base was the hydroxide ion. It should be first, because it is by far the most common base in water solution. Nevertheless, getting hydroxide into solution where it can do its work as a base can be a problem at times. You were told in Chapter 10 that most hydroxides are insoluble in water. The major exceptions are hydroxides of the Group IA metals, such as NaOH and KOH. Solid hydroxides such as $Ca(OH)_2$ and $Mg(OH)_2$ react with acid solutions, but they don't dissolve enough in water to affect some indicators.

11.7 STRONG AND WEAK

In the preceding sections, we have described acids and bases as *strong* or *weak*. This is an important distinction, and we should explain what it means.

We begin by saying what we do *not* mean. In Chapter 10 we spoke of strong and weak coffee and pointed out that what people mean by these terms is what the chemist would call *concentrated* and *dilute*. In everyday speech, "strong" and "weak" generally refer to concentration. In chemistry they do not.

A solution of 0.01 *M* HCl would be considered rather dilute, but it is a solution of a strong acid. A solution of 10 *M* $HC_2H_3O_2$ is very concentrated, but it contains a weak acid. To clarify what we mean by strong and weak, we will discuss the terms in several ways.

"STRONG" AND "WEAK" REACTIONS

When 10 *M* HCl is poured over limestone, CO_2 is produced very rapidly, and the material bubbles like a hot cola opened just after shaking. When 10 *M* $HC_2H_3O_2$ is poured over limestone, bubbles certainly form, but the action is much less violent. The action of the acetic acid is "weaker" than the action of the hydrochloric acid.

Something similar happens when metals are added to these two acids. Hydrogen gas is produced in both cases, but the action is much more vigorous with HCl than with $HC_2H_3O_2$ of the same concentration. Observations of this kind led to the description of one acid as strong and the other acid as weak.

Both reactions involve the H^+ ion, and both HCl and $HC_2H_3O_2$ appear capable of furnishing the same number of these ions (one for each molecule in solution). Why, then, does one acid react more vigorously than the other? The explanation lies in another observation of these two acids. Before we mention it, we want to see whether you remember enough from Chapter 7 to interpret the observation.

11-23. A conductivity apparatus is placed in a solution and the bulb lights, as shown in Figure 11.7 (p. 312). What does this tell you about the particles in solution?

11-24. Would you expect the bulb on a conductivity apparatus to light when the electrodes are placed in a solution of HCl? Of $HC_2H_3O_2$? Why?

11-25. Which would you expect to show more conductivity (brighter bulb), a 0.01 *M* solution of HCl or a 10 *M* solution of HCl? Why?

Figure 11.7
What kind of particles in solution cause the bulb to light? (Photo by Tom Greenbowe and Keith Herron.)

Now that you have refreshed your memory about what conducts electricity in solutions, you are ready to examine conductivity in HCl and $HC_2H_3O_2$. Figures 11.8 and 11.9 show conductivity in 10 *M* HCl and 10 *M* $HC_2H_3O_2$, respectively. The HCl solution is an excellent conductor of electricity. The light is "strong," indicating an abundance of ions in the solution. By contrast, the conductivity in the 10 *M* $HC_2H_3O_2$ is barely detectable. The light is "weak," indicating very few ions in solution.

Even though equal volumes of these solutions contain the same number of solute *molecules,* the number of *ions* in the HCl solution is far greater than the number of ions in the acetic acid solution. Since H^+ ions account for acid properties, HCl is a stronger (more potent) acid than acetic acid.

Figure 11.8
HCl(aq) is an excellent conductor of electricity, indicating that there are many ions in solution. (Photo by Tom Greenbowe and Keith Herron.)

Figure 11.9
$HC_2H_3O_2(aq)$ is a poor conductor of electricity, indicating that there are few ions in solution. (Photo by Tom Greenbowe and Keith Herron.)

If we could magnify the solutions of these acids until the solute particles became visible, we would see something like what is illustrated in Figure 11.10. Virtually all of the molecules of HCl are **dissociated** or broken apart into H^+ and Cl^- ions. By contrast, only a few of the $HC_2H_3O_2$ molecules are dissociated into H^+ and $C_2H_3O_2^-$ ions. Figure 11.10 is actually misleading, because it shows one out of three of the acetic acid molecules dissociated. In fact, only about one out of every thousand molecules breaks apart in a 10 *M* solution of acetic acid. Acetic acid is a weak acid, because it exists in solution as molecules rather than ions.

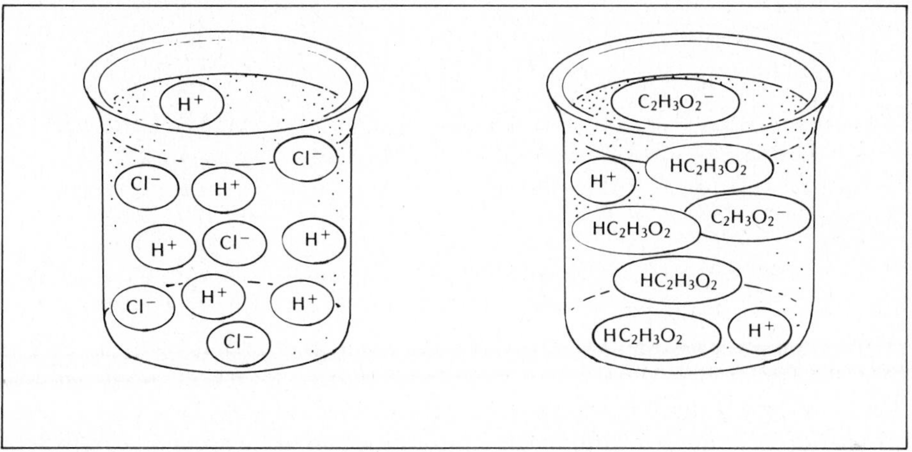

Figure 11.10
Since HCl(aq) is a good electrical conductor, we believe that most of the HCl molecules come apart to form ions, as shown on the left. Since $HC_2H_3O_2(aq)$ is a poor electrical conductor, we believe that just a few of the molecules come apart to form ions, as shown on the right.

When chemists speak of strong or weak acids and bases, they refer to how much the compound dissociates into ions. A **strong** acid or base is one that dissociates completely into ions when dissolved in water. A **weak** acid or base is one that dissociates very little.

Let us now look at "strong" and "weak" in a third way. We have defined an acid as a proton donor and a base as a proton acceptor. We can imagine the proton being pushed and pulled between acids and bases in solution. An acid is "pushing" the proton onto some other molecule or ion. A base is "pulling" the proton away from some other molecule or ion. In this frame of reference, *a strong acid is one that donates protons to other molecules very easily. A weak acid is one that has little tendency to donate protons.* By contrast, *a strong base is one that readily accepts protons. A weak base is one that has little tendency to accept protons.*

Every weak acid has a strong base associated with it. Every strong base has a weak acid associated with it. This will make more sense when we discuss it in terms of something like acetic acid.

11.8 REVERSIBLE REACTIONS AND EQUILIBRIUM

Before looking at the strong base and weak acid in acetic acid, we need to clarify a point that has been avoided so far. We have discussed chemical reactions as though they take place on a one-way street. We have written the reactants on the left of a chemical equation, we have written the products on the right, and we have separated the two by an arrow pointing in the direction of the products. For example, we have represented the burning of hydrogen gas in oxygen like this:

$$2H_2 + O_2 \rightarrow 2H_2O$$

The equation says that we start with hydrogen and oxygen and end up with water. This is a reasonable description of what actually happens.

We have also studied the electrolysis of water, and we have represented it like this:

$$2H_2O \rightarrow 2H_2 + O_2$$

This equation says that we start out with water and end up with hydrogen and oxygen. Once again, it is a reasonable description of what actually happens. Still, it should be obvious that the second equation is just the reverse of the first. Although it would be unconventional to do so, we might represent the second equation like this:

$$2H_2 + O_2 \leftarrow 2H_2O$$

Here we have placed the reactants on the right and changed the direction of the arrow to indicate the direction of the chemical change. We might even represent *both* chemical changes (the burning of hydrogen to give water and the electrolysis of water to give hydrogen and oxygen) by the same equation, like this:

$$2H_2 + O_2 \rightleftarrows 2H_2O$$

The arrow going to the right indicates that hydrogen gas and oxygen gas can combine to form water. The arrow going to the left indicates that water can decompose to form hydrogen gas and oxygen gas, the reverse of the first reaction. The two arrows signal a **reversible reaction.**

The two reactions that we are discussing would not normally be shown together, because the conditions necessary for the two reactions to occur are very different. To get hydrogen to burn, we must ignite it with a match or electric spark, and it continues burning until all of the hydrogen or all of the oxygen is consumed. To get water to decompose, we must add an ionic compound to the water and then pass an electric current through the water solution. The reaction continues only so long as we keep the current going.

Although the reaction of hydrogen and oxygen to give water is not easily reversed, many reactions, including most dissociations, are. When acetic acid is dissolved in water, it dissociates as shown by the equation

$$HC_2H_3O_2(aq) \rightarrow H^+(aq) + C_2H_3O_2^-(aq)$$

This equation says, "One mole of acetic acid dissociates to form one mole of hydrogen ions and one mole of acetate ions." Although this is our normal translation of an equation, it can be misleading. It seems to say that, if I poured a mole of acetic acid molecules into a container of water, all of the molecules would dissociate to give ions. From our observations on the conductivity of acetic acid solutions, we know that this isn't so. Very few of the molecules dissociate. A better way to read the equation would be, "For every mole of acetic acid that dissociates, one mole of hydrogen ions and one mole of acetate ions are produced." This statement says nothing about how many molecules dissociate. It just says that each molecule that *does* dissociate produces one hydrogen ion and one acetate ion.

We can readily cause the reverse of this reaction to take place, simply by mixing the two ions involved. We could get the hydrogen ions from a solution of a strong acid like HCl, and we could get the acetate ions from a solution of sodium acetate. The reaction that would take place is described by the following equation:

$$\mathbf{H^+}(aq) + Cl^-(aq) + Na^+(aq) + \mathbf{C_2H_3O_2^-}(aq) \rightarrow \mathbf{HC_2H_3O_2}(aq) + Na^+(aq) + Cl^-(aq)$$

It is easier to see that this is the reverse of the dissociation of acetic acid when the reaction is described by a net ionic equation:

$$H^+(aq) + C_2H_3O_2^-(aq) \rightarrow HC_2H_3O_2(aq)$$

It does not matter where the hydrogen and acetate ions come from. If they are present in the same solution, they can combine to form hydrogen acetate. They could come from the dissociation of hydrogen acetate itself. In other words, when hydrogen acetate dissociates to form hydrogen and acetate ions, some of those hydrogen and acetate ions recombine to form hydrogen acetate molecules. We say that the reaction is reversible, and we write the equation with arrows pointing in both directions.

$$HC_2H_3O_2(aq) \rightleftarrows H^+(aq) + C_2H_3O_2^-(aq)$$

When hydrogen acetate dissolves in water, some of the molecules begin to dissociate immediately, as shown in Figure 11.11 (p. 316). Since there are no

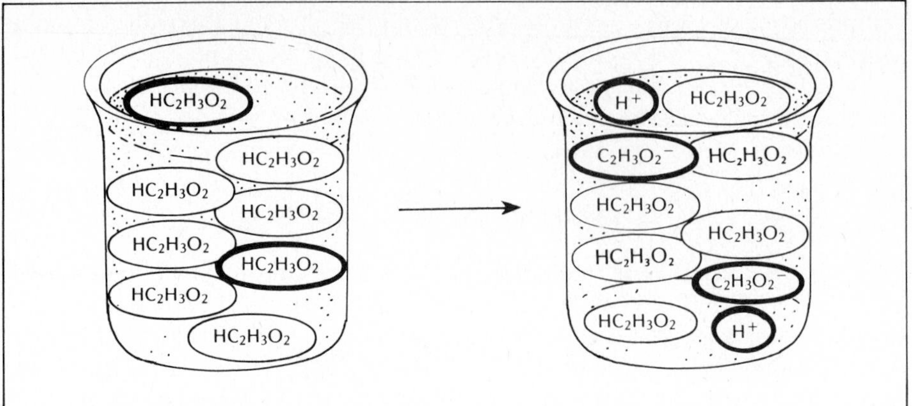

Figure 11.11 *When hydrogen acetate dissolves in water, some of the molecules dissociate to form ions.*

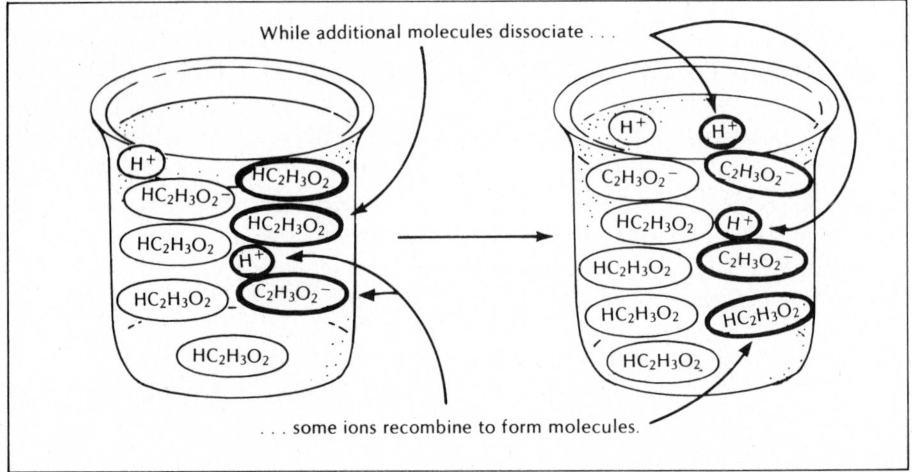

Figure 11.12 *As more molecules dissociate, some of the ions recombine to form molecules again.*

ions in solution at the beginning, the reverse reaction can't occur. However, ions are produced as molecules dissociate, and both reactions can take place, as shown in Figure 11.12.

Eventually the rate at which molecules break apart to form ions is the same as the rate at which ions recombine to form molecules, and the chemical system is described as being in **equilibrium.** At equilibrium, there is no apparent change taking place, because one change exactly balances the other.

When a system finally comes to equilibrium, there may be a large number of ions present or only a few. In the case of acetic acid, we know that there are only a few, because a solution of acetic acid is a poor conductor of electricity. We sometimes represent this fact by drawing one of the arrows in the equation longer than the other:

$$HC_2H_3O_2 \longleftarrow\!\rightharpoonup H^+ + C_2H_3O_2^-$$

When HCl dissolves in water, virtually all of the molecules dissociate to form ions. This fact is emphasized by the equation

$$HCl \leftarrow\!\!\longrightarrow H^+ + Cl^-$$

HCl is such a strong acid that it is usually considered totally dissociated in water, and the arrow pointing to the left is omitted.

When we observe that very few $HC_2H_3O_2$ molecules dissociate, we conclude that the acetate ion is a stronger base than $HC_2H_3O_2$ is an acid.

$$\underset{\text{weak acid}}{HC_2H_3O_2(aq)} \longleftarrow\!\!\rightarrow \underset{\text{strong base}}{H^+(aq) + C_2H_3O_2^-(aq)}$$

Molecules and ions that differ only by a proton (hydrogen ion) are sometimes referred to as **conjugate acid-base pairs.** "Conjugate" means "joined together; coupled; or having a common derivation." Conjugate acid-base pairs are acids and bases that go together. The acid $HC_2H_3O_2$ and the base $C_2H_3O_2^-$ go together. The acid form of the pair contains a proton that is missing in the base form of the pair. That is the only difference between them, and it is the only difference between the members of any conjugate acid-base pair.

If a proton attaches firmly to the base form of an acid-base pair, the pair contains a strong base and a weak acid, as illustrated by $HC_2H_3O_2$ and $C_2H_3O_2^-$. If the proton does not attach firmly to the base form of an acid-base pair, the pair contains a strong acid and a weak base. An example is HCl and Cl^-. Here the acid is almost totally dissociated in water solution, as illustrated by the following equation.

$$\underset{\text{strong acid}}{HCl} \leftarrow\!\!\longrightarrow \underset{\text{weak base}}{H^+ + Cl^-}$$

Table 11.2 shows the relative strength of several acid-base pairs. In the table, the strongest acids are at the top and the strongest bases are at the bottom.

Table 11.2. Relative Strength of Common Acids and Bases

	Acid name	Acid formula		Base formula		Base name
STRONGER ACIDS ↑	hydrochloric	HCl	$\rightleftarrows$	$H^+ + Cl^-$	STRONGER BASES ↓	chloride ion
	nitric	HNO_3	$\rightleftarrows$	$H^+ + NO_3^-$		nitrate ion
	sulfuric	H_2SO_4	$\rightleftarrows$	$H^+ + HSO_4^-$		hydrogen sulfate ion
	hydronium ion	H_3O^+	$\rightleftarrows$	$H^+ + H_2O$		water
	oxalic	$H_2C_2O_4$	$\rightleftarrows$	$H^+ + HC_2H_4^-$		hydrogen oxalate ion

Table 11.2. Relative Strength of Common Acids and Bases (Continued)

	Acid name	Acid formula		Base formula		Base name
↑ STRONGER ACIDS	hydrogen sulfate ion	HSO_4^-	⇄	$H^+ + SO_4^{2-}$	STRONGER BASES ↓	sulfate ion
	phosphoric	H_3PO_4	⇄	$H^+ + H_2PO_4^-$		dihydrogen phosphate ion
	formic	$HCHO_2$	⇄	$H^+ + CHO_2^-$		formate ion
	hydrogen oxalate ion	$HC_2O_4^-$	⇄	$H^+ + C_2O_4^{2-}$		oxalate ion
	acetic	$HC_2H_3O_2$	⇄	$H^+ + C_2H_3O_2^-$		acetate ion
	carbonic	H_2CO_3	⇄	$H^+ + HCO_3^-$		bicarbonate ion
	dihydrogen phosphate ion	$H_2PO_4^-$	⇄	$H^+ + HPO_4^{2-}$		hydrogen phosphate ion
	hypochlorus	$HClO$	⇄	$H^+ + ClO^-$		hypochlorite ion
	boric	H_3BO_3	⇄	$H^+ + H_2BO_3^-$		dihydrogen borate ion
	ammonium ion	NH_4^+	⇄	$H^+ + NH_3$		ammonia
	hydrocyanic	HCN	⇄	$H^+ + CN^-$		cyanide ion
	hydrogen carbonate ion	HCO_3^-	⇄	$H^+ + CO_3^{2-}$		carbonate ion
	monohydrogen phosphate ion	HPO_4^{2-}	⇄	$H^+ + PO_4^{3-}$		phosphate ion
	water	H_2O	⇄	$H^+ + OH^-$		hydroxide ion

11.9 STRONG AND WEAK: A SUMMARY

We have discussed three ways to interpret the terms "strong" and "weak." First, we said that strong acids are those that react vigorously with carbonates and metals, while weak acids are those that produce a "weaker" reaction. Similar comments could be made about the vigor of reactions involving strong and weak bases.

Next, we pointed out that reactions of acids involve the proton or hydrogen ion, and we noted that the difference between strong acids and weak acids, when viewed at the microscopic level, is that strong acids dissociate completely (or nearly so) to form hydrogen ions, whereas weak acids dissociate very little. A strong acid is a strong electrolyte; a weak acid is a weak electrolyte. Weak bases would also be expected to be weak electrolytes.

Finally, we pointed out that acid-base reactions can be considered as competitions between two molecules or ions for protons. A strong base such as OH^- readily accepts protons to form water, and it is difficult to take the proton away from water to form the OH^- ion again. We say that OH^- is a strong base but that water is a very weak acid. By contrast, Cl^- has very little tendency to accept a proton to form HCl. We say that the Cl^- ion is a very weak base but that HCl is a very strong acid. "Strong" and "weak" refer to how tightly the proton is attached to the rest of the molecule at hand.

Before going on, be sure that you understand these ideas by answering the following questions.

11-26. When magnesium metal is added to dilute HNO_3, a very rapid reaction occurs to produce hydrogen gas and magnesium nitrate. On the basis of this information, would you guess that nitric acid is a strong acid or a weak acid?

11-27. Would a solution of nitric acid contain mostly HNO_3 molecules or mostly H^+ ions and NO_3^- ions?

11-28. Would a solution of HNO_3 be a good conductor of electricity or a poor conductor of electricity?

11-29. A solution of HBr is an excellent conductor of electricity. Is HBr a strong acid or a weak acid?

11-30. When oxalic acid and the fluoride ion are in solution, the following reaction takes place:

$$H_2C_2O_4 + F^- \rightleftarrows HC_2O_4^- + HF$$

Which is the stronger base, $HC_2O_4^-$ or F^-?

11-31. Which is the stronger acid, $H_2C_2O_4$ or HF?

11-32. On the basis of the following equation, which is the stronger acid, $H_2PO_4^-$ or $HC_3H_5O_2$?

$$HPO_4^{2-} + HC_3H_5O_2 \rightleftarrows H_2PO_4^- + C_3H_5O_2^-$$

11-33. HPO_4^{2-} can also act as an acid, even though it acts as a base in the previous reaction. Is HPO_4^{2-} a stronger acid than $HC_3H_5O_2$? How do you know?

11-34. When zinc is added to a saturated solution of oxalic acid, hydrogen gas evolves very slowly. Would you guess that oxalic acid is completely dissociated in water? Would you guess that oxalic acid is strong or weak?

11-35. Solutions of strong acids are very irritating to the skin and mucous membranes of the body. Boric acid is used to wash out eyes. Do you think that boric acid is completely dissociated in water? Why?

11.10 INDICATORS

Acid-base indicators are large molecules that can act as weak acids or weak bases like the compounds and ions listed in Table 11.2. Some of them contain hydrogen atoms that can dissociate to give hydrogen ions in much the same way that HCl dissociates in water to give hydrogen ions. In the following equation, HInd represents such an indicator.

$$\underset{\text{acid form}}{HInd} \rightleftarrows H^+ + \underset{\text{base form}}{Ind^-}$$

Ind is not the symbol for an element. It is used here to represent any indicator species. The H in the formula represents the hydrogen atom that becomes an ion when the indicator acts as an acid.

Some indicators are weak bases, and the neutral molecule acts as the proton acceptor rather than the proton donor. This is shown by the following equation:

$$\underset{\text{acid form}}{HInd^+} \rightleftarrows \underset{\text{base form}}{H^+ + Ind}$$

Note that this equation is very similar to the equation showing the NH_4^+ and NH_3 acid-base pair in Table 11.2 (pp. 317–318).

Whether an indicator is a weak acid (as shown by the first equation) or a weak base (as shown by the second equation), the molecule can exist in two forms that have a different color. For example, the acid form of methyl orange ($HInd^+$) is red, and the base form (Ind) is yellow. The acid form of phenolphthalein (HInd) is colorless, and the base form (Ind^-) is red.

When an indicator is in a solution with a high concentration of hydrogen ions, most of the molecules exist in the acid form and the solution takes on the color of the acid form of the indicator. When the indicator is in a solution containing a stronger base, most of the indicator molecules give up their proton and exist in the base form. The solution then takes on the color of the base form of the indicator. Thus the indicator provides information about the concentration of hydrogen ions in the solution.

11.11 HYDROGEN ION CONCENTRATION

Every water solution contains some hydrogen ions. As shown in the last equation in Table 11.2 (pp. 317–318), water can dissociate to give hydrogen ions and hydroxide ions. Even though this *can* happen, few water molecules are dissociated in pure water. At equilibrium, only about 2 molecules of water in 1 billion break apart to form ions. This is not enough ions to conduct much electricity, and the bulb in a simple conductivity apparatus does not light when water is the conducting liquid. However, more sensitive instruments show that the concentration of H^+ and OH^- in pure water is about 10^{-7} *M*. Hence the product of the hydrogen ion concentration (normally represented as $[H^+]$) and the hydroxide ion concentration (represented as $[OH^-]$) is equal to 10^{-14}.

This product remains constant in water solutions. When acids are added to increase the $[H^+]$, some of these added hydrogen ions combine with the hydroxide ions to form water molecules so that the $[OH^-]$ is reduced. The product of $[H^+]$ and $[OH^-]$ is still 10^{-14}. When a base such as NaOH is added to water, the additional hydroxide ions combine with some of the hydrogen ions, lowering the $[H^+]$ concentration so that the product of hydrogen ion concentration and hydroxide ion concentration is still 10^{-14}.

11-36. Write a mathematical equation that says, "The product of $[H^+]$ and $[OH^-]$ is always equal to 10^{-14}."

11-37. When the hydrogen ion concentration in a water solution is 0.01 *M*, what is the hydroxide ion concentration?

11-38. What must the hydrogen ion concentration be in order for the hydroxide ion concentration in a water solution to be 10^{-5}?

Virtually all reactions in living systems take place in water solutions, and the hydrogen ion concentration influences the reactions that can take place. For this reason, chemists have found that it is important to measure hydrogen ion concentrations, even when they are very low.

As suggested by the $[H^+]$ in pure water (0.0000001 *M*), hydrogen ion concentrations do seem small. They can be much smaller. In a saturated solution of NaOH, the hydrogen ion concentration may be as low as 0.000000000000001 *M* (10^{-15} *M*). In a saturated solution of HCl, $[H^+]$ may be as high as 10 *M*. As you can see, the concentration of hydrogen ions in water solutions can vary a great deal. In fact, it varies so much that it is difficult to imagine.

Early in this course you found that a millionth of a sheet of typing paper was about the size of the period at the end of this sentence. The difference between the size of a sheet of paper and a period is great, but it is nothing compared to the differences we are talking about now. A millionth is 10^{-6}. The hydrogen ion concentration in water is 10^{-7}, only $\frac{1}{10}$ as much. Can you imagine $\frac{1}{10}$ of the size of a period? Well, that area compared to the area covered by a whole sheet of paper is similar to the comparison between the hydrogen ion concentration in water and that in a 10 *M* solution of HCl. It is a big difference, but it is only half the story.

Remember that in a saturated solution of NaOH, the hydrogen ion concentration can be as low as 10^{-15} *M*. That is 100 million times less than $[H^+]$ in pure water. To get some idea of the range of concentration we are talking about, try to imagine spreading out 100 million sheets of typing paper. (They would cover 2.33 square miles or, said another way, a square measuring about 1.53 miles on each side.) Now compare the size of $\frac{1}{10}$ of a period with that huge square of typing paper. The difference in size is comparable to the difference in hydrogen ion concentration that can occur in water solutions.

11.12 A SCALE FOR LARGE CHANGES: pH

We face difficulties when we try to represent something that varies over a wide range of values. To illustrate the problem, we will look at some data representing the $[H^+]$ during a chemical reaction. The reaction that we will discuss is the neutralization of a 1 *M* solution of NaOH with a 1 *M* solution of HCl. However, we will focus on the change in the $[H^+]$ during the reaction.

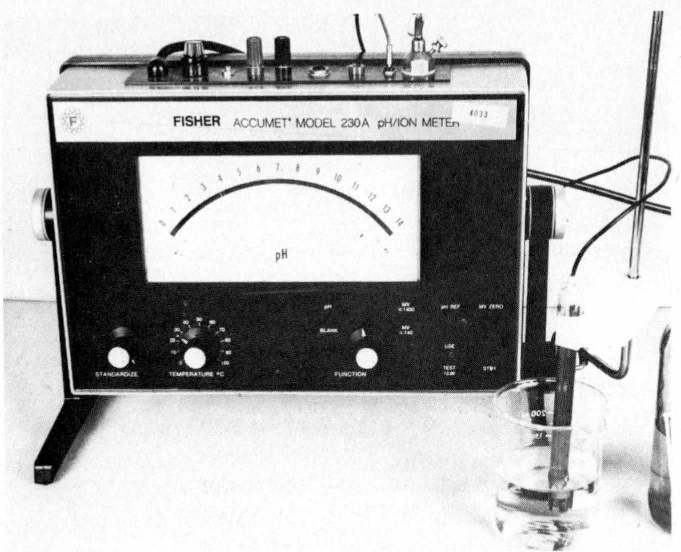

Figure 11.13
A pH meter. This electronic instrument can be used to measure the hydrogen ion concentration of a solution. (*Photo by Tom Greenbowe and Keith Herron.*)

We will start with 100 mL of the 1 *M* NaOH solution and will add 1 *M* HCl, a little at a time, until all of the NaOH has reacted and there is an excess of HCl. Each time we add some of the acid, we will measure the $[H^+]$ of the solution with the instrument shown in Figure 11.13 to see how it changes. Table 11.3 shows the $[H^+]$ at several steps along the way.

Table 11.3. $[H^+]$ as HCl Is Added to 100 mL of 1 *M* NaOH

Volume of HCl added (mL)	*Hydrogen ion concentration*	*Log $[H^+]$*
0	1.0×10^{-14}	−14.0
10	1.2×10^{-14}	−13.9
20	1.4×10^{-14}	−13.8
30	1.8×10^{-14}	−13.7
40	2.3×10^{-14}	−13.6
50	3.0×10^{-14}	−13.5
60	4.0×10^{-14}	−13.4
70	5.6×10^{-14}	−13.3
80	9.1×10^{-14}	−13.0
90	2.0×10^{-13}	−12.7
95	3.8×10^{-13}	−12.4
99	2.0×10^{-12}	−11.7
99.9	2.0×10^{-11}	−10.7
99.99	2.0×10^{-10}	− 9.7
100	1.0×10^{-7}	− 7.0
100.01	5.0×10^{-5}	− 4.3
100.1	5.0×10^{-4}	− 3.3
101	5.0×10^{-3}	− 2.3
105	2.4×10^{-2}	− 1.6
110	4.8×10^{-2}	− 1.4
120	9.1×10^{-2}	− 1.0
130	1.3×10^{-1}	− 0.88
140	1.7×10^{-1}	− 0.77
150	2.0×10^{-1}	− 0.70
200	3.3×10^{-1}	− 0.64

Now try to plot a graph of these data. Show the $[H^+]$ along the y axis and the volume of HCl added along the x axis. Try it . . . but you won't like it.

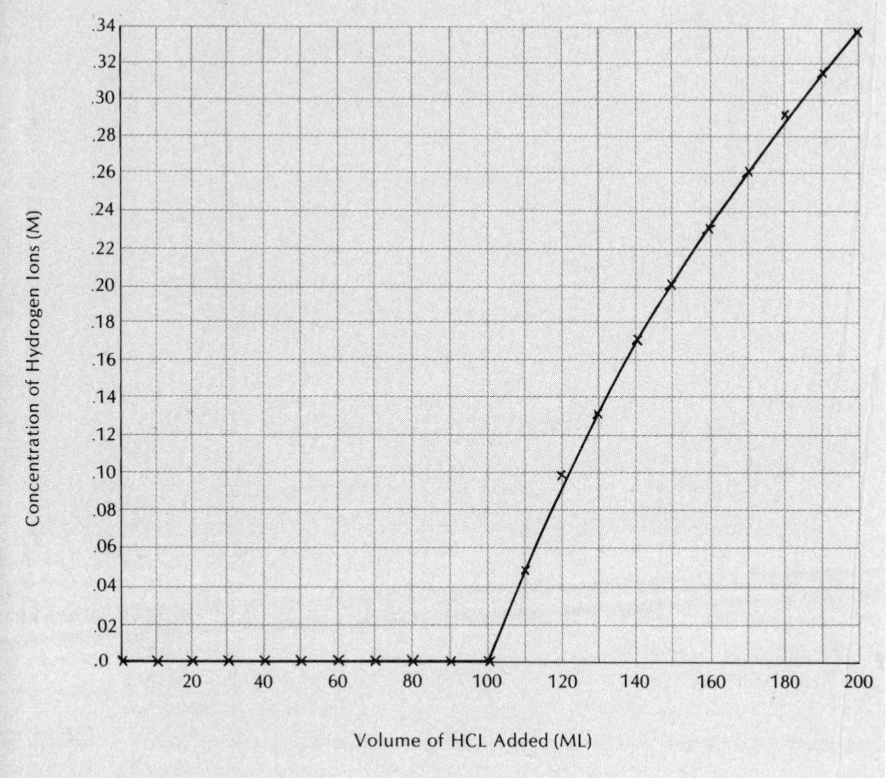

Figure 11.14
A graph showing the hydrogen ion concentration of the solution formed by adding 1 M HC1 to 100 ml of 1 M NaOH. Because of the scale used to represent the hydrogen ion concentration, it appears that there is no change in concentration until 100 ml of the acid have been added.

The problem is that you can't select a scale for the $[H^+]$ that will allow you to place all of the data on a single graph and still give an accurate picture of what is taking place. Figure 11.14 shows a graph that includes all of the data points, but the scale is so small that there appears to be no change in the $[H^+]$ until 100 mL of HCl have been added. Looking at Table 11.3, you know that this isn't true. The $[H^+]$ had doubled by the time 30 mL of acid had been added, and it had increased tenfold by the time 90 mL of acid had been added. This change doesn't show on the graph in Figure 11.14. The only way to make it show is to pick a different scale.

Figure 11.15 (p. 324) shows the data from Table 11.3 plotted with a new scale. Now the change in $[H^+]$ that occurs during the addition of the first 99 mL of acid can be seen. The problem is that the *rest* of the data can't be shown. It just isn't possible to select a scale that shows the changes taking place and still get all the data on the graph.

One solution to this problem is to transform the wide range of values to a logarithmic scale. The logarithm of a number is the power to which 10 must be raised to equal that number. (See *Study Guide* for a discussion of

Figure 11.15
A portion of the data shown in Figure 11.14 plotted with a different scale for the hydrogen ion concentration. Note that now the concentration appears to change very rapidly when 90–100 mL of the acid are added. Using this scale, however, it is impossible to get all the data on the graph.

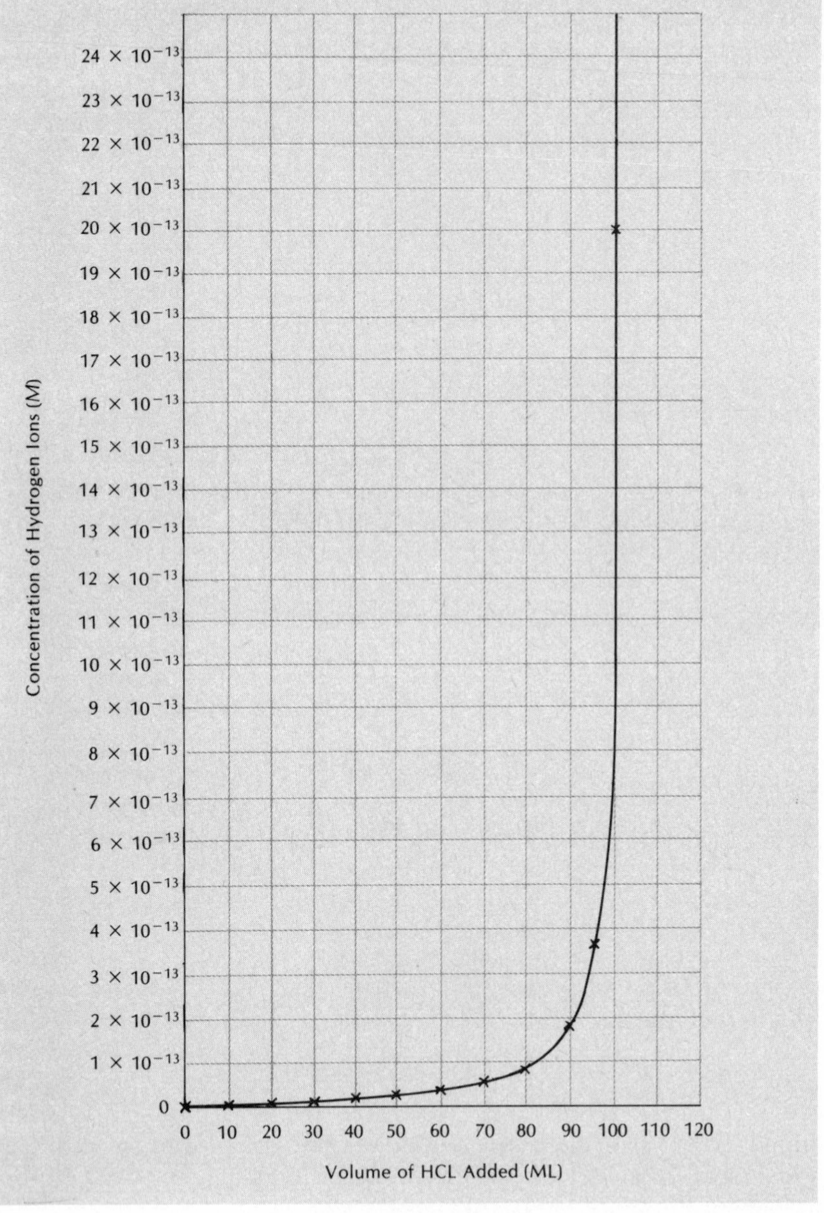

logarithms.) An increase of 1 in the logarithm of a number represents a tenfold increase in the number. This can be seen in Table 11.3 (p. 322).

The $[H^+]$ after 90 mL of acid were added is shown as 2.0×10^{-13}, and the logarithm of that number is shown as -12.7. When 99 mL of acid had been added, the $[H^+]$ had increased 10 times to a value of 2.0×10^{-12}. However, the logarithm increased only 1 to a value of -11.7.

While the $[H^+]$ recorded in Table 11.3 varies from 0.00000000000001 to 0.33 (a change of over 10 million million) the log $[H^+]$ changes only from -14 to near 0. Although the $[H^+]$ can't be represented conveniently on a graph, the log $[H^+]$ can be. The data are plotted in Figure 11.16. Note that the changes in $[H^+]$ during the early stages of the addition of acid are shown *and* that all of the data are plotted.

There is really only one thing that is inconvenient about the graph shown in Figure 11.16. All of the values for the log $[H^+]$ are negative. We are more accustomed to working with positive numbers.

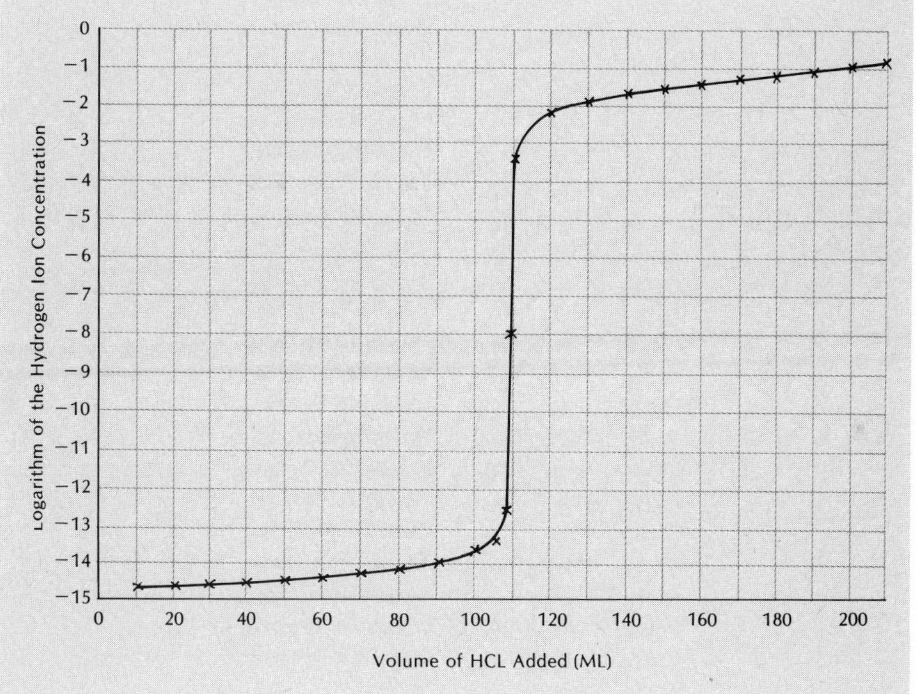

Figure 11.16
Here the logarithm of the hydrogen ion concentration is plotted in place of the molarity itself. This enables us to get all the data on the graph. However, since the logarithms are all negative, the graph does not have its origin in the lower left corner, as is customary.

Any time we represent the log of the hydrogen ion concentration, we are likely to get a negative number. In water solution, the hydrogen ion concentration is seldom greater than 1 *M*, so the log $[H^+]$ is almost always negative. (The logarithm of any number between 1 and 0 is negative.) Not willing to bother with those negative numbers all the time, chemists have defined a new term. The term is **pH.**

$$pH = -\log [H^+]$$

Since the logarithm of the hydrogen ion concentration is almost always negative, minus the logarithm of the hydrogen ion concentration is almost always positive. The log $[H^+]$ and pH are the same number but with opposite

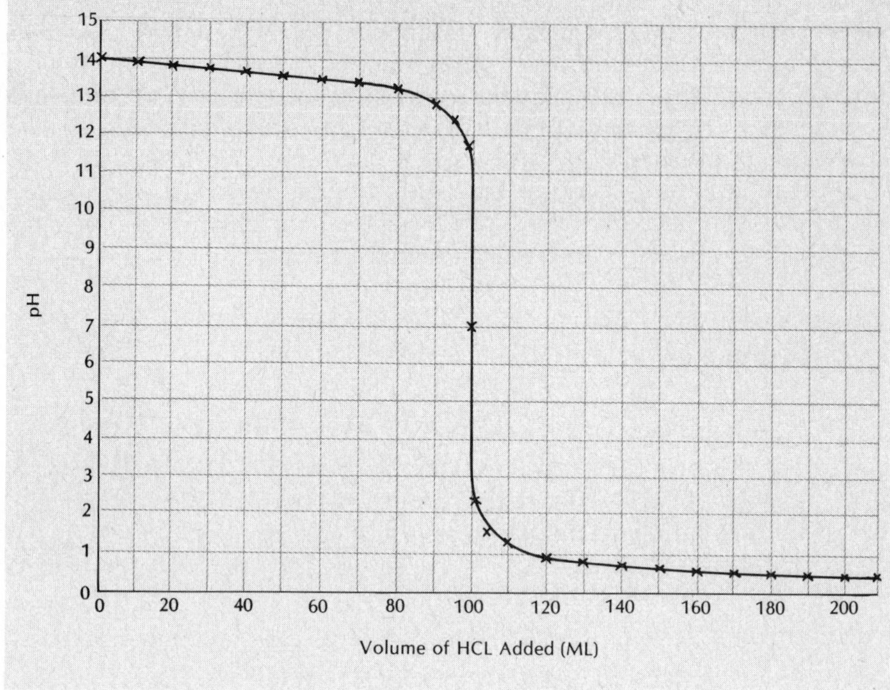

Figure 11.17 *By defining pH as −log [H$^+$], we are able to plot all the data in the customary fashion with the origin in the lower left corner of the graph.*

sign. As a result, as [H$^+$] *increases,* the pH value *decreases* (−5 is larger than −6, but +5 is less than +6). With pH, we can use positive numbers and plot graphs in the upper right quadrant, where we are more accustomed to seeing them.

Figure 11.17 shows the data from Table 11.3 expressed as pH. It is just like Figure 11.16 but rotated 180° on the *x* axis. (In other words, it is what you would see in a mirror placed along the *x* axis of Figure 11.16.)

11.13 TITRATION

The most dramatic thing that Figure 11.17 reveals is the very rapid change in pH (hydrogen ion concentration) when 100 mL of the HCl solution have been added. Something dramatic seems to be taking place at that point. Answering the following review question on stoichiometry should tell you what it is.

11-39. What volume of 1 *M* HCl is needed to react with 100 mL of 1 *M* NaOH?

Perhaps you didn't even need to do the calculations. Since the HCl and NaOH are of the same concentration, equal volumes contain equal numbers of solute molecules. Furthermore, the equation indicates that one molecule of NaOH combines with one molecule of HCl. When 100 mL of 1 *M* HCl are added to 100 mL of 1 *M* NaOH, the resulting solution should consist of only the products of the reaction. The proportion of reactants in the mixture is exactly the same as the proportion described by the balanced equation. This point is called the **equivalence point.** It is the point at which the proportions of reactants in the beaker are equivalent to the proportions described by the equation.

In any reaction between an acid and a base in water solution, there is a dramatic change in the pH of the solution at the equivalence point. This fact provides a way to find the equivalence point and is the basis for one of chemistry's oldest and most important analytical tools, **titration.**

INDICATORS AND TITRATION

You have already seen that indicators are weak acids or weak bases that change color as the $[H^+]$ of a solution changes. Figure 11.18 shows the color of some common indicators at various pH values. As the shaded areas on the graph suggest, indicators do not change color instantaneously at a given pH. The color change is due to the proportion of the indicator molecules in the acid or base form.

Below a pH of 3.2, virtually all of the methyl orange molecules have hydrogen ions attached and are in the acid form, which is red. Above that

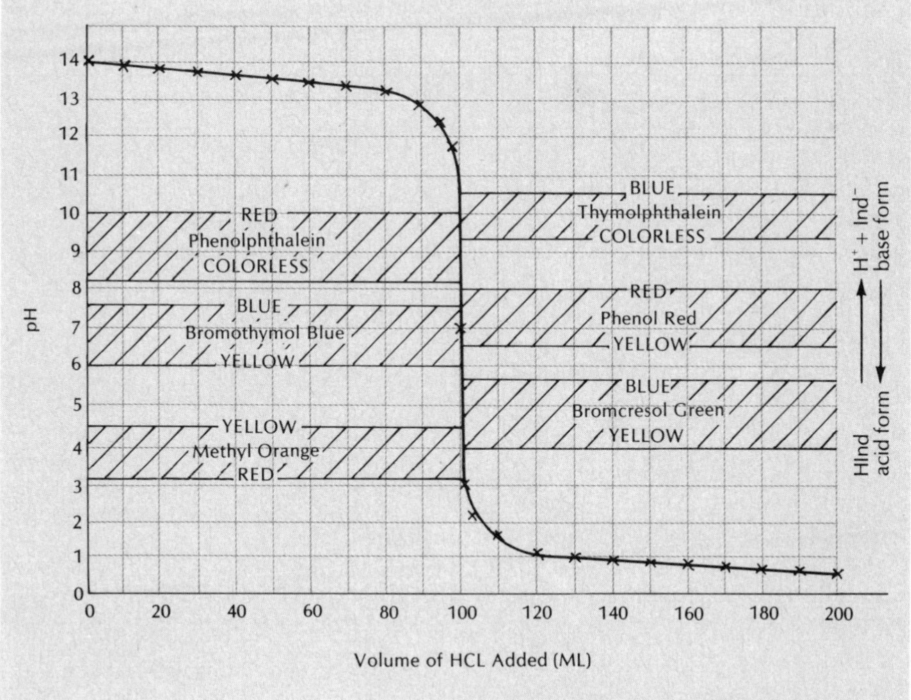

Figure 11.18
The pH range over which six common indicators change color. Note that all the indicators change color in the region of very rapid change in pH. If the proper indicator is selected, a single drop of solution is enough to change the indicator from one color to another.

pH, the methyl orange molecules begin to donate protons to other bases in the solution, leaving the methyl orange in the base form, which is yellow. By the time the hydrogen ion concentration has *decreased* to a pH of 4.4, virtually all of the methyl orange molecules have lost protons and exist solely in the base form of the molecule. In similar fashion, phenol red is almost entirely in the acid form when the pH of the solution is below 6.6. As the hydrogen ion concentration falls (pH rises), molecules in the yellow, acid form dissociate to form the red, base form of the molecule. This transition is essentially complete at a pH of 8.0. Other color changes are shown in the figure.

The major thing to notice about Figure 11.18 is that all six of the indicators change color between the point where 99.9 mL of HCl have been added to the 100 mL of NaOH and the point where 100.1 mL of HCl have been added. In other words, all of these indicators change color at or very near the equivalence point for the reaction. Therefore we can use this color change to indicate the equivalence point. (That's why they are called indicators.)

Titration has many applications. It can be used to find the concentration of solutions that cannot be made accurately from dry solids. Another application might be quality control in the production of household vinegar. Ordinary vinegar is a 5 percent solution of acetic acid. This is equivalent to about 0.83 *M*. We can use titration to be sure that the vinegar produced has this concentration.

In order to explain the procedure used in titration, we will discuss the neutralization reaction between NaOH and acetic acid.

11-40. Write the equation for the reaction.

Note that, in this reaction, 1 mol of NaOH reacts with 1 mol of $HC_2H_3O_2$. We could calculate the acetic acid concentration in vinegar if we knew how many moles of NaOH are needed to react with the acetic acid in a known volume of vinegar. According to the equation, the number of moles of NaOH that react is equal to the number of moles of acetic acid.

The key to the operation is determining when we have added enough NaOH to the solution of vinegar to *just* react with the acetic acid present—in other words, finding the equivalence point. We can do this with indicators.

Phenolphthalein (fee-nol-tha'-lene) is commonly used for this purpose, because it is easy to see the change from colorless in acid solution to red (faint pink when only a drop or two are used) in a base solution. This color change is very dramatic and must be seen to be appreciated.

Add a drop or two of phenolphthalein to a few milliliters of any acid solution. Then, using a medicine dropper, add base to the solution. You will find that the solution remains colorless until you add enough base to neutralize all of the acid. Then the solution turns red, as shown in Figure 11.19. The amazing thing is that a single drop of base causes the color change. Adding

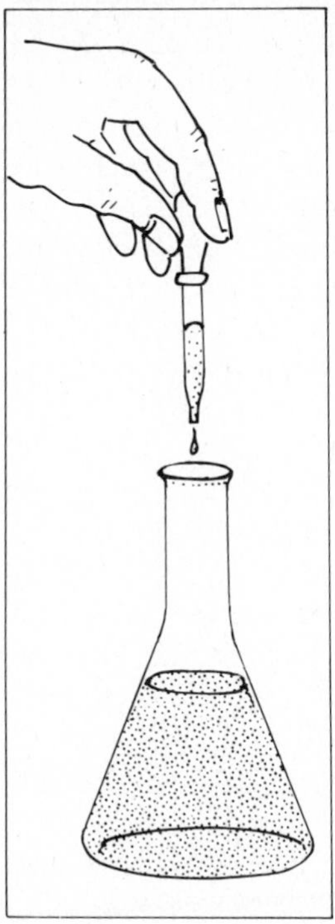

Figure 11.19
A single drop of base is enough to change phenolphthalein from colorless to red when the solution to which it is added is at the equivalence point.

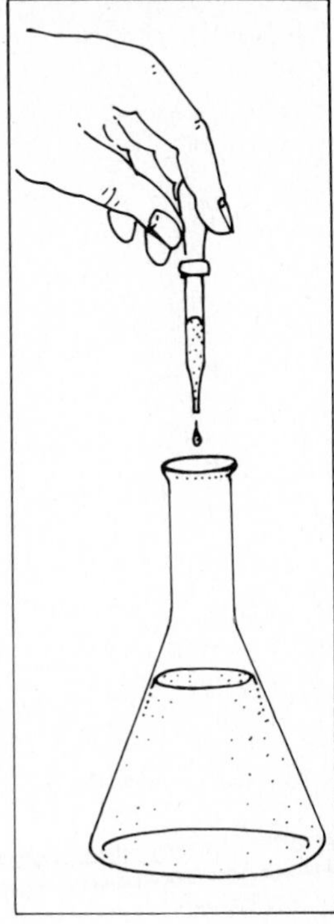

Figure 11.20
A single drop of acid is enough to change phenolphthalein from red to colorless when the solution to which it is added is at the equivalence point.

a drop or two of acid changes the solution to colorless again, as shown in Figure 11-20. You can shift the color from colorless to pink and back again as long as you like by simply adding a drop or two of base and then a drop or two of acid.

Now let us see how we can use this procedure to check the concentration of a vinegar solution.

First, we place 50 mL of vinegar in a flask. (We could use any amount, but 50 mL is a convenient sample.) Next, we add one or two drops of phenolphthalein.

11-41. What is the color of the solution at this point?

Finally we add NaOH of known concentration until the solution just turns pink. We usually add the NaOH from a graduated piece of glassware called a buret so that the volume can be measured accurately. The apparatus is shown in Figure 11-21 (p. 330).

When the solution changes color, we know that the equivalence point has been reached. We can use the equation and the known concentration of the base to calculate the concentration of the acid. The procedure will become clear as you work through the following example.

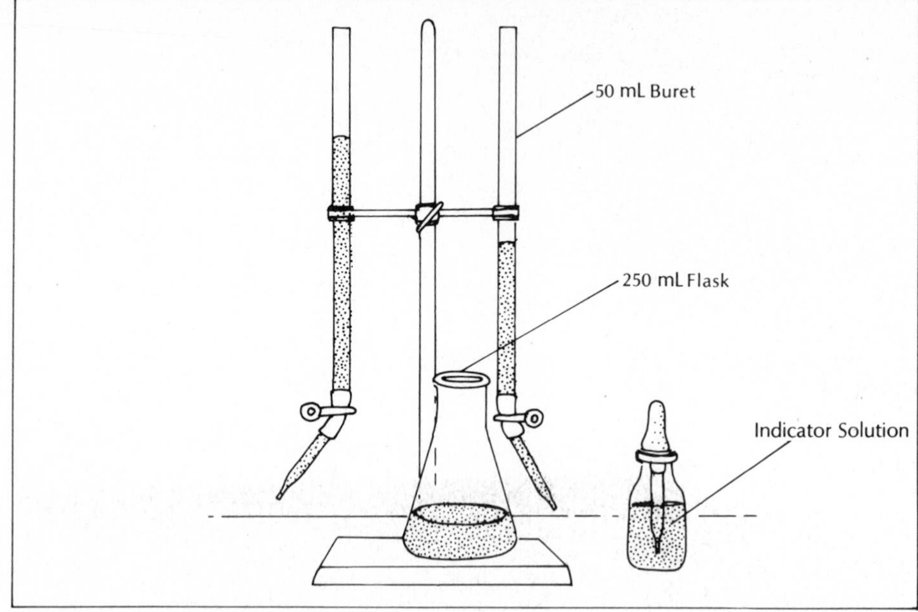

Figure 11.21 *Apparatus for performing a titration. The acid and base are placed in a buret, and the volumes required for the reaction are carefully measured.*

Example 11.1

It is found that 42.5 mL of 1.02 *M* NaOH have been added to 50.0 mL of vinegar when the phenolphthalein in the solution just turns pink. What is the concentration of the vinegar?

By now you know that quantities involved in a reaction are always calculated in terms of moles. First, calculate the number of moles of NaOH that were involved in the reaction.

$$? \text{ mol NaOH} = 42.5\,\cancel{\text{mL NaOH sol}} \times \frac{1.02 \text{ mol NaOH}}{1000\,\cancel{\text{mL NaOH sol}}} = 0.0434 \text{ mol NaOH}$$

The color change occurs at the equivalence point—when the moles of base added are just enough to react with the moles of acetic acid in the vinegar. Therefore we know that it takes 0.0434 mol of NaOH to neutralize the moles of acetic acid present in 50.0 mL of vinegar. The equation tells us the proportions of acid and base that react.

Write a balanced equation for this reaction.

$$NaOH + HC_2H_3O_2 \rightarrow H_2O + NaC_2H_3O_2$$

It is clear from the equation that the number of moles of acid that react is the same as the number of moles of base. However, this is not always true. Show the conversion factor derived from the equation that would be used to complete the following calculation.

$$? \text{ mol } HC_2H_3O_2 = 0.0434 \text{ mol NaOH} \times \underline{\qquad} =$$

$$\frac{1 \text{ mol } HC_2H_3O_2}{1 \text{ mol NaOH}}$$

The equation indicates that 1 mol of sodium hydroxide reacts for each mole of acetic acid present. Since the ratio is one-to-one in this case, the calculation is trivial. We know that there must have been 0.0434 mol of acetic acid in the vinegar, because 0.0434 mol of NaOH was required to neutralize the acid.

The sample of vinegar used in this titration was 50.0 mL. Then we know that the vinegar contains 0.0434 mol of acetic acid in 50.0 mL of solution. What is the molarity of the vinegar?

$$? \frac{\text{mol } HC_2H_3O_2}{1 \text{ L sol}} = \frac{0.0434 \text{ mol } HC_2H_3O_2}{50.0 \cancel{\text{mL}} \text{ sol}} \times \frac{1000 \cancel{\text{mL}}}{\text{L}} = \frac{0.868 \text{ mol}}{\text{L sol}} = 0.868\,M$$

This sample of vinegar is a little more concentrated than the normal 5 percent solution, which is 0.83 *M*.

Using the same procedure, solve the following problems. Try each problem and check your solution with that given at the end of the chapter. Then go to the next problem. Be sure that you write a balanced equation for each reaction. You *must* know the number of moles of acid that react with 1 mol of base. The ratio will not always be one-to-one, as it was in the example we just worked.

11-42. What is the concentration of a NaOH solution when 30 mL of 0.5 *M* HCl are needed to neutralize 50 mL of the base?

11-43. What is the concentration of HCl when 25 mL of this HCl solution neutralize 25 mL of 0.5 *M* $Ca(OH)_2$?

11-44. What is the concentration of acetic acid in vinegar when 32.5 mL of 0.56 *M* NaOH are required to neutralize 15 mL of the vinegar?

11-45. What is the concentration of NH_3 in household ammonia when 48.25 mL of 0.5246 *M* HCl are needed to neutralize 22.00 mL of the ammonia solution?

11-46. Oxalic acid ($H_2C_2O_4$) is a solid that is only slightly soluble in water. However, as it reacts with KOH or any other strong base, it continues to dissolve. When 6.25 g of oxalic acid are placed in water containing phenol red, the suspension is yellow. A solution of KOH is added until the color changes to red. What is the molarity of the KOH solution if 32.2 mL are required to change the phenol red to red?

11-47. A 5.0-g tablet of $Mg(OH)_2$ neutralizes 450 mL of stomach acid. What is the molarity of the HCl in the solution?

SUMMARY

Acids and bases are among the most important classes of chemicals. There are many compounds that act as acids and bases, and many reactions can be described as acid-base reactions. You should be aware that the number of chemicals classified as acids and the number of reactions called acid-base reactions depend on who is doing the classifying and for what purpose.

In this chapter, we have defined an acid as a proton donor and a base as a proton acceptor. This concept of acids and bases was derived independently in 1923 by Thomas Martin Lowry in England and Johannes Nicholas Brønsted in Denmark. It is usually called the Brønsted-Lowry theory of acids and bases, or just the Brønsted acid-base theory. This idea has been very useful and has essentially replaced a theory proposed by Arrhenius in 1884. Arrhenius described bases as compounds that produce hydroxide ions in water solution—a definition that is now considered too limited.

In future courses you will be introduced to another concept of acids and bases proposed by G. N. Lewis in 1923 but ignored until about 1938. Lewis' concept is broader than the one introduced here and is particularly useful in organic chemistry. However, use of Lewis' theory requires knowledge of the distribution of electrons in molecules—something that is not obvious from the formula and difficult for beginning students to determine. Since Brønsted's theory of acids and bases can explain all the acid-base chemistry that most of us encounter, it is the appropriate one to introduce here. If you continue your study of chemistry, you can expand that concept later.

Keep in mind that the Brønsted notion of acid and base is a little different from the meaning intended when the farmer asks about the acidity of his soil or the swimming pool attendant wonders whether the pool is too basic. In everyday use, substances are considered acidic if they have a hydrogen ion concentration greater than that found in pure water. They are considered basic if the hydrogen ion concentration is less than that found in pure water.

As you have seen, hydrogen ion concentration varies over such a wide range that it is normally expressed on a logarithmic scale called pH. Pure water has a pH of 7. Any pH *above* 7 corresponds to a hydrogen ion concentration *less* than that of water, and the solution is *basic.* Any pH *below* 7 corresponds to a hydrogen ion concentration *above* that of water, and the solution is *acidic.* (Don't be confused by the fact that *higher* pH corresponds to *lower* $[H^+]$. Remember that pH is defined as the *negative* of the logarithm of the $[H^+]$. Consequently, the scale is inverted, as seen in Figures 11.16 and 11.17.) Although water can act as a Brønsted acid or a Brønsted base, we wouldn't normally describe it as either an acid or a base.

We made quite a point of "strong" and "weak" in order to emphasize several things. First, we wanted to remind you that common words often have special meaning in science. You should be alert to this problem. When a familiar word is used in a strange way, ask what it means. Chemists and other professionals who have used words in a special way most of their lives forget that the meaning *is* special and not likely to be understood by the novice.

Even in reference to acids, "strong" and "weak" have different but related meanings. The macroscopic observation that leads to the notion of a strong or a weak acid is the observation that some acids react very vigorously, while other acids of the same concentration do not. The former appear to be "stronger" than the latter. An explanation of this observation was found at the microscopic level when it was observed that strong acids are completely dissociated to give ions, while weak acids are only partially dissociated into ions. The Brønsted theory of acids and bases provides a third view. A strong acid donates protons very readily; a weak acid gives up protons reluctantly.

All of these explanations are compatible, and all are different from the common use of the terms. Strong coffee is concentrated coffee, but concentration is *not* what we are talking about when we describe an acid as strong or weak.

Equilibrium was mentioned in this chapter but not discussed fully. As you continue your study of chemistry, you will find that equilibrium is an important idea. Few chemical reactions go to completion. There are reactants left over. It is often important to know the proportion of reactants and products present when the reaction stops and how this proportion is affected by temperature, pressure, and other conditions that we can manipulate when we carry out a reaction. These questions involve the study of equilibrium, which can become quite complex. At this stage, you should know that a system at equilibrium shows no observable change even though changes are occurring at the microscopic level. You should know that the proportion of reactants and products present at equilibrium can be changed by removing or adding one of the products or tampering with the system in some other way.

Analytical chemistry is the branch of chemistry that focuses on the questions "What do I have?" and "How much do I have?" More and more, these questions are answered with the aid of mechanical instruments, but the old technique of titration introduced in this chapter is still used routinely to determine the amount of a chemical that is present in solution.

The success of any titration depends on having some kind of indicator to show when the equivalence point for the reaction is reached. In this chapter you have learned how compounds called acid-base indicators signal the equivalence point in an acid-base reaction. Other chemicals can be used to signal the equivalence point for other reactions. Mechanical devices can be used for still others.

Acids and bases are such an important class of compounds that it is worthwhile to remember some typical reactions of acids and bases. Many of these can be described by generalized equations or net ionic equations. Even though there are exceptions to the generalizations that have been made about reactions between metals and acids, between carbonates and acids, between metal oxides and acids, and between acids and bases, the exceptions are few. Remembering these generalizations will enable you to predict a large number of very useful reactions.

Questions and Problems

11-48. Classify each of the following reactions as a replacement, metathesis, or acid-base reaction. (It may be possible to classify some reactions in two categories.)

a. $2HBr(aq) + Cl_2(aq) \rightarrow 2HCl(aq) + Br_2(g)$
b. $2HBr(aq) + Zn(s) \rightarrow ZnBr_2(aq) + H_2(g)$
c. $HBr(aq) + KOH(aq) \rightarrow KBr(aq) + H_2O(\ell)$
d. $HBr(aq) + AgNO_3(aq) \rightarrow AgBr(s) + HNO_3(aq)$
e. $HCHO_2(aq) + NH_3(aq) \rightarrow NH_4CHO_2(aq)$
f. $C_3H_5O_2^-(aq) + HI(aq) \rightarrow HC_2H_5O_2(aq) + I^-(aq)$
g. $BaCl_2(aq) + H_2SO_4 \rightarrow BaSO_4(s) + 2HCl(aq)$
h. $NaOH(aq) + H_2SO_4(aq) \rightarrow NaHSO_4(aq) + H_2O(\ell)$
i. $Fe(s) + 2H_3PO_4(aq) \rightarrow Fe(HPO_4)_2(aq) + H_2(g)$
j. $FeO(s) + 2H_3PO_4(aq) \rightarrow Fe(HPO_4)_2(aq) + H_2O(\ell)$

11-49. Write a balanced chemical equation for the reaction between any acid and each of the following substances.

a. Ca
b. CaO
c. $Ca(OH)_2$
d. $CaCO_3$
e. Na
f. Na_2O
g. NaOH
h. $NaHCO_3$

11-50. Give the conjugate acid for each of the following bases.

a. OH^-
b. HCO_3^-
c. ClO_4^-
d. S^{2-}
e. CN^-
f. SO_3^{2-}
g. HSO_3^-
h. $C_7H_5O_2^-$
i. F^-
j. NO_2^-

11-51. Give the conjugate base for each of the following acids.

a. HCO_3^-
b. HClO
c. HBr
d. H_3O^+
e. HSO_4^-
f. NH_4^+
g. H_2S
h. $H_2PO_4^-$
i. HSO_3^-
j. H_2O

11-52. For each of the following equations, identify the compound on each side of the equation that acts as an acid and the compound on each side that acts as a base.

a. $HNO_2 + CN^- \leftrightarrows NO_2^- + HCN$
b. $HSO_4^- + C_2O_4^{2-} \leftrightarrows HC_2O_4^- + SO_4^{2-}$
c. $H_2PO_4^- + HCO_3^- \leftrightarrows H_2CO_3 + HPO_4^{2-}$
d. $NH_4^+ + HPO_4^{2-} \leftrightarrows NH_3 + H_2PO_4^-$
e. $OH^- + HSO_4^- \leftrightarrows H_2O + SO_4^{2-}$
f. $HF + SO_3^{2-} \leftrightarrows F^- + HSO_3^-$
g. $HCO_3^- + ClO^- \leftrightarrows HClO + CO_3^{2-}$

11-53. When the pH is given, write the $[H^+]$; when the $[H^+]$ is given, write the pH. Indicate whether each solution is acidic or basic.

a. pH = 4
b. $[H^+] = 10^{-11}$
c. $[H^+] = 0.001$
d. pH = 12
e. pH = 0
f. $[H^+] = 10^{-1}$
g. pH = 7
h. $[H^+] = 0.1$
i. $[H^+] = 0.000000001$
j. pH = 8

11-54. The pH or $[H^+]$ for two solutions is given. For each pair, indicate which solution is more acidic.

a. pH = 4; pH = 6
b. $[H^+] = 0.001$; $[H^+] = 0.0001$

c. pH = 2; $[H^+] = 10^{-3}$
d. pH = 9; $[H^+] = 10^{-8}$
e. pH = 7; pH = 5
f. $[H^+] = 10^{-4}$; $[H^+] = 0.001$

11-55. Calculate the concentration of the unknown solution.
a. What is the concentration of KOH when 34 mL of it react with 42 mL of 1.5 *M* H_2SO_4 solution?
b. 51.68 mL of 0.1 *M* HCl are needed to titrate 27.36 mL of aqueous ammonia. What is the concentration of the ammonia?
c. When phenolphthalein is added to a slurry containing 0.15 mol $Ca(OH)_2$, it turns red. When 46.21 mL of HCl(aq) are added, the solution turns colorless. What is the concentration of the HCl solution?
d. 0.583 g of $Mg(OH)_2$ neutralizes 50.0 mL of liquid taken from a human stomach. Assuming that the only acid in the stomach is HCl, what is the concentration of the stomach acid?

More Challenging Problems

11-56. Find the pH or $[H^+]$ corresponding to each $[H^+]$ or pH.
a. pH = 11.33
b. $[H^+] = 1.2 \times 10^{-3}$ *M*
c. $[H^+] = 4.5 \times 10^{-8}$ *M*
d. pH = 0.234
e. pH = 7.68
f. $[H^+] = 0.00204$ *M*

11-57. 5.25 g of $Ca(OH)_2$ are mixed with 50 mL of water to form a suspension. After mixing, two drops of phenolphthalein are added. What volume of 1.2 *M* HNO_3 is needed to turn the suspension colorless?

11-58. Acids can be neutralized by ordinary limestone ($CaCO_3$), quicklime (CaO) or slaked lime ($Ca(OH)_2$). If the three compounds sell for the same price per ton, which would be the cheapest base with which to neutralize acid wastes?

11-59. 25.0 mL of an acid solution neutralize 46.2 mL of a 1.33 *M* NaOH solution. What volume of the acid is needed to neutralize 3.5 g of $Mg(OH)_2$?

11-60. Which reactant will be left over when 25.0 mL of 0.33 *M* HCl are mixed with 32 mL of 0.13 *M* $Ca(OH)_2$?

11-61. What is the pH of the solution formed when 35 mL of 0.22 *M* HCl are mixed with 55 mL of 0.15 *M* NaOH?

11-62. In Example 11.1 it was mentioned that a 5 percent acetic acid solution corresponds to a molarity of 0.83 *M*. Assuming that the density of the solution is the same as the density of water, do the calculations to show that this value is correct.

11-63. When hydrochloric acid is purchased commercially, the molarity of the acid is not indicated on the bottle. However, the label does indicate that the solution is about 37 percent HCl by weight and that the density of the solution is 1.18 g/cm³. Using this information, calculate the molarity of the acid.

Answers to Questions in Chapter 11

11-1. The CO_3^{2-} ion

11-2. Elements are composed of only one kind of atom; compounds contain at least two kinds of atoms.

11-3. The reactants would consist of two compounds; the products would consist of two compounds. In the products, the electropositive part of one reactant would be combined with the electronegative part of the other reactant, and vice versa.

11-4. A replacement reaction involves an element and a compound. During reaction, the free element replaces an element in the compound.

11-5. One of the best tests would be the addition of an acid. Bubbles of carbon dioxide gas would indicate either a carbonate or a bicarbonate.

11.6. This would be very difficult. Heating it strongly to see whether it decomposes to form two or more products weighing less than the original sample would be a good start. Melting it (if possible) and passing an electric current through the melt would be another approach. If neither test produced evidence of decomposition, I would suspect that it is an element.

11-7. Once again, this would be very difficult. Many types of reactions could produce a yellow solid. Unless you could identify the reactants and the product, it would be impossible to classify the reaction.

11-8. Metals shine, conduct electricity, conduct heat, and can be hammered into sheets and pulled into wires. These are properties to look for. However, a given metal might lack one or more of these properties, and a nonmetal may possess one or more of them.

11-9. Look at the periodic table. Metals are on the left.

11-10. Pure substances look homogeneous and have constant boiling and melting points. Impure substances may possess any one of these properties, but they are unlikely to possess all three.

11-11. It is neither a replacement reaction nor a metathesis reaction.

11-12. Pour the liquid on baking soda ($NaHCO_3$) and see whether bubbles of CO_2 form. Pour the liquid on zinc and see whether bubbles of H_2 form. Pour the liquid over a dull copper penny and see whether it removes the copper oxide coating. Add a few drops of an indicator to see whether the liquid gives the color characteristic of an acid. Although sour taste is characteristic of acids, do not taste the liquid; that is too dangerous.

11-13. One (atomic number)

11-14. One (atomic number)

11-15. Remove the electron.

11-16. A proton

11-17. HNO_3 is the acid; KOH is the base.

11-18. NH_4^+ is the acid; NH_2^- is the base.

11-19. HCl is the acid; NH_3 is the base.

11-20. H_2SO_4 is the acid; H_2O is the base.

11-21. HCO_3^- is the acid; PO_4^{3-} is the base.

11-22. $HC_2H_3O_2$ is the acid; HCO_3^- is the base.

11-23. There are ions in the solution.

11-24. Since both HCl and $HC_2H_3O_2$ are acids, and acids are compounds that furnish hydrogen ions, both solutions should show conductivity.

11-25. The more concentrated solution should have more ions. The bulb should be brighter when the 10 *M* solution is tested.

11-26. Strong

11-27. Mostly the ions

11-28. A good conductor

11-29. Strong

11-30. F^- (The arrows indicate that the products are favored at equilibrium.)

11-31. $H_2C_2O_4$

11-32. $HC_3H_5O_2$ (The arrows indicate that the products are favored at equilibrium.)

11-33. No. If it were, it could not act as a base in relation to acetic acid.

11-34. No; weak

11-35. No. It must be a weak acid. If it dissociated to give H^+ in high concentrations, it would irritate the eyes.

11-36. $[H^+] \times [OH^-] = 10^{-14}$

11-37. $10^{-2} \times [OH^-] = 10^{-14}$

$$[OH^-] = \frac{10^{-14}}{10^{-2}} = 10^{-12}$$

11-38. $[H^+] \times 10^{-5} = 10^{-14}$

$$[H^+] = \frac{10^{-14}}{10^{-5}} = 10^{-9}$$

11-39. 100 mL

11-40. $NaOH + HC_2H_3O_2 \rightarrow NaC_2H_3O_2 + H_2O$

11-41. Colorless

11-42. $? \text{ mol HCl} = 30\,\cancel{\text{mL HCl sol}} \times \dfrac{0.5 \text{ mol HCl}}{1000\,\cancel{\text{mL HCl sol}}}$

$$= 0.015 \text{ mol HCl}$$

$$? \text{ mol NaOH} = 0.015\,\cancel{\text{mol HCl}} \times \frac{1 \text{ mol NaOH}}{1\,\cancel{\text{mol HCl}}}$$

$$= 0.015 \text{ mol NaOH}$$

$$? \frac{\text{mol NaOH}}{\text{L sol}} = \frac{0.015 \text{ mol NaOH}}{50\,\cancel{\text{mL}} \text{ sol}} \times \frac{1000\,\cancel{\text{mL}}}{\text{L}}$$

$$= 0.3\,M \text{ NaOH}$$

11-43. $2HCl + Ca(OH)_2 \rightarrow CaCl_2 + 2H_2O$

$$? \text{ mol Ca(OH)}_2 = 25\,\cancel{\text{mL Ca(OH)}_2 \text{ sol}}$$

$$\times \frac{0.5 \text{ mol Ca(OH)}_2}{1000\,\cancel{\text{mL sol}}}$$

$$= 0.0125 \text{ mol Ca(OH)}_2$$

$$? \text{ mol HCl} = 0.0125\,\cancel{\text{mol Ca(OH)}_2}$$

$$\times \frac{2 \text{ mol HCl}}{1\,\cancel{\text{mol Ca(OH)}_2}}$$

$$= 0.0250 \text{ mol HCl}$$

$$\frac{0.0250 \text{ mol HCl}}{25\,\cancel{\text{mL}} \text{ sol}} \times \frac{1{,}000\,\cancel{\text{mL}}}{\text{L}} = \frac{1 \text{ mol HCl}}{\text{L sol}} = 1\,M$$

11-44. $NaOH + HC_2H_3O_2 \rightarrow H_2O + NaC_2H_3O_2$

$$? \text{ mol NaOH} = 32.5\,\cancel{\text{mL NaOH sol}}$$

$$\times \frac{0.56 \text{ mol NaOH}}{1000\,\cancel{\text{mL sol}}}$$

$$= 0.0182 \text{ mol NaOH}$$

Without doing the calculation, we can see from the equation that the moles of acid must equal the moles of base used. Then

$$\frac{0.0182 \text{ mol } HC_2H_3O_2}{15 \cancel{\text{mL}} \text{ sol}} \times \frac{1{,}000 \cancel{\text{mL}}}{\text{L}} = 1.2\,M\ HC_2H_3O_2$$

11-45. $NH_3 + HCl \rightarrow NH_4Cl$

$$? \text{ mol HCl} = 48.25 \cancel{\text{mL HCl sol}} \times \frac{0.5246 \text{ mol HCl}}{1{,}000 \cancel{\text{mL sol}}} = 0.02531 \text{ mol HCl}$$

Again, the equation shows a one-to-one ratio, so without calculations we can say that

$$\frac{0.02531 \text{ mol } NH_3}{22.00 \cancel{\text{mL}} \text{ sol}} \times \frac{1{,}000 \cancel{\text{mL}}}{\text{L}} = 1.150\,M\ NH_3$$

11-46. $H_2C_2O_4 + 2KOH \rightarrow K_2C_2O_4 + 2H_2O$

$$? \text{ mol } H_2C_2O_4 = 6.25 \cancel{\text{g } H_2C_2O_4} \times \frac{1 \text{ mol } H_2C_2O_4}{90.04 \cancel{\text{g } H_2C_2O_4}} = 0.0694 \text{ mol } H_2C_2O_4$$

$$? \text{ mol KOH} = 0.0694 \cancel{\text{mol } H_2C_2O_4} \times \frac{2 \text{ mol KOH}}{1 \cancel{\text{mol } H_2C_2O_4}} = 0.1388 \text{ mol KOH}$$

$$\frac{0.1388 \text{ mol KOH}}{32.2 \cancel{\text{mL}} \text{ sol}} \times \frac{1{,}000 \cancel{\text{mL}}}{\text{L}} = 4.31\,M \text{ KOH}$$

11-47. $Mg(OH)_2 + 2HCl \rightarrow MgCl_2 + 2H_2O$

$$? \text{ mol HCl} = 5.0 \cancel{\text{g } Mg(OH)_2} \times \frac{1 \cancel{\text{mol } Mg(OH)_2}}{58.33 \cancel{\text{g } Mg(OH)_2}} \times \frac{2 \text{ mol HCl}}{1 \cancel{\text{mol } Mg(OH)_2}} = 0.17 \text{ mol HCl}$$

$$\frac{0.17 \text{ mol HCl}}{450 \cancel{\text{mL}} \text{ sol}} \times \frac{1{,}000 \cancel{\text{mL}}}{\text{L}} = 0.38\,M \text{ HCl}$$

READING

There are so many articles in periodicals that relate to acids and bases that the reading section of this chapter could easily be larger than the entire book. But only one representative article has been selected. It deals with the problem of acid mine wastes.

PHOSPHATE PROCESS TREATS ACID MINE DRAINAGE

A quartet of scientists from Wright State University, Dayton, Ohio, has turned a sewage treatment technique upside down and developed a new process for treating stream waters contaminated by acid mine drainage. Ordinary phosphate rock is a major ingredient in the method, which was described by Dr. Michael J. Smith, associate professor of chemistry, in a presentation to the Division of Environmental Chemistry. Graduate student Michael F. Cox, former graduate student David M. Haile, and staff hydrogeologist Brent E. Huntsman worked with Smith in developing the process. According to the Dayton team, treatment with phosphate before lime neutralization greatly reduces the sludge handling problem and also is more effective in removing iron.

Acid mine drainage occurs when iron disulfide minerals found in many coal deposits are exposed to air and water by mining operations. The minerals are oxidized to sulfuric acid and iron(II) sulfate that, leached by natural drainage waters, trickle into streams. The acid flow can destroy vegetation, kill fish, and render streams useless for recreation or as water supplies. In Appalachia alone, some 5700 miles of streams are considered to be seriously contaminated. The deadly flow can continue for decades, even from mines long abandoned, if nothing is done to counteract it.

Technology exists to purify the contaminated waters, Smith notes. But in many cases it is too expensive for wide application. The most common approach is treatment with limestone or lime, which neutralizes the acid and precipitates the iron. Limestone, although cheap, is not very efficient. As the pH goes up during neutralization, its reactivity diminishes. Lime is more efficient, and effective at higher pH, but it costs more. Also, lime neutralization results in a colloidal sludge that is slow to settle and difficult to dispose of.

Another serious environmental problem—in some

Source: The American Chemical Society, *Chemical and Engineering News,* Vol. 55, No. 14 (April 4, 1977), p. 27.

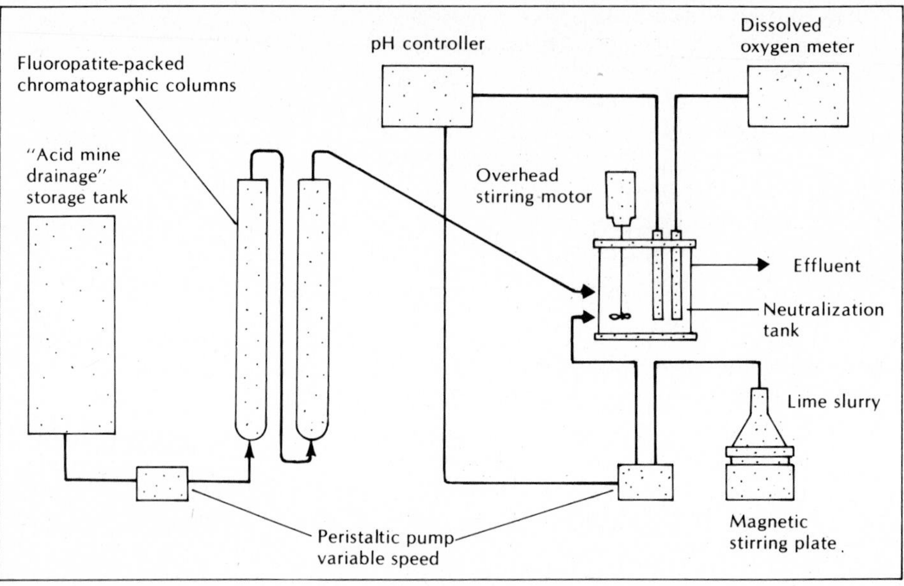

regions—is phosphate pollution of domestic sewage. One method of fighting this problem is to treat the sewage with iron(II) sulfate, then aerate it. Iron(II) is oxidized to iron(III) and phosphates are precipitated almost completely when the effluent is neutralized with lime. It occurred to Smith and his associates that this same phenomenon could be put to use in treating waters contaminated by acid mine drainage. After all, Smith notes, such waters can be grossly described as solutions of iron(II) sulfate in sulfuric acid. Thus, addition of phosphate, followed by aeration and neutralization, might be an economically feasible process.

So, the team set up a miniature treatment plant, concocted some synthetic acid mine drainage (the real thing can vary too much from day to day), and ran some experiments. They found an inexpensive source of phosphate in fluoroapatite, or ordinary phosphate rock. It's a good reagent, Smith notes—soluble in acid solutions but insoluble in neutral solutions. Overtreatment is almost impossible. Since phosphate "loading" is proportional to influent acidity, no metering device is needed; the influent can simply run through a bed of phosphate pebbles.

The phosphate-lime treatment was found to remove contaminants more effectively than neutralization alone, Smith says. For example, iron(II) removal efficiency was 99.9% at the optimum neutralization pH of 6.0. When neutralized, the phosphate-loaded effluent produced a dense, heavily flocked precipitate that settled rapidly and had a much higher solids content than the colloidal precipitate produced in conventional lime neutralization. According to Smith, the high (29%) concentration of phosphate in the sludge dilutes toxic contaminants; thus, it should be possible to use the material as a soil conditioner.

Another advantage of the technique is that it substantially reduces the amount of reagent needed for effective treatment. In conventional lime neutralization, iron removal takes place in two steps. First, iron(II) is oxidized to iron(III):

$$Fe^{2+} + \tfrac{1}{4}O_2 + H^+ \rightarrow Fe^{3+} + \tfrac{1}{2}H_2O$$

which consumes a mole of H^+ for every mole of Fe^{2+} oxidized. Then iron(III) is precipitated as the hydroxide:

$$Fe^{3+} + 3H_2O \rightarrow \underline{Fe(OH)_3} + 3H^+$$

generating three moles of H^+ for every mole of Fe^{3+} precipitated. In other words, enough lime is required to neutralize not only the acid originally present but that produced by the oxidation, hydrolysis, and precipitation of iron.

In the phosphate-lime method, the initial oxidation

proceeds in the same way as in simple lime neutralization. But the iron is precipitated as phosphate, not as hydroxide:

$$Fe^{3+} + PO_4^{-3} \rightarrow \underline{FePO_4}$$

In this case, H^+ is not generated in the precipitation step. Overall, in fact, one mole of H^+ is consumed for every mole of $FePO_4$ precipitated, reducing some of the free mineral acidity originally present in the acid mine drainage. In the Wright State experiments, 44% less lime was needed for neutralization after phosphate pretreatment, compared to the amount required for conventional neutralization. Also, since phosphate has a catalytic effect on the rate of iron(II) oxidation in air-saturated solutions, aeration time could be shorter.

Materials costs for phosphate-lime treatment likely would be comparable to those for conventional lime neutralization—about 1.1 cents per 1000 gal, Smith says. However, sludge disposal costs, which range from 1.8 to 2.0 cents per 1000 gal, probably would be reduced substantially.

Work now under way at Wright State suggests that phosphate pretreatment greatly improves the efficiency of limestone neutralization, Smith notes, adding that substitution of limestone for the more expensive lime would further reduce treatment costs.

12

The Behavior of Gases

OBJECTIVES

Direct and Indirect Proportionalities

If you think algebra is useless, get ready to change your mind. You will use algebra in this chapter. More important, you will learn more about the significance of some algebraic relationships. To begin, you will need to accomplish these objectives:

1. Given two variables (such as mass and volume) that are directly proportional, write a mathematical statement of that fact. (See Problem 12-15.)
2. Given two variables (such as pressure and volume of a gas) that are inversely proportional, write a mathematical statement of that fact. (See Problem 12-20.)
3. Given a mathematical statement of the relationship between two variables, indicate whether the variables are directly proportional, inversely proportional, or neither. (See Problem 12-14.)
4. Given a graph of the relationship between two variables, indicate whether the relationship is a direct proportion, an inverse proportion, or neither. If the relationship is a direct proportion, write an equation for the relationship. (See Problems 12-12 and 12-13.)

Gas Laws

The mathematical relationships we have mentioned are used to describe the behavior of gases. When you have accomplished the following objectives, you will have a tool that enables you to find the pressure, temperature, or volume of a gas when it expands or contracts.

5. Given the volume of a gas at one pressure, find the volume at another pressure, or, given the volume, find the pressure (temperature and moles constant). (See Problem 12-24.)
6. Given the volume of a gas at one temperature, find the volume at another temperature, or, given the volume, find the temperature (pressure and moles constant). (See Problem 12-40.)
7. Given the values for any three of the variables that describe a gas (pressure, volume, temperature, and number of moles), calculate the corresponding value for the fourth variable. (See Problem 12-47.)

Temperature and Pressure

Since the volume of a gas depends on temperature and pressure, you need to know a few things about these two properties.

8. Convert pressure from units of torr, kilopascals, or atmospheres to another of these units.
9. Convert temperature given in degrees Celsius to Kelvin, and vice versa. (See Problems 12-32 and 12-33.)
10. State the values for standard temperature and pressure.
11. Given the total pressure for a mixture of gases and the pressure exerted by all but one, find the pressure exerted by the other gas. (See Problem 12-66.)

Molecular Mass

Finally, you should be able to use your knowledge of gases to find the molecular mass of a compound. The following objectives cover this skill.

12. Find the volume of any number of moles of an ideal gas at any temperature and pressure. (See Problem 12-53.)
13. Given the mass of a given volume of gas at a specified temperature and pressure, find the molecular mass of the gas. (See Problem 12-65.)

Microscopic Model

We have encouraged you all along to "think molecules." We continue to do so here. We hope that, by the end of the chapter, you will be able to do the following:

14. Given a description of the behavior of a gas when the pressure is changed, explain the behavior in terms of moving molecules. (See Problem 12-29.)
15. Given a description of the behavior of a gas when the temperature is changed, explain the behavior in terms of moving molecules. (See Problem 12-43.)

Words You Should Know

directly proportional
constant
inversely proportional
newton
pascal
barometer
torr
Boyle's law
Kelvin
ideal gas
Charles' law
ideal gas equation
ideal gas constant (R)
molar volume
STP (standard temperature and pressure)
Dalton's law of partial pressure
diffusion
vapor pressure

12.1 READING MATHEMATICAL SENTENCES

A large part of teaching is communicating. If you have learned to read English sentences, then I should be able to communicate to you through the symbols on this page. As efficient as the English language may be, we often wish to communicate ideas that are awkward to express in words. We use the language of mathematics. Fine—if you can read mathematics. And you can. At least you can read such statements as "24 + 16 = 40," "24 ÷ 4 = 6," and a host of more complex mathematical sentences.

DIRECT PROPORTIONALITY

In this section we will focus on one kind of mathematical sentence that is used often in science. This is a sentence that says that one quantity is directly proportional to some other quantity. You will recall that "the weight of an object is directly proportional to its mass." This is an example of the kind of statement to which I refer. There are many others: "The volume of a gas is directly proportional to the absolute temperature of the gas." "The acceleration of an object is directly proportional to the force acting on the object." "The rate of a chemical reaction is directly proportional to the concentration of a given reactant." In this exercise we will explore what is meant by **directly proportional.**

The more mass an object has, the more it weighs. We might begin our exploration of directly proportional by making a tentative definition. We might say, "If A increases as B increases, then A is directly proportional to B." As we will see, this is not strictly correct, but it will do for a start.

We begin by discussing a relationship that you already know: the relationship between the volume of water in a cylinder and the height of the water in the cylinder. You know that when the height of the water increases, the volume of the water also increases.

Table 12.1 shows data taken when a cylinder was filled to different heights, as shown in Figure 12.1. We will discuss these data, but if you have a cylinder handy, collect data for your cylinder and go through the same analysis with those data as well.

Table 12.1. Volume of a Cylinder as a Function of Height

Height (cm, ±0.2)	*Volume (cm^3, ±2)*	*Volume divided by height (cm^2)*
3.1	64	________
6.1	128	________
10.0	211	________
14.6	307	________

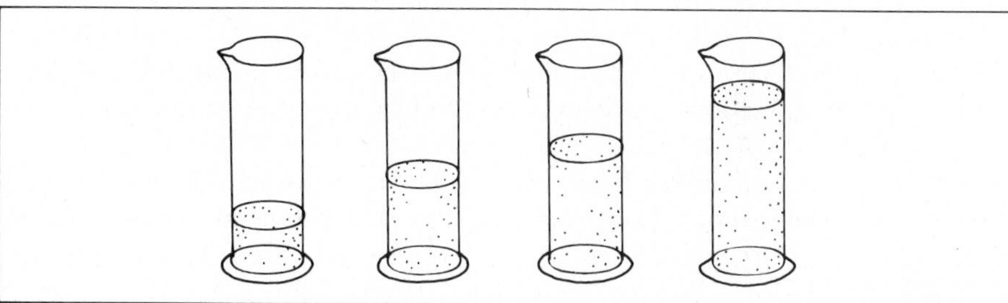

Figure 12.1 *The same cylinder filled to different heights.*

12-1. Do the division called for, and fill in the third column of Table 12.1.
12-2. Did you get the same number each time you divided the volume of water in the cylinder by the height of the water? (In other words, are all of the values in column 3 of Table 12.1 the same?)

If you remembered to use what you learned in Chapter 2 about significant digits, you should have found that all values in column 3 of the table are the same. The first two calculations justify only two significant digits in the answer, and the value obtained is 21. The third and fourth calculations provide three digits for the answer: 21.1 and 21.0. Expressed to two digits, all four values are the same.

No matter what cylinder you selected, you would find that the volume of water in the cylinder divided by the height of the water gives the same number. This is the fact that we express when we say, "The volume of water in the cylinder is directly proportional to the height of the water in the cylinder." This relationship is expressed mathematically as:

$$\frac{V}{h} = k$$

We have used the letter V to represent the volume, h to represent height, and k to represent the constant number that results from the division. (The value of k remains the same for a particular cylinder but changes when we change cylinders.)

Multiplying both sides of this equation by h, we get

$$\left(\frac{V}{h}\right)(h) = (k)(h)$$

$$V = kh$$

Translating this into ordinary English, we would read, "The volume of water in a cylinder is equal to some number that is constant times the height of the water in the cylinder." The last equation that we wrote is the normal way in which we state that "V is directly proportional to h."

In discussing the relationship between the volume of a cylinder and its height, we have left out an important point. All of our measurements were made on the *same* cylinder, which means that the diameter of the cylinder we used was always the same; that is, it was constant. Therefore, the mathematical statement that we made is correct *only* when the diameter of the cylinder is constant. Volume would *not* be directly proportional to height when we filled cylinders of different diameters or vessels like those shown in Figure 12.2. We might make the following statement to be perfectly clear.

Figure 12.2 *Containers that do not have a constant diameter.*

$$V = kh \quad (d \text{ constant})$$

We have added the parenthetical statement to indicate that the equality holds only under the condition that we measure the height of the water in a cylinder that has the same diameter from top to bottom.

There are many statements in science that are true only when some added condition holds. In Chapter 11 you learned that acids react with zinc to produce hydrogen gas. We can say that "the volume of hydrogen produced is directly proportional to the weight of zinc that reacts," but this is true only when there are no side reactions (such as the one with the nitrate ion when HNO_3 is used as the acid). You also learned in Chapter 11 that strong acids react more rapidly than weak acids. As a matter of fact, we can say that "the rate of reaction is directly proportional to the concentration of hydrogen ions in solution," but this is true only when the temperature of the solution is kept constant.

It is a good idea to state conditions that must hold in order for a relationship to be true. This is not always done. You must be alert to unstated conditions that may be important.

We began our discussion of direct proportion with the intuitive notion that two things (such as the volume and height of a cylinder) are directly proportional when one increases as the other increases. Using a cylinder as an example, we developed a mathematical statement of this relationship. For the cylinder, it was expressed in two ways:

$$\frac{V}{h} = k \quad (d \text{ constant})$$

$$V = kh \quad (d \text{ constant})$$

In general we can say that "A is directly proportional to B if $A/B = k$."

Before going on to another topic, consider whether the volume of a cylinder is proportional to the *diameter* when the *height* is kept constant. You can't stretch a cylinder to different diameters, but several cylinders can be filled to the same height, as shown in Figure 12.3.

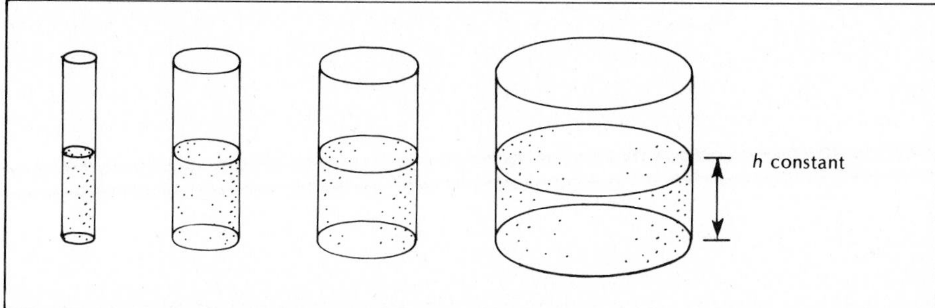

Figure 12.3 *Cylinders with different diameters filled to the same height.*

12-3. Do you think that the volume of a cylinder is directly proportional to the diameter when the height is kept constant?

12-4. Write a mathematical sentence that says, "The volume of a cylinder is directly proportional to the diameter."

12-5. How could you check to see whether the statement is true?

Table 12.2 shows data obtained by measuring the diameter and volume of several cylinders filled to the same height. Complete column 3 of the table.

Table 12.2. Volume of a Cylinder as a Function of Diameter

Diameter (cm, ±0.2)	*Volume (cm^3, ±2)*	*Volume divided by diameter (cm^2)*
2.3	15.3	________
3.4	34.2	________
4.4	56.0	________
5.3	82.0	________
6.8	132.0	________
10.3	313.5	________

12-6. Are all values in column 3 of Table 12.2 the same?
12-7. Is the volume of a cylinder directly proportional to the diameter?

If you paid attention to significant digits when you did the calculations to complete Table 12.2, you know that the numbers differed even in the places that contained certain digits. $V/d \neq k$. (This is read, "The volume divided by the diameter is *not equal* to a constant.") Now we know that our original statement, "If A increases as B increases, then A is directly proportional to B," is incorrect, because we have found a case in which V increases and d increases but V is *not* directly proportional to d.

It is difficult to write a simple sentence that defines "directly proportional" without using a mathematical expression. You might like to try making such a statement. Perhaps you can see why we use mathematical sentences in science. Once you learn to read them, they are much more precise and easier to interpret than the English sentences that express the same ideas.

We hope that you are learning to read and write mathematical sentences to express important relationships in physical science. To give you more practice, let us return to the volume of a cylinder.

You may remember from your study of geometry that the formula for finding the volume of *any* cylinder is

$$V = \pi r^2 h$$

In words this reads, "The volume of a cylinder is equal to pi (a constant) times the radius of the cylinder squared times the height of the cylinder." Since the radius of a circle is half of the diameter, we could also write the formula for the volume of a cylinder as

$$V = \pi \left(\frac{d}{2}\right)^2 h \quad \text{or} \quad V = \pi \frac{d^2}{4} h$$

Looking at this formula, we can see immediately—well, almost immediately—that the volume of a cylinder is *not* directly proportional to the diameter when we keep the height constant. Rearranging the last equation, we get

$$V = (\pi h/4)d^2$$

Pi is a constant and the number 4 is a constant. (*Any* number is always the same.) If we do not change the height, the value of what is represented inside the parentheses remains unchanged (is constant) and we can write

$$V = kd^2 \quad (h \text{ constant})$$

Clearly, the volume of a cylinder is directly proportional to the *square* of the diameter. It is *not* directly proportional to the diameter itself. If you go back to Table 12.2 and square all the numbers that you have for the diameter of the cylinders, you will find that the volume divided by the diameter squared *is* a constant. Try it.

From the data in Table 12.1, we found that the volume of a cylinder *is* directly proportional to the height when we keep the diameter constant. We wrote

$$V/h = k \quad (d \text{ constant})$$

$$V = kh \quad (d \text{ constant})$$

We might ask, "What is the value of the constant, k, and what does V/h represent?" Since the formula for the volume of a cylinder is $V = \pi r^2 h$, it should be obvious that the value of k in the foregoing expression is simply pi times the radius of the cylinder squared; that is, $k = \pi r^2$. The area of a circle is πr^2. V/h is equal to the area of the circular cross section of the cylinder.

12-8. Using the data from Table 12.1, find the radius of the cylinder that you used to make the measurements.

12-9. What is the area of the circular cross section of the cylinder?

INVERSE PROPORTIONALITY

We have talked a great deal about direct proportionality. Another important relationship in science is inverse proportionality. When two measurements are **inversely proportional,** one gets smaller as the other gets larger. You already know the limitations of these simple English statements. A mathematical statement is more precise. A is inversely proportional to B if A times B is a constant, or $AB = k$.

12-10. Write two other mathematical expressions that say, "A is inversely proportional to B." (Note: The new statements must be algebraically equivalent to $AB = k$.)

12-11. Name a relationship that you have already studied in this course that represents a direct proportion or an inverse proportion.

12-12. Pick any numbers for A and B such that A/B is always the same number. (For example, pick several pairs of numbers such that A/B is equal to 3.) Using these numbers, plot a graph with the values of A along the y axis and the values of B along the x axis. What is the shape of the resulting plot? When one variable is directly proportional to another, what does the plot of those two variables look like?

12-13. Now pick numbers for A and B such that AB is always the same number. (For example, pick pairs of numbers such that $AB = 12$.) Using these values for A and B, plot a graph. When one variable is inversely proportional to another, what does the plot of those two variables look like?

The mathematical exercises in this section were included because both direct and inverse proportionalities occur frequently in science. Examples of both types of relationships will be found in the following sections on gases. Although the letters used to represent variables will change, you should be able to recognize the general form of each mathematical relationship as a direct or inverse proportionality.

12.2 INTRODUCTION TO GASES

Of the three states of matter (solid, liquid, and gas), the gaseous state is easiest to describe. The study of gases has played an important role in our understanding of matter. It is time for us to look at gases more closely.

In Chapter 3 we described gases at the microscopic level. We imagined that gases such as air are composed of molecules or individual atoms that are far apart, as shown in Figure 12.4. We said that the molecules are in constant random motion, as suggested by Figure 12.5. The random motion is due to the fact that a molecule moving through space continues in a straight line

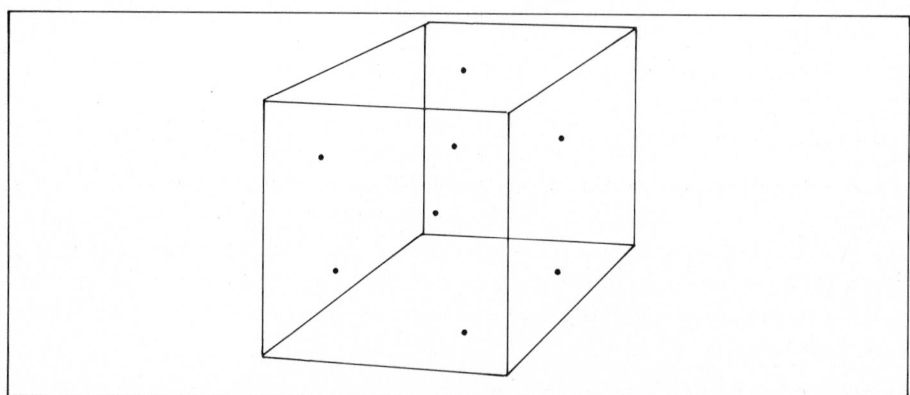

Figure 12.4 *The molecules of a gas are very far apart with nothing between them.*

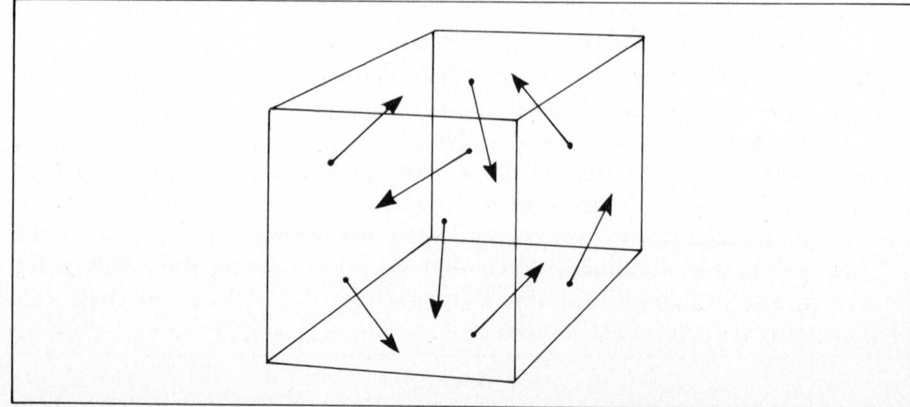

Figure 12.5 *The molecules of a gas are in constant random motion.*

until it hits something. When it does, it ricochets off in a new direction and at a different speed, which depends on the angle of impact and the momentum of the particles that collide.

Since gas molecules move in a straight line until they strike something, they are said to occupy all of the space available to them. Whether you put 100 molecules or 100 trillion molecules in a jar, the molecules fill the jar. The molecules don't literally occupy all of the space, but they do move around in all of the space in the jar. That space is considered to be the volume occupied by the gas.

As the molecules move about in a jar or in a balloon or in a bicycle tire, they hit the sides of the container and bounce back. These molecules exert pressure. The more molecules, the more hits; the more hits, the more pressure.

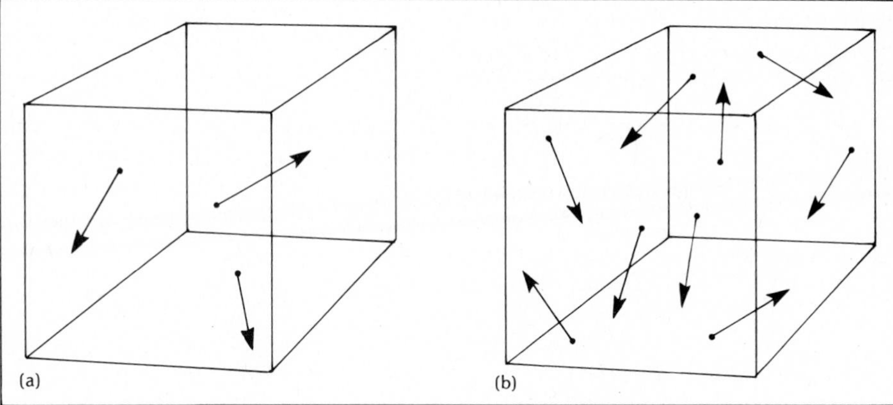

Figure 12.6 *Gases exert pressure when the molecules collide with the walls of the container. (a) Few molecules produce little pressure. (b) Many molecules produce high pressure.*

Before continuing our discussion of gases, we will digress to be sure that you understand pressure.

12.3 PRESSURE

Like density and temperature, pressure is an intensive property. It is a measure of the force exerted on a given area. For example, a pressure of 10 newtons/m^2 indicates that a force of 10 newtons is exerted on an area of 1 square meter. This is the same as a force of 100 newtons exerted on an area of 10 square meters or 1,000 newtons exerted on an area of 100 square meters. In all cases, the force on each square meter is the same.

You may not be familiar with the term "newton." The **newton** (N) is an SI unit of force. It is named for Sir Isaac Newton and is used extensively in physics. It is just a little more than the force exerted by a 100-g mass held in your hand. (Perhaps you haven't held a 100-g mass lately. An ordinary flashlight battery is close to the same mass.)

The standard unit of pressure in SI units is the newton per square meter (N/m^2) which is called a **pascal** (Pa) in honor of the Frenchman Blaise Pascal. Pascal studied pressure in liquids and discovered that pressure in a liquid is exerted equally in all directions.

As you can tell by holding a flashlight battery, a force of 1 N is not very great. When that force is spread over an area of 1 m^2, the pressure exerted is very small indeed. A more convenient unit is 1,000 times this much pressure, the kilopascal (kPa), or 1,000 N/m^2.

12-14. A force of 10 N exerted on an area of 1 m^2, a force of 100 N exerted on an area of 10 m^2, and a force of 1,000 N exerted on an area of 100 m^2 all give a constant pressure of 10 N/m^2. How would you describe the relationship between the total force exerted and the area over which the force is exerted, when pressure is constant?

12-15. Write a mathematical statement of this relationship, using F to represent force, A to represent area, and P to represent the constant pressure.

Traditionally, chemists have not expressed pressure in pascals or kilopascals. This may change with the adoption of SI units, but it is still more common for chemists to express pressure by comparison to the atmosphere. When we say that "the gas exerts a pressure of two atmospheres" we mean that the pressure of the gas is twice that exerted by the atmosphere.

ATMOSPHERIC PRESSURE

The pressure exerted by the atmosphere isn't constant. You know this from listening to weather reports. You are told that "the barometer is falling" or "the barometer is rising." We trust that the barometer is really staying put, but that the pressure of the atmosphere measured by the barometer is rising or falling.

Atmospheric pressure changes from day to day at any one place and is less at high altitudes than at low altitudes. If we are going to speak of "one atmosphere" (atm) as a standard pressure, we must define just what atmosphere we are talking about.

Standard atmospheric pressure is the pressure exerted by a column of mercury 760 mm high. Why mercury? Because, using a column of mercury, it is easy to make an instrument to measure pressure.

Fill with mercury a glass tube that is closed on one end and then invert the column so that the open end is in a pool of mercury. Presto! You have a **barometer.** A picture of such an instrument is shown in Figure 12.7.

When you look at the drawing of the barometer in Figure 12.7, you will see that the mercury does not go all the way to the top of the tube. Why? After

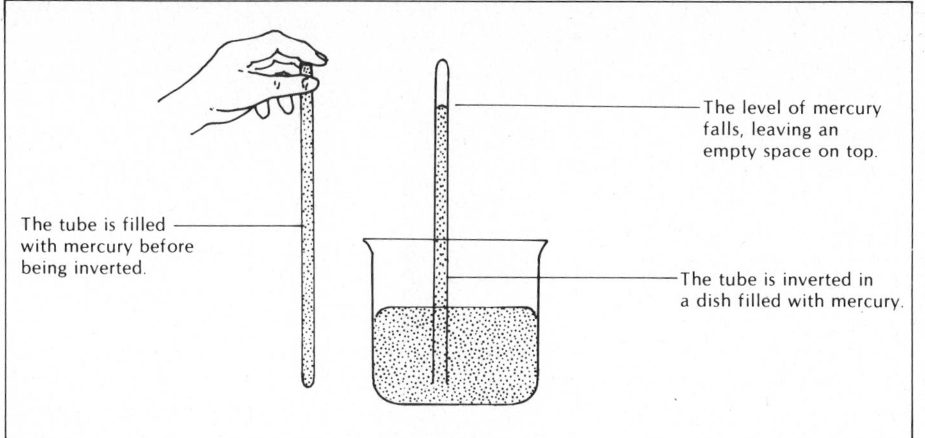

Figure 12.7
A mercury barometer is made by filling a tube with mercury and placing the open end of the tube in a pool of mercury. The length of the column of mercury in the tube is directly proportional to atmospheric pressure.

all, the tube *was* full when it was placed in the pool of mercury. Actually, a better question is "Why didn't *all* of the mercury come out of the tube when it was turned upside down in the pool of mercury?"

If you swim, you know that the underwater pressure on your body depends only on your depth below the surface of the water. The pressure that you feel is caused by the weight of whatever is above you. Generally all that is above you is water and air.

Figure 12.8 shows a lake with two swimmers under the water. They are both shown at the same depth below the surface. Even though one swimmer is under a boat, he feels no more pressure than the other swimmer at the same depth.* This is true at the surface of the water as well.

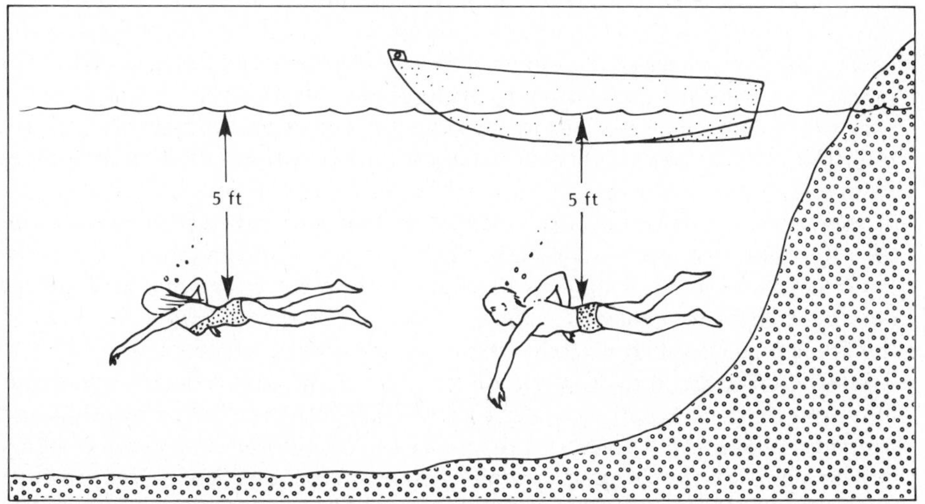

Figure 12.8
In any fluid, the pressure is equal at all points that are the same depth below the surface of the fluid.

*Even though the weight of the boat is a force on the swimmer, the boat has displaced water weighing the same amount.

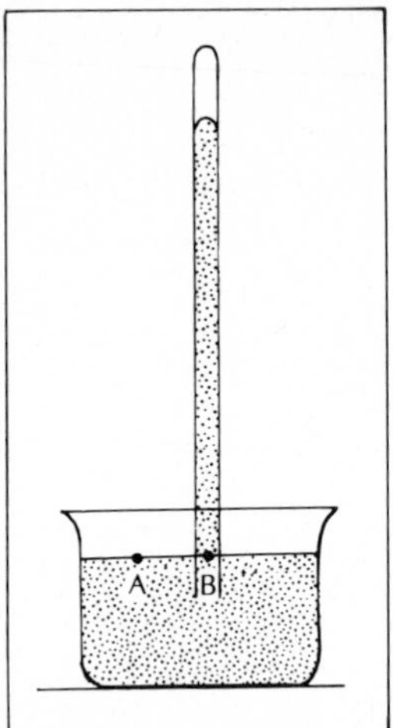

Figure 12.9
The pressure at point A and the pressure at point B are the same.

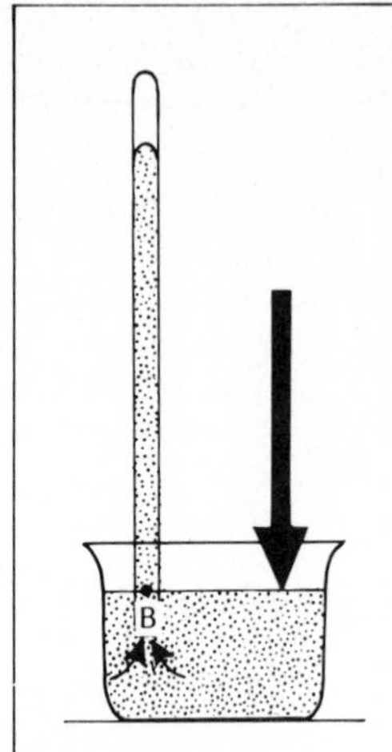

Figure 12.10
When atmospheric pressure increases, mercury is forced into the tube.

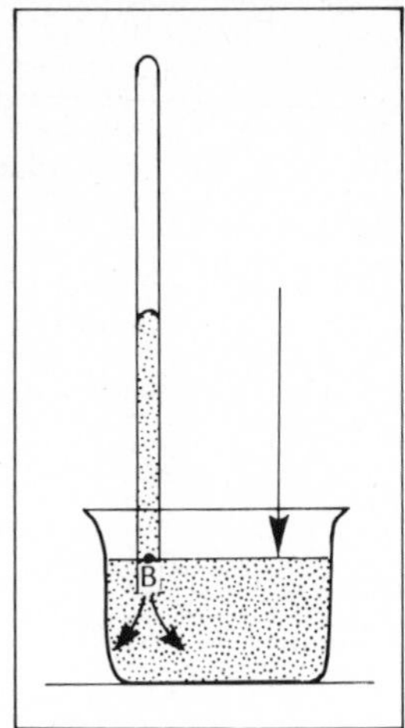

Figure 12.11
When atmospheric pressure decreases, mercury flows out of the tube.

Figure 12.9 shows a point on the surface of the pool of mercury in a barometer (point A) and another point at the same level under the mercury-filled tube (point B). These two points are at the same pressure. The pressure in both cases is due to the weight of the material above. The only thing above point A is air. The pressure at point A is atmospheric pressure. The only thing above point B is a column of mercury. (There is air *outside* the tube, but the tube is sealed so that air cannot get inside. The space above the column of mercury in the tube is empty.) The pressure exerted by the column of mercury in the tube is the same as atmospheric pressure.

If you make a mercury barometer like the one shown in Figures 12.7 (p. 351) and 12.9, you will find that the height of the column of mercury changes from day to day. It changes because the pressure of the atmosphere changes. When the atmospheric pressure increases, mercury is forced into the glass tube on the barometer, as shown in Figure 12.10, until the pressure at point B is the same as atmospheric pressure. When the atmospheric pressure decreases, mercury runs out of the glass tube, as shown in Figure 12.11, until the pressure at point B is again equal to atmospheric pressure.

Since the height of the column of mercury in the barometer changes with atmospheric pressure, the height gives an indirect measure of atmospheric pressure. When the weather report says that "the barometer reading is 29.4 in.," it is indicating that the atmospheric pressure is the same as the pressure exerted by a column of mercury 29.4 in. high. If the weather report gave the length of the column in metric units, it would state that "the barometer reading is 747 mm."

We should point out that the height being measured is the vertical height from the surface of the pool of mercury to the top of the column of mercury. It is an experimental fact that the shape of the glass column, its diameter, and the angle at which the tube is inserted in the pool of mercury make no difference. The vertical height is all that matters. All of the barometer tubes shown in Figure 12.12 would register the same pressure. For each, the height of the mercury above the surface of the pool of mercury would be the same.

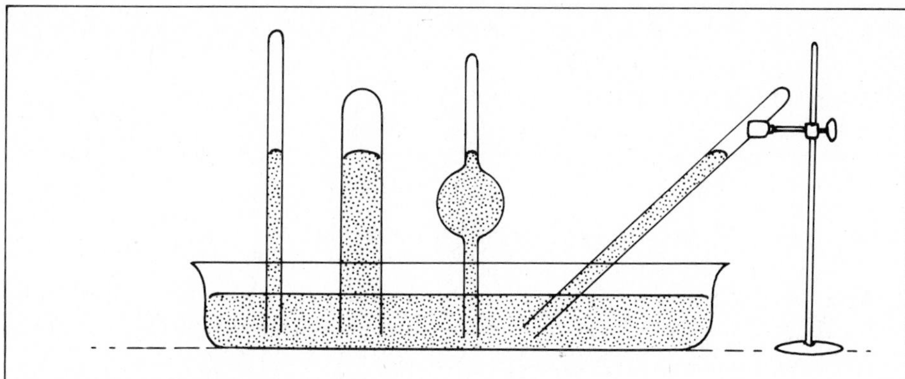

Figure 12.12 *The height of a column of mercury in a barometer does not depend on the size or shape of the tube. It depends only on the atmospheric pressure.*

UNITS FOR PRESSURE

When you read other chemistry texts or use a reference book, you are likely to see gas pressures expressed in "mm of mercury" or "torr." A **torr** is the pressure exerted by a column of mercury 1 mm high. The unit is named in honor of Evangelista *Torr*icelli (1608–1647), who invented the mercury barometer. A pressure of 740 torr is the same as a pressure of 740 mm of Hg, the pressure exerted by a column of mercury 740 mm high.

Neither torr nor millimeters of mercury nor the common English unit of inches of mercury shows the relationship between pressure, force, and area. Pressure is a measure of the force exerted over a given area. It has units of force per unit area. As we have said, the correct SI unit is newtons per square meter (N/m^2), or pascals (Pa). As SI units become more common, most chemists will switch to the pascal or, more commonly, the kilopascal (kPa) to express gas pressures. We will use the kilopascal in this book, but you are likely to encounter other pressure units in other courses.

The various units of pressure used to express standard atmospheric pressure are as follows:

$$\mathbf{1\ atm} = \mathbf{760\ torr} = 760\text{ mm Hg} = \mathbf{101.3\ kPa} = 1.013 \times 10^5\ N/m^2$$
$$= 1.013\text{ bar} = 29.92\text{ in. Hg} = 14.68\text{ lb/in}^2$$

The value of 1 atm and the value of 760 torr (or 760 mm Hg) are *exact* numbers and have an infinite number of significant digits, because the standard atmosphere is defined as the pressure exerted by a column of mercury 760 mm high. All other values are derived from this value and are given here to four significant digits. Units commonly used in chemistry are shown in bold type.

12-16. The liquid used in a barometer is mercury, but any other liquid would work. Why do you think mercury is used? (Hint: The density of mercury is 13.6 g/cm³, and it freezes at −39°C.)

12-17. How tall would you need to make a column of water so that it would weigh as much as a column of mercury 760 mm tall?

12-18. Convert 742 torr to kilopascals and to atmospheres.

12-19. Convert 122 kPa to torr and newtons per square meter.

12.4 THE SPRING IN AIR: BOYLE'S LAW

All of you know that air can be compressed. You compress it every time you use a bicycle pump. By pushing on the handle of the pump, you force air into a smaller space than it occupied before.

You also know that, when you push on the air in the pump, it pushes back. This spring in air interested Robert Boyle (1627–1691), and he investigated it quantitatively. He found that, when the temperature of a gas is kept constant, the volume of the gas decreases in an orderly way when the pressure is increased. Doubling the pressure cuts the volume in half; tripling the pressure cuts the volume to one-third; quadrupling the pressure cuts the volume to one-fourth; and so on. In other words, the volume of the gas was found to be *inversely proportional* to the pressure applied to the gas.

This inverse proportionality was not apparent to Boyle when he collected his data and, reportedly, became apparent only when someone else pointed it out. Nevertheless, the quantitative relationship is known as **Boyle's law** in honor of the man who collected the data without appreciating their full meaning.

12-20. In Section 12.1, you learned to write a mathematical expression that represents an inverse proportionality. Write a mathematical expression that says, "The volume (V) of a gas is inversely proportional to the pressure (P) of the gas."

Figure 12.13 shows a syringe connected to a pressure gauge with a thin piece of tubing. By pushing on the plunger of the syringe, we can compress the gas in the syringe into a smaller volume. The pressure of the gas in the syringe can be read from the gauge. Table 12.3 shows several values taken with this instrument for volume and corresponding pressure. The pressure is shown in the familiar English units of pounds per square inch and in the SI units of kilopascals.

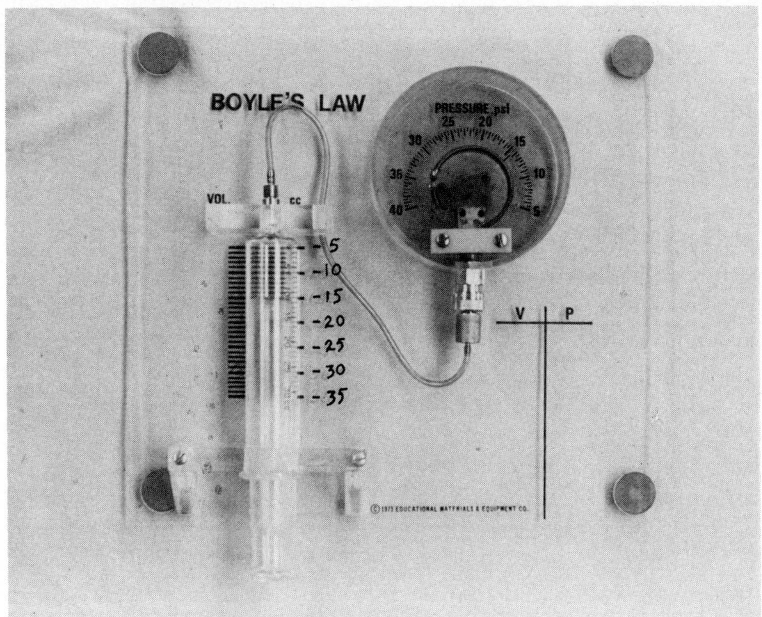

Figure 12.13
Apparatus for demonstrating Boyle's law. As we change the volume of gas in the syringe by pushing on the plunger, the gauge records the pressure exerted by the gas.

12-21. Table 12.3 shows actual experimental data recorded to the proper number of significant digits. Based on these data, are you willing to say that $PV = k$, where P represents the pressure of the gas, V represents the volume of the gas, and k represents a constant number?

12-22. What is the equivalent of 102 kPa in units of newtons per square meter?

12-23. Using your answer to question 12-22 as the pressure exerted by the gas when the volume was 15.0 cm³, calculate a value for k. What are the units for k?

The particular value of k that you just calculated represents the constant number that is obtained when the volume of this particular sample of gas is multiplied by the pressure exerted by the gas. The sample of gas used in this experiment was 15 cm³ of air placed in the syringe at the prevailing

Table 12.3. Pressure Exerted by a Gas When Volume Is Changed

	Pressure	
Volume (cm³)	*Pounds per square inch (psi)*	*Kilopascals (kPa)*
5.0	45.2	311
10.0	22.1	152
15.0	14.8	102
20.0	11.2	77.0
25.0	9.0	62
30.0	7.6	52

atmospheric pressure. (In other words, 15 cm^3 of air was drawn into the syringe and the syringe was then connected to the pressure gauge as it read atmospheric pressure.) If we had started the experiment by placing 30 cm^3 of air in the syringe rather than 15 cm^3, PV would be a constant during the experiment, but it would not be the same constant that you calculated in Question 12-23. This can be a little confusing. When is this constant constant?

When we investigated how the volume of a cylinder changes with the height of a cylinder, we found that $V = kh$—provided that the diameter of the cylinder remained constant. We added a parenthetical expression after the equation to remind us of this condition. We need to do the same thing with Boyle's law. It should be written

$$PV = k \quad (n, T \text{ constant})$$

The equation now says, "The pressure of a gas multiplied by the corresponding volume of that gas will always give you the same number, provided you don't change the amount of gas (the number of gas molecules, n) or the temperature (T) of the gas."

If the data shown in Table 12.3 had been taken with a leaky syringe, PV would not have been constant, because n, the moles of gas molecules in the syringe, would not have remained constant. $PV = k$ only when n is kept constant.

If the data shown in Table 12.3 had been taken while the syringe was held in my hand, PV would not have been constant. My hand would have gradually warmed the gas in the syringe, causing it to expand or exert more pressure. $PV = k$ only if temperature is kept constant.

The relationship that we have derived—namely, that the pressure times the volume of a gas is constant—does not hold exactly for *every* gas at *all* temperatures and pressures, but it is amazingly constant for ordinary gases at moderate temperatures and pressures. It is a convenient relationship to use if you know the volume of a gas at one pressure and want to know the volume that the gas would occupy at some new pressure or, conversely, the pressure that the gas would exert at some new volume. Chemists do not need to work problems of this kind very often, but engineers do. Work through the following example with me, and you will see how the relationship can be used.

Example 12.1

What volume of air at standard atmospheric pressure would you have to force into a bicycle tire that has a volume of 1.00 liter in order to pressurize the tire to 65 lb/in^2 (abbreviated on the tire as psi)?

The question asks about a relationship between pressure and volume. What is the mathematical statement of this relationship?

$$PV = k \quad (T, n \text{ constant})$$

We normally don't write down the conditions shown in parentheses, but you should remember them. What are they telling you must be true in order for us to use the mathematical relationship that you wrote?

The air must not change temperature as it is put inside the tire, and the number of molecules we are talking about when the air is outside the tire must be the same as the number of molecules we are talking about when the air is inside the tire; in other words, the tire can't spring a leak.

$PV = k$ says that I can multiply the pressure and volume of a sample of gas, getting a number I have called k, and then put that sample of gas under more pressure, and multiply the new pressure times the new volume, and *still* get that same number, k. In other words, the product PV is constant.

From the data given in the problem, calculate a numerical value for k.

$$k = 65 \frac{\text{L lb}}{\text{in}^2}$$

The problem says that we will put enough air in the tire that has a volume of 1 liter to give a pressure of 65 lb/in². Then,

$$PV = k$$

$$65 \frac{\text{lb}}{\text{in}^2} \times 1.00 \text{ L} = k$$

$$65 \frac{\text{L lb}}{\text{in}^2} = k$$

The problem could be worked without calculating a value for k, but finding k may help you see the reasoning that we are using. $PV = k$ says that the product of the pressure and volume of the gas that we put in the bike tire will always be 65 L lb/in², no matter what the pressure and volume of the gas may be. When the air is *outside* the tire, PV must also equal 65 L lb/in².

The problem says that the gas is at normal atmospheric pressure before it is placed in the tire. What is the value of normal atmospheric pressure?

$$1 \text{ atm} = 760 \text{ torr} = 101.3 \text{ kPa} = 29.92 \text{ in. Hg} = 14.68 \text{ lb/in}^2$$

All of these values represent normal (standard) atmospheric pressure expressed in different units. Which of these values would you need to use with the value of k we found (65 L lb/in²) in order to calculate the volume of air outside the tire in units of liters?

$$14.68 \text{ lb/in}^2$$

The value of k that we calculated has pounds per square inch as part of the units. If we want to get rid of these units to leave only units of liters, the pressure must be expressed in pounds per square inch.

Now use the relationship $PV = 65 \text{ L lb/in}^2$ to calculate the volume of the air before it is put into the bike tire.

$$PV = 65 \text{ L lb/in}^2$$

$$14.68 \text{ lb/in}^2 \times V = 65 \text{ L lb/in}^2$$

$$V = \frac{65 \text{ L } \cancel{\text{lb}}}{\cancel{\text{in}^2}} \times \frac{1 \ \cancel{\text{in}^2}}{14.68 \ \cancel{\text{lb}}}$$

$$V = 4.4 \text{ L}$$

As mentioned earlier, it really isn't necessary to calculate a value for k in order to solve this problem. We have said that the pressure exerted by the air in the tire times the volume of the air in the tire is equal to k, or

$$P_{\text{tire}}V_{\text{tire}} = k$$

We have also said that the pressure of that air in the atmosphere times the volume it occupies in the atmosphere is equal to the same number, which we have represented by k:

$$P_{\text{atm}}V_{\text{atm}} = k$$

A basic rule of algebra is that two quantities equal to the same quantity are equal to each other. Hence we can write

$$P_{\text{tire}}V_{\text{tire}} = P_{\text{atm}}V_{\text{atm}}$$

Using this relationship, substitute the numerical values given in the statement of the problem and find the volume that the gas will occupy at atmospheric pressure.

$$P_{\text{tire}}V_{\text{tire}} = P_{\text{atm}}V_{\text{atm}}$$

$$\frac{65 \text{ lb}}{\text{in}^2} \times 1.00 \text{ L} = \frac{14.68 \text{ lb}}{\text{in}^2} \times V_{\text{atm}}$$

$$\frac{1 \ \cancel{\text{in}^2}}{14.68 \ \cancel{\text{lb}}} \times \frac{65 \ \cancel{\text{lb}}}{\cancel{\text{in}^2}} \times 1.00 \text{ L} = \frac{1 \ \cancel{\text{in}^2}}{\cancel{14.68 \text{ lb}}} \times \frac{\cancel{14.68 \text{ lb}}}{\cancel{\text{in}^2}} \times V_{\text{atm}}$$

$$4.4 \text{ L} = V_{\text{atm}}$$

Sometimes Boyle's law is expressed like this:

$$P_1V_1 = P_2V_2$$

The subscripts are used to identify two different sets of conditions. In our example, the 1 could represent the pressure and volume of the air when it is in the tire, and the 2 could represent the pressure and volume of the air when it is outside the tire.

You can use either expression for Boyle's law to solve the following problems. About the only difficulty that you should encounter is in the units. If the two pressures are not expressed in the same units or the two volumes are not expressed in the same units, the answer you get will not make sense.

12-24. A sample of propane gas occupies a volume of 2.5 liters at 1 atm pressure. What volume would it occupy at 0.2 atm pressure?

12-25. What pressure would be required to place 12 liters of air at 1 atm pressure into a 3-liter bottle?

12-26. A 1-liter bottle of oxygen gas is under a pressure of 100 kPa. What would the pressure be if the gas were placed in a 250-mL bottle?

12-27. A certain gas occupies 100 mL at a pressure of 2 atm. What volume would it occupy at a pressure of 79 kPa?

12-28. By what factor must the pressure be changed to compress 450 mL of gas to a volume of 100 mL?

12.5 THE MICROSCOPIC EXPLANATION OF PRESSURE

In the previous section, we discussed the relationship between the pressure and volume of a gas. Nothing was said about what the molecules of gas were doing. Now you are invited to think about what is happening at the microscopic level. We believe that this is important, because changes that occur in matter can best be understood by thinking about atoms and molecules. Thinking about things you can't see requires practice. This is a good time to get some of that practice.

Recall that gases are made up of molecules or atoms, which can be represented by spheres. These particles are very far apart in gases. Each molecule is moving. It moves in a straight line until it hits something. Then it bounces off and continues in a straight line until it hits something else. (See Figures 12.4 through 12.6.)

Pressure is exerted on the walls of a container when the molecules strike the wall; the more hits, the more pressure. In addition, a molecule moving fast hits harder than a molecule moving slowly. The average speed of the molecules in a gas is measured by the temperature. The higher the temperature, the faster the molecules are moving.

With these ideas in mind, see how well you can explain the following macroscopic observations in terms of what the molecules of a gas are doing.

12-29. You know that the pressure of a gas times its volume is constant. This is illustrated by the syringe shown in Figure 12.14. Talking in terms of moving molecules, explain why the pressure inside the syringe shown in Figure 12.14(b) is twice the pressure inside the syringe shown in Figure 12.14(a).

Figure 12.14

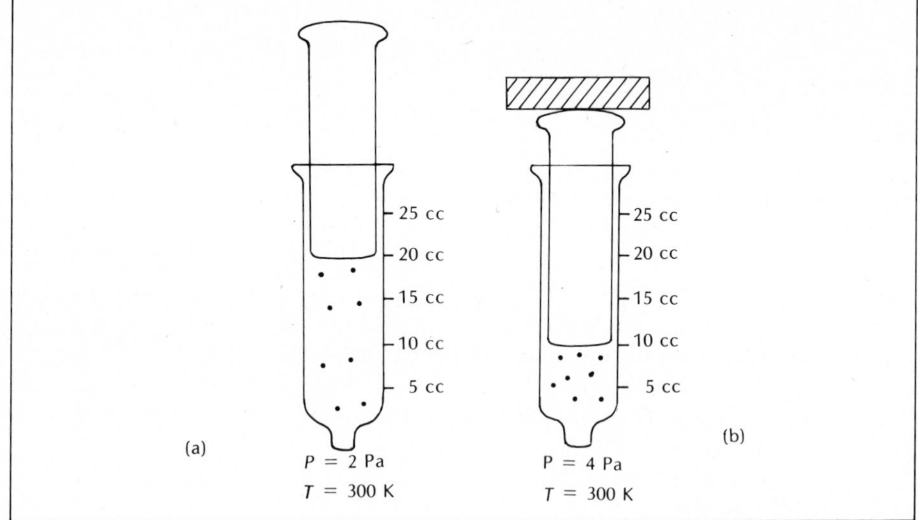

12-30. We have emphasized that $PV = k$ only when n and T are constant. Figure 12.15 illustrates a change in pressure and volume in a leaky syringe. (Note that there are fewer molecules shown in Figure 12.15(b) than in Figure 12.15(a). Some have leaked out of the syringe.) Talking in terms of moving molecules, explain why the pressure inside the syringe shown in Figure 12.15(b) is *not* twice the pressure inside the syringe shown in Figure 12.15(a).

Figure 12.15

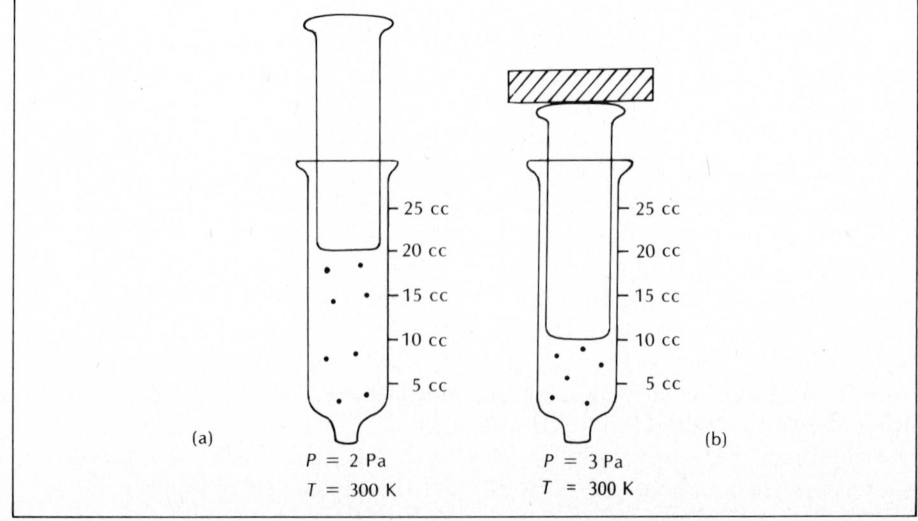

12-31. Talking in terms of moving molecules, explain why the pressure inside the syringe shown in Figure 12.14(b) would be greater than 4 Pa if the temperature of the gas increased to 400 K.

12.6 TEMPERATURE EFFECTS: CHARLES' LAW

In the preceding discussion, we were careful to point out that $PV = k$ only when the temperature of the gas is not changed. Changing the temperature of a gas results in a change in volume when the gas is in a container that allows for expansion. When the gas is in a rigid container, the pressure changes. This is a fact that you all know very well. Discarded aerosol cans explode when they are placed in a fire. The increased temperature of the gas creates enough pressure to rupture the can. Balloons burst for the same reason when they get hot, and automobile tires register a higher pressure when they heat up during driving. We now want to look at the quantitative relationship between the volume of a gas and its temperature.

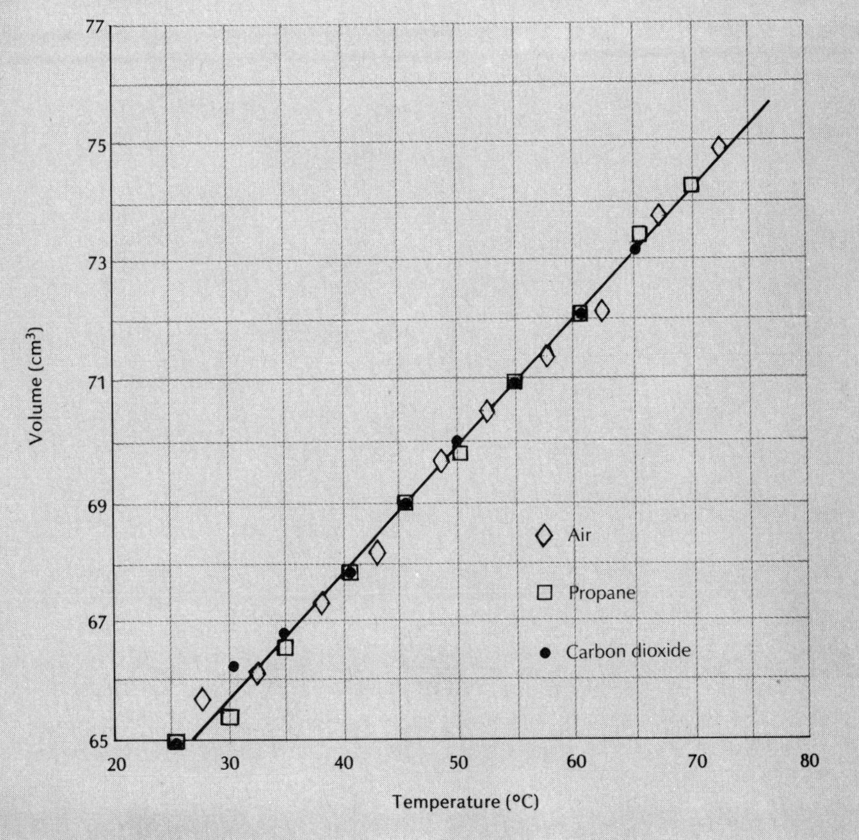

Figure 12.16 *Graph showing the volume of three different gases at various temperatures.* (*As adapted from Figures 3.6 and 3.7 in* Introductory Physical Science, *IPS Group of ESI.* © *1967 by ESI.*)

Figure 12.16 (p. 361) shows a graph obtained by plotting the volume of various gases as their temperature increased. These data were obtained by placing a sample of each gas in a syringe and then heating the gas to various temperatures in a pan of water. A simplified model of the apparatus is shown in Figure 12.17. As the temperature was changed, data were recorded for the volume of the gas at the various temperatures (Table 12.4).

As Figure 12.16 shows, the data seem to fall on the same straight line, regardless of the gas used in the experiment. Similar experiments have been performed on a number of gases, and all seem to yield this same straight line when the temperature is varied over a moderate range. When the temperature approaches the condensation point of a gas, the gas begins to turn to a liquid and the points plotted no longer fall on the line shown.

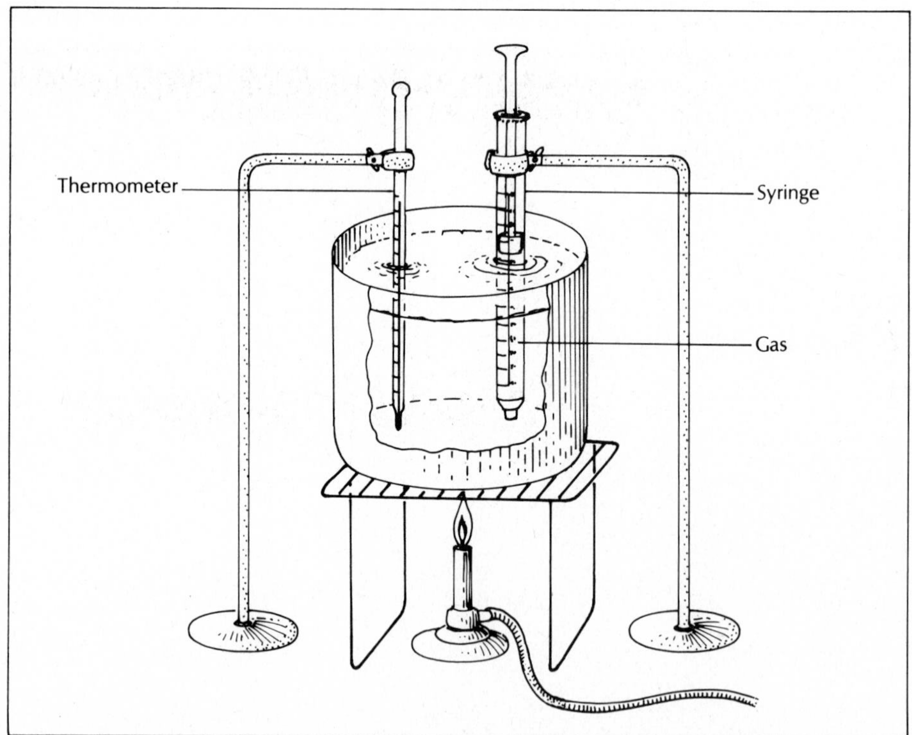

Figure 12.17 *An apparatus similar to this was used to collect the data plotted in Figure 12.16.* (*As adapted from Figures 3.6 and 3.7 in* Introductory Physical Science, *IPS Group of ESI. © 1967 by ESI.*)

Some of you will recall that a fairly simple equation can be written to describe any straight-line graph. The general equation for a straight line is $y = ax + b$. Here y represents what is plotted along the y axis, x represents what is plotted along the x axis, a is the slope of the line, and b is the value of y where the line crosses the y axis (the y intercept). In the graph shown in Figure 12.16, we have plotted the volume of the gas along the y axis and the

Table 12.4. Volume of Gas at Various Temperatures

Air		*Propane*		*Carbon dioxide*	
Temperature (°C)	*Volume* (cm^3)	*Temperature* (°C)	*Volume* (cm^3)	*Temperature* (°C)	*Volume* (cm^3)
25.1	65.0	25.1	65.0	25.1	65.0
27.4	65.8	30.1	65.9	30.3	66.2
32.7	66.2	35.3	66.6	35.0	67.0
37.9	67.4	40.2	67.8	40.1	67.8
42.9	68.2	45.1	69.0	45.0	69.0
48.1	69.8	50.0	69.8	50.1	70.2
52.3	70.6	54.9	71.0	54.8	71.0
57.6	71.5	60.2	72.2	59.9	72.2
62.3	72.5	65.0	73.4	64.5	73.2
67.1	73.8	70.0	74.2	—	—
72.5	75.0	—	—	—	—

As adapted from Figures 3.6 and 3.7 in *Introductory Physical Science,* IPS Group of ESI. © 1967 by ESI.

temperature in degrees Celsius along the x axis. We can write an equation for the relationship between the volume and temperature of a gas as follows:

$$y = ax + b$$

$$V = at + b$$

The value of a and b depends on the moles of gas that we place in the syringe when we begin the experiment.

Note that the equation is one that would describe a direct proportionality between the volume and temperature of a gas except for the value of b added to the expression. If the y intercept were zero, the equation would simply be "$V = at$," which says that "the volume of a gas is directly proportional to the temperature." The proportionality constant, k, is represented by a in the equation as written.

Direct proportionalities are easy to work with. It would be nice if we could change the graph shown in Figure 12.16 so that the value of b would be zero. Then the relationship between the volume of a gas and the temperature of the gas could be represented as a direct proportion, such as $V = at$.

What we have proposed can be done rather easily. All we need to do is to define a new temperature scale that has −273°C as its zero point. Such a scale was defined in Chapter 2. It is the Kelvin temperature scale. The size of the Kelvin (the degree itself is called a **Kelvin**) is exactly the same as the size of the Celsius degree. The only difference between the two scales is the temperature that is assigned a value of zero.

Figure 12.18 (p. 364) shows the data from Table 12.4 plotted on a smaller scale with the line through the data points extended to a temperature of 0 Kelvin. The boxed area of Figure 12.18 shows the area covered by Figure 12.16.

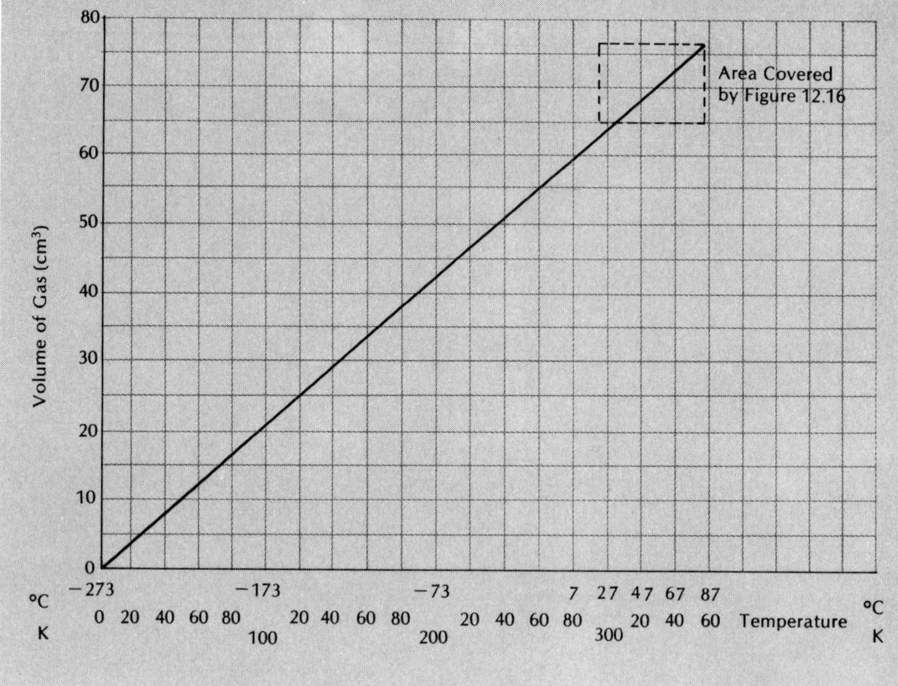

Figure 12.18
Graph showing the data from Figure 12.16 extended to the point of zero volume. Any real gas would change to a liquid before reaching this theoretical value at "absolute zero."

Because we have changed the temperature scale to Kelvin, the line through the data points passes through the origin on the graph and the y intercept is zero. We can now say that the volume of a gas is directly proportional to the Kelvin temperature. This can be expressed by the following equation:

$$V = kT$$

V stands for the volume of the gas, k stands for the proportionality constant, and the capital T stands for the temperature in Kelvin. Kelvin is abbreviated as K, and no degree sign is used. Thus, 100°C is equal to 373 K. In words, "one hundred degrees Celsius is equal to three hundred and seventy-three Kelvin."

To refresh your memory, convert the following temperatures from degrees Celsius to Kelvin or from Kelvin to degrees Celsius.

12-32. 27°C
12-33. 27 K
12-34. 0°C
12-35. 0 K
12-36. 373 K
12-37. 37°C (normal body temperature)

Before leaving Figure 12.18 and discussing the macroscopic behavior of gases, we should make one further point. As you cool a sample of gas, the volume of the gas gets smaller, as shown by the data for real gases in Figure 12.16. However, Figure 12.18 suggests that cooling a sample of a gas to 0 Kelvin would result in zero volume for the gas. This isn't true. First of all, no substance exists that remains a gas at such low temperatures. Long before this temperature is reached, any gaseous substance would turn to a liquid or solid. The graph shows a kind of "ideal" arrangement in which we have a gas that will never become liquid. The line on the graph in Figure 12.18 describes the behavior of this "ideal" gas at all temperatures, but it describes the behavior of real gases only at moderately high temperatures, such as those shown in the boxed area on the graph.

The relationship between the temperature and volume of a gas was first noticed by Jacques Charles (1746–1823). It is known as **Charles' law.** Just as Boyle's law can be used to find the change in volume when the pressure is varied, Charles' law can be used to find the change in volume when the temperature is varied. Since $V = kT$, the ratio of volume to temperature (V/T) is always the same for a given sample of gas. Using the same notation that we used in discussing Boyle's law, we can write

$$\frac{V}{T} = k \quad (P, n \text{ constant})$$

$$\frac{V_1}{T_1} = \frac{V_2}{T_2}$$

As before, the parenthetical expression is a reminder that the relationship holds only when you do not change the pressure of the gas or the amount of gas (number of moles) that you are considering. You should also remember that, in working with Charles' law, *we must express temperature in Kelvin.*

Before proceeding to the next section, work the following problems to be sure that you understand how this relationship is used.

12-38. A plastic bag containing 1 liter of air at 27°C is heated to 100°C. What is the new volume of the gas? (Note: Did you remember to change from Celsius to Kelvin? If you did not, your answer is wrong. The equation is true *only* when temperature is expressed in K.)

12-39. To what temperature (in °C) would you need to heat a liter of air at 27°C in order to just double the volume?

12.40. An "empty" aerosol can explodes when thrown into a fire, because the increase in temperature increases the pressure of the gas inside the can. The only alternative would be for the can to get larger. How large would a 250-mL aerosol can have to get in order to contain the gas inside it with no increase in pressure when the can is heated from 25°C (a reasonable room temperature) to 450°C (a rather "cool" fire)?

12-41. You have 125 mL of gas at 27°C and want to put the gas into a 100-mL bottle without changing the pressure. How much do you have to cool the gas?

12-42. When the volume of a gas is increased from 100 mL to 250 mL by an increase in temperature, how does the final temperature of the gas compare to the initial temperature of the gas?

12.7 THE MICROSCOPIC EXPLANATION OF TEMPERATURE

Once again you are invited to explain what has been described at the macroscopic level by talking about the molecules that make up a gas. Remember that temperature is a measure of the average kinetic energy of the gas molecules. The kinetic energy of any moving particle can be described as one-half of its mass times its velocity squared:

$$K = \tfrac{1}{2}mv^2$$

As the temperature of a gas increases, the molecules move faster, resulting in an increase in kinetic energy.

12-43. The volume of a gas divided by its temperature is constant, as illustrated by Figure 12-19. Talking in terms of the motion of molecules, explain how the pressure inside the syringe shown in Figure 12-19(b) can be the same as the pressure inside the syringe shown in Figure 12-19(a).

Figure 12.19

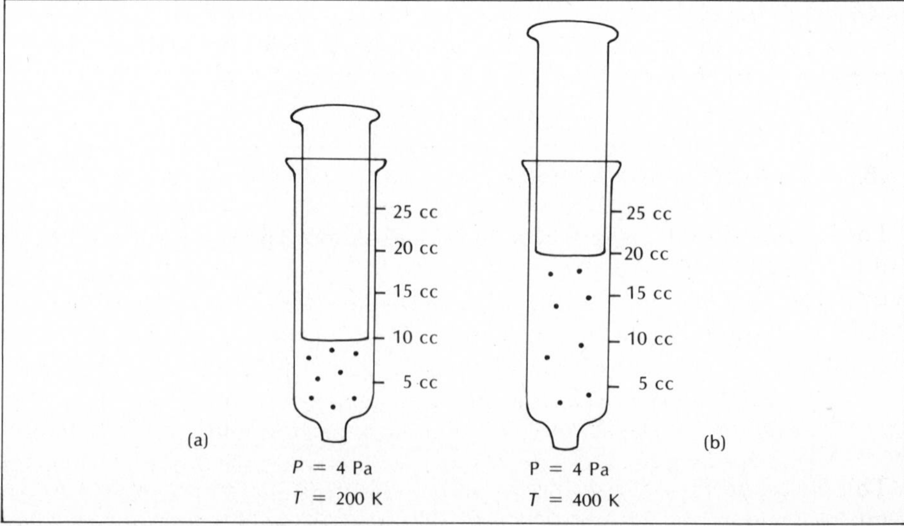

12-44. We have emphasized that V/T is constant only when we do not change the number of molecules. Talking in terms of moving molecules, describe what would happen if more molecules were added to the syringe shown in Figure 12.19(b).

12-45. We have emphasized that V/T is constant only when we do not change the pressure of the gas. Figure 12.20 shows a syringe before and after heating, but the position of the plunger is not shown in Figure 12.20(b). Draw a sketch to show where you think the end of the plunger would be (that is, what volume of gas would be contained in the syringe).

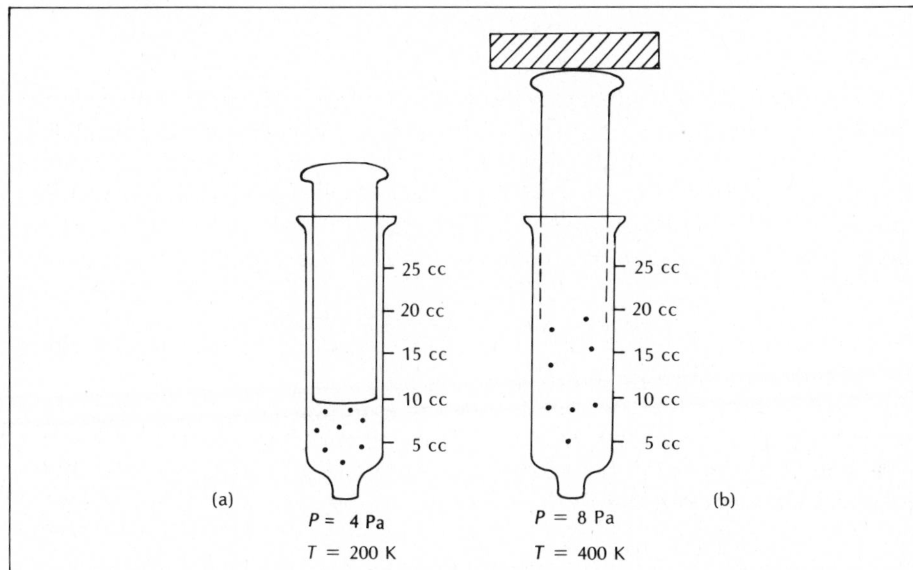

Figure 12.20

12-46. Talking in terms of moving molecules, explain why the plunger would be at a different level in Figure 12.20(b) than in Figure 12.19(b).

12.8 COMBINED GAS LAWS

The situation described in Figure 12.20 is one in which *both* the temperature and the pressure of a gas are changed. It is possible to combine Boyle's law and Charles' law into a form that makes it convenient to solve such problems exactly. This combined form is usually written as

$$\frac{P_1V_1}{T_1} = \frac{P_2V_2}{T_2} \quad (n \text{ constant})$$

This equation is convenient for solving many problems encountered in engineering, but none of the equations that we have discussed is particularly useful in chemistry. In chemistry we are usually interested in knowing the number of moles of gas that we have. It is difficult to weigh gases and then use the molecular mass to calculate the number of moles of gas present. However, it is fairly easy to measure the volume of a gas. A relationship that could be used to calculate the number of moles of gas in a given volume would be a useful tool. With it, we could measure the volume of gas produced when $NaHCO_3$ decomposes, for example, and use that volume to calculate the moles of soda that decomposed. Such a relationship does exist. It is known as the **ideal gas equation,** and it is represented by

$$PV = nRT$$

P, *V*, and *T* have their usual meaning. *R* represents a constant known as the **ideal gas constant,** and *n* represents the number of moles of gas considered.

A derivation of the ideal gas equation is too involved for this course, but it is not difficult to show that the ideal gas equation is equivalent to Boyle's law when we keep temperature constant and to Charles' law when we keep pressure constant. To see this, compare the ideal gas equation to Boyle's law.

Boyle's law:	$PV = k$ (n, T constant)
Ideal gas equation:	$PV = nRT$

For Boyle's law to hold, *n* and *T* must be kept constant. If we keep them constant, *nRT* is a constant. *R*, *n*, and *T* are all constant, and the product of constants is also a constant. Under the conditions of constant *n* and *T*, the ideal gas equation reduces to

$$PV = nRT = k$$

This is Boyle's law.

Now let us compare the ideal gas equation to Charles' law.

Charles' law:	$V/T = k$ (n, P constant)
Ideal gas equation:	$PV = nRT$
Rearranging:	$V/T = nR/P$

Now if we do not change the value for *n* and *P* (they remain constant), *n* times *R* divided by *P* does not change either. We could just as well write

$$V/T = nR/P = k$$

When *P* and *n* are held constant, the ideal gas equation is identical to Charles' law.

The advantage of the ideal gas equation is that we have a single equation that holds for any gas under moderate conditions of temperature and pressure. We do not have to make any assumptions about keeping temperature, pressure, or the number of moles of gas constant.

The value of R is the same regardless of the temperature of the gas. It is the same for all pressures. It is the same no matter how many moles of gas we consider. The value of R can be calculated from experimental data. This is done in the following example.

Example 12.2

Exactly 1 mol (32.0 g) of oxygen gas was collected in the laboratory at a temperature of 24°C and a pressure of exactly 100.0 kPa. The volume occupied by the mole of gas at this temperature and pressure was measured as 24,686 mL.

The measurements made in the laboratory provide values for each variable in the ideal gas equation except R. These values are as follows:

$$V = 24.686 \text{ L}$$
$$P = 100.0 \text{ kPa}$$
$$T = 297 \text{ K } (24°\text{C})$$
$$n = 1.00 \text{ mol}$$

Use these values and the ideal gas equation to calculate a value for R.

$$R = 8.31 \frac{\text{L kPa}}{\text{mol K}}$$

We obtained this value simply by substituting in the equation as follows:

$$PV = nRT$$

$$100.0 \text{ kPa} \times 24.686 \text{ L} = 1.00 \text{ mol} \times R \times 297 \text{ K}$$

$$\frac{100.0 \text{ kPa} \times 24.686 \text{ L}}{1.00 \text{ mol} \times 297 \text{ K}} = R$$

$$8.31 \frac{\text{L kPa}}{\text{mol K}} = R$$

Notice the units for R. Liter appears, indicating that volume has been expressed in liters. The kPa in the units indicates that pressure has been measured in kilopascals. The mol and K in the denominator indicate that the amount of substance has been measured in moles and that the temperature has been measured in Kelvin. Since the relationship is true *only* when temperature is in Kelvin, and since we measure amount of substance in moles, the units shown in the denominator do not change. However, a value for R can be calculated with pressure and volume expressed in other units.

Convert the value of R that we have just calculated to units of L atm/mol K.

$$R = 0.0820 \frac{\text{L atm}}{\text{mol K}}$$

$$1.00 \text{ atm} = 101.316 \text{ kPa}$$

Then,

$$8.31 \frac{\text{L } \cancel{\text{kPa}}}{\text{mol K}} \times \frac{1.00 \text{ atm}}{101.316 \ \cancel{\text{kPa}}} = 0.0820 \frac{\text{L atm}}{\text{mol K}}$$

Just for practice, calculate R in units of mL torr/mol K. (1 atm is equal to 760 torr.)

$$R = 6.24 \times 10^4 \frac{\text{mL torr}}{\text{mol K}} \quad \text{(actual value)}$$

$$0.0820 \frac{\cancel{\text{L}} \ \cancel{\text{atm}}}{\text{mol K}} \times \frac{1000 \text{ mL}}{1 \ \cancel{\text{L}}} \times \frac{760 \text{ torr}}{1 \ \cancel{\text{atm}}} = 6.23 \times 10^4 \frac{\text{mL torr}}{\text{mol K}}$$

Values of R

$$8.31 \frac{\text{L kPa}}{\text{mol K}}$$

$$0.082 \frac{\text{L atm}}{\text{mol K}}$$

$$62.4 \frac{\text{L torr}}{\text{mol K}}$$

(The difference between the actual value and the value calculated is due to rounding error compounded during three separate calculations.)

Once you know the value for R, you can use it in calculations to find other variables in the equation. The following problems can be solved in the same way in which you just solved for R. The only difference is that you now know a value for R and will solve the equation for some other variable. Be careful with the units to be sure that the answer is sensible.

12-47. How many moles of gas are contained in 250 mL of oxygen at a temperature of 27°C and 1.0 atm pressure?

12-48. When 1 mol of a gas is contained in a 20-liter bottle at a temperature of 27°C, what is the pressure of the gas?

12-49. What is the temperature of 1 mol of gas in a 20-L container when the pressure of the container is 1.0 atm?

12-50. How many moles of air molecules are there in an open 1.00 L bottle when the temperature of the room is 25°C and the barometer reading is 750 mm Hg?

REMEMBERING EQUATIONS VS. REGENERATING THEM

You have seen several equations in this chapter. You may be wondering whether you have to remember them all. It really isn't necessary. If you remember $PV = nRT$ and think carefully about what you are being asked to do, you can solve any problem you are likely to face. Work through the following examples with me and you will see how it is done.

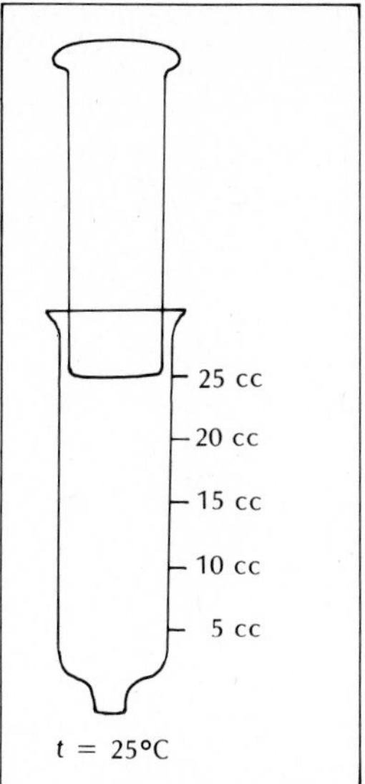

Figure 12.21

Example 12.3

The syringe shown in Figure 12.21 has 25.0 cm³ of air in it when it is at room temperature and pressure. How much pressure would you need to exert on the plunger to force all of the air into a space of 11.0 cm³?

As you read this question, what do you assume the pressure of the 25 cm³ of gas to be?

We aren't told. It is the same as room pressure, but we aren't told what that is. If we were in the room with a barometer, we could measure it. Assume that the barometer reading is 98.6 kPa.

As you read the problem, is the temperature of the air or the moles of gas in the syringe going to change?

Evidently not. There is no mention of either changing.

As I understand this question, we want to know the change in volume when we change the pressure on the gas and nothing else. This could be done with Boyle's law, but we will use the ideal gas law.

$$PV = nRT$$

In this problem, neither n nor T changes. If n doesn't change, R doesn't change, and T doesn't change, then nRT doesn't change and we can say:

$$PV = nRT = \text{a constant}$$

This means that

$$P_1V_1 = P_2V_2$$

Using this relationship, find the pressure that you would have to exert in order to reduce the volume from 25.0 cm^3 to 11.0 cm^3.

$$P_2 = 224 \text{ kPa}$$

Here is my solution:

$$P_1V_1 = P_2V_2$$

$$98.6 \text{ kPa} \times 25.0 \text{ cm}^3 = P_2 \times 11.0 \text{ cm}^3$$

$$\frac{98.6 \text{ kPa} \times 25.0 \cancel{\text{cm}^3}}{11.0 \cancel{\text{cm}^3}} = P_2$$

$$224 \text{ kPa} = P_2$$

Example 12.4

When the temperature of the room is 25°C, how much would you have to cool the 25.0 cm^3 of air in the syringe shown in Figure 12.21 in order to get the volume down to 11.0 cm^3 without changing the pressure?

What is to be changed and what is to be held constant in this problem?

The pressure and the number of moles of gas are constant. Temperature and volume are changing.

Rearrange the ideal gas equation so that the things that change are shown on the left and the things that remain constant are shown on the right of the equals sign.

$$V/T = nR/P$$

Since everything on the right is constant, we can write

$$\frac{V_1}{T_1} = \frac{V_2}{T_2}$$

Now use this expression to find the temperature of the air when it has a volume of 11.0 cm^3.

$T_2 = 131$ K (Did you remember to convert 25°C to 298 K?)

Here is my solution:

$$\frac{V_1}{T_1} = \frac{V_2}{T_2}$$

$$\frac{25.0 \text{ cm}^3}{298 \text{ K}} = \frac{11.0 \text{ cm}^3}{T_2}$$

$$T_2 \times \frac{25.0\ \text{cm}^3}{298\ \text{K}} = 11.0\ \text{cm}^3$$

$$T_2 = 11.0\ \cancel{\text{cm}^3} \times \frac{298\ \text{K}}{25.0\ \cancel{\text{cm}^3}}$$

$$T_2 = 131\ \text{K or } -142°\text{C}$$

The problems that follow are similar to the last two examples in that some conditions of the gas are not changed while other conditions are. By rearranging the ideal gas equation, you can find what is constant and solve the problems. These would be good problems to work with a partner. Think out loud as you solve the problem, and have your partner check your strategy as you go along.

12-51. A 500-mL bottle contains 1 mol of gas at 27°C. What will the pressure of the gas be when an additional 0.5 mol of gas is added to the bottle?

12-52. What change must be made in the temperature of a gas to increase the volume from 125 mL to 150 mL without changing the pressure?

12-53. A student collected two gases under the same conditions of temperature and pressure. The volume of the first gas was 125 mL, and the volume of the second gas was 225 mL. How do the number of moles of gas in the two samples compare?

12.9 MOLAR VOLUME AND DALTON'S LAW OF PARTIAL PRESSURE

When you calculated the value for R in Example 12.2, you were told that the volume of 1 mol of oxygen gas was measured at 24°C and 100.0 kPa. The volume was found to be about 24.7 liters. The volume of exactly 1 mol of hydrogen gas at this same temperature and pressure would also be about 24.7 liters.

In Chapter 6 you learned that Avogadro suggested many years ago that equal volumes of gases (at the same temperature and pressure) contain equal numbers of particles. The converse must also be true. Equal numbers of gas particles (such as 1 mol) must occupy equal volumes, *provided* that they are measured at the same temperature and pressure.

The volume of 1 mol of gas is called the **molar volume.** The molar volume at 24°C and 100 kPa is 24.7 liters. The molar volume is usually expressed at standard temperature and pressure (**STP**). The standard temperature is 0°C. The standard pressure is 760 torr, or 101.3 kPa. At this temperature and pressure—and *only* at this temperature and pressure—1 mol of gas occupies a volume of 22.4 liters.

It is rather amazing that equal numbers of gas molecules occupy the same volume, exerting the same pressure at the same temperature. It is instructive to think about this fact a little more.

Figure 12.22 shows two containers of gas with equal numbers of molecules. If the containers are of the same size, the pressure inside the two containers must be the same. (Equal volumes of gas *at the same temperature and pressure* contain equal numbers of molecules.) This means that the average pressure exerted by the large molecules is the same as the average pressure exerted by the small ones. This can be true only if the small molecules move much faster than the large molecules. There is evidence that they do. When a bottle of hydrogen gas is opened on one side of a room and an instrument to detect the gas is placed on the other side of the room, it is not long before some of the hydrogen gas is detected. However, the time required for carbon dioxide (or some other heavier gas) to move across the room is much greater. This movement of a gas through other gases is called **diffusion.** Light gases diffuse much faster than heavy ones, because the molecules of light gases move faster than the molecules of heavy gases.

Now let us consider another idea. When we connect the two boxes shown in Figure 12.22 for a while, the gases mix. The mixtures shown in Figure 12.23 might result if we keep the boxes connected for just a short time.

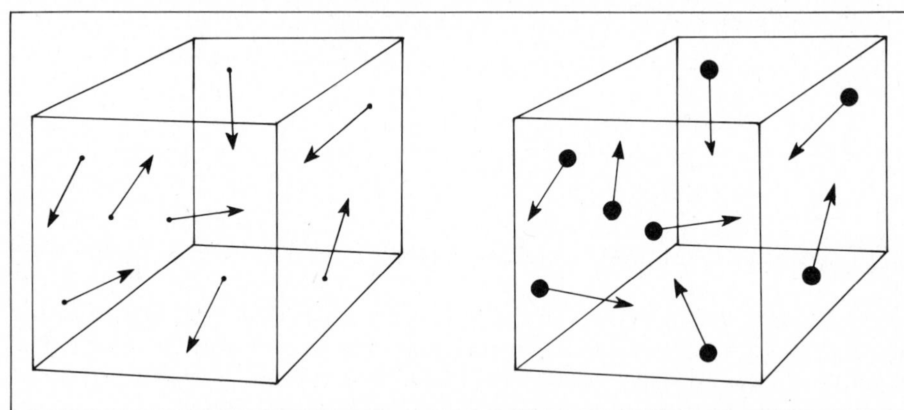

Figure 12.22 *Boxes containing two different gases.*

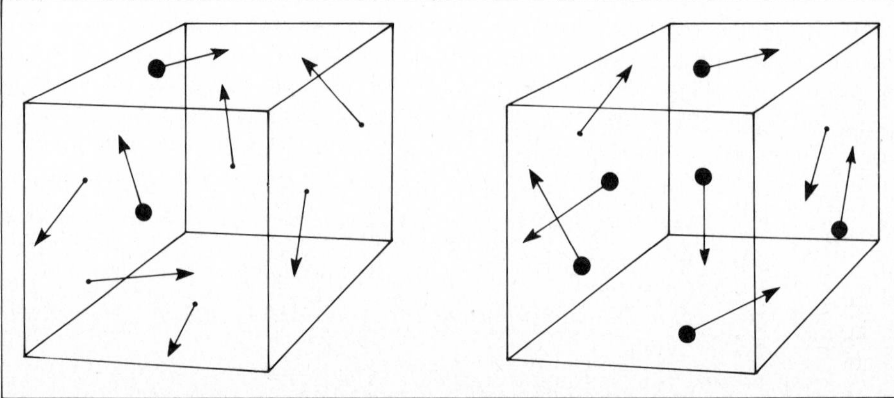

Figure 12.23 *The same boxes after being connected for a short time.*

The mixtures that are formed still contain the same number of molecules as before. The total pressure must be the same. The average pressure exerted by each molecule (large or small) must still be the same. This means that the total pressure of the gas in each box is simply the sum of the pressures exerted by individual molecules in the box. This fact is called **Dalton's law of partial pressure** and is usually written as

$$P_{\text{total}} = P_1 + P_2 + P_3 + \cdots$$

P_{total} represents the total pressure of a mixture of gases, and P_1, P_2, etc., represent the pressure exerted by the molecules of gas 1, gas 2, and so on. P_1 is also the pressure that gas 1 would exert if it were alone in the container. P_2 is the pressure that gas 2 would exert if *it* were alone in the container, and so on. A mixture of gases can have any number of components, and the pressure exerted by each gas in the mixture is not influenced by the other components.

Air is primarily a mixture of nitrogen and oxygen gas. The total pressure of the atmosphere is due to the pressure exerted by the nitrogen plus the pressure exerted by the oxygen. We could express this mathematically by saying that

$$P_{\text{atm}} = P_{O_2} + P_{N_2}$$

Since about 79 percent of the molecules in air are nitrogen and about 21 percent are oyxgen, we could even calculate the partial pressure of each gas.

$$P_{N_2} = 0.79 \times P_{\text{atm}}$$

$$P_{O_2} = 0.21 \times P_{\text{atm}}$$

If you know the composition of any gas mixture, you can calculate the partial pressure due to any one component in a similar way.

The most common application of Dalton's law of partial pressure is to make corrections for data obtained when gases are collected over water. Gases that do not dissolve in water are frequently collected by filling a bottle with water, turning it upside down in a pan of water, and then letting the gas bubble into the bottle to displace the water. A typical collection arrangement is shown in Figure 12.24, page 376. Since the gas passes through water, some water evaporates. Hence the bottle of gas contains the gas being collected plus some water molecules, as illustrated in Figure 12.25, page 376. The amount of water that evaporates depends on the temperature of the water.

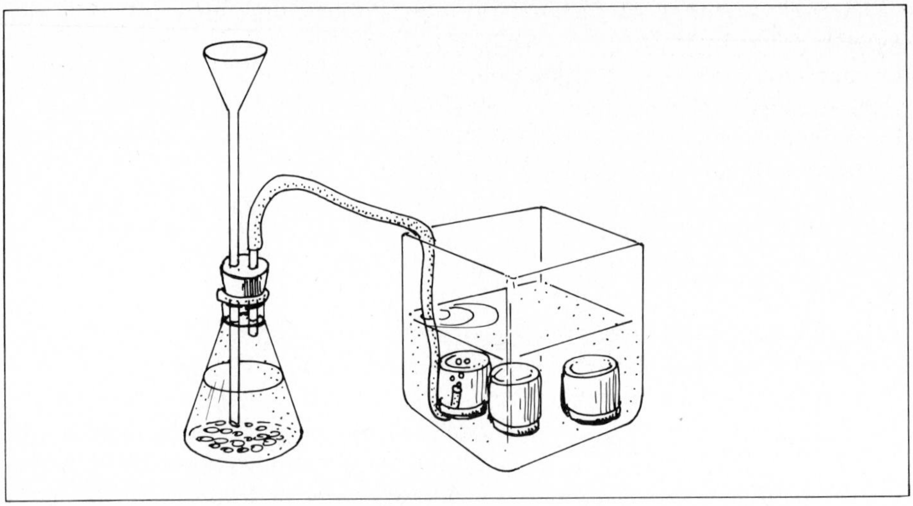

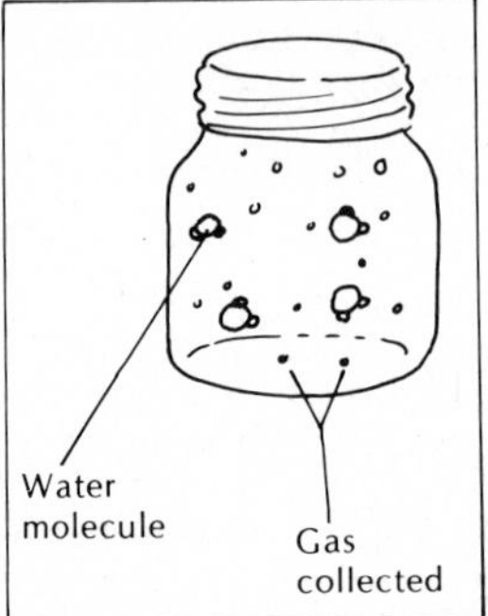

Figure 12.24
A typical apparatus for preparing and collecting gases that are not soluble in water.

Figure 12.25
Gases collected over water always contain some water molecules as an impurity.

12.10 VAPOR PRESSURE

We normally think of a liquid changing to a gas at its boiling point, but this isn't quite true. Any liquid evaporates to produce some gaseous molecules, even at low temperatures, and these molecules bounce around and exert pressure. The pressure exerted by the gaseous molecules that come from the liquid is called the **vapor pressure** of the liquid material. Vapor pressure increases as the temperature of the liquid increases. When the vapor pressure is the same as atmospheric pressure, bubbles of gas form in the liquid, and we say that the liquid is boiling.

Table 12.5 shows the vapor pressure of water at several temperatures. Tables such as this can be used to find the vapor pressure of water at the temperature at which a gas is collected. By subtracting the vapor pressure of water from the total pressure of the gas in the collecting bottle, we find the pressure due to the gas we are interested in.

For example, the total pressure inside the bottle shown in Figure 12.25 is

Table 12.5. Vapor Pressure of Water at Various Temperatures

Temperature (°C)	*Vapor pressure* torr	kPa
0	4.58	0.61
10	9.21	1.23
20	17.54	2.34
21	18.65	2.49
22	19.83	2.64
23	21.07	2.81
24	22.38	2.98
25	23.76	3.17
26	25.21	3.36
27	26.74	3.56
28	28.35	3.78
29	30.04	4.00
30	31.82	4.24
40	55.32	7.37
50	92.51	12.33
60	149.38	19.91
70	233.7	31.15
80	355.1	47.33
90	525.76	70.08
100	760.00	101.32

due to the gas collected plus the water molecules that evaporated when the gas bubbled through the water. We can say that

$$P_{\text{total}} = P_{\text{gas}} + P_{\text{water}}$$

If we know the total pressure of the gas in the bottle and the temperature of the water that the gas bubbled through, we can find the pressure due to the gas we collected.

The following example illustrates the use of Dalton's law of partial pressure.

Example 12.5

A student produced CO_2 by adding acid to some marble chips in a flask, as shown in Figure 12.24, and collected the gas by bubbling it into jars filled with water. When a jar was filled, it was raised to the surface of the water and capped with a lid. The student figured that the CO_2 in the jar should be at about the same temperature as the water, because it had bubbled through the water. She measured the temperature and found that it was 24.0°C.

The student also assumed that the pressure of the gas in the jar was the same as atmospheric pressure in the room, because the lid was placed on the jar at the surface of the water. (If the gas were at a higher pressure, it would escape into the room.) She read the barometer and found the pressure to be 98.6 kPa. Before the gas was produced, she measured the volume of the jar by filling it with water from a graduated cylinder. The volume was 250 mL to the nearest milliliter.

What the student really wanted to know was the number of moles of CO_2 in the jar. How can she answer her question?

Using the ideal gas equation, $PV = nRT$, the student should be able to calculate the number of moles of gas in the jar. She knows the volume of the gas in the jar, the temperature of the gas in the jar, and the pressure of the gas in the jar. Since R is a constant, the only unknown in the equation appears to be n, the number of moles. There is one problem, however. Do you see what it is?

The problem is that the pressure in the jar is due to the molecules of the CO_2 *plus* the molecules of the water that evaporated during collection. The pressure used in the equation must be the pressure due to the CO_2 alone. This problem can be eliminated with Dalton's law of partial pressure.

$$P_{\text{total}} = P_{CO_2} + P_{H_2O}$$

What is the pressure due to the CO_2 molecules in the jar?

95.6 kPa

Table 12.5 gives the vapor pressure of water at 24°C as 2.98 kPa. The total pressure in the jar was 98.6 kPa. Then,

$$P_{\text{total}} = P_{CO_2} + P_{H_2O}$$

$$98.6 \text{ kPa} = P_{CO_2} + 2.98 \text{ kPa}$$

$$95.6 \text{ kPa} = P_{CO_2}$$

Now that you know the pressure exerted by the CO_2, calculate the moles of CO_2 collected.

9.68×10^{-3} mol CO_2

I first list the values that I know. These are

$$P = 95.6 \text{ kPa} \qquad R = 8.31 \frac{\text{L kPa}}{\text{mol K}}$$

$$V = 0.250 \text{ L} \qquad T = 297 \text{ K } (24°\text{C})$$

$$n = ?$$

Substituting these values into the ideal gas equation, we have

$$PV = nRT$$

$$95.6 \text{ kPa} \times 0.250 \text{ L} = n \times 8.31 \frac{\text{L kPa}}{\text{mol K}} \times 297 \text{ K}$$

$$\frac{95.6 \text{ kPa} \times 0.250 \text{ L}}{8.31 \frac{\text{L kPa}}{\text{mol K}} \times 297 \text{ K}} = n$$

$$9.68 \times 10^{-3} \text{ mol} = n$$

The following problems are similar to the example that you just finished.

12-54. A student prepared oxygen gas by decomposing $KClO_3$ and collecting the gas by displacement of water. The barometric pressure is 102 kPa, and the temperature is 27°C. How many moles of oxygen were collected if the volume of the gas above the water was 2.5 liters?

12-55. A second student did the same experiment and collected only 150 mL of gas. The temperature and pressure were the same as in Question 12-54. How many moles of oxygen did he collect?

12-56. In this experiment, 1.5 mol of oxygen gas are produced from 1 mol of $KClO_3$. What volume of gas is collected when 1 mol of $KClO_3$ is decomposed and the oxygen gas is collected at the temperature and pressure given in Question 12-54?

12-57. How many moles of $KClO_3$ would you have to decompose in order for us to collect 500 mL of gas over water at 30°C and a pressure of 98.3 kPa?

12.11 FINDING MOLECULAR WEIGHT

In Chapter 6 you learned to find the empirical formula of a compound. And you learned that the empirical formula may not be the true formula. Acetylene (C_2H_2) and benzene (C_6H_6) both have the same empirical formula (CH), and it is the true formula of neither. However, the molecular weights of acetylene and benzene are very different. When we know the molecular weight of a compound with the empirical formula CH, we can easily find the true formula of the compound.

The gas laws provide a tool for finding the molecular weight of gaseous compounds. By measuring the volume of a gas at some known temperature and pressure, we can find the number of moles of gas in the sample. If we can then weigh the sample of gas, we have the mass of a known number of moles. From that, we can find the mass of 1 mol of the gas and the molecular weight. The following example shows how this is done.

Example 12.6

A 10.2-g piece of dry ice is placed in the corner of a trash bag. The bag is quickly flattened and sealed to contain the gas produced when the dry ice changes from a solid to a gas. After several minutes, the volume of gas in the bag is measured and found to be 5.8 liters. The barometer reading is 99.6 kPa, and the temperature of the gas is measured at 25°C. Find the molecular weight of the gas produced.

As a first step, use the ideal gas equation to find the number of moles of gas produced.

$$n = 0.23 \text{ mol}$$

The 0.23 mol of gas came from the dry ice. It weighed 10.2 g. If 0.23 mol of gas weighs 10.2 g, how much does 1 mol weigh?

---

44 g

Here is my solution:

$$? \text{ g} = 1 \cancel{\text{mol gas}} \times \frac{10.2 \text{ g gas}}{0.23 \cancel{\text{mol gas}}} = 44 \text{ g gas}$$

If 1 mole of the gas weighs 44 g, then the molecular weight is 44. Dry ice is solid carbon dioxide, which has an empirical formula of CO_2 and a true formula of CO_2.

It is difficult to weigh gases accurately, and molecular weights calculated in this way are seldom as accurate as the one we obtained in the example. However, even approximate values for the molecular weight may be sufficient to indicate the true formula of a compound.

If you knew that the empirical formula for carbon dioxide was CO_2 but were not sure whether the molecular formula was CO_2 or C_2O_4, you would probably get the correct formula even if you calculated a molecular weight of 55—a 25 percent error.

In Example 12.6, the number of moles of gas was calculated in one step and the molecular weight of the gas was calculated in a second step. Some people do it in one step by modifying the ideal gas equation to read like this:

$$PV = \frac{g}{MW}RT$$

The symbol for the number of moles, *n*, has been replaced by something new. The *g* in the formula stands for the mass of the gas in grams, and *MW* stands for the molecular weight of the gas. Since we calculate the number of moles of a compound by dividing the mass of the compound by its molecular weight, $n = g/MW$. This new expression has been substituted for *n*, and the molecular weight (*MW*) can be calculated directly. If you find this shortcut easier, use it. If you do not, continue to work the following problems in two or more steps.

In some problems you may be given more information than you need. If so, simply ignore the nonessential data. In other problems you may be given less information than you need. In that case, indicate what additional information you must have in order to answer the question.

12-58. An oxide of nitrogen with an empirical formula of NO is collected in a 50-mL syringe at a temperature of 27°C and a pressure of 100 kPa. The sample of gas weighs 0.102 g. What is the formula of the gas?

12-59. Moth flakes weighing 106 g are placed in a container and heated to 227°C. The flakes totally vaporize to produce a pressure of 150 kPa in the 20 L container. What is the molecular weight of the moth flakes? What is the true formula?

12-60. A compound of sulfur and oxygen is 50 percent S and 50 percent O. As a gas, 3.2 g of the compound occupy a volume of 1.12 L at a pressure of 101 kPa and a temperature of 273 K. The empirical formula of the compound is SO_2. What is the true formula of the compound?

12-61. 21.8 g of calcium carbide (CaC_2) react with water to produce 0.34 mol of gas at a temperature of 27°C and a pressure of 105 kPa. The empirical formula of the gas is CH. What is the molecular formula of the gas?

SUMMARY

This chapter covered a lot of territory. We began with two relationships that are commonly encountered in science: direct proportionality and inverse proportionality. *A* is directly proportional to *B* if the ratio *A*/*B* is constant. *A* is inversely proportional to *B* if the product *AB* is constant.

A simple example of inverse proportionality is the relationship between the pressure and volume of a gas. A simple example of direct proportionality is the relationship between the Kelvin temperature and volume of a gas. These relationships, known as Boyle's law and Charles' law, are used to predict the volume of a gas when the pressure or the temperature, respectively, is changed.

It was clear very early in the chapter that one must be careful in describing proportional relationships. The common-sense notion of direct proportion says that *A* and *B* are directly proportional if *B* increases as *A* increases. This isn't quite right. The two must increase *proportionally;* if one doubles, the other must double; if one is cut to one-third, the other must be cut to one-third. As we discussed gases, we found that the volume of a gas is *not* directly proportional to the Celsius temperature but *is* directly proportional to the Kelvin temperature. For that reason, we *must* use Kelvin temperature in all calculations involving gases.

Boyle's law and Charles' law are good examples of inverse and direct proportionality, but they are not the most useful relationships for chemists.

Chemists usually ask questions about the number of moles of gas present in a sample, and neither of these laws includes moles in the formula. The ideal gas equation, $PV = nRT$, was presented as a relationship that includes moles and combines the information found in Boyle's and Charles' laws. It is a relationship that can be used to solve any problem involving gases.

Only $PV = nRT$ is called an *ideal* gas equation, but all of the relationships described in this chapter actually hold only for "ideal" gases. An ideal gas, like an ideal professor or an ideal student, exists only in our imagination. Real people and real gases may approach that ideal under *certain* conditions but not under *all* conditions. No *real* gas obeys the mathematical relationships in this chapter at every temperature and pressure. We can imagine a gas that would, and we call it "ideal." Still, we work with *real* gases, and you should be aware that the ideal gas equation describes the behavior of real gases only under moderate conditions of temperature and pressure. Don't be surprised if experimental results differ from predictions when gases are subjected to very high pressures or very low temperatures.

The ideal gas equation provides one means of finding the molecular weight of a compound. If you can weigh a compound and then turn it into a gas and find its volume, you can calculate its molecular weight. This can be important in finding the true formula of a compound. In Chapter 6 you learned how to calculate the empirical formula of a compound. Now you have a tool that enables you to find the true formula as well.

In actual laboratory work, gases are often collected by water displacement. This causes some problems, because the gas collected isn't pure. It contains water molecules. Dalton's law of partial pressure provides a way to handle the problem. It says that the total pressure of a gas is the sum of the pressures due to each component. The vapor pressure of water has been measured at many temperatures. We can make corrections by looking up these values in a table.

The pressure of a gas makes little sense unless you know the meaning of "pressure." Early in this chapter, we pointed out that pressure is the force exerted over a given area. In SI units, the standard unit is the newton per square meter (N/m^2), commonly called the pascal (Pa). This is such a small pressure that the kilopascal (kPa) is more often used. Actually, even these pressure units look odd to older chemists. Traditionally, gas pressures have been measured in millimeters of mercury or torr. These units derive from the instrument used to measure atmospheric pressure, the mercury barometer.

Questions and Problems

The following questions review skills you learned in previous chapters and provide practice with the ideal gas equation. In some questions you are given more information than you need; in others you are not given enough. This is much closer to real life situations than problems that include just what you need. Knowing what you need to know in order to solve a problem is just as important as the skill of solving the problem once you have the information. These questions give practice in both areas.

12-62. What volume of CO_2 is produced by the complete combustion of carbon when the temperature of the CO_2 collected is 300 K and the pressure is 1 atm?

12-63. $2\ Na(s) + H_2O(\ell) \rightarrow 2NaOH(aq) + H_2(g)$
According to this equation, how many moles of hydrogen gas can be produced from 5 g of Na when the hydrogen is collected at 30°C and 100 kPa?

12-64. Two identical syringes are filled to the same volume at the same temperature and pressure. One syringe is filled with nitrogen and the other with oxygen.
   a. How do the number of moles of the two gases compare?
   b. How do the number of molecules of the two gases compare?
   c. The mass of the nitrogen is 0.700 g. What is the mass of the oxygen?

12-65. A mole of propane (C_3H_8) occupies a volume of 22.4 liters at STP. What is the molecular weight of propane?

12-66. A sample of hydrogen collected over water at 25°C and 101 kPa occupies 0.100 L. What would be the volume of dry hydrogen at the same temperature and pressure?

12-67. How many moles of hydrogen gas were collected in question 12-66?

12-68. 4.0 L of a gas are stored in a cylinder at a temperature of 15°C and a pressure of 39 kPa. With the temperature kept constant, a piston moves in the cylinder until the volume of the cylinder is 3.0 L. What is the pressure at this new volume?

12-69. Hydrochloric acid reacts with zinc to produce hydrogen gas. What volume of 0.25 *M* HCl would be needed to produce 0.25 g of hydrogen gas?

12-70. The pressure in an automobile tire is 30 psi, and you want to increase it to 40 psi. What change must take place in the number of moles of gas? (Assume that the temperature and volume of the tire remain constant.)

12-71. Ammonia gas is made by the Haber process by the direct combination of nitrogen and hydrogen gases at a temperature of 450°C and a pressure of 15,000 kPa. How many moles of ammonia can be made from 1 tonn (10^6 g) of hydrogen gas by this process?

12-72. 25 mL of a gas at 27°C and 2 atm pressure weigh 0.018 g. What does 1 mol of the gas weigh?

12-73. How much do 20.0 L of oxygen gas weigh when the temperature is 30°C and the pressure is 98 kPa?

12-74. How much do 5.0 L of hydrogen gas weigh at a normal pressure of 99.3 kPa and a comfortable temperature of 23°C? (The gas obviously weighs something. Why do you think a hydrogen-filled balloon rises in air rather than falling? Why does wood float in water but not in air?)

More Difficult Problems

12-75. 70 L of a gas that is at normal atmospheric pressure and weighs 48 g are decomposed to produce 36 g of carbon and 12 g of hydrogen. What is the empirical formula of the gas?

12-76. CO_2 cartridges contain liquid carbon dioxide. If the density of liquid CO_2 is 0.80 g/mL, what volume of liquid CO_2 would be required to inflate a life jacket of 4.0 L when the temperature is 27°C and the pressure is 101 kPa?

12-77. How large a raft could be inflated with a 25-mL cartridge of liquid CO_2 when the CO_2 filling the raft is at STP?
12-78. Ammonia gas is very soluble in water. If 2.5 L of ammonia gas at 30°C and 112 kPa dissolve in water to form 125 mL of solution, what is the concentration of the aqueous ammonia?
12-79. Ammonia gas reacts with hydrogen chloride gas to form solid ammonium chloride. When 5 L of each gas at identical temperature and pressure are mixed, what remains after the reaction is finished?
12-80. Which weighs more, 10 L of carbon monoxide gas or 10 L of oxygen gas? (Assume that both gases are at the same temperature and pressure.)
12-81. What volume is occupied by a cup (about 250 g) of water after it is changed to steam at 120°C and normal atmospheric pressure (101 kPa)?
12-82. What is the density of hydrogen gas at 30°C and 98 kPa?
12-83. Prove that the pressure exerted by a column of liquid depends only on the height of the liquid. That is, prove that the pressure in a barometer tube with a 1-cm³ cross section is the same as the pressure in a tube with a 10-cm³ cross section.

Answers to Questions in Chapter 12

12-1. 21, 21, 21.1, 21.0
12-2. Yes. Considering that the last digit is uncertain in each case, you must assume that the numbers are the same.
12-3. ???
12-4. $V/d = k$ (h constant)
12-5. Measure the diameter and volume of water in cylinders like those shown in Figure 12.3 and do the indicated division. If V/d always gives the same number, the statement is true.
12-6. Definitely not. The values calculated are 6.6, 10, 13, 15, 19, and 30.4.
12-7. Evidently not
12-8. Table 12.1 shows that k has a value of 21 cm². If $k = \pi r^2$, then

$$21\ \text{cm}^2 = \pi r^2$$

$$r^2 = \frac{21\ \text{cm}^2}{3.14159}$$

$$r^2 = 6.6845\ \text{cm}^2$$

$$r = 2.6\ \text{cm}$$

12-9. 21 cm² (Remember that $V/h = \pi r^2$.)
12-10. $A = k/B$ $B = k/A$
12-11. (1) "Density" is the name given to the direct proportionality observed between the mass and the volume of a substance. $D = M/V$ D is the constant; M and V are directly proportional.

(2) The mass of an element (or compound) is directly proportional to the number of moles of the substance. The proportionality constant (k) is equal to the atomic mass of the element or the molecular mass of the compound. (You may think of other examples.)

12-12.

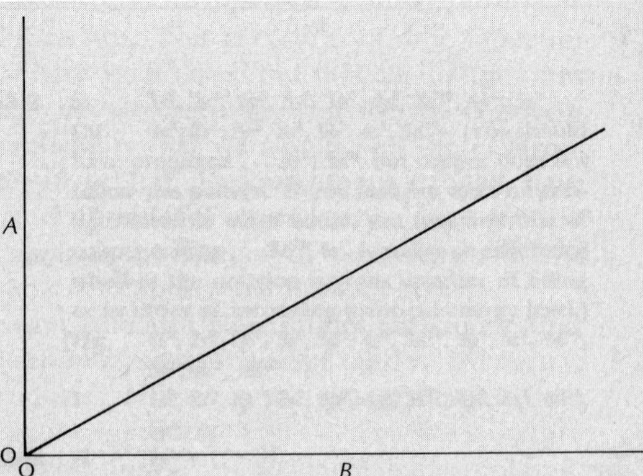

Figure 12.26
When values for two variables that are directly proportional are plotted, the graph is a straight line that passes through the 0,0 point.

12-13.

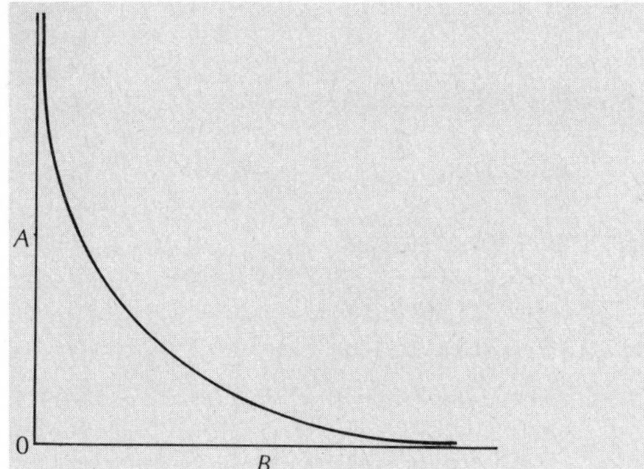

Figure 12.27
When values for two variables that are inversely proportional are plotted, the graph is a hyperbola. As values for A and B become large, the graph comes closer and closer to the two axes but never reaches them.

12-14. Pressure is equal to force divided by area: $P = F/A$. When pressure is constant, the force is directly proportional to the area over which the force is exerted.

12-15. $F/A = P$

12-16. If a less dense liquid were used, the column would have to be very tall to hold enough liquid to equal atmospheric pressure. For example, if the liquid were water, the column would have to be about 34 ft tall. Another advantage of mercury is that it is unlikely to freeze in normal use. Water would, if it were left outside.

12-17. 33.89 ft

12-18. 742 torr = 98.9 kPa = 0.976 atm

12-19. 122 kPa = 915 torr = 122,000 N/m^2

12-20. $PV = k$

12-21. Yes. The values I get when I multiply the pressure in kPa by the volume in cm^3 are 1.6×10^3, 1.52×10^3, 1.53×10^3, 1.54×10^3, 1.5×10^3, 1.6×10^3.

12-22. 102,000 N/m^2

12-23. $$\frac{102{,}000\ \text{N}}{\cancel{\text{m}^2}} \times 15.0\ \cancel{\text{cm}^3} \times \frac{1\ \overset{\text{m}}{\cancel{\text{m}^3}}}{10^6\ \cancel{\text{cm}^3}} = 1.53\ \text{N m}$$

(The units are newtons times meters or "newton-meters.")

12-24. 10 liters (12.5 ignoring significant digits)

12-25. 4 atm

12-26. 400 kPa

12-27. 300 mL (255.7 ignoring significant digits)

12-28. The new pressure must be 4.5 times the old pressure.

12-29. The same number of molecules have been pushed into half the space. If they are still moving at the same average speed (the temperature hasn't changed), they should hit the walls twice as often. Hence the pressure would be doubled.

12-30. Now we do not have so many molecules. With fewer molecules to occupy the space, there are fewer hits on the walls of the container. The space is still half of the original, but there are not twice as many hits now.

12-31. If the temperature were increased, the molecules would be moving faster. If they move faster, they hit the walls more often and have more impact when they hit. Both factors increase the pressure.

12-32. 300 K

12-33. −246°C

12-34. 273 K

12-35. −273°C

12-36. 100°C (the normal boiling point of water)

12-37. 310 K

12-38. 1 L (1.243 L ignoring significant digits)

12-39. 600 K or 327°C (To double the volume, you must double the *Kelvin* temperature.)

12-40. 610 mL (606.5 ignoring significant digits)

12-41. The new temperature is 240 K, or 60 less than the original temperature of 300 K (27°C).

12-42. $$V_1/T_1 = V_2/T_2$$

$$\frac{100\ \text{ml}}{T_1} = \frac{250\ \text{ml}}{T_2}$$

$$T_2 = \frac{250}{100}T_1$$

Thus the new temperature must be 2.5 times the old temperature. (Remember, that is 2.5 times the *Kelvin* temperature!)

12-43. Increasing the temperature increases the speed of the molecules. The molecules hit the walls more often and with more momentum, thus exerting more pressure.

12-44. One of two things could happen. If something keeps the plunger on the syringe from moving, the pressure inside the syringe will increase because of the additional molecules hitting the walls of the syringe. If nothing keeps the plunger from moving, the plunger will be pushed back by the increased pressure until the pressure exerted by the molecules inside the syringe is the same as the pressure exerted by the molecules of air on the outside. The volume occupied by the gas would then be larger.

12.45.

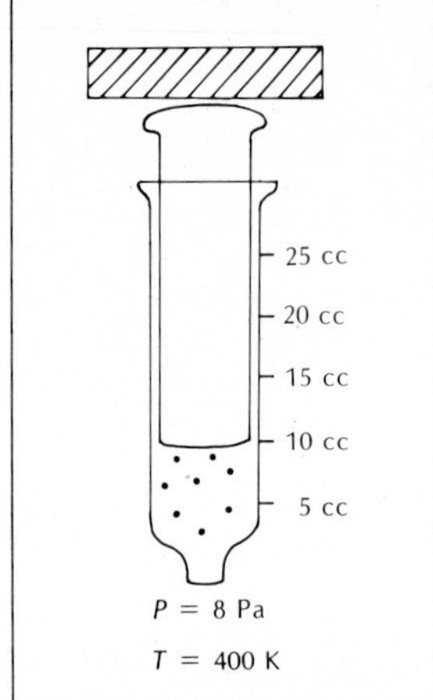

12-46. The only way that the same number of molecules can exert a greater pressure at the temperature of 400 K is for those molecules to be squeezed into a smaller space. This way they will strike the walls more often, exerting more pressure.

12-47. n = 0.010 mol (Did you express temperature in *Kelvin* and use a value for R that has the same units for volume and pressure as the units used to express the volume and pressure of the gas?)

12-48. 1 atm (1.23 ignoring significant digits) If you used the proper value for R and were careful with your units, only units of atmospheres remained for the answer.

12-49. 200 K (243.9 ignoring significant digits)

12-50. Because we have not calculated a value for R using millimeters of Hg as a pressure unit, it is easier to convert 750 mm of Hg to kPa or atm and then do the problem. Since 760 mm of Hg = 1 atm or 101 kPa, 750 mm of Hg = 0.987 atm or 99.7 kPa. Using either of these values and the appropriate value for R, you should find that n = 0.040 mol.

12-51. P = ?; V = 0.5 L; n = 1.5 mol; T = 300 K (The value used for R depends on the units that you want for your answer.)

P = 70 atm (73.8 ignoring significant digits) when R has units of atmospheres.

P = 7000 kPa (7,454 ignoring significant digits) when R has units of kilopascals.

12-52. When P is constant and we assume that n is constant, we can say that $V_1/T_1 = V_2/T_2$.

$$\frac{125 \text{ mL}}{T_1} = \frac{150 \text{ mL}}{T_2}$$

$$T_2 = \frac{150}{125} T_1 = 1.2T_1$$

Thus the Kelvin temperature must be increased by 20 percent. If questions like this confuse you, you might try working them by *assuming* some initial temperature, calculating the final temperature, and then finding the ratio of final temperature to initial temperature. Try this strategy using different initial temperatures, and you will see that you always get the same value for T_2/T_1.

12-53. n_2 = 225/125 n_1 = 1.8n_1. The volume of the second gas is 1.8 times the volume of the first gas. Since equal volumes of gas contain an equal number of molecules, the second gas must have 1.8 times as many moles as the first gas.

12-54. Did you remember to subtract the vapor pressure of water at 27°C to get the pressure due to the O_2 collected? If you did, you should have calculated that 0.099 mol of O_2 was collected.

12-55. 5.9×10^{-3} mol. You can work the problem again or multiply your answer to Question 12-54 by 150/2500. Do you see why the latter procedure works?

12-56. 38 L

12-57. We must first find the moles of oxygen gas collected. Knowing that, we can use the equation for the reaction to calculate the moles of $KClO^3$ that must react.

P = 94.1 kPa (98.3 − vapor pressure of water)

V = 0.5 L; n = ?

R = 8.31 L kPa/mol K

T = 303 K

Using these values and solving the ideal gas equation for n, I get 0.02 mol as the volume of O_2 collected (0.0187 mol ignoring significant digits). Now, using the equation, we can find the moles of $KClO_3$ that decomposed.

$$2KClO_3 \rightarrow 2KCl + 3O_2$$

$$? \text{ mol } KClO_3 = 0.02 \text{ mol } O_2 \times \frac{2 \text{ mol } KClO_3}{3 \text{ mol } O_2}$$

$$= 0.01 \text{ mol } KClO_3 \text{ (0.0125 mol ignoring significant digits)}$$

12-58. Probably N_2O_2. Using the ideal gas equation, I calculated that the 50 mL of gas was 0.002 mol. That weighed 0.102 g, so 1 mol would weigh 51 g. Hence the estimated molecular mass is 51. NO has a molecular mass of 30, and N_2O_2 has a molecular mass of 60. The molecular mass I calculated (51) is much closer to 60 than to 30.

12-59. The 106 g of moth flakes produced 0.72 mol of gas. Hence the mole mass is 106 g ÷ 0.72 mol =

147 g/mol. Without the empirical formula, we have no way of finding the true formula. Paradichlorobenzene is an organic compound used for moth flakes. It has a molecular mass of 147. This is probably what the compound is.

12-60. 1.12 L of the gas contain 0.0498 mol. This amount weighs 3.2 g, so the weight of 1 mol is 3.2 g ÷ 0.0498 mol = 64 g/mol. SO_2 has a molecular mass of 64, so the true formula must be SO_2.

12-61. We are told the number of moles of gas produced, so we do not have to use the ideal gas equation. What we need to know is the mass of the 0.34 mol of gas produced. We aren't told. We may be able to solve the problem indirectly from the mass of CaC_2 that reacts. The 21.8 g of CaC_2 represent 0.34 mol of CaC_2. Since there are two C atoms in one molecule of CaC_2,

$$0.34 \text{ mol } CaC_2 \rightarrow 0.68 \text{ mol CH}$$
$$0.34 \text{ mol } CaC_2 \rightarrow 0.34 \text{ mol } C_2H_2$$

The correct formula must be C_2H_2, acetylene.

READING

Oversimplification appears to be a necessary part of beginning courses. The whole story is complex, and understanding it requires knowledge that can only be developed after the basics are mastered. One of the places in which this text oversimplifies is the discussion of chemical and physical change in Chapter 4. The following article will help you see why the earlier discussion is inadequate.

PHYSICAL VERSUS CHEMICAL CHANGE

Walter J. Gensler

Most elementary chemistry texts, generally in the first few pages, present a short discussion of "physical" and "chemical" change. Often this includes a definition that reads something like: Physical changes are those that involve a change in state or form of a substance but do not produce a new substance; chemical changes are those that result in a new chemical entity. Examples of physical processes or changes are given as: boiling, melting, and other phase changes; breaking one piece of material into two; thermal expansion; solution; deformation; etc. Most texts also say something about physical properties, an incomplete list of which would include: size, shape, state, melting point, temperature, color, density, taste, smell, ductility, viscosity, hardness, thermal conductivity, electrical conductivity, etc.

The present note argues that the distinctions drawn between "physical" and "chemical" are neither warranted nor wise. This position plus the prevalent attitude that trying to define a distinction is more a pastime for idle moments than a matter of deep significance leads to the conclusion that such distinctions should find no place in first-year textbooks.

The phase changes of water provide a favorite illustration of physical change. Thus: "Water boils; in this change no new substance is formed, but water changes from the liquid state to the gaseous state." "Steam is still water and may be condensed to the common liquid." "In freezing water to ice, only the physical aspects of the matter are changed. It is still water." Let us examine this and some of the other illustrations more closely. Through first hand experience, everybody knows that, in fact, ice is *not* water; to maintain otherwise smacks of double talk. In effect, in a discipline where experiment is paramount, the novice is being asked to distrust and discard his own experimental results and to place his faith in authority. Some texts develop the idea of a phase change as "physical" change by stressing the identity of the "chemical composition" in the different phases. But this is either the beginning of an unproductive circular argument or it forces the introduction of terms that at this stage have little meaning. Even from a more sophisticated point of view, it is not clear why phase changes are nonchemical. A detailed description of the processes—i.e., the mechanism of phase changes—is surely best given in terms of

Source: *Journal of Chemical Education,* Vol. 47, No. 2 (February 1970), pp. 154–155.

changes in intermolecular "chemical" bonding. Polymorphism, for example, can be treated logically by specifying the changes in intermolecular or interionic bonding. Some may hold that the intermolecular (as against the intramolecular) nature of these changes provides the basis for classification as "physical." In the first place, it would seem that this kind of distinction is pointless. In the second place, even accepting this differentiation, some phase changes would still emerge as "chemical" (e.g., graphite to diamond; gaseous P_4 to black polymeric P_n), since they involve what is unreservedly "chemical," namely, the covalent bond.

Solution, or its converse, crystallization from solution, is often presented as an example of physical change. Here too, only by asking for blind faith can the text expect a novice to accept the statement that a piece of sugar is the "same" before and after it dissolves in water. The evidence of the student's repeated, first-hand observations (i.e., experiment) strongly suggests that this is not true. Actually, the student's reasonable conclusion has much to commend it, since a detailed picture for dissolution or for crystallization will inevitably involve changes in "chemical" bonding between submicroscopic particles. Where ions are involved, additional processes of ionization, ion aggregation, and ion dissociation—all "chemical" processes in good standing—come into consideration.

In some texts, recovery of the unchanged solute when solvent is removed is offered as evidence that "the solution process does not produce a different substance" and hence is a "physical process." More generally, "Physical changes are those changes which do not permanently alter the material and which allow the material to be recovered unchanged by restoring the original conditions." So far as recovery from solution is concerned, what is observed is the macroscopic end result of a series of submicroscopic steps, none of which can be defended as "physical." Thus, why the appearance of unchanged material on "restoring the original conditions" is diagnostic of a "physical" process is not clear. Reversibility turns out to be a precarious and unsatisfactory criterion for the absence of chemical change, for were this criterion to be applied, the following changes would have to be classed as "physical"

$$H_2 \underset{\text{cool}}{\overset{\text{heat}}{\rightleftharpoons}} 2H$$

$$Na_{(metal)} + NH_{3(l)} \rightleftharpoons Na^+_{(solvated)} - e^-_{(solvated)}$$

$$[\text{racemic acid} + \text{dextro alkaloid}]_{solution} \underset{\text{warm}}{\overset{\text{cool}}{\rightleftharpoons}}$$

$$[\text{dextro acid} \cdot \text{dextro alkaloid}]_{crystal} + [\text{levo acid}]_{solution}$$

Breaking a solid into two pieces is a traditionally reliable example of a "purely physical" change. Yet, as soon as the events at the rupture point are examined closely, one realizes that either covalent bonds are broken or that intermolecular forces (i.e., chemical bonds) are overcome. Thus, even disregarding the high chemical reactivity of a new "clean" surface, fragmentation is by no means devoid of "chemistry." Speculation on the process taking place at the contact point when one piece of solid touches another would probably lead into some interesting areas of chemistry. It is true that the amount of material undergoing "chemical change" on fracture or contact is so small as to make little difference macroscopically or thermodynamically. But to base a distinction on the percentage of the material affected is not satisfying. When fragmentation is repeated and specific surface goes up, the extent of change is no longer negligible and a recognized branch of chemistry emerges. In this connection, the distinction between physical adsorption (a "physical process") and chemical adsorption or chemisorption is not sharp; it is made more for convenience than because of any fundamental difference. Similar considerations apply to the "physical" process of liquid subdivision as in emulsification or in fog formation from bulk liquid.

The deformation of a solid by drawing or forging (and the related physical properties of ductility, elasticity, malleability, hardness and brittleness) are given as familiar "physical changes." But in the last analysis, these macroscopic changes must be described in terms of changes in "chemical" bonding. To classify the macroscopic aspect of a phenomenon as "physical" and the submicroscopic aspect of the same phenomenon as "chemical" is again not particularly appealing.

Placing taste and odor under "physical properties" is inappropriate. Working hypotheses of these imperfectly understood processes involve preliminary solution of the substances followed by interactions between the molecules and receptor sites that are just as "chemical" as the Michaelis-Menten interactions between substrate and enzyme. Fascinating and probably very lengthy chapters on the detailed chemistry of taste and smell are yet to be written.

Fluorescence, or perhaps more often, the more familiar phosphorescence, is sometimes given as an example of a "physical" process. This brings up the whole area of the interactions of radiation and matter. To classify the production and the behavior of the photochemically excited state as "physical" will be supported by some and decried by others. Most chemists, in a sensible way, would avoid the question as barren. Although not presented in the early sections of texts, the "physical" or "chemical" nature of the processes leading to, and away from, the transition state comes to mind as a similar kind of concern. Another would be labeling conformational changes as "physical" or "chemical."

Perhaps a logical, appealing and useful distinction between "physical" and "chemical" processes will be (or possibly already has been) formulated. If so, the definitions would probably involve concepts of model systems, submicroscopic particles, interatomic bonds, intermolecular bonds, etc., whose understanding and appreciation call for more than a few introductory pages in a general chemistry text. Since the distinction is not really meaningful to the beginner—in fact, as now presented it is frequently indefensible and confusing—and since the way in which such an empty distinction helps the author in his exposition of chemistry is not obvious, why not forget the traditional distinction and leave it out altogether?

13

The Structure of Atoms

OBJECTIVES

Although we do not expect you to know all there is to know about atoms when you finish this course, we do think you should be able to do the following:

1. Given the mass number and the atomic number of an isotope, indicate the number of protons and neutrons in the nucleus of the atom. (See Problem 13-34.)
2. Given the atomic number or the name of an element, write the electron configuration for the element. (See Problem 13-35.)
3. Given the name of and the charge on an ion, write the electron configuration for the ion. (See Problem 13-36.)

There is a definite relationship between the periodic table and the electron configuration of the elements. You should have a good idea of what that relationship is when you can do the following:

4. Indicate the number of valence electrons for any element shown in the periodic table. (See Problem 13-37.)
5. By referring to the periodic table, indicate the energy sublevels occupied by the valence electrons of an element in any A group. (See Problem 13-39.)
6. By referring to the periodic table, write the shorthand version of the electron configuration for any element.
7. Write the electron-dot structure for any element in an A group of the periodic table. (See Problem 13-38.)

Words You Should Know

proton	energy level	valence electron
neutron	principal energy level	alkali metal
electron	quantized	halogen
mass number	principal quantum number	noble gas
isotope	*aufbau* principle	transition metal
line spectrum	electron configuration	electron-dot structure
emission spectrum		Lewis structure
diffraction grating		

I once asked a student what chemistry was about and the response was "Atoms and molecules." It is not a *bad* response, and it is certainly understandable. Atoms and molecules are so much a part of the chemist's explanation of the natural world that one might think that the study of atoms and molecules *is* chemistry. Still, I prefer to think of chemistry as the study of matter—what it is and how it changes from one form to another. The notion of atoms and molecules is just a tool for increasing that understanding. Still, it is an important tool. Consequently, in this chapter we return to atoms in an effort to gain greater understanding of why some changes in matter take place while others do not.

13.1 A REVIEW OF ATOMIC STRUCTURE

Before developing our model of atoms further, let's recall what has already been said.

First, we explained the behavior of matter by assuming that all matter is made up of a few elements. Each element contains only one kind of atom, and the atoms of one element are different from the atoms of all other elements. These atoms of different elements can combine in definite proportions to form new substances called compounds. The fundamental particle of the compound is the molecule.

We found that most of the mass of the atom is in the nucleus at its center, that the two fundamental particles in the nucleus are protons and neutrons, and that the protons have a positive electrical charge. The number of protons in the nucleus distinguishes the atoms of one element from the atoms of a second element. The number of protons determines the magnitude of the positive charge on the nucleus of the atom. The number of protons is called the atomic number of the element.

Atoms normally have no net electrical charge, so there must be negatively charged particles to balance the positive charge of the protons. We called these negatively charged particles electrons and cited evidence that the electrons are outside of the nucleus and occupy most of the space taken up by the atom. Electrons contribute very little to the mass of the atom. The number of electrons in a neutral atom is the same as the number of protons.

Except for nuclear reactions like those that take place in atom bombs, protons and neutrons cannot be removed from, or added to, the nucleus. However, electrons *can* be exchanged between atoms. When an atom loses an electron, it produces a positively charged ion; when an atom gains an electron, it produces a negatively charged ion. Such ions are attracted to one another and are found in crystals of ionic compounds.

Although we will not be able to complete the picture of atoms in this course, two additional pieces of information will be added. The first piece of information involves the nucleus; the second involves the electrons around the nucleus.

13.2 MASS NUMBER

We have said that the mass of an atom is due almost entirely to the protons and neutrons in the nucleus of the atom. Hence the atomic mass of an element is very nearly equal to the mass of protons and neutrons in the nucleus. Hydrogen has a mass of about 1.00. The only thing in the nucleus of a normal hydrogen atom is one proton. Helium has an atomic mass of about 4 and an atomic number of 2. Helium has two protons and two neutrons in its nucleus. We could estimate the number of neutrons in any element in a similar way, except for one complication—not all atoms of a given element have the same number of neutrons.

ISOTOPES

Up to now, we have implied that all atoms of an element are identical. This is not true. For example, 99.985 percent of the hydrogen atoms on the face of the earth have nothing but a proton in the nucleus, but 0.015 percent (about 3 out of every 20,000 atoms of hydrogen) have a neutron in the nucleus as well. These atoms are still hydrogen. They have one proton and one electron. They react chemically just like any other hydrogen atom, but they are about twice as heavy as normal hydrogen. This additional mass has led to the term "heavy water" to describe the substance created when these atoms of hydrogen are combined with oxygen.

Atoms of the same element that have different numbers of neutrons are called **isotopes.** The prefix *iso* means "equal," and *tope* comes from the Greek word *topos,* which means "place." Isotopes have an "equal place" in the periodic table, and they have almost identical properties.

Chemically, all isotopes of an element behave in the same way, but the existence of these isotopes complicates the atomic mass that we calculate. Most elements have two or more isotopes that occur in nature. The atomic mass shown on the periodic table is an average of the masses of the isotopes as they occur in nature.

Most carbon atoms that occur in nature have six protons and six neutrons, giving a **mass number** of 12. This isotope of carbon is the currently accepted standard for atomic mass, but the periodic table shows an atomic mass for

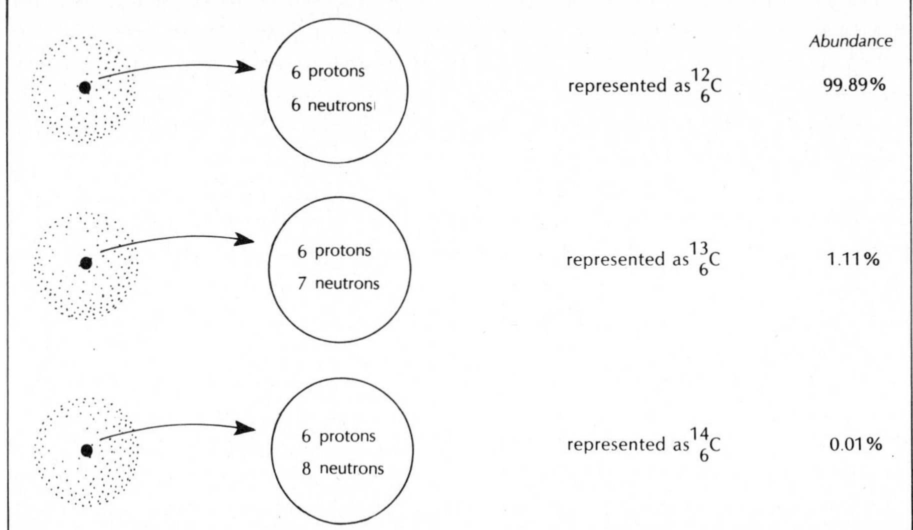

Figure 13.1 *The composition and abundance of the three naturally occurring isotopes of carbon.*

carbon of 12.01115. The higher atomic mass is due to the fact that some atoms of carbon have seven neutrons and others have eight. Adding these neutrons to the six protons in all carbon atoms, we find that these isotopes have mass numbers of 13 and 14 (Figure 13.1). Even though these isotopes are not common in nature, enough of them exist to raise the average atomic mass of carbon above 12.000. Other factors that affect the atomic mass of atoms complicate the picture so that it is *not* possible to subtract the atomic number from the *atomic mass* to find the number of neutrons in an atom. (It *is* possible to subtract the atomic number from the *mass number* to find the number of neutrons in an atom.)

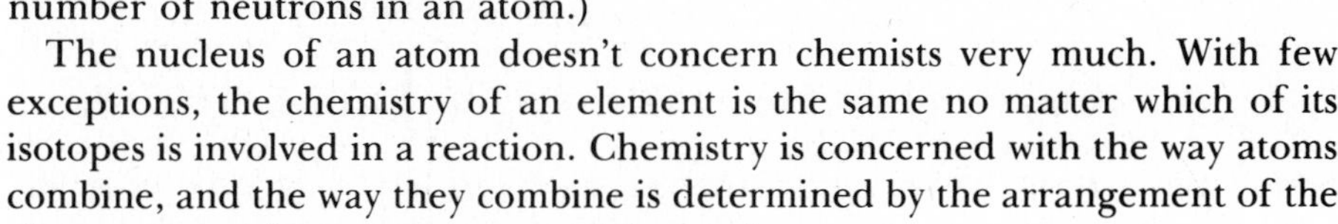

The nucleus of an atom doesn't concern chemists very much. With few exceptions, the chemistry of an element is the same no matter which of its isotopes is involved in a reaction. Chemistry is concerned with the way atoms combine, and the way they combine is determined by the arrangement of the electrons, not the arrangement of the nucleus.

13.3 ELECTRON ARRANGEMENT

The discussion that follows is a mixture of fact and fiction. The factual part is the observations of matter that have been used as a basis for the fiction. The fiction is the mental pictures dreamed up to explain the facts. A number of these fictional stories have been published over time, and there are likely to be more. Each is an attempt to develop a mental picture that is more consistent with the factual observations, more convenient to use, or both. So far, none has been perfect. The models that seem closest to reality are difficult to use, because they rely on complex mathematics. Those that are easiest to use fall short of the truth.

This presents us with a problem. To talk about the model that appears closest to the truth involves complex ideas that you are not prepared for. On the other hand, to present a much simpler model would not prepare you for discussions that you are likely to encounter in later courses. Our decision is to compromise. We will stick to the more sophisticated model but tell only the simpler parts of the story.

WHY WE CARE ABOUT ELECTRON ARRANGEMENT

You already know that electrons get involved when ions are formed. We actually believe that electrons are involved when *any* chemical change occurs. Electron arrangement is the key to explaining many chemical facts that you have already learned.

Consider water. You know that it has a formula of H_2O rather than HO, as Dalton imagined. Its formula reflects the combining numbers of 1 for hydrogen and 2 for oxygen. But why do these elements have *these* combining numbers rather than some other?

You know that water boils at 100°C. You probably see nothing strange about that, but it is really quite remarkable. Water has a molecular mass of 18. Most substances with massess that small boil at temperatures below 0° Celsius. The unusually high boiling point of water is partly due to its shape. Water is bent, as shown in Figure 13.2. But why is water bent?

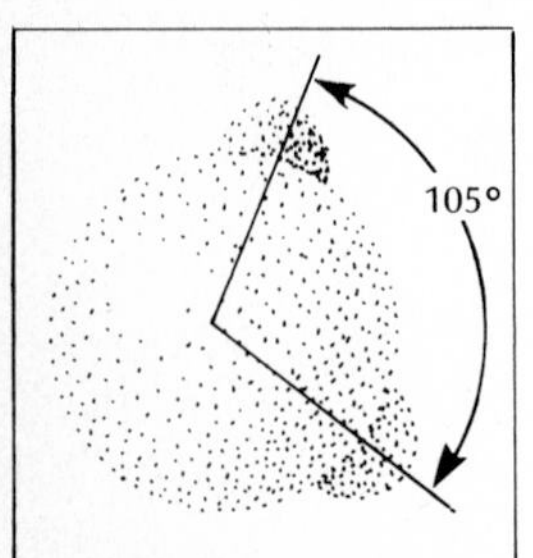

Figure 13.2
The water molecule is usually described as "bent." Lines drawn from the oxygen nucleus through the hydrogen nuclei form an angle of about 105°.

You know that water is an excellent solvent—a fact that also results from the shape of the water molecule. You also know that substances like HCl dissociate to form ions when they dissolve in water but that other compounds dissolve in water without forming ions. Why?

Although we cannot give full answers to these questions, we can begin. In each case, the answer depends on the arrangement of electrons in the atom.

13.4 EVIDENCE FOR ELECTRON ARRANGEMENT: LINE SPECTRA

Most of what we believe about the arrangement of electrons is based on observations of the light that is given off when an electric current passes through a glass tube containing a small amount of a gaseous element. An example of such a tube is the familiar fluorescent light bulb, shown in Figure 13.3.

Fluorescent lights contain mercury vapor. When the tube is connected to a source of high-voltage electricity, electrons flow across the tube and strike the atoms of mercury that are in their path. When the mercury atoms are hit by the electrons, the electrons within the atom gain energy and are said to be excited. These electrons quickly return to their original unexcited or "ground" state by emitting energy in the form of light.

When you look at the light emitted by atoms that have been excited, you do not see white light. An uncoated fluorescent lamp emits a pale blue light. The coating on the inside of the bulb changes this pale blue light into the white light that you normally see from such bulbs. Neon lights used for

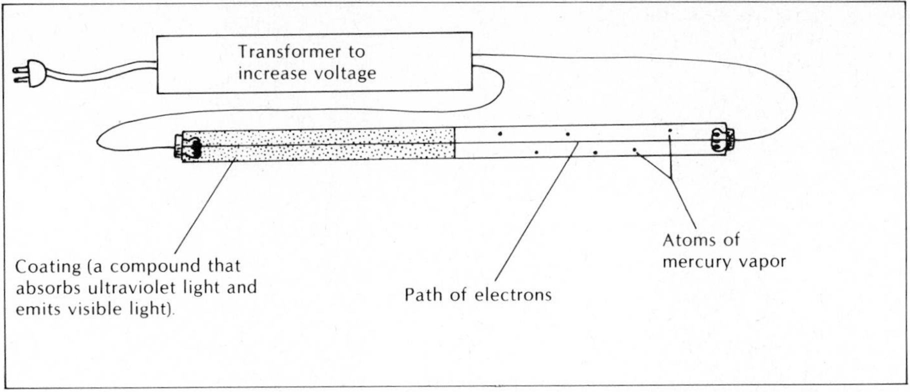

Figure 13.3
A fluorescent light bulb. Electrons traveling across the tube strike mercury atoms, which are "excited" and then emit energy primarily as ultraviolet light. A coating on the inside of the bulb absorbs the ultraviolet light and emits light that appears white.

advertising are uncoated glass tubes filled with neon gas, and these bulbs emit a red glow. Other gases produce still other colors.

We can see the exact colors produced by a gas-discharge tube by looking at the tube through a piece of glass or plastic called a diffraction grating (Figure 13.4). A **diffraction grating** is a transparent material with many evenly spaced lines scratched on the surface. The more lines the better. Most common gratings have between 4,000 and 8,000 lines per centimeter.

Figure 13.4
Diffraction gratings. These thin plastic sheets (slides enclosed in cardboard) have thousands of grooves in the surface. As light passes through the grating these grooves break it into the colors of the rainbow. (*Photo by Tom Greenbowe.*)

When you look at the sun or an incandescent light through a diffraction grating, you see a rainbow. When you look at a neon light through a diffraction grating, you see only those sections of the rainbow that contain the colors emitted by the neon gas in the tube. Each element has a unique pattern, and this pattern can be used to identify the element in the tube. Your instructor will probably demonstrate this for you. It must be seen to be appreciated.

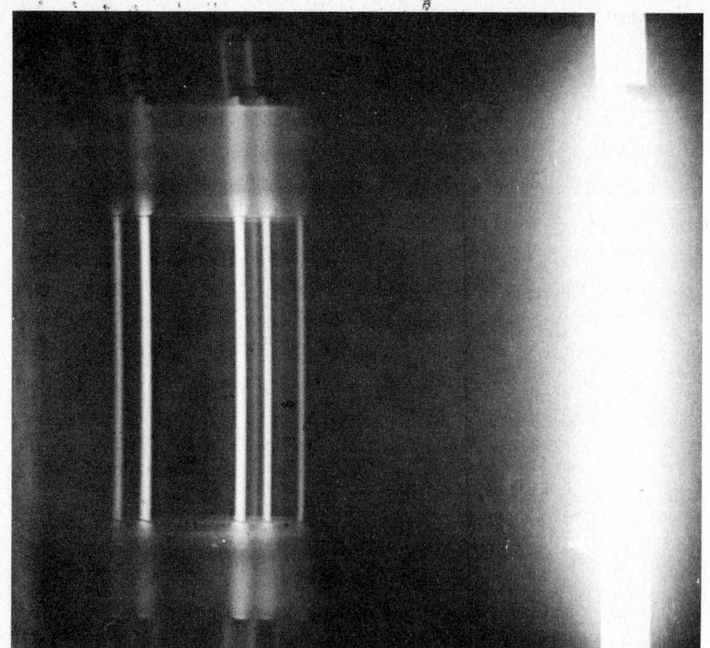

Figure 13.5
A picture of a gas-discharge tube taken through a diffraction grating. The overexposed area on the right is the actual tube. Each image of the tube shown on the left is formed by one particular wavelength of light emitted by the gas in the tube. (Photo by Dudley Herron.)

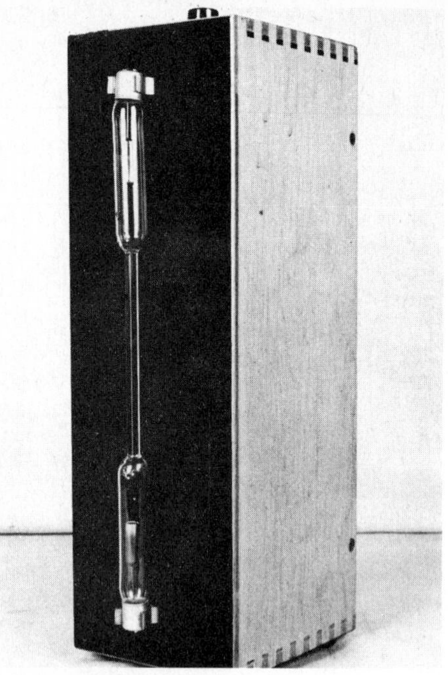

Figure 13.6
A picture of the same gas-discharge tube taken without the diffraction grating over the lens of the camera. (Photo by Tom Greenbowe.)

To give you some idea of what you will see when you look through a diffraction grating at a gas-discharge tube, I taped a diffraction grating over the lens of a camera and took the picture reproduced as Figure 13.5. Figure 13.6 shows the discharge tube that is being photographed. Note that the picture taken through the diffraction grating produces multiple images of the tube. Each of the images formed is a different color and corresponds to one of the wavelengths of light that is given off by the particular gas in the tube. The gas used to produce Figure 13.5 was helium. If hydrogen had been used, a different series of images would have been observed; if neon had been the gas in the tube, still different images—corresponding to other wavelengths of light—would be seen. The particular series of images (lines) produced by a gas in the tube is known as the line spectrum for that gas. A spectrum is "an array of entities." A **line spectrum** is an array of lines showing the various wavelengths of light emitted by a substance.

Each wavelength of light corresponds to a particular amount of energy. Light of two different wavelengths (colors) is light of two different energies. Light is apparently emitted by atoms in a gas when electrons in the atom change from one level of energy to another. The fact that atoms emit only particular colors (wavelengths) of light suggests that the electrons in the atoms can exist only at certain, discrete energy levels. The following analogy should help clarify what we are trying to say about the electrons in atoms.

13.5 ENERGY LEVELS: THE BOOKCASE ANALOGY

When you hold a book above the floor and drop it, the book loses energy as it falls. It can do work when it hits the floor. The higher you hold the book, the more energy it has and the more work it can do when it hits the floor. You can give the book *any* amount of energy by simply changing its initial height.

Suppose that you place a book on a shelf of a bookcase rather than holding it. Now the book can have only certain increments of energy corresponding to the distance from the various shelves to the floor. We might say that the energy of the book is **quantized** when we put it on the bookcase. The energy of a book on a high shelf is greater than the energy of a book on a lower shelf. A book can lose energy by being moved from a high shelf to a lower shelf or by being moved from a shelf to the floor. However, it can lose or gain only *certain amounts* of energy as it is moved from one shelf to another.

We might think of the electrons in an atom as being on various "energy shelves," or **energy levels.** When the electron moves from a higher energy shelf to a lower shelf, it loses energy, and we see this energy in the form of light of a particular wavelength. Since the electron can move only from one shelf to another, only a few energy changes can result from this shifting of electrons. Consequently, we see only a few lines in the spectrum of light emitted, rather than a continuous rainbow. A bookcase like the one in our analogy appears in Figure 13.7.

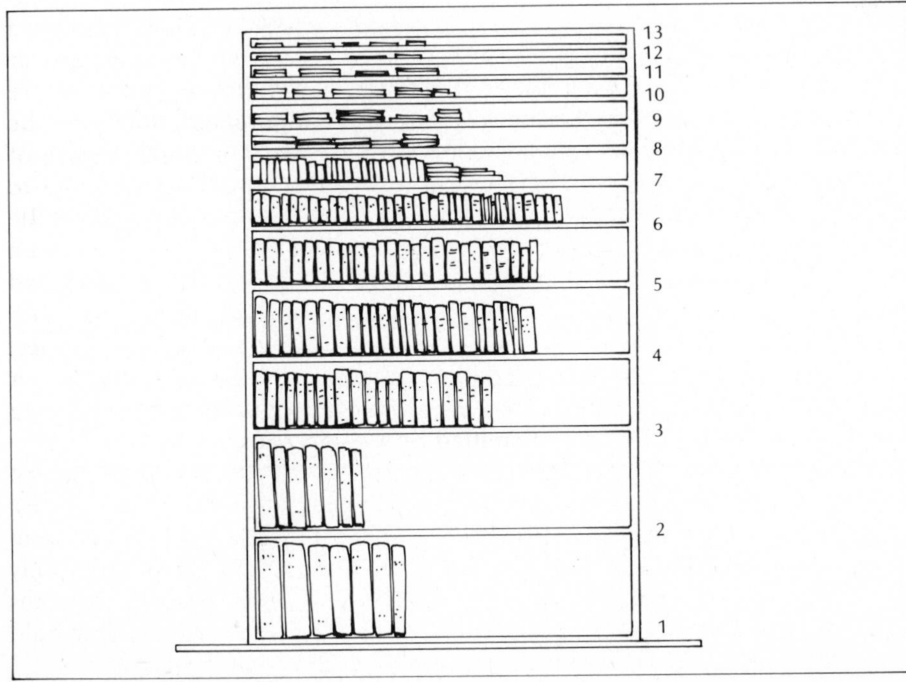

Figure 13.7
A bookcase. Each shelf holds books at a different energy level.

We should not carry the analogy too far. Older pictures of atoms suggested that the energy levels of electrons correspond to particular distances from the nucleus of the atom in much the same way in which books in a bookcase are at different distances from the floor. More recent observations suggest that this picture is too simple. It is not possible to predict how far an electron of a particular energy is from the nucleus. The electron moves about in some way that we do not fully understand. At times, an electron of a given energy seems to be very close to the nucleus, and at other times it appears to be very far away (Figure 13.8). We are wiser to focus on the differences in the electron's energy level and not to worry about the exact position of the electron at any given time.

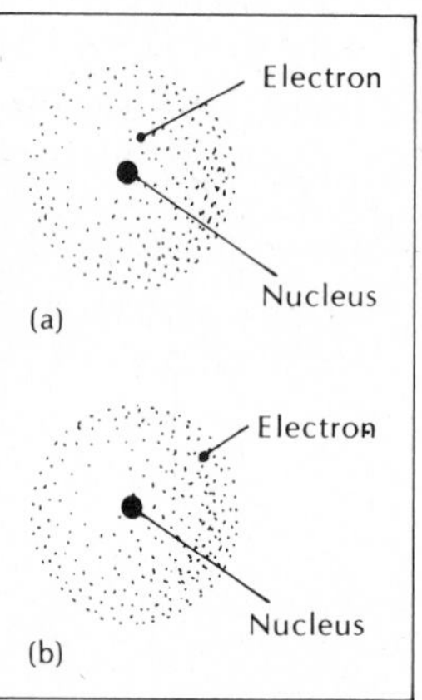

Figure 13.8
Current models of atoms do not tell us where the electron is. A given electron may be near the nucleus or far away at any given time. (a) If we could stop the electron in an atom of hydrogen, we would sometimes find it close to the nucleus. (b) At other times, this same electron would be very far from the nucleus.

We can say that, when we compare an electron at a higher energy level with an electron at a lower energy level, *on the average* the electron of higher energy is farther from the nucleus than the electron of lower energy, as suggested by Figure 13.9. This is not the same as saying that an electron of higher energy is *always* farther from the nucleus. At some instant it may actually be

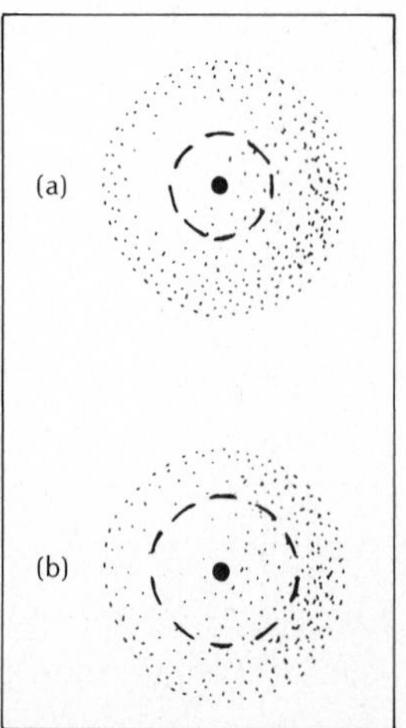

Figure 13.9
On the average, electrons of high energy are farther from the nucleus than electrons of low energy. (a) The dashed line shows the average distance from the nucleus for an electron of low energy. (b) Average distance from the nucleus for an electron of high energy.

closer to the nucleus than the electron of lower energy, but over a period of time, the electron of higher energy is far from the nucleus more often than the electron of lower energy.

This idea of energy level is the cornerstone of the model that we want to present. We need some way to talk about electrons at different energy levels. In the system commonly used, numbers are assigned to the **principal energy levels** in an atom, beginning with 1 for the lowest possible level. These numbers are called **principal quantum numbers** and have values of 1, 2, 3, 4, ...

Certain experimental observations can be explained when we assume that the principal energy levels in atoms are divided into sublevels. We might illustrate this by our bookcase analogy.

Imagine that each shelf of a bookcase has on it another bookcase with smaller shelves, as shown in Figure 13.10. This would divide the energy levels into smaller increments.

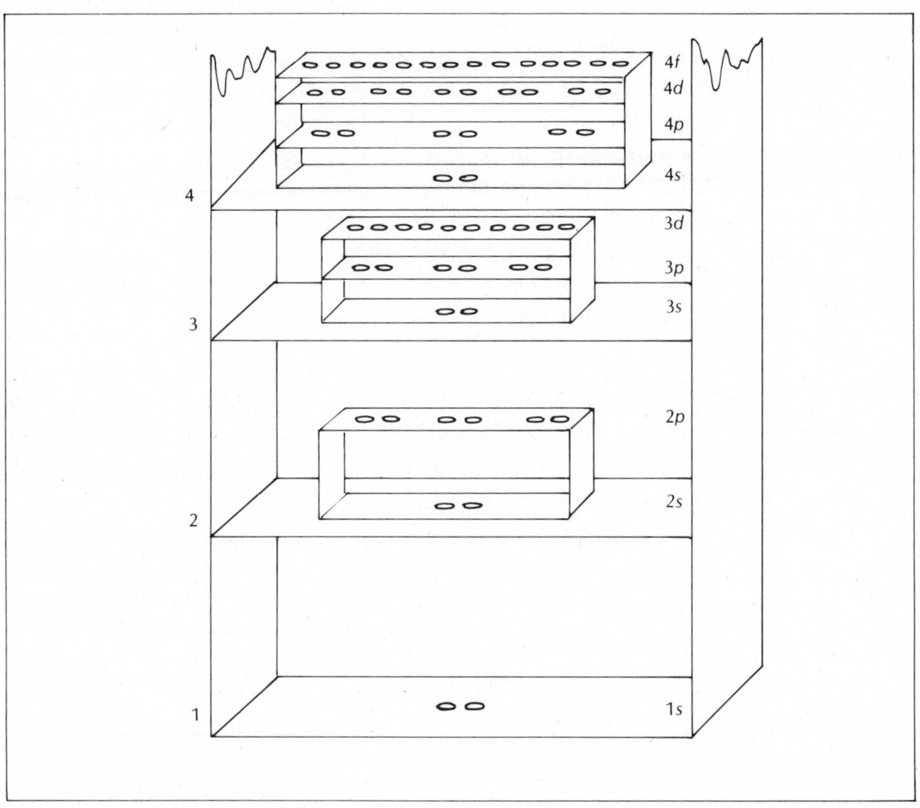

Figure 13.10 *The principal energy levels of atoms are divided into sublevels.*

To be consistent with experimental observations, our model does not have the same number of sublevels for each principal energy level. The first energy level has only one sublevel, the second energy level has two sublevels, the third has three sublevels, and so on.

In a complete treatment of this model, the sublevels are designated by numbers, but in qualitative discussions of electron arrangement, letters are used. The first (lowest) sublevel of any energy level is designated by *s*, the second by *p*, the third by *d*, and the fourth by *f*. These letters derive from early observations of line spectra. Some lines were described as *s*harp, others as *p*rincipal, others as *d*iffuse, and still others as *f*ine. The first letters of these descriptive terms were later adopted to designate electron energy sublevels.

We might mention that other sublevels are possible at higher principal energy levels. When encountered, these sublevels are lettered alphabetically beginning with *g*. The only sublevels commonly considered are the *s*, *p*, *d*, and *f* sublevels that we have mentioned. These are labeled in Figure 13.10.

Note that the energy "subshelves" in Figure 13.10 have indentations: two for the *s* sublevel, six for the *p* sublevel, ten for the *d* sublevel, and fourteen for the *f* sublevel. These represent "energy positions" where we could place an electron. And they represent the *only* positions where we could place an electron. What we are saying is that no more than two electrons can occupy an *s* sublevel, no more than six electrons can occupy a *p* sublevel, no more than ten can occupy a *d* sublevel, and no more than fourteen can occupy an *f* sublevel. This rule is essential for predicting the electron arrangement in an atom, and you should remember it. (It's easy. Did you notice that each sublevel holds four more than the previous sublevel?)

With one additional rule, you can begin to predict the electron arrangement for various atoms. This rule, known as the ***aufbau*** **principle,** states that *an electron normally occupies the lowest available energy level.* Before going on to describe the electron arrangement in atoms, let us summarize what we have said about **electron configuration.**

1. Electrons can exist in atoms only at certain, discrete energy levels.
2. The principal energy levels in an atom are described by the principal quantum numbers. The lowest principal energy level is numbered 1, the next is 2, and so on until the electron receives so much energy that it is totally removed from the atom.
3. Each principal energy level has sublevels. The number of sublevels is the same as the principal quantum number; that is, the number-1 level has only one sublevel, the number-2 level has two sublevels, and so on.
4. Sublevels are usually designated by letters, as indicated in Table 13.1.

Table 13.1. Energy Sublevels of Electrons

Sublevel	*Designation*	*Number of electrons held*
First	*s*	2
Second	*p*	6
Third	*d*	10
Fourth	*f*	14
Fifth	*g*	18

5. Electrons normally occupy the lowest available energy level. If an atom has only one or two electrons, they would occupy the *s* sublevel of the first principal energy level. If an atom has 3 electrons, the first two would occupy the 1*s* level and the third would go into the *s* sublevel of the second principal energy level, and so on.

This last point will become clearer as you work through the following example with me. Remember to cover each step of the problem until you formulate your answers.

Example 13.1

What energy levels would be occupied by electrons in an oxygen atom?

First, how many electrons are there in an oxygen atom?

Eight (The atomic number of oxygen is 8.)

Look at Figure 13.10 and indicate where these electrons would be located.

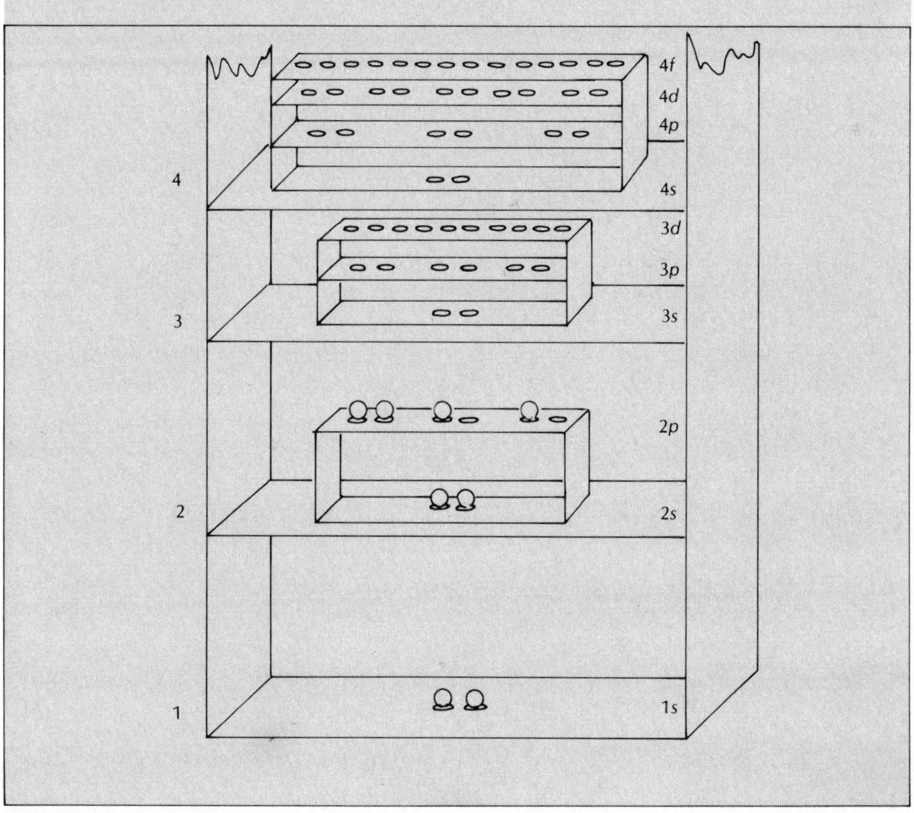

According to the *aufbau* principle, the eight electrons in an oxygen atom are found in energy levels beginning at the lowest possible level. Two electrons occupy the *s* sublevel of the first energy level. Since there are no more "notches" at this energy level, the other electrons go to the second energy level. Here two electrons can occupy an *s* sublevel, and the remaining four electrons go to the *p* sublevel. Note that, when the eight electrons in oxygen have been assigned, there are still two empty "notches" in this *p* sublevel. We know that oxygen has a combining number of 2. Later we will see that these facts are related.

It is awkward to draw pictures, so notations have been developed to represent the electron configuration we have just described. For the electron configuration of oxygen, we would write

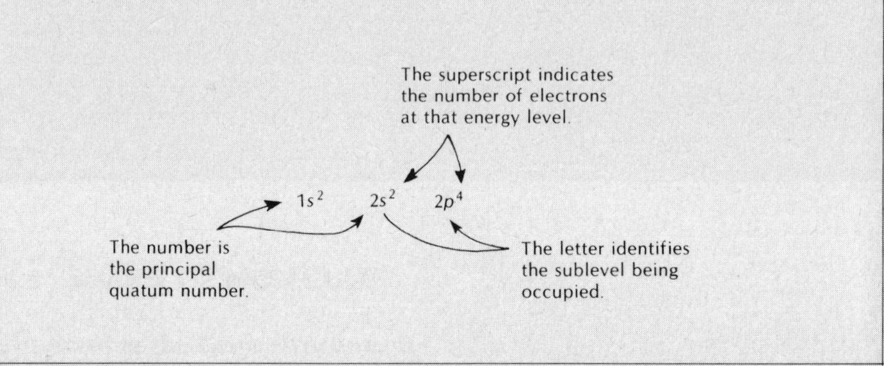

This notation says, "There are two electrons in the *s* sublevel of the first principal energy level, two electrons in the *s* sublevel of the second principal energy level, and four electrons in the *p* sublevel of the second principal energy level."

There are, of course, other energy levels in the oxygen atom. We *could* write something like the following for the electron configuration of oxygen:

$$1s^2,\ 2s^2,\ 2p^4,\ 3s^0,\ 3p^0,\ 3d^0,\ \ldots$$

We do not normally include energy levels that are empty when we show electron configurations.

Look at Figure 13.10 and show where you would expect to find the electrons in a chlorine atom.

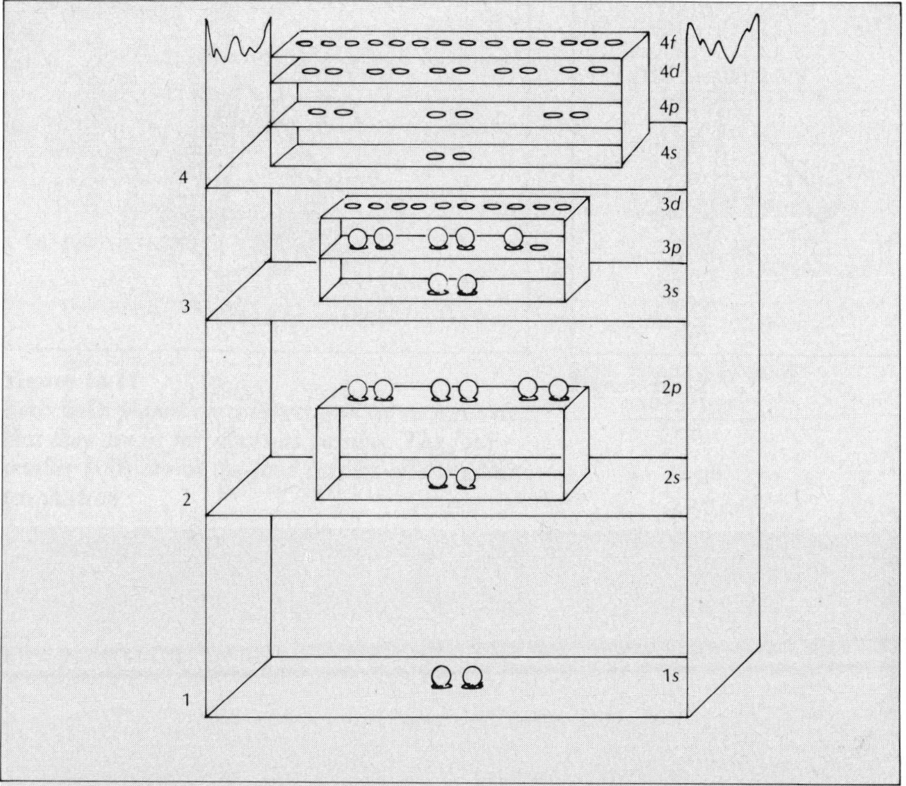

Now use the shorthand notation to indicate the electron configuration for chlorine.

$1s^2$, $2s^2$, $2p^6$, $3s^2$, $3p^5$

Note that the electron configuration for chlorine leaves one unoccupied space in the last sublevel that has electrons. Recall that the combining number of chlorine is also 1. We will explore the relationship between combining number and electron configuration after you have written the electron configuration for several elements.

To help you remember, the various sublevels are listed in Figure 13.11.

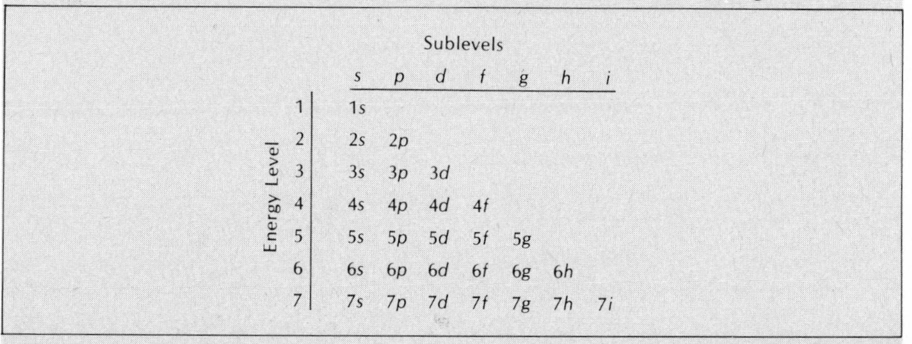

Energy Level	Sublevels s	p	d	f	g	h	i
1	1s						
2	2s	2p					
3	3s	3p	3d				
4	4s	4p	4d	4f			
5	5s	5p	5d	5f	5g		
6	6s	6p	6d	6f	6g	6h	
7	7s	7p	7d	7f	7g	7h	7i

Figure 13.11 *Energy sublevels.*

13.6 WRITING ELECTRON CONFIGURATIONS

13-1. In the space supplied, write the electron configuration for the first eighteen elements. The first three are done as examples.

H $1s^1$

He $1s^2$

Li $1s^2$, $2s^1$

Be

B

C

N

O

F

Ne

Na

Mg

Al

Si

P

S

Cl

Ar

If you remembered that an *s* sublevel can hold only two electrons and that a *p* sublevel can hold only six, you probably had no trouble with this exercise. However, if you continue with the nineteenth element, potassium, you will almost certainly make an error. From what has been said, we would expect the next electron (number 19) to go into the 3*d* sublevel. However, experimental observation suggests that this is not the case. It goes into the 4*s* sublevel instead. We will not concern ourselves with *why* this is so but will concentrate on how you can make the right prediction for most elements. We will do this by using Figure 13.11. Figure 13.12 shows the same diagram with diagonal lines added.

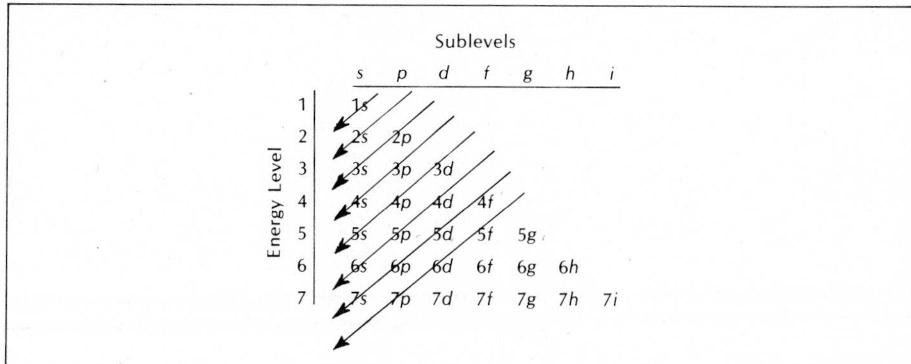

Figure 13.12 *Normal order of filling energy sublevels.*

If you follow the diagonal lines in Figure 13.12, you can predict the order in which the various sublevels are filled. As shown, the correct order of filling is 1*s*, 2*s*, 2*p*, 3*s*, 3*p*, 4*s*, 3*d*, 4*p*, 5*s*, 4*d*, 5*p*, 6*s*, 4*f*, 5*d*, 6*p*, 7*s*, 5*f*, 6*d* and 7*p*. Although other sublevels are thought to exist, no known elements have electrons in these sublevels when the atom is in an unexcited state.

13-2. Using the diagonal rule shown in Figure 13.12, write the electron configuration for the following elements.

Sr ____________________

Cu ____________________

Hg ____________________

I ____________________

If you used the rule correctly, you should have come up with the correct electron configuration for three of the four elements, but not for the fourth. The rule that you have used does not work perfectly. It predicts an electron configuration for copper of $1s^2$, $2s^2$, $2p^6$, $3s^2$, $3p^6$, $4s^2$, $3d^9$. The correct configuration is the same up to the $4s$ sublevel. However, copper has only one electron in the $4s$ sublevel, and it has ten electrons in the $3d$ sublevel, giving . . . $4s^1$, $3d^{10}$. We will not try to explain such exceptions in this course, but you should be aware that they exist.

13-3. Following the rules that you have been given, write the electron configurations for each of these elements.

H __________	B __________
Li __________	Al __________
Na __________	Ga __________
K __________	In __________
Rb __________	

VALENCE ELECTRONS

Now that you have written the electron configuration for these elements, look at what you have done. First, note that the elements on the left are the first five elements in Group IA of the periodic table and that the elements on the right are the first four elements in Group IIIA of the periodic table. You should also see a pattern in the electron configurations that you have written.

All of the elements in Group IA have a *single* electron in the *highest principal energy level* that is occupied (level 1 for H, level 2 for Li, level 3 for Na, level 4 for K, and level 5 for Rb). In a similar fashion, all of the elements in Group IIIA have *three* electrons in the highest principal energy level that is occupied. (Two of these three electrons are in an s sublevel, and one electron is in a p sublevel.) These electrons at the highest principal energy level are the electrons that are most often near the surface of the atom, and they are the electrons that are normally involved in bonding between atoms. Those electrons that may be involved in bonding are called the **valence electrons.** *For all of the A groups in the periodic table, the group number indicates the number of electrons in the highest principal energy level. This is also the number of valence electrons in the element.*

13.7 PERIODICITY AND ELECTRON ARRANGEMENT

GROUP IA(s^1): THE ALKALI METALS

Hydrogen, normally shown in Group IA of the periodic table, is unusual and should not be considered a member of this group. It is a gas at normal temperatures, is not a good conductor of heat or electricity, and does not have any of the other properties that we associate with metals. All other elements in this group, the **alkali metals,** have these properties. Furthermore, each of these elements reacts with water to form a hydroxide and hydrogen gas, each reacts rapidly with oxygen in the atmosphere, each reacts with hydrogen gas to give a hydride, and each forms ions with a single positive charge. This similarity in properties is generally explained by the similarity in the electron configuration of these elements—specifically, the fact that each has a single valence electron in an *s* sublevel.

Although there are similarities in the properties of these elements, they are not identical. This too is largely due to differences in electron configuration. Lithium has its valence electron in the second principal energy level; sodium has its valence electron in the third principal energy level, and so on down the column of the periodic table. Consequently, the size of the atoms increases from top to bottom in the column. This difference in size causes differences in such properties as density, melting point, and heat of vaporization.

GROUP IIA(s^2): BERYLLIUM, MAGNESIUM, CALCIUM, STRONTIUM, AND BARIUM

Group IIA elements have two valence electrons, both in an *s* sublevel. Elements in this group all have a combining number of 2 and form positive ions with a charge of 2+. However, the properties of elements in this group differ more than the properties of elements in Group IA. Beryllium in particular, with only two electrons between the valence electrons and the nucleus, melts and boils at temperatures almost twice as high as the melting and boiling points of other members of the group, and it is considerably harder than the larger atoms below it.

GROUPS IIIA(s^2p^1), IVA(s^2p^2), VA(s^2p^3), AND VIA(s^2p^4)

The similarity among the elements within a group, so evident in Group IA and still apparent in Group IIA, is less apparent in these groups. Most of you have seen carbon, which heads Group IVA, in both its forms: diamond and graphite. You would scarcely think of carbon in either form as being similar to tin and lead at the bottom of the same group. Carbon is a nonmetal. It is brittle and does not have a metallic luster. It reacts with oxygen to form a gaseous oxide. By contrast, tin and lead have those properties normally associated with metals.

In a similar fashion, nitrogen, at the top of Group VA, is very different from bismuth at the bottom. Oxygen, at the top of Group VIA, is quite different from tellurium, even though the shift in properties is not quite so dramatic as in the previous group. (Polonium, the last element in the group, is radioactive and virtually nonexistent. Although many of its properties are known, there seems to be little point in discussing them.)

The combining numbers of elements in these groups vary considerably as well. Although oxygen may have a combining number of 1, 2 is by far the most common combining number for oxygen. Tellurium may have combining numbers of 2, 4, or 6. Carbon has common combining numbers of 2, 3, and 4; tin and lead have common combining numbers of 2 and 4. The most common combining number for elements in Group IIIA is 3, the same as the number of valence electrons; but gallium also has a combining number of 2, and both indium and thallium have a combining number of 1 in some compounds.

Although it is possible to find relationships between the electron configurations of these elements and their properties, the relationships are complex. You will learn about them in a later course when you have more chemical facts available to you.

GROUP VIIA(s^2p^5): THE HALOGENS

By the time we reach this group to the right of the periodic table, similarities among the elements in the group are again clear. Astatine does not exist in nature, and little is known about it. The others in this group (fluorine, chlorine, bromine, and iodine) form diatomic molecules, react readily with metals to form binary compounds in which the halogen has a combining number of 1, and react with hydrogen to form gaseous compounds that dissolve in water to form strong acids.

The common combining number of 1 for the **halogens** may seem strange. When you write the electron configuration for these elements, you see that each has seven valence electrons: two in an *s* sublevel and five in a *p* sublevel. The combining number of 1 seems to be related to the fact that one additional electron will fill the *p* sublevel, producing the same electron configuration as the noble gas found to the right of the element.

GROUP O(s^2p^6): THE NOBLE GASES

Until the 1960s, the elements to the right of the periodic table (which we now call the **noble gases**) were called the inert gases, signifying little or no tendency to react with other elements. For years it was assumed that these elements could not combine chemically. However, as the model of electron configuration that we are introducing here began to develop, there seemed to be no reason why elements near the bottom of the group should not form some stable compounds.

Radon is a highly radioactive gas that is extremely dangerous to work with, and it disintegrates very rapidly to form isotopes of other elements. It is not an element to experiment with. Xenon, the next element from the bottom, is not radioactive but it is very rare. Both cost and limited availability discouraged experimentation, and compounds of xenon were not made until recent years. Although several compounds of xenon have been produced, none of the Group O gases is very reactive, and most of them have no known compounds. The electron configurations of these elements appear to be very stable.

The combining numbers of the elements in Groups IA, IIA, and VIIA are often described in terms of the noble gases. In the previous section, we pointed out that the halogens take on the electron configuration of their neighboring noble gas when they acquire an additional electron to form halide ions with a charge of -1. This arrangement appears to be quite stable. There is no evidence of a halogen acquiring more than one additional electron.

Write the electron configurations for Ne, Na, and Mg; Ar, K, and Ca; Kr, Rb, and Sr; and Xe, Cs, and Ba. Now look at the triads together. It is clear that Na has the same electron configuration as the noble gas Ne when it loses one electron to form a Na^+ ion. In similar fashion, Mg acquires the Ne configuration when it forms the Mg^{2+} ion. The alkali metals acquire the same configuration as a noble gas when they lose a single electron, and the elements in Group IIA acquire a noble gas configuration when they lose two electrons. Once again, the fact that the elements in Group IA always form $1+$ ions and that elements in Group IIA always form $2+$ ions suggests that the noble gas arrangement of electrons is particularly stable.

This generalization about the stability of the noble gas configuration has proved useful in predicting the formulas of many compounds, but it should not be taken too far. Lead and tin form many compounds containing ions with a charge of $2+$ as well as compounds containing ions with a charge of $4+$. Only the $4+$ ion has a noble gas configuration. There are *many* other exceptions.

13.8 VALENCE ELECTRONS, ELECTRON ARRANGEMENT, AND THE PERIODIC TABLE

We have discussed the relationship between the electron configuration of the elements, their properties, and their position in the periodic table. All elements in Group IA of the periodic table have a single valence electron in an *s* sublevel. All elements in Group IIA have two valence electrons in an *s* sublevel. Beginning with Group IIIA, the *p* sublevels, along with the *s* sublevel, contain valence electrons. Group IIIA elements have two valence electrons in the *s* sublevel and one in the *p* sublevel, Group IVA elements have two *p* electrons along with the two in the *s* sublevel, and so on across the table. In Figure 13.13 (p. 410), the periodic table is broken into sections to emphasize the sublevels that are being filled. You may want to refer to Figure 13.13 as you answer the following questions.

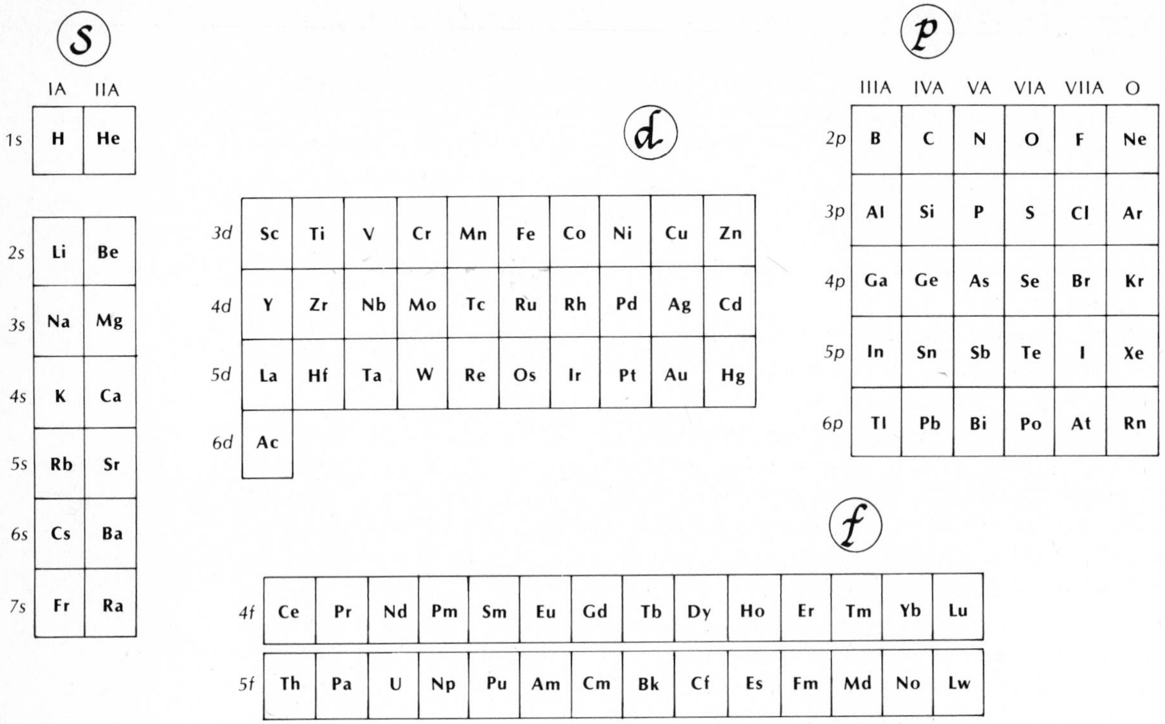

s

	IA	IIA
1s	H	He
2s	Li	Be
3s	Na	Mg
4s	K	Ca
5s	Rb	Sr
6s	Cs	Ba
7s	Fr	Ra

d

3d	Sc	Ti	V	Cr	Mn	Fe	Co	Ni	Cu	Zn
4d	Y	Zr	Nb	Mo	Tc	Ru	Rh	Pd	Ag	Cd
5d	La	Hf	Ta	W	Re	Os	Ir	Pt	Au	Hg
6d	Ac									

p

	IIIA	IVA	VA	VIA	VIIA	O
2p	B	C	N	O	F	Ne
3p	Al	Si	P	S	Cl	Ar
4p	Ga	Ge	As	Se	Br	Kr
5p	In	Sn	Sb	Te	I	Xe
6p	Tl	Pb	Bi	Po	At	Rn

f

4f	Ce	Pr	Nd	Pm	Sm	Eu	Gd	Tb	Dy	Ho	Er	Tm	Yb	Lu
5f	Th	Pa	U	Np	Pu	Am	Cm	Bk	Cf	Es	Fm	Md	No	Lw

Figure 13.13
Periodic table sectioned to show the energy sublevels that are being filled.

13-4. How many valence electrons do elements in Group IA have? What sublevel is the valence electron in?

13-5. Elements in Group VIA have six valence electrons. What sublevel(s) are the valence electrons in?

13-6. What group in the periodic table has elements with three valence electrons in the *p* sublevel? Do these elements have other valence electrons?

13-7. Without writing the electron configuration for tin, indicate the sublevel(s) that contain(s) the valence electrons.

13-8. In what sublevel do you find the valence electrons of iron?

You probably answered 3*d* for the last question, but 4*s* is a better response. The valence electrons are the electrons that are available for bonding, and they *always* include those at the highest principal energy level. For iron, that means the 4*s* electrons. Let's look at the electron configuration for iron.

$$\text{Fe} \quad 1s^2,\ 2s^2,\ 2p^6,\ 3s^2,\ 3p^6,\ 4s^2,\ 3d^6$$

If you look at the periodic table at the front of the book, you will see that iron is in row 4. The row of the periodic table indicates the principal energy level containing valence electrons. Every element in row 4, including iron, has valence electrons in the fourth principal energy level. However, this does not mean that the lower principal energy levels are completely occupied.

Figure 13.12 shows that the order of filling the various sublevels in an atom is regular through the $3p$ sublevel. At this point, however, electrons fill the $4s$ sublevel before filling the $3d$. In Figure 13.13, iron is among those elements in the block labeled *d*. These elements are known as the **transition metals.** Note that there are ten columns in this block of elements, corresponding to the ten electrons that fit into a *d* sublevel.

Scandium (Sc) and yttrium (Y) in the first column of this block have a single electron in a *d* sublevel; titanium (Ti), zirconium (Zr), and hafnium (Hf) have two *d* electrons; vanadium (V), niobium (Nb), and tantalum (Ta) have three *d* electrons; and so on across the row. (There are some deviations from this pattern in the fourth and ninth columns, but these need not concern you now.) Keep in mind that the *d* sublevel that is being filled as you move from left to right across a row is the *d* sublevel *below* the normal valence electrons. In row 4 of the periodic table, it is the $3d$ sublevel that is being filled. There are already electrons in the $4s$ sublevel.

Because transition metals have two electrons in the highest principal energy level, a common combining number for the transition elements is 2. However, the electrons in the *d* sublevel may also become involved in bonding. These electrons can be regarded as valence electrons along with those electrons at the highest principal energy level. It is the general rule that transition elements have more than one combining number. Two exceptions to the rule are Zn, which has a combining number of 2, and Ag, which has a combining number of 1.

13.9 ELECTRON CONFIGURATION OF TRANSITION METAL IONS

When students are asked to write the electron configuration for ions formed from elements in one of the B Groups of the periodic table, they often make errors. The error occurs when the student calculates the number of electrons in the ion and then follows Figure 13.12 to write the configuration. This leads to the following *incorrect* configuration for Fe^{2+}:

$$Fe^{2+} \quad 1s^2,\ 2s^2,\ 2p^6,\ 3s^2,\ 3p^6,\ 4s^2,\ 3d^4 \quad \text{(incorrect)}$$

Ions are formed when electrons are *removed* from neutral atoms, and the electrons that are most easily removed are those at the highest principal energy level. When electrons are removed from iron to form the Fe^{2+} ion, it is the $4s$ electrons that are removed. The *correct* configuration for the ion is

$$Fe^{2+} \quad 1s^2,\ 2s^2,\ 2p^6,\ 3s^2,\ 3p^6,\ 3d^6 \quad \text{(correct)}$$

13-9. Write the electron configuration for Zn^{2+}.
13-10. Write the electron configuration for Cr^{3+}.
13-11. Write the electron configuration for Ni^{2+}.

13.10 SHORTHAND NOTATION FOR ELECTRON CONFIGURATIONS

As you have been writing electron configurations for various atoms and ions, you have probably noticed that the first part of the notation for several atoms is the same. For example, in the foregoing exercises, the electron configuration for all of the ions began with

$$1s^2, 2s^2, 2p^6, 3s^2, 3p^6$$

This is the electron configuration for argon. In other words, each of the ions considered has its first eighteen electrons arranged just like the electrons in argon, the immediately preceding noble gas. This part of the electron configuration is often indicated by [Ar]. This shorthand method of representing electron configuration is outlined in the following example.

Example 13.2

Write the electron configuration for Hg.

Mercury has eighty electrons, but we need not write the configuration for all of them. Most can be represented by the symbol for the noble gas that precedes mercury in the periodic table. Which element is this?

---

xenon, element 54

Using the periodic table, we go down the column of noble gases to the one whose atomic number is just under that of the element considered. We would then write:

[Xe]

[Xe] represents the arrangement for the first fifty-four electrons in mercury. The only thing left is to decide where the next electrons go.

Mercury is in *row 6* of the periodic table. After writing the symbol for xenon, we continue with those electrons that start the *sixth* principal energy level. Referring to Figure 13.12, we see that these are the $6s$ electrons. By continuing to follow the order of filling described by Figure 13.12 and counting until we have a total of eighty electrons represented, we can complete the electron configuration. Do it yourself before looking at my results.

Hg [Xe]^{54}, $6s^2$, $4f^{14}$, $5d^{10}$

Practice this shorthand notation by writing the electron configuration for each of the following elements and ions.

13-12. Pb	13-14. U	13-16. Ba
13-13. Ag^+	13-15. Sn^{2+}	

13.11 LEWIS STRUCTURES

Since the electrons that are involved in chemical bonding are the valence electrons, it is often convenient to use some kind of diagram to represent these electrons in an atom. The American chemist G. N. Lewis proposed that we use the symbol for the element to represent *the nucleus plus all electrons except the valence electrons* and then use dots around the symbol to represent *the valence electrons.* These structures are called **electron-dot structures** or **Lewis structures.** Every element in Group IA of the periodic table has a single valence electron. The Lewis structure for one of these elements is the symbol for the element and a single dot. H· would represent the hydrogen atom, and Li· would represent the lithium atom. In a similar fashion, oxygen has six valence electrons (it is in Group VIA) and would be represented like this:

Lewis structures are nothing more than a convenient way to show valence electrons. The placement of the dots around the symbol has no special significance and should never be interpreted as an actual location of an electron relative to the nucleus of the atom. H·, ·H, $\dot{\text{H}}$, and $\underset{\cdot}{\text{H}}$ are all equivalent, and none of these symbols indicates the actual location of the electron in the atom. It is customary to write the dots in pairs on the four sides of the symbol, but even this is not necessary.

Use the periodic table to find the number of valence electrons and write the Lewis structure for each of the following elements.

13-17. Aluminum	13-19. Phosphorus	13-21. Calcium
13-18. Sodium	13-20. Chlorine	13-22. Iodine

Lewis structures are not used to represent neutral atoms very often. We use them when we want to represent molecules and complex ions. You will do that in Chapter 14 after you consider bonding. However, before you can draw Lewis structures for molecules or complex ions, you must know the total number of dots to show. In neutral molecules, the valence electrons for all of the atoms in the molecule are shown. For ions, electrons are subtracted or added to give the ion the appropriate charge. Indicate the number of dots that must be shown in the electron-dot structure for each of the following ions and molecules.

13-23. H_2O
13-24. Cl_2
13-25. I^-
13-26. OH^-
13-27. CH_4
13-28. SO_4^{2-}
13-29. NO_3^-
13-30. NO_2
13-31. PCl_5
13-32. CO_3^{2-}
13-33. PO_4^{3-}

SUMMARY

Matter is made up of atoms, and the atoms of all elements are different. Each element is identified by an atomic number, which indicates the number of protons in the nucleus of the atom. The proton is a positively charged particle with a mass of 1.673×10^{-24} g, or a relative mass of 1.007.

Joining the protons in the nucleus of atoms are particles with no electrical charge, called neutrons. Neutrons have a mass of 1.675×10^{-24} g, or a relative mass of 1.009.

Actually, not all atoms of a given element are identical. They all have the same number of protons, but they may have different numbers of neutrons. Such atoms of the same element are called isotopes. The sum of the protons and neutrons in an atom is the mass number for that atom. Various isotopes of an element are frequently represented by the symbol for the element with the atomic number written at the bottom left and the mass number written at the top left. Using this notation, we identify the three naturally occurring iostopes of oxygen as $^{16}_{8}O$, $^{17}_{8}O$, and $^{18}_{8}O$.

The relative (atomic) mass for the elements is based on a particular isotope of carbon, $^{12}_{6}C$. This isotope, called carbon-12, is assigned a mass of exactly 12.000. This same relative mass scale produces a relative mass for the proton of 1.007 and a relative mass for the neutron of 1.009. Strange as it may seem, in the case of nuclear mass, the whole is *not* equal to the sum of the parts. Adding the relative mass of the six protons in carbon-12 and the relative mass of the six neutrons in carbon-12, we get a relative mass of *more than* 12, the relative mass of the atom. The difference between the mass of the separate parts and the mass of the atom is related to the energy required to hold the positively charged protons together in the nucleus of the atom, and it helps to explain how atoms like $^{16}_{8}O$ can have a relative mass less than the mass number for the isotope.

The atomic mass of oxygen-16 is 15.99491. The atomic mass for oxygen recorded in the periodic table is slightly higher than this: 15.9994. The difference is due to the presence of small amounts of oxygen-17 and oxygen-18 in naturally occurring oxygen.

The nucleus of the atom can be interesting, but chemists are primarily concerned with what surrounds the nucleus, because the electrons are the particles involved in chemical bonding.

The currently accepted model of the atom does not describe the position of the electron in the atom. It suggests that the electrons are moving about the nucleus in some manner that we do not understand. However, we do know that the various electrons in the atom differ in their energy. Our model describes the electrons in terms of their energy. Numbers are used to represent the principal energy level of the electron, and letters are used to represent sublevels of energy. Only a limited number of electrons can "occupy" a particular energy sublevel; two can occupy an *s* sublevel, six can occupy a *p* sublevel, ten can occupy a *d* sublevel, and fourteen can occupy an *f* sublevel.

By following a few simple rules, we can describe the energy level of all electrons in an isolated atom. This description is called the electron configuration for the atom.

There appears to be a close relationship between the chemical properties of elements, their positions in the periodic table, and the electron configurations of the elements. For example, all elements in a given group of the periodic table have the same number of valence electrons in the same energy sublevels.

It is the valence electrons that are most often near the surface of the atom, and these electrons are normally involved in bonding between atoms. As a result, valence electrons receive a great deal of attention in chemistry. We represent the nucleus and all electrons except the valence electrons by the symbol of the element and the valence electrons by dots around the symbol. This kind of representation is sometimes called an electron-dot structure or Lewis structure.

Questions and Problems

13-34. Complete the following table.

Symbol	*Number of protons*	*Number of neutrons*	*Number of electrons*
$^{17}_{8}O$	______	______	______
______	6	7	6
$^{238}_{92}U$	______	______	______
$^{23}_{11}Na^{+}$	______	______	______
______	7	7	10
$^{36}_{17}Cl^{-}$	______	______	______

13-35. Select an element at random and write its electron configuration.

13-36. Write the electron configuration for each of the following ions.

a. Sn^{2+} c. Ag^{+} e. Ca^{2+}
b. S^{2-} d. I^{-} f. Pb^{4+}

13-37. How many valence electrons are there in each of the following atoms?

a. At d. Si g. Be
b. Bi e. Cs h. Po
c. Se f. Ga

13-38. Draw the electron-dot structure for each atom listed in Question 13-37.

13-39. For each of the following elements, indicate the sublevels occupied by the valence electrons and the number of valence electrons in each sublevel.

a. As d. Rb g. F
b. Sr e. Ge h. Xe
c. In f. Te

13-40. On the average, which would you expect to find closer to the nucleus, an electron in a 4*p* energy level or an electron in a 3*d* energy level? Explain.

13-41. Explain why the light from an uncoated fluorescent lamp contains only a few distinct wavelengths rather than all the colors of the rainbow. (Hint: A fluorescent lamp is a gas-discharge tube filled with mercury vapor.)

13-42. An older concept of atoms pictures the electrons moving around the nucleus in circular (or eliptical) orbits, much like the planets moving around the sun. How does the idea presented in this chapter differ from that older picture?

13-43. The order given in Figure 13-12 for filling the energy sublevels within an atom does not always hold. In what sections of the periodic table do you find most of the exceptions?

13-44. The bookcase analogy used in this chapter may help us visualize some of the properties of electrons, but all analogies tend to oversimplify. In what respects does the bookcase analogy misrepresent what we believe about electron arrangement?

Answers to Questions in Chapter 13

13-1.

H	$1s^1$
He	$1s^2$
Li	$1s^2, 2s^1$
Be	$1s^2, 2s^2$
B	$1s^2, 2s^2, 2p^1$
C	$1s^2, 2s^2, 2p^2$
N	$1s^2, 2s^2, 2p^3$
O	$1s^2, 2s^2, 2p^4$
F	$1s^2, 2s^2, 2p^5$
Ne	$1s^2, 2s^2, 2p^6$
Na	$1s^2, 2s^2, 2p^6, 3s^1$
Mg	$1s^2, 2s^2, 2p^6, 3s^2$
Al	$1s^2, 2s^2, 2p^6, 3s^2, 3p^1$
Si	$1s^2, 2s^2, 2p^6, 3s^2, 3p^2$
P	$1s^2, 2s^2, 2p^6, 3s^2, 3p^3$
S	$1s^2, 2s^2, 2p^6, 3s^2, 3p^4$
Cl	$1s^2, 2s^2, 2p^6, 3s^2, 3p^5$
Ar	$1s^2, 2s^2, 2p^6, 3s^2, 3p^6$

13-2. Sr $1s^2, 2s^2, 2p^6, 3s^2, 3p^6, 4s^2, 3d^{10}, 4p^6, 5s^2$

Cu $1s^2, 2s^2, 2p^6, 3s^2, 3p^6, 4s^1, 3d^{10}$ (You should have predicted . . . $4s^2, 3d^9$; but copper does not follow the pattern. If you look up electron configurations in other books, you may find that of copper ending . . . $3d^{10}, 4s^1$. It makes no difference whether the notation is given in order of filling or in order of increasing principal energy level.)

Hg $1s^2, 2s^2, 2p^6, 3s^2, 3p^6, 4s^2, 3d^{10}, 4p^6, 5s^2, 4d^{10}, 5p^6, 6s^2, 5d^{10}$

I $1s^2, 2s^2, 2p^6, 3s^2, 3p^6, 4s^2, 3d^{10}, 4p^6, 5s^2, 4d^{10}, 5p^5$

13-3.

H	$1s^1$
Li	$1s^2, 2s^1$
Na	$1s^2, 2s^2, 2p^6, 3s^1$
K	$1s^2, 2s^2, 2p^6, 3s^2, 3p^6, 4s^1$
Rb	$1s^2, 2s^2, 2p^6, 3s^2, 3p^6, 4s^2, 3d^{10}, 4p^6, 5s^1$
B	$1s^2, 2s^2, 2p^1$

Al $1s^2, 2s^2, 2p^6, 3s^2, 3p^1$
Ga $1s^2, 2s^2, 2p^6, 3s^2, 3p^6, 4s^2, 3d^{10}, 4p^1$
In $1s^2, 2s^2, 2p^6, 3s^2, 3p^6, 4s^2, 3d^{10}, 4p^6, 5s^2, 4d^{10}, 5p^1$

13-4. One in an s sublevel
13-5. Two are in the s sublevel, and four are in the p sublevel.
13-6. Group VA. These elements also have two electrons in an s sublevel.
13-7. $5s^2, 5p^2$
13-8. $4s$ (and possibly $3d$)
13-9. Zn^{2+} $1s^2, 2s^2, 2p^6, 3s^2, 3p^6, 3d^{10}$
13-10. Cr^{3+} $1s^2, 2s^2, 2p^6, 3s^2, 3p^6, 3d^3$ (Once the $4s$ electrons have been removed, electrons will be removed from the $3d$ sublevel.)
13-11. Ni^{2+} $1s^2, 2s^2, 2p^6, 3s^2, 3p^6, 3d^8$
13-12. Pb [Xe] $6s^2, 4f^{14}, 5d^{10}, 6p^2$
13-13. Ag^+ [Kr] $4d^{10}$ (You would predict [Kr] $4d^9$, $5s^1$.)
13-14. U [Rn] $7s^2, 5f^3, 6d^1$ (From the rules given, you would predict . . . $5f^4$.)
13-15. Sn^{2+} [Kr] $5s^2, 4d^{10}, 5p^2$
13-16. Ba [Xe] $6s^2$
13-17. ·Al:
13-18. Na·
13-19. :P̣:
13-20. :Ċ̣l:
13-21. Ca:
13-22. :İ̤:
13-23. 8
13-24. 14
13-25. 8
13-26. 8
13-27. 8
13-28. 32
13-29. 24
13-30. 17
13-31. 40
13-32. 24
13-33. 32

READING

Modern theories of atomic structure are built around the idea of atomic orbitals. Sanderson's delightful article elaborates on the use of orbitals in explaining chemical bonds.

NOTHING IS VERY IMPORTANT

R. T. Sanderson

Gracious, I don't mean that you're not very important—or, for that matter, that mothers, pies, peace, love, and chemistry are not very important! I mean precisely that *nothing* can also be very important.

But this nothing is a special kind. It's not what fills (or fails to fill) the answer sheet when your memory runs dry. Instead, it's the kind in great demand for making holes in donuts and macaroni. It's that nothing which occupies empty space. In an atom, it is where an electron could or might be, but isn't.

That's the beauty of quantum mechanics. It dictates that we not hope to focus on a sharp image of an atom. The image remains diffuse, electrons being spread out through certain *regions* in space about the nucleus. Atoms can't accommodate electrons just any old place—they're found only in certain regions, called orbitals. Each orbital can accommodate only two electrons, with opposing magnetic spins. Quantum mechanics not only labels each electron, but conveniently assigns quantum numbers to empty orbitals so that we can tell one nothing from another.

Each successive electron added to the existing electron cloud is repulsed by all other electrons but is attracted to the nucleus. Hence electrons can't be added in any stable way, unless the available region provides a net attraction. In other words, nuclear attraction must exceed interelectron repulsion. This can happen only in the region occupied by nothing.

Source: *Chemistry,* Vol. 49, No. 8 (October 1976), pp. 6–7.

This particular kind of nothing, called orbital vacancy, is like the space in a two-car garage for a one-car family. So long as garden tools and rummage are kept out, a future second car *could* be accommodated in the empty slot.

Why Is Nothing So Very Important?

Most substances, except the noble gases, consist of atoms joined together. These compounds could not exist unless binding forces, called chemical bonds, held the atoms together. Because the same electrons are attracted simultaneously to two atomic nuclei, a chemical bond can form.

Important to bond formation is electrostatic attraction between opposite charges—between the nucleus and electrons. But an atom without an outermost orbital vacancy lacks a region where another atom's electron is close enough to be attracted by the nucleus. No vacancies—no bonds. No bonds—no substances, except single atoms. You and I would be nothing but mixtures of gases. That's exactly why the noble elements are gases—with few exceptions, they exist as single atoms.

Nothing Spelled Out

If we wish, we can spell out the importance of nothing as it applies to outermost orbital vacancies. First, what does it tell us about different types of bonding? In metallic bonding there are more outermost vacancies than electrons. This permits any shared electrons to spread out—become delocalized, as quantum mechanics people say—giving rise to familiar metallic properties.

In covalent bonding there are as many or more electrons than vacancies in the outermost level. This does not permit delocalization, but confines electrons in a covalent bond to localized sharing. In coordinate-covalent bonding one atom must provide double nothing—a vacant orbital capable of accommodating an electron *pair* from another source. In hydrogen bonding there is no actual vacancy at the hydrogen. However, its electron cloud is withdrawn toward the more electronegative atom . . . to provide at least a fraction of nothing. This is in contrast to ionic bonding where the metal atom has plenty of nothing around it. Remember, it's the empty spaces that unmask nuclear attractions.

Perhaps we should avoid explaining what nothing means, lest we come to argue over nothing. But—inescapably—there are different qualities of nothing—for example, vacancies about atoms of different elements. Even an imaginary electron might sense a net attraction, corresponding to the "effective nuclear charge," over the distance of the atomic radius. The effective nuclear charge is the original nuclear charge minus what the other electrons have blocked off. It's equivalent to subtracting the electron repulsions from the attraction exerted by the nucleus. The net attractive force between the electron (if it were there) and the effective nuclear charge is called "electronegativity."

In the periodic table, from left to right across the elements in a group, effective nuclear charge increases because the outer electrons' ability to block off positive charge decreases. A large effective nuclear charge results in the electron cloud being held closer, thus reducing the atomic radius. A larger effective nuclear charge operating over a shorter distance increases electronegativity. This is why electronegativity ranges from very low in alkali metals to very high in halogens.

When like atoms form bonds, electronegativity influences bond strength by determining the attraction of each nucleus for the bonding electrons. When unlike atoms form bonds, electronegativity exerts a similar influence on the covalent portion of total bond energy. In addition, it determines the relative magnitude of the covalent and ionic contributions to total bond energy. Bond energies, in turn, contribute vitally to the relative stabilities of different substances—and there lurks an important incentive for chemical change. Reactions tend to go in the direction of forming stronger bonds.

This is why vacancies are at least as important as electrons in atomic structures. One might *almost* say that nothing is more important than anything.

14

Chemical Bonding

Most of us hate theory. Give us tantalizing tastes, seductive aromas, and alluring colors. *That's* the kind of chemistry we can appreciate. We may not *understand* it, but we can certainly appreciate it.

There comes a time when it is worth the effort to understand. If we can understand the chemistry of photosynthesis, we may increase our food supply. If we analyze the molecules that carry genetic codes, we may find a cure for disease. If we can determine the mechanism of reactions occurring in the environment, we may protect it for our children's children as well as for ourselves. This kind of understanding will not come unless we probe the nature of molecules—how they come together, how they don't, and why. What is the glue that holds atoms of less than 100 kinds of elements together in millions of living, breathing forms that we see and are each day?

In this chapter, we begin to look at bonding and develop a simple model that we can use to talk about differences between molecules. As you continue your study of chemistry, you will learn more. You will still enjoy sights and smells, but you will be proud that you have begun to understand why.

OBJECTIVES

1. Give a qualitative description of what causes a chemical bond to form.
2. Distinguish between an orbit, an orbital, and an energy level.
3. Use the octet rule to predict structures for molecules involving elements in row 2 of the periodic table.
4. Given an electron-dot structure, identify single bonds, double bonds, triple bonds, and nonbonding electrons. (See Problem 14-49.)
5. Given the formula of simple molecules or ions, draw an electron-dot structure for the molecule or ion. (See Problem 14-48.)
6. Given the electron-dot structure for a molecule or ion, predict the shape of the molecule or ion.

7. Given a table of electronegativities, predict whether a bond between two atoms will be covalent, polar-covalent, or ionic. (See Problem 14-50.)
8. State the relationship between combining number and oxidation number.
9. Given the formula of a molecule, state the oxidation number for each element in the molecule. (See Problems 14-51 and 14-52.)

Words You Should Know

bond energy
orbital
octet rule
bonding electron
nonbonding electron
single bond
double bond
triple bond
bond angle
tetrahedron
covalent bond
polar
ionic bond
electronegativity
polar-covalent bond
oxidation number

14.1 WHAT WE NEED TO KNOW ABOUT BONDING

Several important questions arise as we attempt to understand nature. Many can be answered through a study of bonding. Before developing a simple model of bonding, we will examine some of these questions.

HOW MANY ATOMS ARE CONNECTED?

In the final analysis, the only way to know the formula of a compound is to determine the amount of each reactant in the compound and calculate the proportions, as discussed in Chapter 6. Still, it would be nice to be able to predict the *most likely* formula of a compound. A good theory of bonding helps us explain the chemical fact that LiCl is more likely to form than Li_2Cl or $LiCl_2$.

We can't predict the formula for *all* compounds. From the law of multiple proportions, we know that more than one compound of the same two elements are possible. We have H_2O and H_2O_2, CO and CO_2, CuCl and $CuCl_2$. Predicting formulas is dangerous at best, but a good theory of bonding allows intelligent guesses about the structure of molecules.

WHICH ATOMS ARE CONNECTED?

The structure of molecules implies two things: what atoms are connected and what shape they produce in the molecule. Information about the atoms that are connected can be important in predicting chemical properties. (Conversely, information about chemical properties provides important clues to structure.) Phosphoric acid has the formula H_3PO_4, and all *three* hydrogen protons can be donated to a strong base. Phosphorous acid has the formula H_3PO_3, and only *two* of the hydrogen protons can be donated to a strong base. Why?

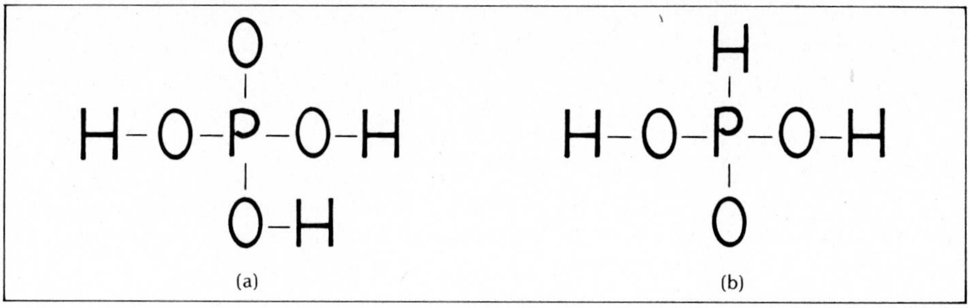

```
      O                      H
      |                      |
H-O-P-O-H              H-O-P-O-H
      |                      |
      O-H                    O

     (a)                    (b)
```

Figure 14.1
(a) Phosphoric acid, H_3PO_4. (b) Phosphorous acid. H_3PO_3.

Figure 14.1 shows how atoms are connected in these two compounds. In phosphoric acid, all hydrogen atoms are connected to oxygen, and all three of the hydrogen protons can be removed by a strong base. In phosphorous acid, two hydrogen atoms are attached to oxygen, but one is attached to phosphorus. The hydrogen atoms attached to oxygen can be removed by a strong base, but the hydrogen attached to phosphorus cannot. This is one example of a chemical property that can be understood through knowledge of the atoms that are connected.

WHAT IS THE SHAPE OF THE MOLECULE?

Shape is the other aspect of structure, and it helps determine chemical properties. If you have worked a jigsaw puzzle, you know that pieces of the wrong shape will not fit. The same is true of molecules. Many chemical reactions—especially those in living systems—depend on the ability of two molecules to get close to each other. If a molecule has the correct shape, it can come close to another molecule and react; if the molecule has the wrong shape, it can't get close enough to react. The action of drugs, the digestion of food, the production of food in green plants, and the reproduction of cells in your body involve chemical reactions in which molecular shape plays an important role. A good theory of bonding predicts the shape of molecules.

HOW STRONG IS THE BOND?

When two atoms bond, energy is released; when a bond is broken, the same amount of energy must be supplied to tear the atoms apart. The energy involved in these processes of bond forming and bond breaking is known as the **bond energy.** It is a measure of bond strength and can be very important. It can tell you how many calories of energy you get from a gram of sugar and how many calories of energy are stored when a tree produces cellulose. The strength of a bond can indicate which bonds in a molecule come apart easily and which stubbornly remain stuck. A good model of bonding provides a basis for understanding these differences in bond energy.

As you can see, a good theory of bonding offers better understanding of many chemical facts. The simple theory that we introduce here can do some of these things. The more sophisticated theory that you will study if you continue in chemistry can do much more.

14.2 WHAT CAUSES A BOND?

We know that atoms join to form molecules. The question is why. What is the glue that causes them to stick together? A complete answer is beyond the scope of this book, but we can make a start.

Atoms are made of particles that have a positive electrical charge and particles that have a negative electrical charge. Unlike electrical charges attract each other. These facts provide an explanation for all bonding.

We begin with a simple example: two hydrogen atoms combining to form one molecule of H_2.

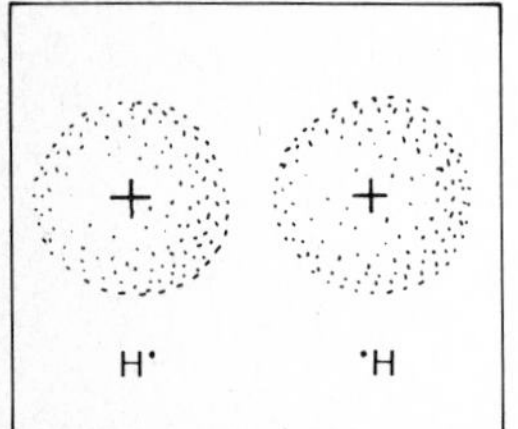

Figure 14.2
Two hydrogen atoms far apart.

Figure 14.2 represents two hydrogen atoms separated from one another. The + in the center represents the single proton in the nucleus of each atom, and the fuzzy sphere around it represents the single electron. Below each picture the electron-dot structure for the hydrogen atom appears.

When we imagine the two atoms moving very close to one another, we end up with a picture similar to the one shown in Figure 14.3. Once again, the + represents the proton at the center of each atom, and the fuzzy area represents the electrons. In this case two electrons are represented, because there are two electrons associated with the two atoms. Below the diagram is the electron-dot structure for the H_2 molecule.

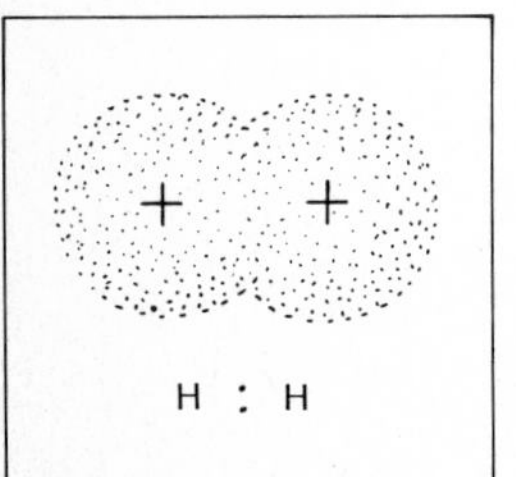

Figure 14.3
A hydrogen molecule, H_2.

What are we trying to say has happened? When the atoms are far apart, the only force of attraction is between the proton on one atom and the electron on the *same* atom. As long as the atoms are far apart, there are no other forces large enough to have any major effect. As the atoms approach each other, however, other forces become important. There are repulsive forces between the electrons on the two atoms, and there are repulsive forces between the protons on the two atoms. There are also new attractive forces. The proton on one of the atoms can attract the electron on the other atom, and vice versa. If these new attractive forces are greater than the new repulsive forces, the two atoms stick together to form a molecule. If the new repulsive forces are greater than the attractive forces, a bond does not form and the two atoms move apart again. The question, then, is whether the new attractive forces or the new repulsive forces are greater. How do we know?

Figure 14.4
Attractions and repulsions come into play when two hydrogen atoms come close.

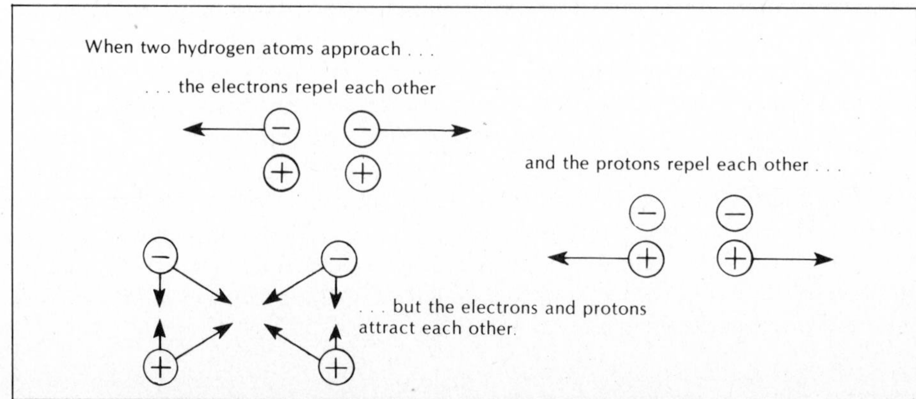

Accurate calculation of the attractive and repulsive forces that occur when two atoms come close is complex. Consequently, simplified procedures have been invented for discussing bond formation.

14.3 ORBITALS

In most explanations of bonding, the arrangement of electrons in the atom is important. In Chapter 13, we said that the modern concept of atoms tells us the energy of the electron but not its location. That is not exactly true.

Although we can't determine where an electron of a particular energy is at any given time, the mathematical model on which electron configurations are based does allow us to calculate the *probability* of finding the electron in some *region* of space.

The fuzzy sphere representing the electron in Figure 14.2 is actually the spherical region of space around the nucleus of a hydrogen atom wherein the electron moves about 90 percent of the time. This region of space is referred to as the 1*s* **orbital.**

The term "orbital" should not be confused with "orbit." An older atomic model described the electrons as circling the nucleus in orbits much as planets orbit the sun. That model has proved unsatisfactory. In the model that we are using, no effort is made to describe the motion of the electron, and no effort is made to determine the *exact* location of the electron at any time. What *is* described is the region of space in which there is a 90 percent chance of finding the electron. This *region of space* is called an orbital.

An orbital can contain no more than two electrons. There is a single *s* orbital at each principal energy level, and it has the spherical shape shown for the hydrogen atom. The *p* sublevel can hold a total of six electrons. Since each orbital can hold no more than two electrons, it follows that there are three *p* orbitals. The three *p* orbitals together form a spherical region of space much like the spherical region described by the *s* orbital. However, when the three *p* orbitals are examined separately, it becomes apparent that an individual *p* orbital is not spherical. Rather, it has two lobes on either side of the nucleus. The three *p* orbitals are shown in Figure 14.5.

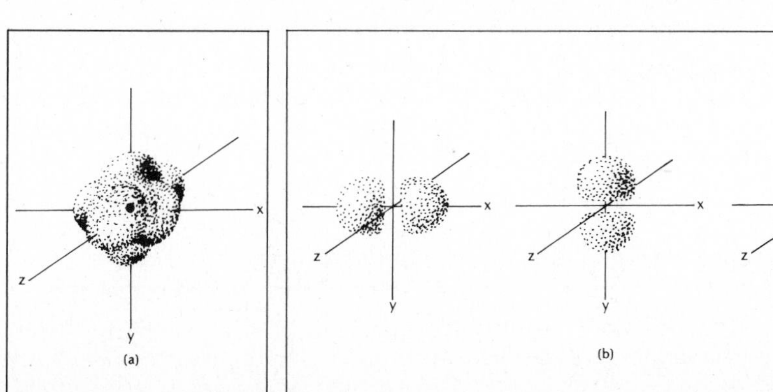

Figure 14.5
(a) The three p *orbitals form a spherical charge cloud surrounding the nucleus. (b). The two lobes along each axis represent a* single *orbital that can contain one or two electrons.*

There are five orbitals for the ten *d* electrons and seven orbitals to accommodate the fourteen electrons at the *f* energy sublevel. The shapes of the *d* and *f* orbitals are more complex and will be left for a later course.

The idea of orbitals is derived from an atomic model based on the assumption that the electron in an atom behaves as a wave. This model has proved valuable in predicting or explaining the shape of molecules, the existence of compounds not previously known, and the strength of chemical bonds. However, careful development of this notion requires an understanding of wave phenomena (wave mechanics) and mathematics that is well beyond the scope of this book. We will settle for a simple picture that can provide an approximate explanation for most of the questions raised at the beginning of the chapter.

Before going on to this simple picture, let's summarize our basic understanding of how bonds form. The following statements are consistent with the more sophisticated model of bonding that we have alluded to, as well as with the simplified model that follows.

1. Bonds form between atoms when nuclei from two atoms simultaneously attract the same pair of electrons. (Bonds involving one electron are possible but very rare.)
2. Simultaneous attraction of two nuclei for the same electrons is not sufficient for bonds to form. These attractive forces are always accompanied by repulsive forces between the two nuclei and between the electrons. For bonds to form, the forces of attraction must be greater than the forces of repulsion.
3. Another way of saying that the forces of attraction must be greater than the forces of repulsion is to say that bonds form when the energy level of the molecule is lower than the energy level of the isolated atoms.
4. As a practical matter, prediction of whether a bond can form between atoms is based on one of several approximate procedures. One of the simplest of these is to examine the electron configuration of the atoms and see whether there are "unfilled" orbitals that can overlap as the two atoms approach, thus increasing the concentration of electrons in the region of space between the two nuclei. This procedure leads to correct predictions for the number of bonds formed between many atoms.
5. "Number of bonds" refers to the number of pairs of electrons simultaneously attracted to the same two nuclei. One pair constitutes a "single" bond, two pairs constitute a "double" bond, and three pairs constitute a "triple" bond.

14.4 LEWIS STRUCTURES AND THE OCTET RULE

In Chapter 13 we called attention to the fact that noble gases do not form many compounds. There are no known compounds of helium, neon, or argon and only a few compounds of krypton and xenon. It was suggested that the electron configuration of these gases is particularly stable. In discussing the combining number of Group IA, Group IIA, and Group VIIA elements, we pointed out that these elements form stable ions by losing or gaining electrons to produce an electron configuration that is the same as the electron configuration of one of the noble gases. These observations have led to the "rule of eight" and the simple picture of bonding that we want to introduce.

The rule of eight or **octet rule** is popular because it is a simple picture that allows us to predict the bonding between certain atoms. The only assumptions are that

1. No more than eight valence electrons can be placed around the nucleus of any atom.
2. Bonds form until each atom acquires an electron configuration like that of one of the noble gases.

Hydrogen, lithium, and sometimes beryllium assume the helium configuration with only two valence electrons. Most other elements assume a configuration with eight valence electrons; hence the "octet" rule.

By working through the following example with me, you will see how the octet rule works.

Example 14.1

How are the atoms bonded in the Cl_2 molecule?

The structures that we will draw to represent molecules are based on the Lewis structures introduced in Chapter 13. Draw the electron-dot structure for two chlorine atoms.

- -

$$:\ddot{\underset{\cdot\cdot}{Cl}}\cdot \quad \cdot\ddot{\underset{\cdot\cdot}{Cl}}:$$

Each chlorine atom should be represented by the symbol Cl, and there should be seven dots around the symbol to represent the seven valence electrons. How the dots are arranged is not important, but it is conventional to show them in pairs on four sides of the symbol.

You must not attach too much importance to the placement of the dots. Remember, our model of atoms cannot reveal the exact location of electrons. Figure 14.6 is a better representation of what we would expect to see if we could look at these two chlorine atoms.

Figure 14.6 *Two chlorine atoms far apart.*

Figure 14.6 shows the spherical region of space containing all of the chlorine electrons. Five of these electrons are in 3*p* orbitals, which means that one electron must be alone in a 3*p* orbital. When the two atoms approach each other, the two 3*p* orbitals (one on each atom) overlap so that the electrons in these half-filled 3*p* orbitals are simultaneously attracted to the nuclei of both atoms. The two atoms stick together, heat is released, and we say that a bond has formed. This process is illustrated in Figure 14.7.

Under the diagram of each atom we have shown its electron-dot structure. The dots are drawn differently on the two atoms to make them easier to distinguish. All electrons are alike, however, and it is impossible to know whether the electrons located between the nuclei came from different atoms, as suggested by the drawing, or from just one of the atoms. What we do know is that the final arrangement of electrons is symmetrical, with the equivalent of two electrons located between the two atoms.

The dots shown between the two symbols for chlorine atoms represent **bonding electrons;** the other dots represent **nonbonding electrons.**

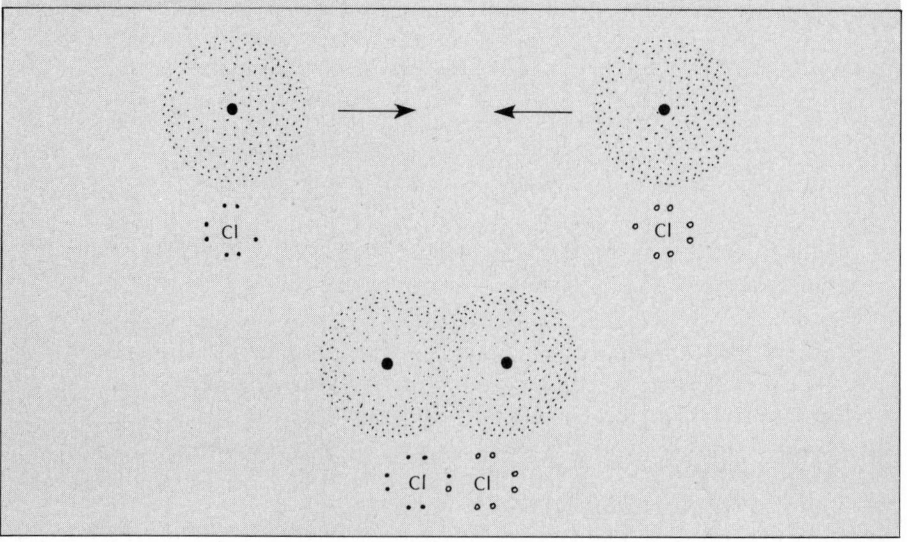

Figure 14.7 *Chlorine atoms bonding to form a molecule, Cl_2.*

These nonbonding electrons are sometimes referred to as lone pairs of electrons. One pair of bonding electrons represents a **single bond,** two pairs represent a **double bond,** and three pairs represent a **triple bond.**

On the basis of what we have done so far, we would say that two chlorine atoms bond together to form a single bond. This can be predicted from the octet rule. Bonding electrons are considered part of the octet around *both* atoms. They are counted twice, as shown in Figure 14.8.

Figure 14.8

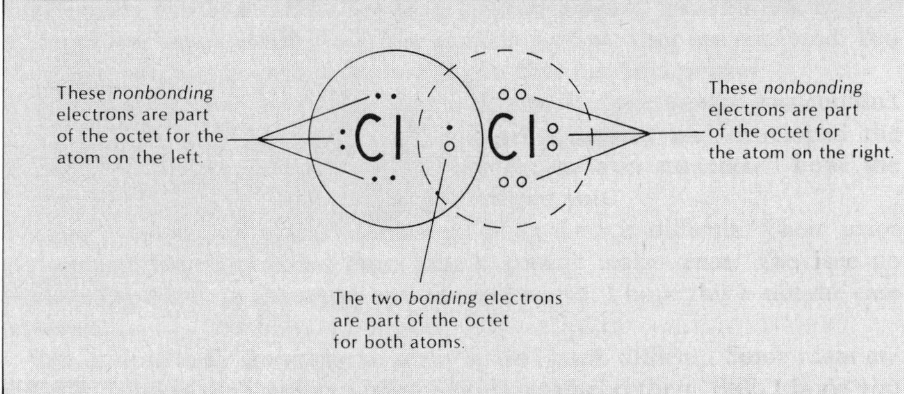

See whether you can draw an electron-dot structure for Cl_2 that satisfies the octet rule (has eight valence electrons around each atom) but *does not* have a single bond.

You should have found this impossible to do. If the proper number of electrons are shown with a double bond, you get something like this:

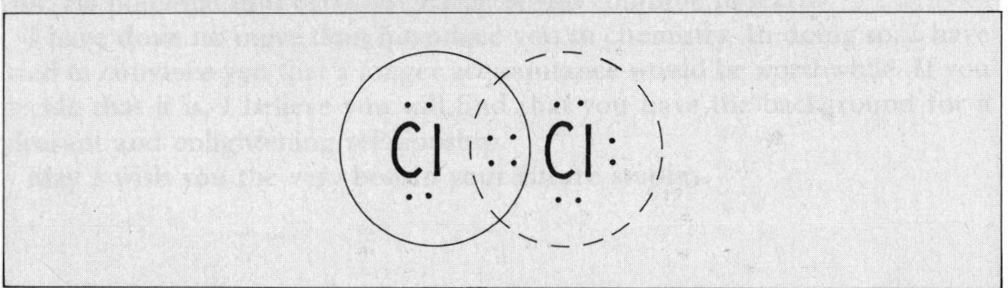

Since the bonding electrons are counted as part of the octet around both atoms, this structure produces too many electrons around one of the atoms. In some molecules, by contrast, it is necessary to predict a double or even a triple bond in order to have eight valence electrons around each atom.

Draw an electron-dot structure for two isolated nitrogen atoms.

$$\cdot\overset{\cdot}{\underset{\cdot\cdot}{N}}\cdot \qquad \overset{\cdot}{\underset{\cdot\cdot}{N}}:$$

You should show five dots around each symbol for nitrogen. The arrangement of the dots is not important, however, because these structures are not intended to indicate where electrons would be found in the atom. I have purposely drawn the dots around the two atoms in different ways to emphasize this point.

Now arrange these ten dots around the two symbols in a structure that could represent N_2 and satisfies the octet rule.

$$:N\vdots\vdots N:$$

Only a structure showing six electrons between the two atoms satisfies the octet rule. N_2 contains three bonds between the two nitrogen atoms. This structure is consistent with the experimental fact that the bond between two nitrogen atoms is very strong. It is much more difficult to break apart the nitrogen molecule than to break the bond between two chlorine atoms, which we predicted would be connected by a single bond. You can see that our simple model leads to predictions that conform to experimental facts.

Still, this model doesn't predict many things that interest us. For example, the octet rule cannot tell us that the formula of the nitrogen molecule is N_2 rather than N_4 or N_6. Figure 14.9 shows electron-dot structures that satisfy the octet rule, but neither of these structures represents a stable form of nitrogen.

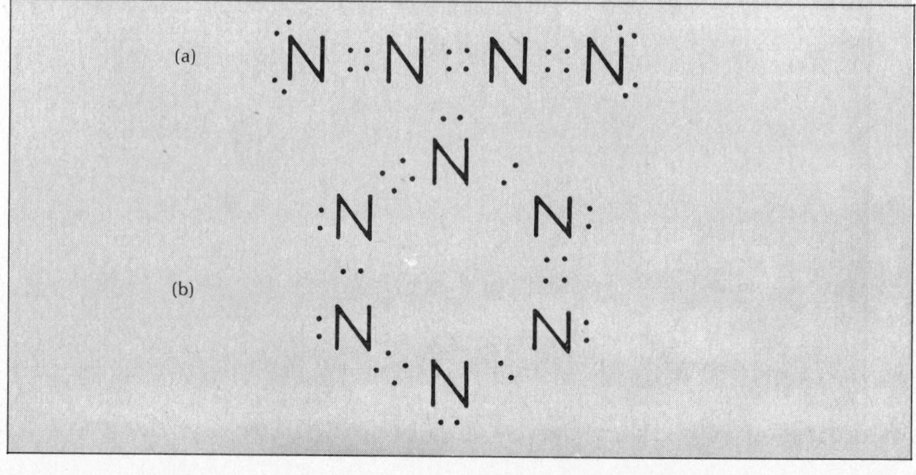

Figure 14.9
Lewis structures for nonexistent molecules. (a) A hypothetical structure for N_4. (b) A hypothetical structure for N_6.

Note that the structure proposed for N_6 has the six atoms bonded together in a ring. Although this structure does not exist, there are many compounds that contain carbon atoms bonded together in a ring like the one shown.

As one final example, draw an electron-dot structure for O_2.

$$\ddot{\underset{\cdot\cdot}{O}}::\ddot{\underset{\cdot\cdot}{O}}$$

The only electron-dot structure for O_2 that satisfies the octet rule has a double bond between the two oxygen atoms. A double bond is stronger than a single bond. Since four electrons are being simultaneously attracted to two nuclei, it takes more energy to pull the atoms apart. Furthermore, atoms connected by a double bond are closer together than atoms connected by a single bond. Experimental measurement of bond lengths and bond strengths suggests that the atoms in O_2 are connected by a double bond. There is a problem, though. Liquid oxygen is pulled into a magnetic field, as shown in Figure 14.10. Only substances with unpaired

Figure 14.10
Liquid oxygen in a test tube is drawn into the magnetic field of a strong magnet. This behavior is characteristic of substances that contain unpaired electrons. (Photos by *Tom Greenbowe and Keith Herron.*)

electrons in the molecule behave like this. The electron-dot structure that was drawn for O_2 does not show that there are unpaired electrons. Once again, we see that there are limitations to the electron-dot structures that we are drawing.

To be perfectly honest, the octet rule holds only for ionic compounds of elements in Groups IA and IIA, binary compounds of elements in Group VIIA, and compounds of carbon, nitrogen, and oxygen. Although many compounds of other elements *do* obey the octet rule, the exceptions are so numerous that predictions based on the rule are very dangerous for those elements.

Since there are so many exceptions to the octet rule, you may wonder why we bother with it at all. The answer is that, even though there are few elements that obey the octet rule all the time, these few elements form many compounds. *Millions* of important molecules contain only carbon, oxygen, and hydrogen. The octet rule predicts the electron-dot structures for these compounds very well, and the electron-dot structures tell us a great deal about the compounds. In spite of its many limitations, it is a useful rule.

Use the octet rule to draw electron-dot structures for the following molecules, and indicate the number of bonds joining the two atoms.

14-1. CO
14-2. Br_2
14-3. HCl
14-4. H_2O
14-5. H_2S

You probably had no trouble with these structures. As molecules become larger, predicting the electron-dot structure becomes more difficult. It is a good idea to approach the task in a systematic way. Here are some suggestions that have helped others.

RULES FOR WRITING LEWIS STRUCTURES

1. The first problem is to decide *which atoms are connected.* For example, which arrangement is correct for CO_2?

$$\text{C—O—O} \quad \text{or} \quad \text{O—C—O}$$

Only a sophisticated chemical analysis of compounds can tell us for *certain* how atoms are connected, but the following generalizations usually lead to correct predictions.

a. When several atoms of one kind are connected to one atom of another kind, the single atom is in the center. In other words, in CO_2 we predict that C is between the two oxygen atoms, and in SO_3 we predict that S is in the center of the molecule with each oxygen connected to the sulfur.

b. If one arrangement is more symmetrical than another, the more symmetrical arrangement is more likely. O—C—O is more symmetrical than C—O—O, so O—C—O is more likely.

Using these rules, decide which of the following arrangements is more likely.

14-6. (a) S—O—O with O above and O below the central O or (b) O—S—O with O above and O below S

```
         O                  O
         |                  |
(a) S—O—O      or   (b) O—S—O
         |                  |
         O                  O
```

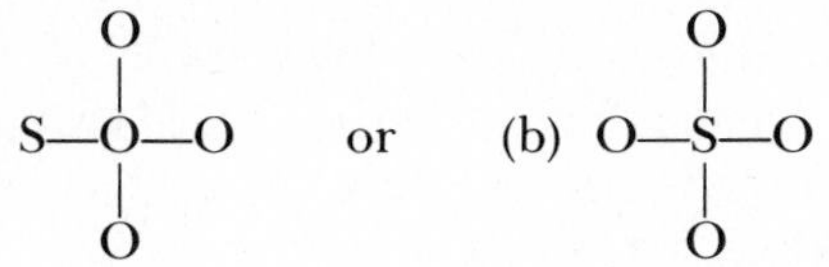

14-7. (a) O—C—O—O or (b) C bonded to three O atoms

```
                              O     O
                               \   /
(a) O—C—O—O     or     (b)      C
                                |
                                O
```

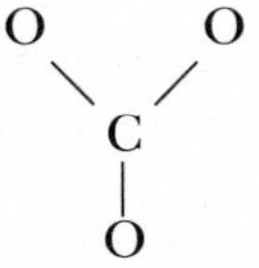

14-8. (a) H—O—O—H or (b) H—O—H—O

14-9. (a) N—N—O—O or (b) O—N—N—O

14-10. (a) Cl—P—Cl—Cl with Cl above and Cl below P or (b) P bonded to five Cl atoms

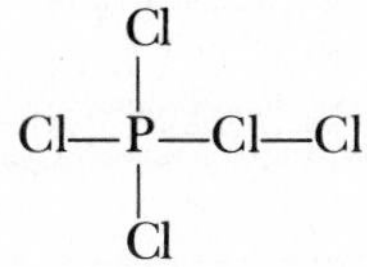

```
          Cl
          |
(a) Cl—P—Cl—Cl     or    (b)  PCl5 (P surrounded by Cl, Cl, Cl, Cl, Cl)
          |
          Cl
```

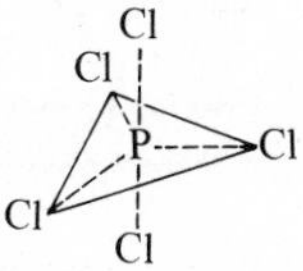

2. After deciding which atoms are connected, the next problem is to decide how many dots should appear in the electron-dot structure. For neutral molecules, the number of dots is the sum of the *valence* electrons for all atoms in the molecule. In CO_2 we would show sixteen dots: four for the carbon atom and six for each of the oxygen atoms.

How many dots should appear in the electron-dot structure for each of the following molecules?

14-11. SO_3

14-12. CH_4

14-13. H_2SO_4

14-14. PH_3

In the case of ions, dots must be added or subtracted to reflect the charge on the ion. For the sulfate ion (SO_4^{2-}), we would need to show thirty-two dots: six for sulfur, six for each oxygen, and two more for the two electrons that give the ion its 2− charge.

How many dots appear in the electron-dot structure for each of the following ions?

14-15. NH_4^+

14-16. CO_3^{2-}

14-17. NO_3^-

14-18. PO_4^{3-}

---

3. After calculating the total number of dots that must appear in the structure, you must decide where to place the dots so that all atoms obey the octet rule. This can be done by trial and error, but for more complicated structures, it is better to do it systematically. The entire procedure will become clearer as you work through the following example.

Example 14.2

Show the electron-dot structure for carbon monoxide.

The first step is to show which atoms are connected. There are only two atoms in our example, so there is no doubt about this point. What is the second step in writing the structure?

- - -

You must count the valence electrons in each atom to determine the number of dots in the structure. How many valence electrons must you show for CO?

- - -

Ten. Carbon has four valence electrons and oxygen has six. See whether you can arrange these ten electrons so that both carbon and oxygen have an octet around them.

- - -

:C:::O:

Did you have difficulty in arranging the electrons? Here is a systematic procedure that will work.

After you have counted the valence electrons, decide how many electrons would have to be shown to obey the octet rule *without any bonds*. In this case there are two atoms in the structure, and each requires eight electrons. You would need sixteen electrons to place an octet around the two atoms.

Now find the difference between the actual number of valence electrons and the number needed for an octet around each atom. What would this difference be for CO?

- - -

$$16 - 10 = 6$$

This difference is the *number of bonding electrons*. There must be six electrons located between the two atoms. This is what was shown in the structure.

Let's go through the entire procedure with a second example. What is the electron-dot structure for the phosphate ion?

First, predict how the atoms will be connected.

- - -

```
    O
    |
O — P — O
    |
    O
```

In this case, several atoms of one kind are connected to one atom of a different kind. We predict that the single atom is the central atom and that the like atoms are arranged symmetrically around this central atom.

Now that you have decided how the atoms should be connected, how many electrons must be shown in the structure?

Thirty-two. Each oxygen atom has six valence electrons for a total of twenty-four. The phosphorus atom has five valence electrons, and the 3− charge on the ion indicates that there are three additional electrons in the structure. The twenty-four for the oxygen atoms, five for the phosphorus atom, and three for the charge on the ion give a total of thirty-two.

How many electrons would be needed to place an octet around these five atoms if there were no bonds?

$$5 \times 8 = 40$$

There are thirty-two electrons to be shown in the structure, and forty are needed for the five octets. How many bonding electrons must be shown in the structure?

$$40 - 32 = 8$$

Begin by showing a single bond between each pair of atoms that are connected. If there are bonding electrons left, start forming double bonds.

Show how the bonding electrons should be placed in PO_4^{3-}.

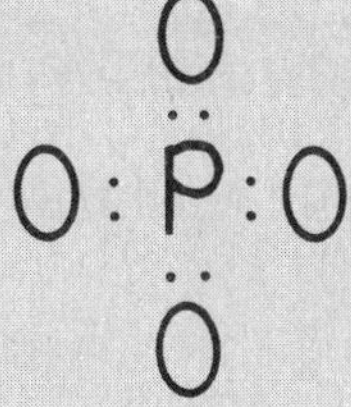

All that remains is to add electrons to complete each octet. Show the final structure, and check to be sure that you have shown the proper number of electrons.

```
       ..
      :O:
   ..  ..  ..
  :O:P:O:
   ..  ..  ..
      :O:
       ..
```

There are thirty-two dots shown, which is the number that should be shown.

Practice by drawing electron-dot structures for the molecules and ions listed in Questions 14-11 through 14-18. Check your drawings with those shown as answers for Questions 14-19 through 14-26.

14.5 SHAPES OF MOLECULES

In drawing the Lewis structures for molecules and ions, you have already dealt with one question that we wanted to answer. With a few simple rules, we predicted the structure for several molecules and ions. Lewis structures tell us nothing about the shape of molecules, but a simple extension of the theory allows us to predict the general shape of many molecules and ions.

In order to predict molecular shape from Lewis structures, we begin with two assumptions. First, we assume that the electrons making up the octet around an atom are as far apart as possible. This is reasonable, because all electrons have the same electrical charge and like charges repel. The second assumption is that electrons remain in pairs. Most of the electrons that we will deal with are electrons involved in bonding. They are held close together by attraction to the positively charged nuclei of the atoms that are bonded.

If you experiment with a ball to represent the core of an atom and smaller spheres to represent electrons, you will find that four pairs of electrons are arranged as shown in Figure 14.11 when they are as far apart as you can get them. This places the nucleus of the atom at the center of a tetrahedron and the four electron pairs at the corners of the tetrahedron, as shown in Figure 14.12.

By assuming that the electrons in an octet are arranged as shown in Figure 14.12, you can predict the shape of many molecules with considerable accuracy. Do this in the following example.

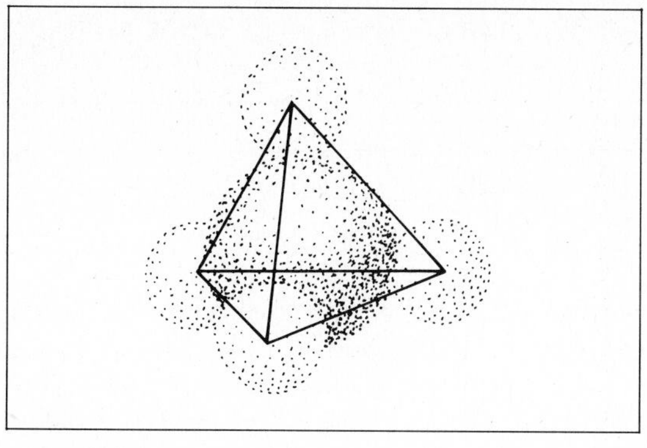

Figure 14.11
Four balls placed on a larger ball in such a way that they are as far apart as possible. The four smaller balls are at the four corners of a regular tetrahedron.

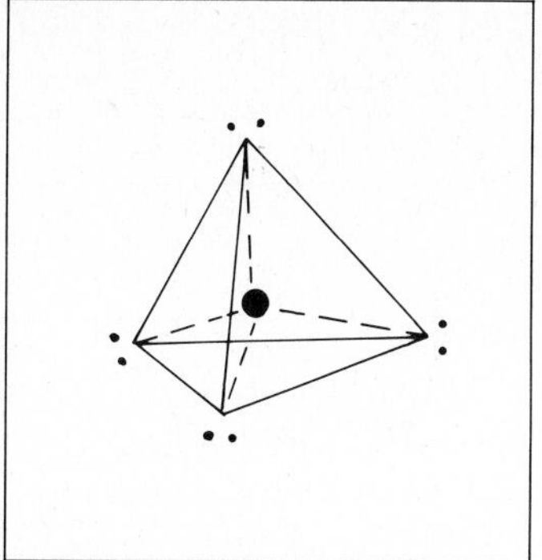

Figure 14.12

Example 14.3

What is the shape of the water molecule?

First, you need the electron-dot structure for water. Draw it.

$$\mathrm{H}:\ddot{\underset{\cdot\cdot}{\mathrm{O}}}:\mathrm{H}$$

Now arrange the pairs of electrons at the corners of a tetrahedron, and predict the shape of the molecule.

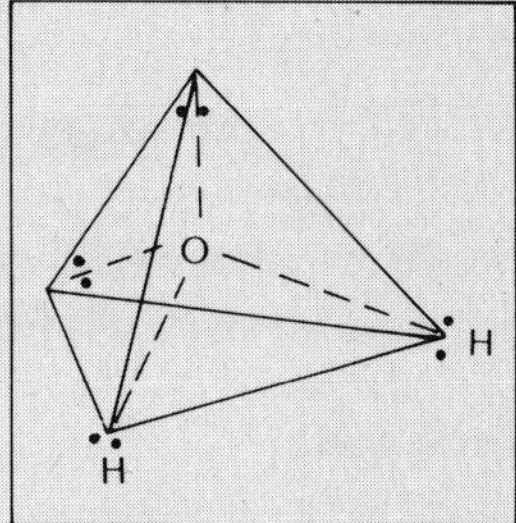

We might describe the water molecule as bent. It does not have the three atoms in a straight line like this: H—O—H. Rather, the three atoms form an angle: H⟋O⟍H. The angle formed is called the **bond angle.** From the shape of a tetrahedron, we would predict that the size of the angle formed is 109°28′. The actual angle is 104°30′. For a simple model, this is very good agreement. (Slight modifications of the model produce even better agreement with experimental data.)

Now use this model to predict the shape of each of the following molecules.

14-27. NH_3
14-28. H_2S
14-29. CH_4
14-30. HCl

All of the structures that you just drew involved molecules with single bonds. The molecule with only two atoms (HCl) must be linear. The molecule with two atoms connected to a central atom (H_2S) must be bent like water. The molecule with three atoms connected to a central atom (NH_3) could be described as a pyramid with the central atom at the apex and the other atoms at each corner of the triangular base. The molecule with four atoms connected to a central atom (CH_4) can be described as a central atom bonded to atoms at the corners of a tetrahedron. Other molecules with the same number of atoms bonded to a central atom by single bonds have the same shape as these molecules. When double or triple bonds are involved, other shapes result. Example 14.4 shows how to predict the shape for molecules that contain multiple bonds.

Example 14.4

What is the shape of the carbon dioxide molecule?

First, draw the electron-dot structure for CO_2.

O::C::O

Carbon is connected to each oxygen atom by a double bond. As long as you keep the four pairs of electrons at the four corners of a tetrahedron, the only way to get four electrons between two atoms is to place one edge of each tetrahedron in contact with an edge of the other tetrahedron, as shown in Figure 14.13.

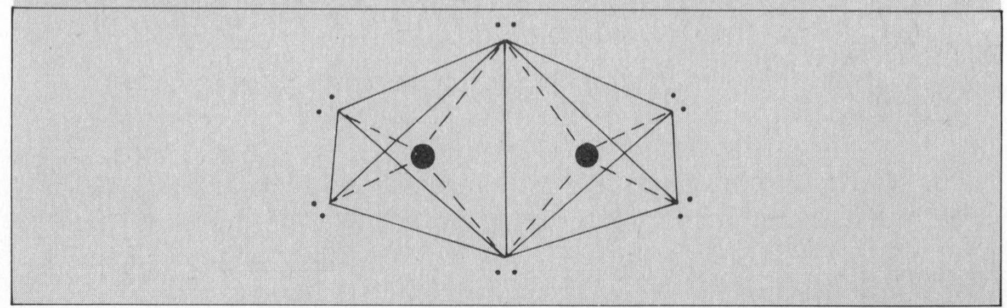

Figure 14.13
Two tetrahedra joined along an edge will place two pairs of electrons between the two atoms, representing a double bond.

What do you predict for the shape of CO_2?

It should be linear: O—C—O. This is apparent in the arrangement of tetrahedra shown in Figure 14.14.

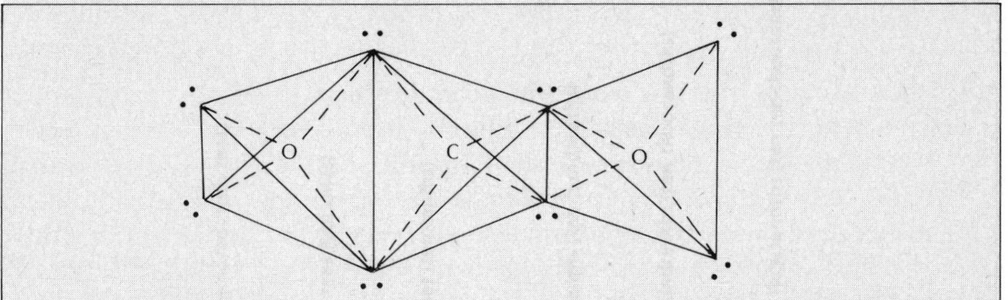

Figure 14.14 *The shape of the carbon dioxide molecule.*

Now consider a way to represent a triple bond, such as that found in CO. You must place three pairs of electrons between the atoms. What parts of the tetrahedra must come into contact to make this possible?

One of the faces of each tetrahedron. The triple bond in CO is represented in Figure 14.15.

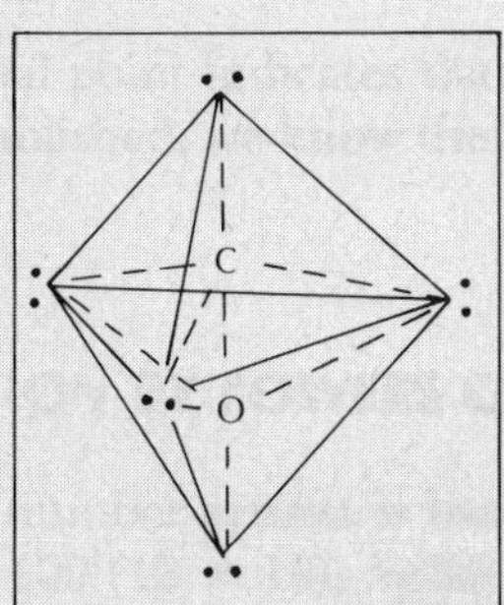

Figure 14.15 *The shape of the carbon monoxide molecule.*

Now predict the shape of each of the following molecules and ions.

14-31. SO_2
14-32. PCl_3
14-33. SO_3
14-34. NH_4^+
14-35. HCN

We have limited our discussion of molecular shape to those molecules that have an octet of valence electrons—those molecules that obey the octet rule. However, the basic notion introduced here can be extended to account for the shape of molecules such as BeH_2, BF_3, PCl_5, ICl_3, SF_6, and XeF_4, none of which obey the octet rule.

14.6 COVALENT AND POLAR-COVALENT BONDS

We have said that any chemical bond results from two nuclei simultaneously attracting the same pair of electrons. We have used Lewis structures to depict these bonding electrons between the nuclei that are bonded. The question that we address now is whether the two nuclei attract the bonding electrons equally.

Consider again the H_2 molecule, with the electron-dot structure

$$H{:}H$$

It seems reasonable to assume that the two electrons in the molecule are equally attracted to each nucleus. After all, the two nuclei are identical. Each nucleus has the same electrical charge, producing the same force of attraction between the bonding electrons and each nucleus. On the average, the electrons should be the same distance from each nucleus. We say that the electrons are equally shared between the two nuclei, and we describe the bond as a **covalent bond.**

Using the same argument, we could predict that the single bond connecting the chlorine atoms in Cl_2 is covalent. Since the two chlorine atoms are identical, the bonding pair of electrons would be equally shared by the two atoms.

Now consider the compound formed between hydrogen and chlorine, HCl. The electron-dot structure for HCl is

$$H{:}\ddot{\underset{..}{Cl}}{:}$$

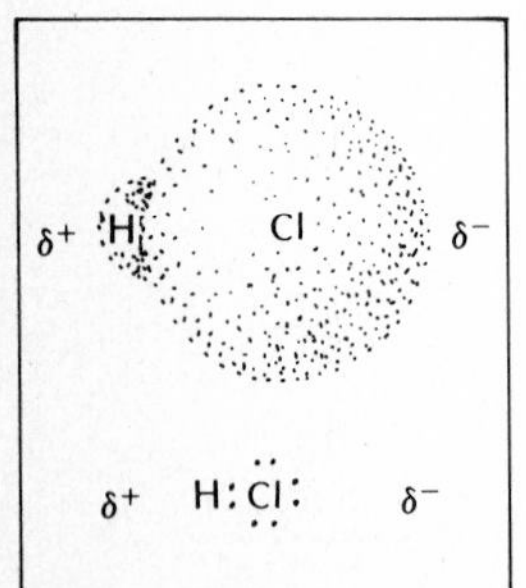

Figure 14.16
The HCl molecule is polar. It has a partial positive charge on the hydrogen end and a partial negative charge on the chlorine end.

Since hydrogen has only one proton in its nucleus and chlorine has seventeen protons in its nucleus, there is little reason to assume that the pair of electrons involved in bonding is shared equally between the two nuclei. Experimental observation suggests that they are not. It appears that the bonding electrons are attracted much more by the chlorine nucleus than by the hydrogen nucleus. On the average, the bonding electrons spend more time near the chlorine nucleus than near the hydrogen nucleus. The electrical charge is not evenly distributed throughout the molecule, and it has positive and negative ends, as suggested by Figure 14.16. We say that the molecule is **polar.**

The lower-case Greek letter delta (δ) is used to mean "partial." Figure 14.16 indicates that the HCl molecule has a partial positive charge on the hydrogen end of the molecule and a partial negative charge on the chlorine end. Because of these partial charges, HCl molecules tend to "line up" as the positive end of one molecule attracts the negative end of another.

Polarity of chemical bonds can produce important chemical properties in molecules. For one thing, if molecules have slightly positive and negative ends, there is more attraction than usual between the molecules. It should take more energy to separate the molecules, and we might expect to see higher boiling points for such substances.

Many of the properties of water are due to its polar character. You have already seen that the water molecule is bent. It is also true that the bonds in the water molecule are polar. As a result, there is a partial positive charge on the side with the hydrogen atoms and a partial negative charge on the oxygen side of the molecule, as shown in Figure 14.17.

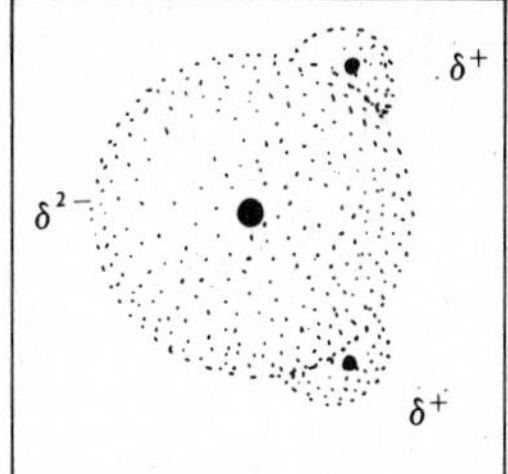

Figure 14.17
The water molecule, showing its polar nature.

The polar nature of water partially accounts for the solubility of ionic compounds in water. The water molecules surround the positive and negative ions and are attracted to them as suggested by Figure 14.18. The attachment of the water molecules is normally not strong, but it is sufficient to supply some of the energy needed to separate the oppositely charged ions that are found in a crystal of an ionic solid.

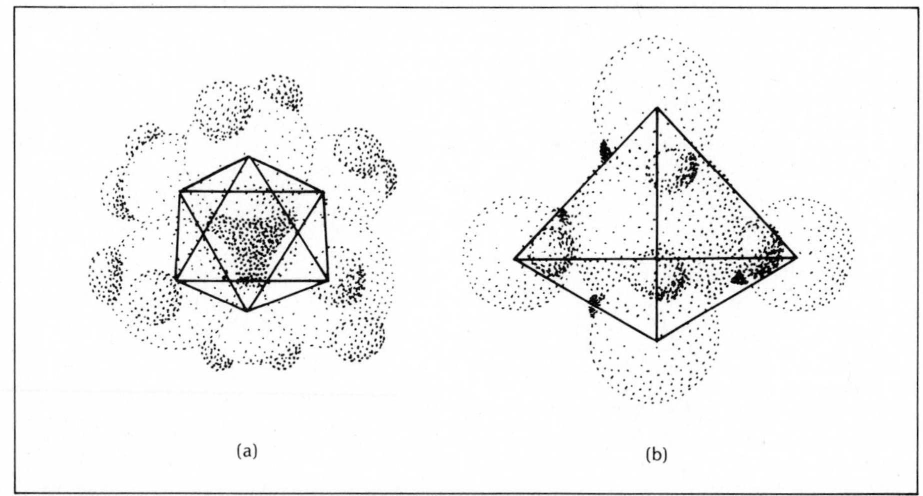

Figure 14.18
(a) In solution, sodium ions (Na^+) are surrounded by six water molecules. (b) Chloride ions (Cl^-) are surrounded by four water molecules. Note that the negative side of the water molecules is attracted to the Na^+ and that the positive side is attracted to the Cl^-.

POLAR BONDS AND POLAR MOLECULES

Some students assume that a molecule that has polar bonds is a polar molecule. That is not necessarily so. A molecule is polar if the center of positive charge does not coincide with the center of negative charge. This is the case for both HCl and H_2O, but CH_4 is not a polar molecule, even though it has four polar bonds. Figure 14.19 may help you see why.

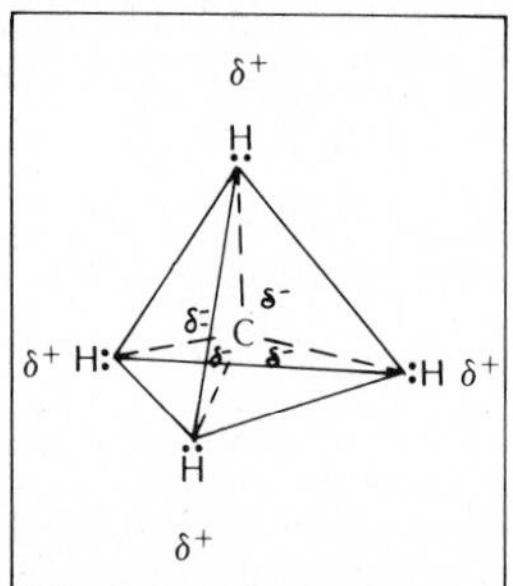

Figure 14.19
All of the bonds in CH_4 are polar, but the molecule itself is nonpolar because of the symmetrical arrangement of the molecule.

Each of the four bonds has a positive and negative end, but they are symmetrically arranged around the center of the molecule. As a result, the center of positive charge and the center of negative charge are both at the center of the molecule. The molecule is not polar.

So far we have not explained how you can tell which end of a polar bond is positive and which end is negative. It is no simple matter to predict which of two nuclei has the greater attraction for bonding electrons. In order to predict which end of a polar bond is positive and which is negative, chemists have devised a scale called electronegativity. The electronegativity of each element is listed in Table 14.1.

Table 14.1. Electronegativities of Elements

H																
2.20																
Li	Be											B	C	N	O	F
0.98	1.57											2.04	2.55	3.04	3.44	3.98
Na	Mg											Al	Si	P	S	Cl
0.93	1.31											1.61	1.90	2.19	2.58	3.16
K	Ca	Sc	Ti	V	Cr	Mn	Fe	Co	Ni	Cu	Zn	Ga	Ge	As	Se	Br
0.82	1.00	1.36	1.54	1.63	1.66	1.55	1.83	1.88	1.91	1.90	1.65	1.81	2.01	2.18	2.55	2.96
Rb	Sr	Y	Zr	Nb	Mo	Tc	Ru	Rh	Pd	Ag	Cd	In	Sn	Sb	Te	I
0.82	0.95	1.22	1.33	1.6*	2.16	1.9*	2.2*	2.28	2.20	1.93	1.69	1.78	1.96	2.05	2.1*	2.66
Cs	Ba	*	Hf	Ta	W	Re	Os	Ir	Pt	Au	Hg	Tl	Pb	Bi	Po	At
0.79	0.89		1.3*	1.5*	2.36	1.9*	2.2*	2.20	2.28	2.54	2.00	2.04	2.33	2.02	2.0*	2.2*
Fr	Ra															
0.7*	0.9*															

*Values indicated by an asterisk are taken from Pauling, L. *Nature of the Chemical Bond,* 3rd ed. Ithaca: Cornell University Press (1960), p. 93. All other values are from Allred, A. L. "Electronegativity Values from Thermochemical Data," *Journal of Inorganic Nuclear Chemistry,* 1961, Vol. 17, pp. 215–222.

14.7 ELECTRONEGATIVITY

The tendency of an atom to attract bonding electrons is called the **electronegativity** of the element. Electronegativity cannot be measured directly, and the values given in Table 14.1 are on a relative scale proposed by Linus Pauling.

Note that the element with the highest electronegativity is fluorine, located in the upper right corner of the periodic table. The element with the lowest electronegativity is francium, located in the lower left corner. As a rule, electronegativity increases as we go from left to right on the periodic table and decreases as we go from top to bottom. Thus it is usually possible to predict the relative electronegativity of two elements just by locating the elements in the periodic table.

The noble gases are not shown in Table 14.1. With few exceptions, these elements do not form compounds. Since electronegativity is a measure of the attraction for *electrons involved in bonding,* a value for an element that does not form bonds does not make sense.

ELECTRONEGATIVITY AND BOND POLARITY

When identical atoms are connected, the bond formed is described as nonpolar. The bonding electrons are equally shared between the two atoms. The two atoms have the same electronegativity, and the two atoms have the same tendency to attract the bonding electrons.

When two elements that have similar electronegativity are bonded, the bonding electrons are almost equally shared. The bond may be described as a covalent bond or a **polar-covalent bond,** depending on how large the difference in electronegativity is. If the bond is described as polar, the element with the higher electronegativity is the negative end of the bond.

14.8 IONIC BONDS

When the difference in electronegativity between two elements is large (greater than 1.7 on the Pauling scale), the influence of the more electronegative element is so great and the influence of the less electronegative element is so small that it appears as though the bonding electrons have actually been transferred to the element of higher electronegativity. When this happens, positive and negative ions form, and the resulting bond between atoms is described as an **ionic bond.**

Sodium has an electronegativity of 0.93, and chlorine has an electronegativity of 3.16. The electron-dot structure for NaCl is

$$\mathrm{Na}\!:\!\underset{\cdot\cdot}{\overset{\cdot\cdot}{\mathrm{Cl}}}\!:$$

However, we know that NaCl exists as ions when the salt is melted, and we know that it forms ions when dissolved in water. The difference in electronegativity between sodium and chlorine is 2.23. This is enough to suggest that the sodium atom has very little influence over the bonding electrons. We generally assume that, even in solid NaCl, the bonding electrons are totally under the control of the chlorine atom, and we represent NaCl as

$$Na^+ \quad :\ddot{\underset{..}{Cl}}:^-$$

There is no sharp boundary between ionic and covalent bonds. All bonds between atoms involve some sharing of electrons, and we can say that there is some covalent character to every bond. Except in the case of identical atoms bonded together, one of the atoms in the bond has more attraction for the electrons than the other atom has. The bond can be described as polar, or it can be described as having some ionic character. When the polarity of the bond becomes very large, the bond is generally described as ionic.

IONIC BONDS AND IONIC COMPOUNDS

We must exercise some caution in discussing ionic and covalent compounds. We are much safer when we discuss ionic and covalent *bonds*. Generally any compound that exists as ions in the liquid state is said to be ionic. This does *not* mean that all *bonds* in the compound are ionic. Sulfuric acid, H_2SO_4, is certainly considered an ionic compound. As a liquid, it exists primarily as H^+ and HSO_4^- ions. Still, the bonds between the oxygen atoms and the sulfur atoms are best described as covalent rather than ionic. Only the bonds between the hydrogen atoms and the oxygen atoms are ionic.

We can't be certain that a substance that conducts an electric current in a liquid state exists as ions in the solid state. Even though melted salt is an excellent conductor of electricity, leading us to conclude that it exists as ions, there is the possibility that these ions are formed as a result of the melting process.

We know that some substances that appear to be covalent compounds can form ions when placed in solution. Ions are formed during the dissolving. Pure HCl is a gas at room temperature, but it can be liquefied. When it is liquefied, it does not conduct an electric current. It is a covalent molecule. When HCl is dissolved in water, the solution is an excellent conductor of electricity. There is evidence that the solution contains $H^+(aq)$ and $Cl^-(aq)$. It would appear that, when polar water molecules and polar HCl molecules come together, the interaction causes the HCl molecule to break apart into ions. (One explanation is that water acts as a base and HCl acts as an acid, as explained in Chapter 11.)

When HCl is pure, it exists as one kind of particle: HCl molecules. When HCl dissolves in water, it exists as different particles: H^+ and Cl^- ions. Evidently a chemical reaction occurs when HCl dissolves in water.

Here we see one of the difficulties in beginning chemistry. We have described the process of something dissolving in water as a *physical* change.

We have said that dissolution involves the mixing of particles rather than the formation of any new particles. In most instances we believe that this is true. In this case, however, the evidence suggests that new particles *are* formed. This particular process of forming a solution is probably best described as a *chemical* change. Nature seems to defy fixed rules. We seem forced—at least in a beginning course—to tell half-truths about nature in order to keep our descriptions simple enough to be understood. We oversimplify by saying that something dissolving in water is undergoing a physical change, even though many examples of dissolution result in new particles. Actually, our whole discussion of chemical and physical change is an oversimplification. The article at the end of Chapter 12 suggests why.

14.9 OXIDATION NUMBER

While in the process of wrapping up loose ends, let us return to "combining number," a term that you learned early in this course. When that term was introduced, some of you may have called it "oxidation number" from a previous course in chemistry. It is true that the combining number and the oxidation number of an element are numerically the same. However, combining numbers are always positive. (How can an element combine with a negative number of hydrogen atoms?) Oxidation numbers for metals are positive, and oxidation numbers for nonmetals and groups that act as nonmetals are negative. Now that we have introduced the idea of electron-dot structures and electronegativity, you can understand how these positive and negative oxidation numbers are derived.

The oxidation number of an atom is determined by pretending. The pretense we make is that *all* bonds are ionic. Playing this game of "pretend-that-all-bonds-are-ionic," we assume that all bonding electrons belong to the atom with the higher electronegativity and determine what the charge on each atom would be. We then define that charge as the **oxidation number** of the element. Let's look at some examples.

Water will do for a start. With circles representing the electrons that were originally on the hydrogen atoms, the electron-dot structure for water is

$$\mathrm{H\,{}^{\circ}_{\bullet}\ddot{\underset{\bullet\bullet}{O}}{}^{\bullet}_{\circ}\,H}$$

Oxygen is more electronegative than hydrogen, so we pretend that the bonding electrons belong to the oxygen:

$$\overset{1+}{\mathrm{H}}\left[{}^{\circ}_{\bullet}\ddot{\underset{\bullet\bullet}{\mathrm{O}}}{}^{\bullet}_{\circ}\right]^{2-}\overset{1+}{\mathrm{H}}$$

You can see that this places eight valence electrons around oxygen. The oxygen atom appears to have gained two electrons, and the charge on the oxygen would be 2−. We say that oxygen has an oxidation number of 2− in water.

Looking at the hydrogen, we see no electrons are assigned to the hydrogen atoms. Each of the hydrogen atoms appears to have lost an electron. The charge on the hydrogen is 1+, and we say that the oxidation number of hydrogen is 1+ in water.

Now let us look at a more complicated example, H_2SO_4 (sulfuric acid). Here is the electron-dot structure, with small circles representing the electrons that were originally on the hydrogen and sulfur atoms:

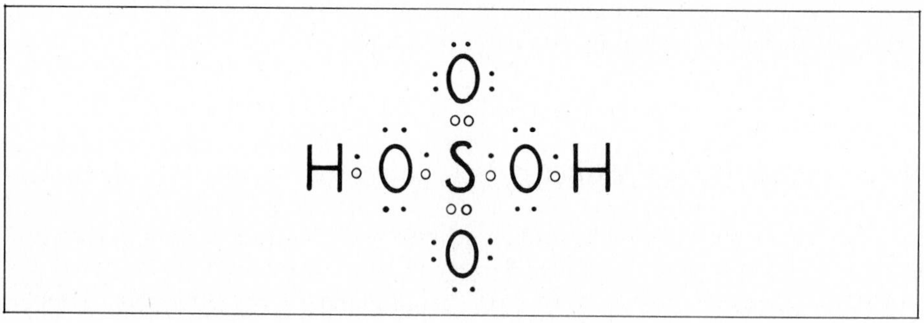

```
          ··
         :O:
          °°
     ··        ··
  H°:O°:S°:O:°H
     ··   °°   ··
         :O:
          ··
```

Oxygen has a higher electronegativity than sulfur, so *all* of the bonding electrons between oxygen and sulfur are assigned to the oxygen atoms. Oxygen also has a higher electronegativity than hydrogen. Consequently, the bonding electrons between hydrogen and oxygen are assigned to the oxygen. This leaves the hydrogen and the sulfur with no valence electrons and the oxygen with eight valence electrons, as indicated here:

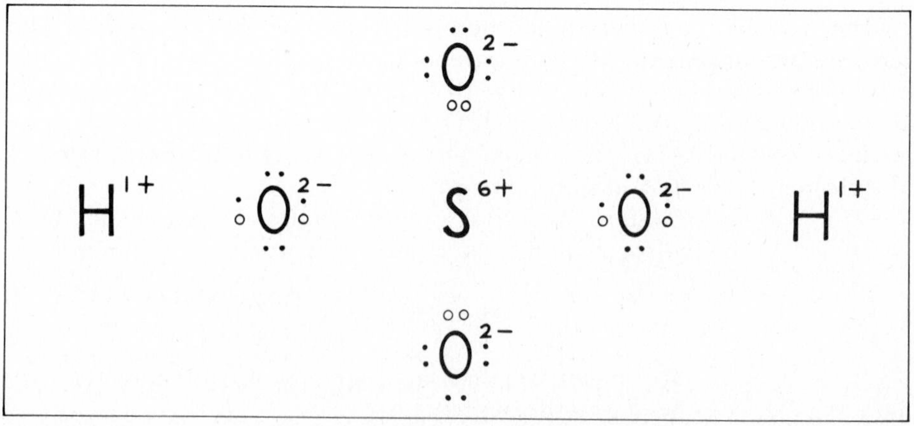

```
                       ··
                      :O:2−
                       °°

  H1+      ··2−                    ··2−        H1+
          °O:         S6+         °O:
           ··                      ··

                       °°
                      :O:2−
                       ··
```

Hydrogen has *lost one electron* and has an oxidation number of 1+. Sulfur has *lost six electrons* and has an oxidation number of 6+. Each oxygen atom has *gained two electrons* and has an oxidation number of 2−.

One might logically ask, "What kind of pretending do we do when identical atoms are joined?" Since there is no difference in electronegativity between identical atoms, we have no reason to assign the bonding electrons to one atom in preference to the other. In H_2, for example, the electron-dot structure is

$$\mathrm{H{:}H}$$

Each hydrogen has equal claim to the bonding electrons, so we distribute them equally, one to each hydrogen:

$$\mathrm{H\,{\cdot}|{\cdot}\,H}$$

In this case, neither atom has lost nor gained an electron, and we conclude that the charge is 0. The oxidation number is also 0. Since all pure elements are composed of identical atoms, all pure elements have an oxidation number of 0.

With few exceptions, hydrogen has an oxidation number of 1+. This is because there are few compounds in which hydrogen is bonded to an atom of lower electronegativity. When hydrogen does bond to elements of lower electronegativity, it has an oxidation number of 1−. This can be illustrated with sodium hydride, NaH. The electron-dot structure is

$$\mathrm{Na{:}H}$$

Hydrogen *does* have a higher electronegativity than sodium, so the bonding electrons are assigned to the hydrogen atom:

$$\overset{1+}{\mathrm{Na}}\ \overset{1-}{\mathrm{{:}H}}$$

This gives hydrogen one more electron than proton, and it has a charge of 1−. The oxidation number is also 1−.

Only one element, fluorine, has an electronegativity greater than that of oxygen. When oxygen is bonded to fluorine, oxygen assumes a positive oxidation number. Such compounds are rare. However, oxygen has an oxidation number of 1− on occasion, rather than its normal value of 2−. This happens when the compound contains oxygen atoms bonded to oxygen atoms. An example is hydrogen peroxide. The formula of hydrogen peroxide is H_2O_2, and the electron-dot structure is

$$\mathrm{H{:}\ddot{\underset{\cdot\cdot}{O}}{:}\ddot{\underset{\cdot\cdot}{O}}{:}H}$$

The electrons bonding hydrogen to oxygen are assigned to the oxygen atom as usual, but the electrons bonding oxygen to oxygen must be split between the two atoms because there is no difference in electronegativity.

$$\begin{array}{cccc} 1+ & 1- & 1- & 1+ \\ \text{H} & :\ddot{\text{O}}: & \ddot{\text{O}}: & \text{H} \end{array}$$

The result is that each oxygen atom gains only one electron and its charge is 1−. All peroxides contain an oxygen atom bonded to an oxygen atom, and the oxidation number of oxygen in these compounds is 1−.

It is inconvenient to write electron-dot structures every time we want to know the oxidation number of an element. Consequently, a number of generalizations are usually memorized to predict oxidation numbers.

RULES FOR ASSIGNING OXIDATION NUMBERS

1. The oxidation number of a monatomic ion is equal to the charge on the ion. (The oxidation number of Cl^- is 1−, that of O^{2-} is 2−, and so on.)
2. The oxidation number of any pure element is 0.
3. The oxidation number of elements in Group IA is 1+ when these elements are in a compound.
4. The oxidation number of elements in Group IIA is 2+ when these elements are in a compound.
5. The normal oxidation number of oxygen in compounds is 2−.
6. The normal oxidation number of hydrogen in compounds is 1+.
7. The oxidation numbers of any other elements in a molecule or ion are selected to make the sum of the oxidation numbers equal to the charge on the molecule or ion.

Use these rules to find the oxidation number of the specified element in each of the following compounds.

14-36. S in H_2SO_3	14-40. C in $HC_2H_3O_2$	14-44. N in NO
14-37. S in SO_3	14-41. C in CO_2	14-45. N in NO_2
14-38. S in S_8	14-42. C in C_2H_2	14-46. N in HNO_3
14-39. C in $MgCO_3$	14-43. N in N_2O	14-47. N in NH_3

SUMMARY

We began this chapter by outlining things that we would like to know about bonding. We proceeded to develop a model that would tell us some of them.

Even though laboratory analysis is the only way to determine the formula of a compound, knowledge of atomic structure and a theory of bonding enable us to predict several formulas. Limited though it is, the octet rule provides an explanation for the formulas of many compounds. In particular, the formulas of compounds between metals in Groups IA and IIA and the halogens in Group VIIA are easily explained by assuming that each atom acquires a noble gas configuration with eight valence electrons (two valence electrons for Li, Be, and H). These compounds are ionic, and our theory of bonding can explain the difference between ionic compounds and covalent compounds.

All bonds result from the simultaneous attraction of two nuclei for the same pair of electrons. When the nuclei are identical the attraction must be equal, but when the two atoms are not the same the attraction is unequal. The electronegativity of an element is an indication of that element's attraction for bonding electrons. The difference in electronegativity between two atoms indicates just how unequal their sharing of electrons is. When the difference is 1.7 or greater, the bonding electrons are almost totally under the influence of the element of higher electronegativity, and the bond is described as ionic. This is the kind of bond formed between the halogens in Group VIIA and most metals.

Only bonds between identical atoms are purely covalent, because only identical atoms have identical electronegativity. All other bonds result in unequal sharing. Such bonds are called polar-covalent bonds. There is a partial positive charge at one end (pole) of the bond and a partial negative charge at the other end.

Polar bonds do not necessarily result in polar molecules. If the shape of the molecule is symmetrical, several polar bonds can be arranged so that the center of negative charge and the center of positive charge coincide. Such molecules are not polar. Clearly the shape of a molecule is important.

Although the octet rule does not predict the shape of molecules, the suggestion that the eight valence electrons are located at the corners of a regular tetrahedron provides a basis for predicting shapes. This simple model gives amazingly accurate predictions. It even explains the experimental fact that a single bond is longer than a double bond, which is longer than a triple bond. A single bond results from tetrahedra joined at the tips, as shown in Figure 14.20(a), page 448. A double bond results from tetrahedra joined along an edge, as shown in Figure 14.20(b), page 448. And a triple bond results from tetrahedra joined at one of the faces, as shown in Figure 14.20(c), page 448. The nuclei of the atoms, located in the center of the two tetrahedra, are closer together in (b) and (c) than they are in (a).

Double and triple bonds are actually an invention to explain the fact that some bonds are more difficult to break than others. It seems reasonable that a bond formed by simultaneous attraction of nuclei for *four* electrons would be stronger than a bond formed by attraction for only *two* electrons. Simultaneous attraction for *six* electrons should be even stronger. Once we know what atoms are connected, the octet rule does a reasonably good job of predicting the number of double and triple bonds in a molecule. A modification of the rule to allow for "resonance" structures will be introduced in a later course. This modification produces even better agreement between theory and experimental results.

Figure 14.20

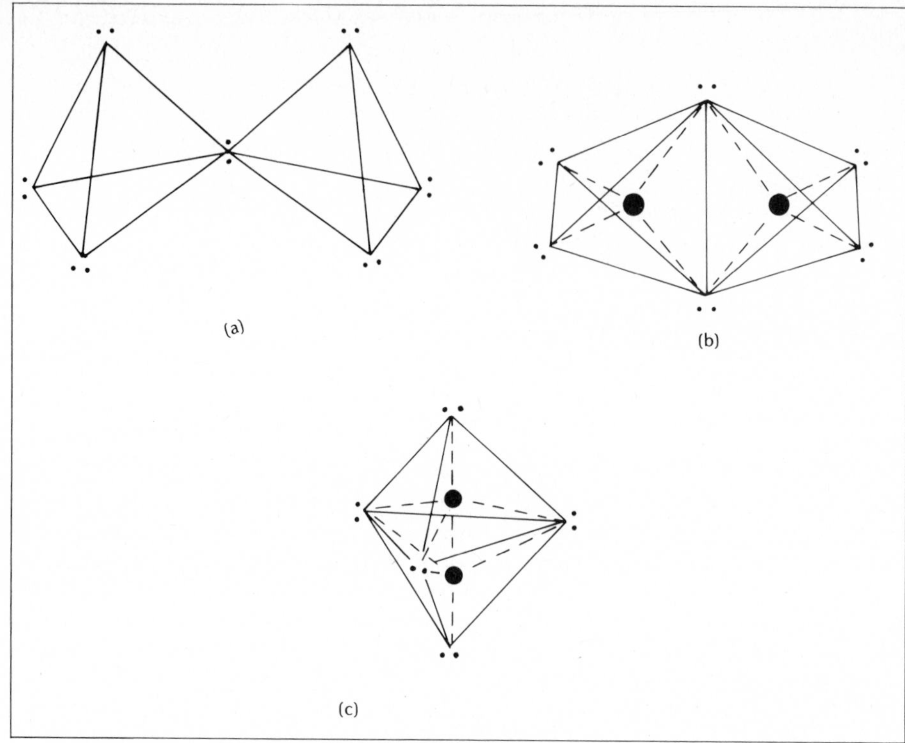

Our description of bonding based on the octet rule is obviously limited. Its primary value is its simplicity. It should not be taken as a literal description of reality. For example, the electron-dot structures based on the octet rule depict electrons in pairs at fixed locations. This is contrary to the model of atoms developed in Chapter 13. There we pointed out that we do not know the location of electrons. The best that we can do is describe the region of space in which there is a high probability that we would find the electron. Such a region of space is called an orbital. Various atomic orbitals have different shapes, and the shape and number of atomic orbitals are used in more sophisticated theories to explain the shapes of molecules and the strength of bonds.

Despite the limitations of the octet rule and electron-dot structures, they are perfectly adequate to explain one important concept used in chemistry: oxidation number. Oxidation number is defined as the charge that an atom would have if all bonds in the molecule were ionic. The electron-dot structure can be used to determine the oxidation number of elements in a compound. However, we can use certain rules to predict the oxidation number of elements even without knowing the structure. Numerically, oxidation numbers and combining numbers are the same, but oxidation numbers are assigned a positive or a negative charge corresponding to the charge that would exist on the atom if all the bonds were ionic.

Questions and Problems

The following ions and molecules are to be used to answer the questions.

a. SO_3^{2-}
b. SO_4^{2-}
c. $SnCl_4$
d. SnO_2
e. SnO
f. CN^-
g. CCl_4
h. ClO_3^-
i. ClO_2^-
j. $SeCl_2$
k. NH_2^-
l. N_2O
m. NO_2
n. N_2O_4
o. NCl_3
p. N_2H_4
q. H_3O^+
r. H_2O_2
s. HNO_3
t. $SiCl_4$

14-48. Draw electron-dot structures for all the molecules and ions given above.

14-49. In the structures you drew for Question 14-48, identify single bonds, double bonds, triple bonds, and nonbonding electrons.

14-50. For the structures drawn in 14-48, identify those bonds that you would expect to be purely covalent, polar-covalent, and ionic.

14-51. Give the oxidation number for each element in each molecule or ion, given in 14-48, using the rules listed on page 446.

14-52. By assigning electrons to the more electronegative element involved in bonding, determine the oxidation number of each atom in the structures you drew in Question 14-48. Are there any cases in which the oxidation number derived in this manner differs from the oxidation number you assigned in Question 14-51?

More Difficult Problems

14-53. Explain why helium does not form a molecule He_2.

14-54. The electron-dot structures for H_2 shows the two electrons located between the hydrogen nuclei. Explain why this representation is misleading.

14-55. Give an example of a nonpolar molecule that has polar bonds, and explain why the molecule is nonpolar even though the bonds are polar.

14-56. Predict the shape for each of the following molecules.

a. C_2H_2 (acetylene)
b. $HCCl_3$ (chloroform)
c. CH_3OH (wood alcohol; methanol)
d. C_2H_6 (ethane)
e. $N(CH_3)_4^+$
f. NH_2OH

14-57. Try to draw electron-dot structures for BeH_2, SF_6, PCl_5, BF_3, ICl_3, and XeF_4. None of these structures obeys the octet rule. All have only single bonds in the molecule. Make your predictions, and discuss the reasons for your predictions with someone else in class. After you finish, look up the actual structures in an advanced chemistry text.

14-58. In the CO_3^{2-} ion, all atoms are in the same plane, all O—C—O bond angles are the same, and the three bonds have the same bond energy. Can a suitable electron-dot structure be drawn for the ion? Explain your answer.

Answers to Questions in Chapter 14

14-1. :C:::O: three bonds

14-2. :Br:Br: one bond

14-3. H:Cl: one bond

14-4. H:O:H one bond between each H and O

14-5. H:S:H one bond between each H and S

14-6. (b)
14-7. (b)
14-8. (a)
14-9. (b) However, this prediction is incorrect. Contrary to expectation, N—N—O—O is the actual arrangement. Such exceptions to the general rule are rare, but they do occur.
14-10. (b)
14-11. 24
14-12. 8
14-13. 32
14-14. 8
14-15. 8
14-16. 24
14-17. 24
14-18. 32
14-19. You must show a double bond between the sulfur and one of the oxygen atoms, but it doesn't matter which one. This implies that one of the oxygen atoms is attached by a double bond and that the other two are attached by single bonds. This is in disagreement with experimental facts. Laboratory measurements show that the three bonds are identical and that the bond length and bond strength are about what one would expect for one and one-third bonds between each pair of atoms. More sophisticated theories of bonding that take into account the shapes of orbitals give a more satisfactory picture of this molecule.

```
:O:S::O:
   :O:
```

14-20.

```
  H
H:C:H
  H
```

14-21. You may show the hydrogens attached to any of the four oxygen atoms. All atoms are connected by single bonds.

```
     :O:
H:O:S:O:H
     :O:
```

14-22.

```
H:P:H
  H
```

14-23.

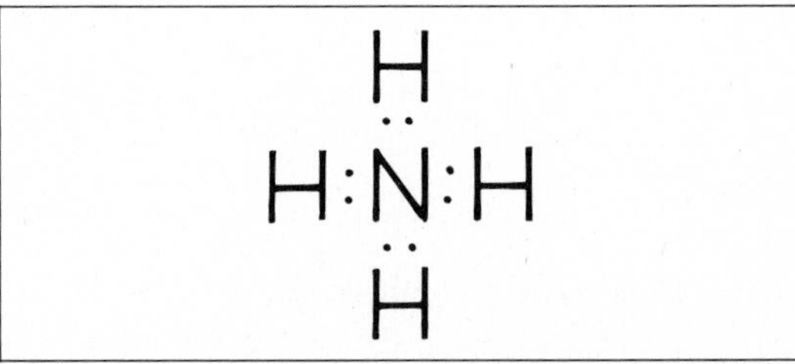

14-24. CO_3^{2-} has the same number of atoms and the same number of valence electrons as SO_3. These structures are identical except that C replaces S. See Question 14-19.
14-25. The structure of NO_3^- is identical to that of CO_3^{2-} and SO_3. See Question 14-19.
14-26.

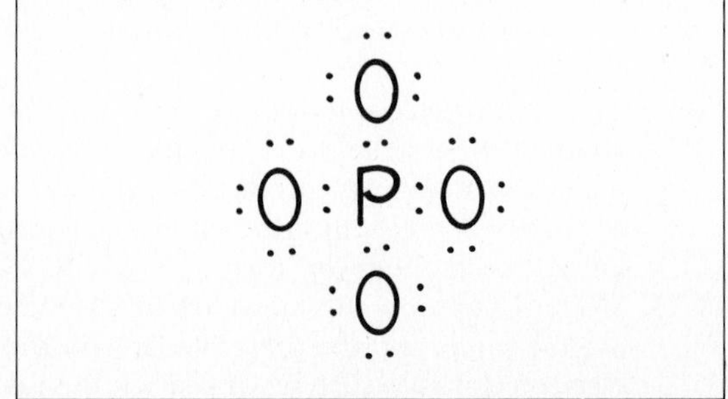

14-27. The shape might be described as a triangular pyramid with nitrogen at the apex and hydrogens at each corner of the base.

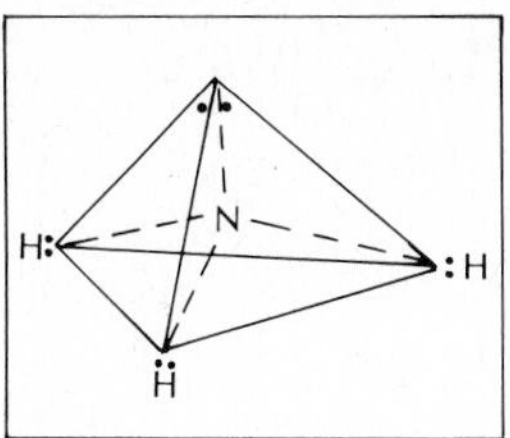

14-28. This is the same shape as that of water. It is usually described as bent.

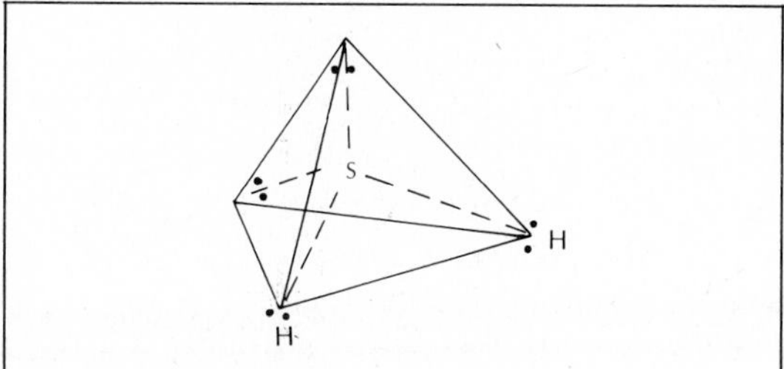

14-29. See Figure 14.19, page 440. The shape is normally described as a tetrahedron with hydrogen atoms at the corners and carbon at the center.

14-30. With only two atoms, the shape must be described as linear.

14-31. Bent. The model predicts one sulfur to oxygen single bond and one sulfur to oxygen double bond. Joining two tetrahedra at a corner and a third tetrahedra along one of the edges would produce a bent shape. Laboratory evidence does indicate that the molecule is bent as predicted, but it also indicates that the two sulfur to oxygen bonds are identical. Theories of bonding based on atomic orbitals provide a better interpretation of this molecule.

14-32. This molecule would be shaped like NH_3. See the answer to Question 14-27.

14-33. All four atoms would lie in the same plane with the angle formed by the single-bonded oxygen atoms and the sulfur being the tetrahedral angle of about 109°. The other oxygen to sulfur to oxygen angles should be larger. This prediction is in reasonably good agreement with experimental data. Experimental data suggest that the molecule is planar with the oxygen atoms at the corners of an equilateral triangle and the sulfur atom in the center.

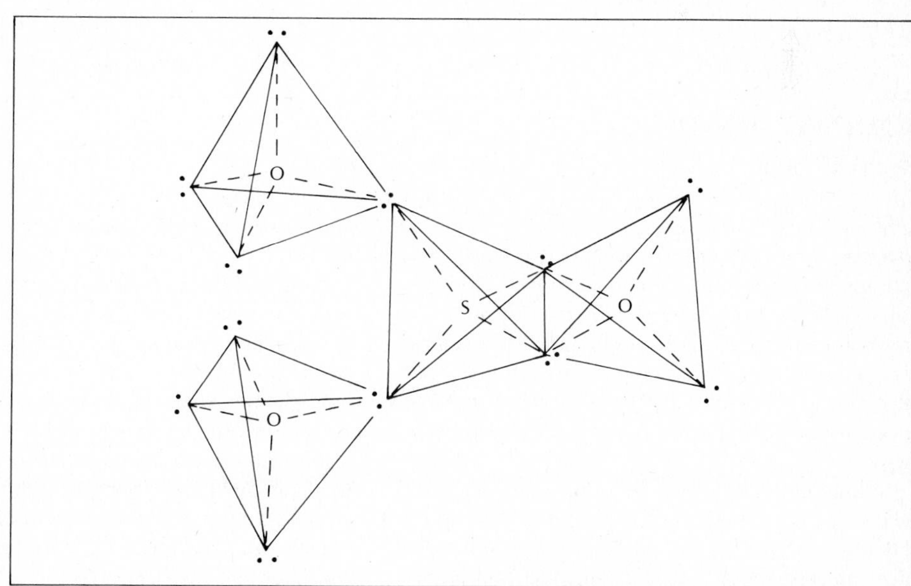

14-34. The hydrogen atoms would be at the corners of a tetrahedron with the nitrogen at the center. (See Figure 14.19, page 440.)

14-35. The molecule would be linear. The triple bond between the carbon and nitrogen would join the tetrahedra along a face. The hydrogen is joined at the apex of the carbon tetrahedron. Thus the three nuclei lie along a straight line. (See Figure 14.15, page 437.)

14-36. 4+
14-37. 6+
14-38. 0
14-39. 4+
14-40. 0 (An oxidation state of 0 for an element in a compound may seem strange, but it can occur when oxidation numbers are assigned using the rules given. If you draw the electron-dot structure for acetic acid and determine the oxidation number of each atom by assigning bonding electrons to the more electronegative element, you will find that one of the carbon atoms has an oxidation number of 3+ and that the other carbon atom has an oxidation number of 3−. The average is 0. At the beginning of the next column is the structure. See for yourself.)

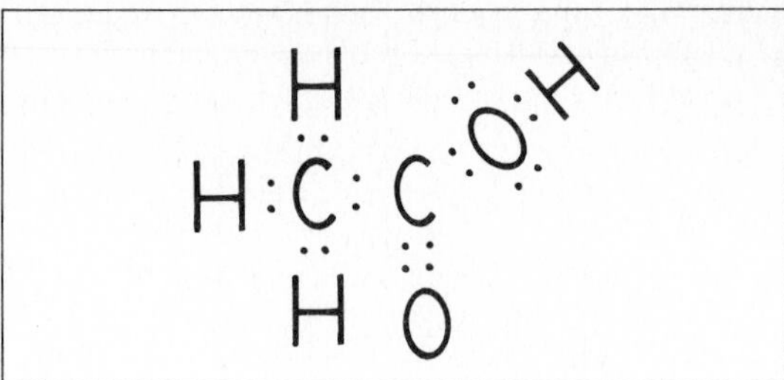

14-41. 4+
14-42. 1−
14-43. 1+
14-44. 2+
14-45. 4+
14-46. 5+
14-47. 3−

READING

In this chapter you were introduced to bonding between atoms within molecules. That isn't the whole story. Molecules of a substance stick together *whenever* a solid forms. Molecules of different kinds stick together when you glue a broken chair or tape an ankle, for instance. The following article suggests some practical applications of bonding between molecules as well as bonding within a molecule.

ADHESIVE BONDING

Myron W. Kelly

Adhesive bonding has many meanings—sealing an envelope, repairing torn paper with cellophane tape, and applying bandages are common uses. The electronics industry fastens components to printed boards, and aluminum honeycomb construction in the aircraft and space industries is another example. In all, solid materials are fastened together by another material, the adhesive, which forms a continuous layer between them.

Various adhesives, made from animals, starch, and cellulosics, have been used for centuries. Within the past four or five decades many synthetic adhesives have replaced those derived from plant and animal sources. Natural adhesives usually lack durability and are sensitive to moisture and heat. Synthetic adhesives are more versatile, and can be "tailor made."

Recent adhesive developments have led to both new composite materials and new applications for old products. Exterior plywood is a relatively new product in which durable, waterproof, synthetic adhesives have replaced animal and vegetable glues. The space program has brought us new lightweight, heat-resistant, and durable adhesives. Many products used widely today could not be produced if synthetic adhesives were not available. Grinding wheels, high-quality sandpaper, wet-strength paper, particle board, and fiber glass are examples.

Advantages of Adhesive Bonding

It's sometimes impossible or, at least, impractical to use other fastening methods. High-pressure laminated table tops and plastic overlays used in furniture

Source: *Chemistry,* Vol. 49, No. 6 (July/August 1976), pp. 14–19.

are examples. Package labeling and wallpaper would also be impractical without adhesives.

Transfer of mechanical stress from one substrate to another is much more uniform with adhesives. Mechanical fasteners concentrate stress at the fastener. But an adhesive bond transfers mechanical stress across the entire glueline from one substrate to the other. It's as if they were fastened by an infinite number of mechanical fasteners.

Because protruding rivets and screwheads are eliminated, smooth surfaces are possible with adhesive bonding. This is critical in some industries. In airplanes, the air resistance of rivets increases drag, and consequently fuel consumption. Adhesive bonding also prevents galvanic corrosion between dissimilar metals. A continuous glueline between substrates prevents physical contact, and hence stops corrosion.

Adhesive bonding led to better utilization of wood resources. The particle board industry uses large quantities of wood waste generated in the conversion of logs to lumber. Laminated wood structures, in which lower quality, smaller wood pieces are glued into larger structures, also contribute significantly to better utilization of wood. Adhesive bonding is vital to both industries.

Many new construction methods and composites depend upon adhesive bonding. Besides laminated structures and particle board, plywood, honeycomb core, stress-skin construction, and strong lightweight materials are only feasible with adhesive bonding.

Disadvantages of Adhesive Bonding

Adhesive bonding does not meet all fastening applications, and it's unrealistic to replace all mechanical fasteners. For one, more elaborate surface preparation is required for adhesive bonding, whereas mechanical fasteners function equally well with both dirty and clean surfaces. For another, maximum mechanical strength of adhesive bonds is not obtained immediately because the adhesive has to solidify. This can take two or three days.

Pressure is required to ensure intimate contact between substrates and adhesive. This limits utility in field applications where it's difficult to bring needed equipment. Also, because no universal adhesive exists, a large adhesive inventory is necessary.

Adhesive Bond Structure and Failure

An adhesive bond can be divided into five regions (Figure 1). Though it may be difficult to separate a particular glueline into its five regions, all adhesive bonds formed (using an adhesive different from either substrate) have these five areas. A bond's maximum strength is limited by the weakest of the five regions.

Adequately designed adhesive bonds should function for their *intended* lifetime, provided the adhesive has been properly used. Failure is caused by improper attention to design, use of the wrong adhesive, or poor application of the proper adhesive.

Adhesive bond failures can be classified broadly into two types—adhesive and cohesive. Adhesive failure occurs when the adhesive does not adhere properly to either one or both substrates. Cohesive failure refers to internal failure within the adhesive or within either of the two substrates. Inspection of the ruptured surfaces usually reveals type of failure: If both faces in the ruptured plane have the same material, *cohesive* failure is the cause; if they are of different material, *adhesive* failure is to blame.

Adhesive failure. Improper surface preparation is often the source of adhesive failures—impurities prevent required intimate contact between substrate and adhesive. Adhesive failure can result when the

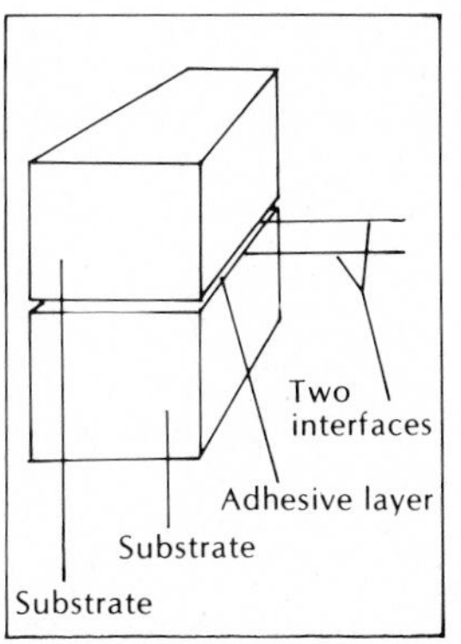

Figure 1

substrate's surface energy is much lower than that of the adhesive. Liquids will only wet and adhere well to solid surfaces with higher surface energies. For example, most polar adhesives with high surface energies do not adhere well to low surface energy solids such as Teflon or polyethylene. Another cause of adhesive failure is corrosion or degradation of the substrate surface by the adhesive. Obviously, destruction of the substrate surface severely weakens or destroys adhesive bonding.

Cohesive failure. Cohesive failure within the *substrates* is good evidence that the adhesive bond is at

Table I. Characteristics of Common Adhesives

Class	*Chemical class*	*Solvent*[a]	*Uses*	*Limitations*
White glues	Polyvinyl acetate	Usually water	Good general glue for porous materials, paper, wood	Low heat resistance, only interior use
Model cement	Cellulosics, butyrates	Organic solvent	Plastic models	Low strength, most are flammable
Epoxy	Epoxies	Water or organic solvents	Strong, versatile adhesive, suitable for many applications, excellent durability	Accurate mixing of components required
Contacts	Elastomerics	Water or organic solvents	High-density overlays for table and counter tops	Low shear strength, adhesive applied to both substrates
Sealants	Silicones	Water or organic solvents	Caulking compounds, excellent water resistance	Low strength
Pressure sensitive	Elastomers on cloth or other backing	None	Decals, tapes	Exceptionally clean surfaces required

[a] Flammability of organic solvents requires caution.

least equal in strength to them. Such failure might be expected when the correct adhesive has been chosen and properly applied. It means, of course, that the adhesive outperforms the materials it's holding together.

Cohesive failure within the *adhesive* layer results either from inherent weakness of the polymer or from discontinuities within the solidified adhesive region. During solidification, many adhesives give off volatile by-products which can be trapped within the adhesive region, forming discontinuities and bubbles in the glueline. These discontinuities are weak regions in which failure first occurs and can then spread. Discontinuities can be reduced by limiting their formation—making a thinner glueline by using less adhesive. The thinnest possible glueline that gives a homogeneous adhesive phase normally results in optimum adhesive bond strength.

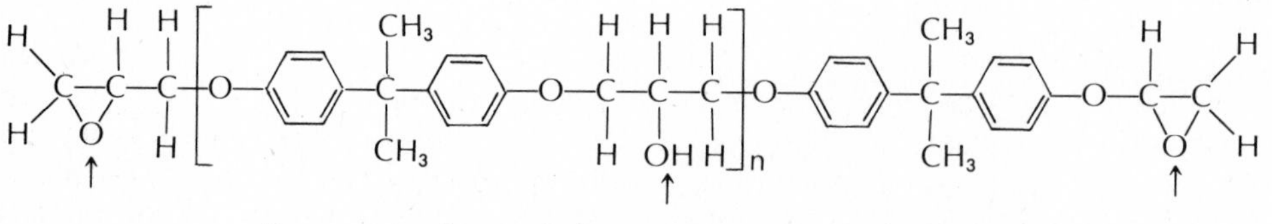

Epoxy resin. Arrows indicate reactive sites for hardener

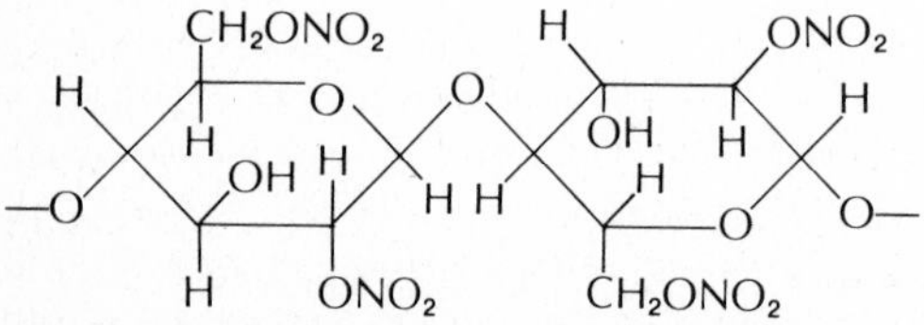

Cellulose nitrate

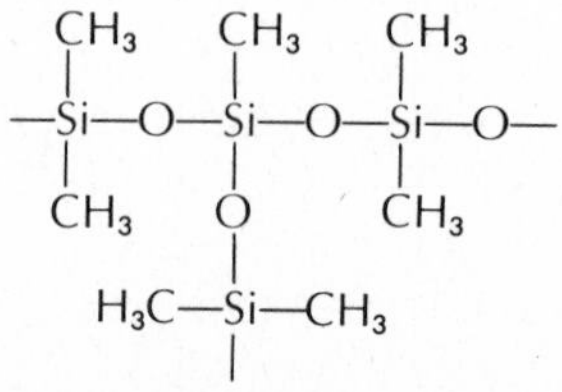

Silicones

```
  H  H  H  H  H  H  H  H  H  H
  |  |  |  |  |  |  |  |  |  |
—C—C=C—C—C—C—C—C=C—C—
  |        |  |  |  |        |
  H        H  H  C6H5 H      H
```

Elastomer,
butadiene-styrene copolymer

Characteristics Required for Adhesives

The many different materials presently used as adhesives range from traditional animal and plant derivatives to recently developed epoxies and hot melts. But four basic requirements must be met: flow, transfer, wetting, and solidification. This is also the correct sequence for optimum bond formation.

Flow. This requirement implies a degree of mobility in the adhesive. Because most adhesives are liquid when applied, they possess the required mobility. The liquid state of hot melts—as the name implies—is reached by heating them above their melting temperatures. Other adhesives, such as the polyvinyl acetates or white glues, are suspended in water or other solvents to attain mobility. Therefore, flow is possible with all adhesives, but why is it necessary?

Most adhesives are not applied to substrates as continuous films but in discrete, noncontinuous areas. Ideally, when sufficient adhesive is applied, it will flow into a continuous film when the two substrates are pressed together. Also, adhesive must fill in substrate surface irregularities, again demanding an adhesive with good flow properties.

However, adhesives can possess too much flow. If applied pressure forces too much adhesive from between the substrates, an inadequate quantity will remain. Reduced pressure and thicker adhesive are the obvious remedies.

Transfer. Adhesive mobility is a prerequisite to this step because the adhesive is applied to only one substrate in most applications. Exposure to pressure and close contact between substrates transfers ad-

Terminology

Adhesive bonding has a language of its own. The more widely used terms will be discussed here with clarifying examples.

- Two solid materials held together by one adhesive are adhesive bonded or glued. The adhesive occupies a specific region, remaining an integral part of the bonded product.
- A composite. The structure formed by bonding two or more materials, called substrates, or adherends.
- The adhesive region within a composite is the glueline. A postage stamp on an envelope exemplifies two substrates separated by a common glueline.
- A material's cohesive strength is determined by the attractive forces between its molecules. Cohesive strength always refers to *inter*molecular attractions within the bulk phase of a material. Polymers, containing large individual molecules, have high intermolecular attractive forces. Consequently, polymers normally possess high cohesive strengths.

When an adhesive has an attractive force high enough to adhere to, or wet, the substrate, its adhesive strength is satisfactory. One can design an adhesive for a given surface. But there are no universal adhesives. Theoretically, adherence of an adhesive to a substrate can be predicted from their surface energies.

- Surface energy is exactly equal to surface tension. More importantly when an adhesive's surface energy is less than or equal to that of the substrate, the two will adhere. Difficulties in determining substrate surface energy frequently limit predictions.

hesive to the second substrate. Poor transfer can occur when the two substrates have poorly fitting surfaces which limit contact under pressure. Here again is emphasized the importance of surface preparation.

Wetting. This term indicates qualitatively the attractive force between the substrate surface and adhesive. Good wetting is the result of strong attractions between adhesive and substrate; poor wetting is the reverse.

Wetting or lack of wetting is normally obvious when a drop of adhesive is placed on the substrate. Without satisfactory wetting, adequate flow and transfer are impossible.

Solidification. The final requirement is hardening or solidification of the adhesive between the substrates. If solidification occurs out of sequence, an unsatisfactory bond is formed.

Penetration. One further requirement for bonding semiporous substrates, such as leather and wood, is the adhesive's ability to penetrate them to a limited degree. This increases the contact area between the substrate and adhesive, and enhances overall bond quality. But penetration must be limited; overpenetration can drain needed adhesive from the glueline.

Solidification Processes—A Closer Look

As we have seen, the final step in satisfactory adhesive bonding is formation of a solid, continuous layer between the two substrates. This process is called curing, hardening, or gelation. Adhesives normally cure by either physical or chemical reactions, or by a combination.

Physical reactions. Cooling and solvent evaporation are two physical reaction cures. No chemical reaction is needed, and no intramolecular bonds are formed or broken. The molecular size of the adhesive polymer remains the same as when applied. Adhesive bonds formed by physical reactions can be readily destroyed by reversing the reaction. Hence, adhesive bonds formed by physical reactions are normally not so durable as chemically cured bonds.

Hot melt bonds are excellent examples of physical cures. These adhesives are applied to substrates in the molten state. Solidification occurs as the temperature falls below the melting point. The adhesive polymer does not change chemically; only a phase change occurs. Unfortunately, elevated temperatures will cause the adhesive to revert to the liquid phase, thus destroying the bond. Hot melts are also called thermoplastic adhesives—slightly elevated temperatures destroy the necessary adhesive characteristics, but lower temperatures regenerate them.

Evaporation of the solvent used to carry an adhesive polymer is also a physical cure. The polymer is not changed during solidification. The solvent is used only as a *carrier* and does not contribute to the adhesive bond. In fact, it must be removed to allow bond formation. Starch, animal and the polyvinyl acetate emulsions, or "white" glues all solidify by water evaporation. These adhesives should not be used in applications where exposure to moisture is likely to occur.

```
  H   H   H   H   H   H
  |   |   |   |   |   |
—C—C—C—C—C—C—
  |   |   |   |   |   |
  O   H   O   H   O   H
  |       |       |
  C=O     C=O     C=O
  |       |       |
  CH3     CH3     CH3
```

Polyvinyl acetate

Chemical reactions. Many adhesives, especially the more durable types, cure by chemical reaction between component polymers. This results in an extremely large and, in most cases, cross-linked polymer network resistant to moderate heat and chemical attack. Solidification by chemical reaction can occur either at room or elevated temperatures, depending on the composition and type of adhesive.

Adhesive systems cured by chemical reaction between component polymers are representative of the thermosetting type. Thermosetting adhesives resist elevated temperatures and usually do not revert to the liquid state upon heating. However, wide variations do exist in the heat resistance of thermosetting adhesives.

Some adhesive systems that solidify by chemical reaction at elevated temperatures can, by suitable catalyst addition, be made to solidify at room temperature. The cyanoacrylate systems, currently receiving wide publicity as a "super" glue, cure rapidly when catalyzed by trace quantities of moisture.

The epoxy adhesives (Table 1, p. 454) are two-component systems in which the user mixes the adhesive polymer with a hardener at the time of use. Hardeners differ from catalysts in that they become an integral part of the hardened adhesive network.

Hardeners are normally either polyfunctional amines or anhydrides that react with the epoxy polymer and solidify into a highly cross-linked structure. Epoxy adhesives result in strong durable bonds normally unaffected by boiling water.

Combinations. There are also many other adhesive systems that solidify by a combination of physical and chemical reactions. These adhesives are applied usually in a solvent medium that must evaporate before the chemical reaction between component

Urea formaldehyde polymer

polymers can occur. By far, this is the most common solidification method.

The normal adhesive is a complex mixture of many components, each component fulfilling a specific function or functions. In addition to the adhesive polymer and solvent, there are fortifiers, extenders, pigments, surfactants, plasticizers, and many other ingredients incorporated into the mix to strengthen, cheapen, and color the final adhesive bond. Table I is a brief summary of some of the more common adhesive systems available for general adhesive bonding. Many brands of each type are available to the consumer.

Plywood Manufacture

The plywood manufacturing industry is one of the largest consumers of adhesives. A description of the manufacturing process exemplifies a large-scale application of adhesive bonding. Plywood is normally made with an odd number of layers—or plies—of wood bonded together with the grain direction of adjacent plies being at right angles to one another. In construction of a three-ply panel, the surface plies have parallel grain orientation, the center-ply, or core, is turned so its grain direction is at right angles to them (Figure 2).

The total thickness of plywood panels can vary from 0.15 to 7.0 cm or more depending upon intended application. Individual plies may vary in

Phenol-formaldehyde polymer

Figure 2

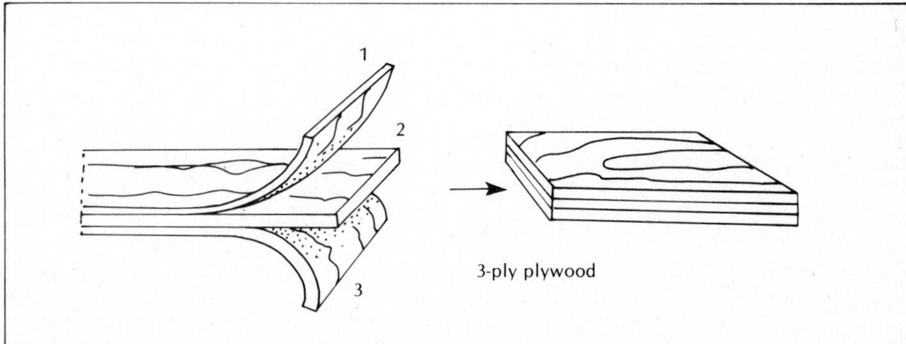

thickness but corresponding plies on opposite sides of the panel's midplane must be the same thickness and have parallel grain.

Plywood panels with unbalanced construction will warp or bend, and thus are easily recognized and culled before they leave the manufacturing site.

Liquid adhesives, normally water solutions of either phenol-formaldehyde or urea-formaldehyde polymers, are applied to both faces of the center ply in three-ply plywood. The veneers are then laid in proper sequence and the panel is placed between two hot platens of a press. While the panel is under pressure, the heat from the hot platens induces a chemical reaction within the adhesive, resulting in a solid continuous adhesive layer between adjacent plies. The panel is removed from the press and allowed to cool, after which it is trimmed to final desired dimensions.

Particle Board Manufacture

Particle board is a wood–polymer composite manufactured from wooden particles held together by a synthetic polymer. A wide range of particle configurations is used because they influence many physical properties, such as strength, surface smoothness, and dimensional stability. Most particle board is produced for interior applications where exposure to moisture and heat is minimal. Hence, urea-formaldehyde-based polymers are the adhesives used for this application. The particle board presently used in this country for outdoor applications is bonded with a phenol-formaldehyde polymer.

Particle board production is highly mechanized, as outlined in the following steps. The wood raw material is reduced to the desired particle shape and size by cutting or grinding. The particles are rapidly dried in large dryers. Next, they're sprayed with liquid adhesive as they are tumbled in a blender. The coated particles are deposited accurately in a mat of proper weight to obtain the desired final density. The mat is then compressed to the desired thickness between two heated platens in a hot press. When subjected to temperatures of 150 to 200°C, the adhesive polymerizes into a hardened, cross-linked polymer which binds the particles together. The pressure ensures intimate contact between the particles, thus facilitating satisfactory bond formation.

Suggested Reading

1. Bickerman, J. J., "The Science of Adhesive Joints," 2nd ed., Academic Press, New York, N.Y., 1968.
2. Cagle, C. V., Ed., "Handbook of Adhesive Bonding," McGraw-Hill, New York, N.Y., 1973.
3. Davies, J. T., Rideal, E. K., "Interfacial Phenomena," 2nd ed., Academic Press, New York, N.Y., 1973.

AN EPILOGUE

It is more fun to teach a class than to write a book. You can talk to your students and they can talk back. You can learn where they are confused. You can tell a joke and see whether they laugh. You can be a person.

I have tried to be a person as I wrote this book. Even though you couldn't tell me where you were confused, I have organized and developed the material on the basis of comments made by my own students. I hope the things that have helped them have also helped you.

Many students don't like chemistry. They find it difficult. Their prior experiences have convinced them that it doesn't make sense. They see no relationship between chemistry and the real world. I hope this is not the case with you.

You undoubtedly found some ideas in this book difficult. Some ideas *are* difficult. It takes time and experience to comprehend them. Still, I hope you found none of the ideas impossible to grasp. In particular, I hope they made sense. I believe that chemistry *is* sensible.

Chemistry is about the real world—the world of delightful colors, delectable aromas, and delicious tastes. I hope you see that it is the source of clean air as well as pollution, that it is the key to good health as well as the source of dangerous drugs, that it will be involved in energy plenty as well as energy shortage. The study of chemistry is the study of you and everything about you. Its potential uses become clearer as you continue to learn.

I have done no more than introduce you to chemistry. In doing so, I have tried to convince you that a longer acquaintance would be worthwhile. If you decide that it is, I believe you will find that you have the background for a pleasant and enlightening relationship.

May I wish you the very best in your future studies.

Appendix 

Basic Math Review

You have studied algebra, but it may have been several years since you used it. This review is to remind you of the basic operations with equalities.

In algebra, x and y are commonly used to represent an unknown number that you want to determine. In science, other letters or even words are used in the same way. You may find, for example,

$$2P + 3 = 11$$

This says, "Two times some number which we represent by P, plus three, is equal to eleven." In order to find the value of the number represented by P, we manipulate the equation in such a way that P stands alone on one side of the equation.

Most simple equations can be solved by remembering that:

1. An equality is not changed when we add the same quantity to both sides.
2. An equality is not changed when we subtract the same quantity from both sides.
3. An equality is not changed when we multiply both sides by the same quantity.
4. An equality is not changed when we divide both sides by the same quantity. (Division by zero is not permissible.)
5. An equality is not changed when we raise both sides to the same power.

Now see how well you can apply these rules by checking the validity of some operations.

Given the equation $2P + 3 = 11$, determine which of the following calculations are correct.

A-1. $2P + 3 = 11$
$2P + 3 - 3 = 11 - 3$
$2P = 8$

A-2. $2P + 3 = 11$
$(2P - 3) + (3 - 3) = 11 - 3$
$2P - 3 = 8$

A-3. $2P + 3 = 11$

$$\frac{2P + 3}{2} = \frac{11}{2}$$

$P + 3 = 5.5$

A-4. $2P + 3 = 11$

$$\frac{2P + 3}{2} = \frac{11}{2}$$

$P + 1.5 = 5.5$

A-5. $2P + 3 = 11$
$2(2P + 3) = 2 \times 11$
$4P + 3 = 22$

A-6. $2P + 3 = 11$

$$\frac{2P + 3}{3} = \frac{11}{3}$$

$2P = 3.66$

Before going on, check your answers with those given at the end of this appendix.

I hope you recognized that only two of the calculations shown are correct. These examples were given to illustrate common errors that students make in solving algebra problems. Students seem to confuse the correct operations for addition and subtraction with the correct operations for multiplication and division. When a number is added or subtracted from one side of an equation, it is added or subtracted only *once*. The process is not applied separately to each term. In multiplication and division, however, the operation *does* affect each term. Division *never* results in a number being "canceled."

To be sure that you understand how to apply these rules of algebra, work the following problems. Work the first five and check your answers. If you miss *any*, work the next set of five and continue until you can work a complete set without error. If you still make mistakes after working all three sets, see your instructor for help.

Set I

A-7. Find the value of V in: $5V - 2 = V + 6$

A-8. Find the value of T in: $\frac{1.5T + 6}{2} = 3$

A-9. Find the value of h in: $A = \frac{1}{2}bh$

A-10. Find the value of n in: $PV = nRT$

A-11. Find the value of X in: $\frac{13}{2.5} = 1.5X + 3$

Set II

A-12. Find the value of t in: $2 + 24t = 26$

A-13. Find the value of y in: $3 + 2x = 3 + 4y$

A-14. Find the value of x in: $3 + 2x = 3 + 4y$

A-15. Find the value of $1.5x$ in: $\frac{3x - 6}{2} = 15$

A-16. Find the value of P in: $\frac{4 - 2P}{1.2} = 3.6$

Set III

A-17. Find the value of n in: $2P + 3n = \frac{0.5P - n}{2}$

A-18. Find the value of P in: $2P + 3n = \frac{0.5P - n}{2}$

A-19. Find the value of z in: $\frac{2 + 3z}{3.2} = 1.8$

A-20. Find the value of x^2 in: $2 - x = x + 4$

A-21. Find the value of F in: $\frac{C}{F - 32} = \frac{5}{9}$

In a beginning chemistry course, you will encounter few problems that are more involved than the ones you have just worked. However, most problems will be worked with numbers that include units. This seems to confuse some students, so it might help to look at a few problems with units.

When units are included in a mathematical problem, they are treated just like variables or unknowns. For example, 2 cm (two centimeters) is treated just like $2x$. Here is a simple example.

Find the value of x if: $100x = 1y$

Dividing by 100, we get: $1x = \frac{1}{100}y$

In the same way, we can ask how many meters are equal to one centimeter.

Find the value of cm if: $100\ \text{cm} = 1\ \text{m}$

Dividing by 100, we get: $1\ \text{cm} = \frac{1}{100}\ \text{m}$

You will note that we treated each unit in this problem as a single variable. That is, we treated cm as a single variable rather than two, c times m. This is the way units are normally treated in scientific calculations. However, you may be interested to know that when metric units are used in calculations, it is possible to treat each letter as a separate variable and get sensible answers. Let's look at the last equation and treat the c and the m in cm as separate variables.

$$100\ \text{cm} = 1\ \text{m}$$

Dividing by m, we get: $100\ \text{c} = 1$

Dividing by 100, we get: $\text{c} = \frac{1}{100}$

You will recall that cm stands for *centi*meters. In the abbreviation, the c stands for the prefix, *centi,* and the m stands for the basic unit, *meter.* In solving the foregoing equation, we find that $\text{c} = \frac{1}{100}$, which is exactly what the prefix *centi* means: one one-hundredth. Metric units and the abbreviations for metric units have been carefully selected so that metric units can be treated in this way.

The following problems involve equations that contain units. Solve the equations just as you did in the previous examples, treating all units as variables. Remember that when the same unit appears in the numerator and the denominator of a fraction, we can "cancel" the units by dividing both the numerator and the denominator of the fraction by that unit. As before, work the problems in sets of five. If you miss any problem in the first set, find out where you made your error and then work the second set. Repeat this process until you are able to work a complete set without making a mistake.

Set I

A-22. Simplify the following expression for x: $x = 25{,}000\ \text{mm} \left(\frac{1\ \text{m}}{1{,}000\ \text{mm}}\right)\left(\frac{1\ \text{km}}{1{,}000\ \text{m}}\right)$

A-23. Find the value of $1\ \text{m}^3$ in: $1\ \text{m} = 100\ \text{cm}$

A-24. Solve for P_2 in: $\frac{760\ \text{torr}}{500\ \text{K}} = \frac{P_2}{100\ \text{K}}$

A-25. Solve for V in: $\frac{2.5\ \text{g}}{\text{cm}^3} = \frac{125\ \text{g}}{V}$

A-26. Solve for n in: $(760\ \text{torr})(22.4\ \text{L}) = n\left(\frac{62.3\ \text{L torr}}{\text{mol K}}\right)(273\ \text{K})$

Set II

A-27. Simplify the following expression for y: $y = 0.02 \text{ kL} \left(\frac{1{,}000 \text{ L}}{1 \text{ kL}}\right)\left(\frac{1{,}000 \text{ mL}}{1 \text{ L}}\right)$

A-28. Find the value of 1 dm³ in: 1 dm = 0.1 m

A-29. Solve for M in: $\frac{1.2 \text{ g}}{\text{cm}^3} = \frac{M}{(2 \text{ cm})(3 \text{ cm})(0.5 \text{ cm})}$

A-30. Solve for h in: $\frac{0.6 \text{ g}}{\text{cm}^3} = \frac{60 \text{ g}}{(5 \text{ cm})(10 \text{ cm})(h)}$

A-31. Solve for T in: $(760 \text{ torr})(11.2 \text{ L}) = 1 \text{ mol} \left(\frac{62.3 \text{ L torr}}{\text{mol K}}\right) T$

Set III

A-32. Simplify the following expression for x: $x = \left(\frac{20 \text{ kg}}{0.2 \text{ m}^3}\right)\left(\frac{1 \text{ m}}{100 \text{ cm}}\right)^3\left(\frac{1{,}000 \text{ g}}{1 \text{ kg}}\right)$

A-33. Solve for h in: $30 \text{ cm}^2 = \frac{1}{2}(5 \text{ cm})\, h$

A-34. Simplify the following expression for H: $H = \left(\frac{0.8 \text{ cal}}{\text{g}}\right)(50 \text{ g}) + \left(\frac{80 \text{ cal}}{\text{g C}^\circ}\right)(500 \text{ g})(0.1 \text{ C}^\circ)$

A-35. Solve for Δt in: $500 \text{ g} \left(\frac{0.25 \text{ cal}}{\text{g}}\right) = \left(\frac{1 \text{ cal}}{\text{g C}^\circ}\right)(50 \text{ g})(\Delta t)$

A-36. Solve for R in: $(1 \text{ atm})(22.4 \text{ L}) = 1 \text{ mol } (R)(273 \text{ K})$

Answers to Questions in Appendix A

A-1. This is legitimate. The same number, 3, was subtracted from both sides of the equation.

A-2. This is incorrect. The net effect of subtracting 3 from both terms on the left was to subtract 6 from the left side of the equation and 3 from the right side. If 6 had been subtracted from the right, the operation would have been correct—though it would not have helped much.

A-3. This is also incorrect. It appears that both sides of the equation are being divided by 2; but when the arithmetic was done, only the first term on the left was divided. The correct way to do the indicated division is shown in Question A-4.

A-4. This is correct. Ordinarily, one would solve this equation by first subtracting 3 from both sides of the equation and then dividing by 2, but the division can be done first. If 1.5 is now subtracted from both sides of the equation, the correct answer of 4 for the value of P is obtained.

A-5. Incorrect. When an equation is multiplied or divided by a number, *all* terms are multiplied or divided by the number. Multiplication by 2 would produce $4P + 6 = 22$. Subtracting 6 from both sides of the equation and dividing by 4 then produces the correct answer of 4 for the value of P.

A-6. Once again, the procedure is incorrect. Division by a number never "cancels" what is in the numerator. Done correctly, division by 3 would yield $0.66P + 1 = 3.66$. If the equation is now solved in the usual way, the answer will be approximately 4. It would be exactly 4 if we had not introduced rounding error in the division.

A-7. $5V - 2 = V + 6$
$4V - 2 = 6$
$4V = 8$
$V = 2$

A-8. $\frac{1.5T + 6}{2} = 3$

$1.5T + 6 = 6$

$1.5T = 0$

$T = 0$

A-9. $A = \frac{1}{2}bh$

$2A = bh$

$h = \frac{2A}{b}$

A-10. $PV = nRT$

$n = \frac{PV}{RT}$

A-11. $\frac{13}{2.5} = 1.5x + 3$

$13 = 3.75x + 7.5$

$3.75x = 5.5$

$x = \frac{5.5}{3.75}$ or 1.46

A-12. $2 + 24t = 26$
$24t = 24$
$t = 1$

A-13. $3 + 2x = 3 + 4y$
$2x = 4y$
$y = \frac{1}{2}x$

A-14. $x = 2y$

A-15. $\frac{3x - 6}{2} = 15$

$3x - 6 = 30$

$3x = 36$

$1.5x = 18$

A-16. $\frac{4 - 2P}{1.2} = 3.6$

$4 - 2P = 4.32$

$-2P = 0.32$

$P = -0.16$

A-17. $2P + 3n = \frac{0.5P - \mathrm{n}}{2}$

$4P + 6n = 0.5P - n$

$4P + 7n = 0.5P$

$7n = -3.5P$

$n = -\frac{1}{2}P$

A-18. $P = -2n$

A-19. $\frac{2 + 3z}{3.2} = 1.8$

$2 + 3z = 5.76$

$3z = 3.76$

$z = \frac{3.76}{3}$ or 1.253

A-20. $2 - x = x + 4$
$-x = x + 2$
$-2x = 2$
$x = -1$
$x^2 = 1$

A-21. $\frac{C}{F - 32} = \frac{5}{9}$

$\frac{9C}{F - 32} = 5$

$9C = 5F - 160$

$5F = 9C + 160$

$F = \frac{9}{5}C + 32$ (You should recognize this equation as the relationship between degrees Fahrenheit, *F*, and degrees Celsius, *C*. The letters are not usually italicized in this equation.)

A-22. $x = 25{,}000 \text{ mm}\left(\frac{1 \text{ m}}{1{,}000 \text{ mm}}\right)\left(\frac{1 \text{ km}}{1{,}000 \text{ m}}\right)$

$x = 0.025$ km (This is the number of kilometers equal to 25,000 millimeters.)

A-23. $1 \text{ m} = 100 \text{ cm}$
$(1 \text{ m})^3 = (100 \text{ cm})^3$
$1 \text{ m}^3 = 1{,}000{,}000 \text{ cm}^3$ (There are one million cubic centimeters in one cubic meter.)

A-24. $\frac{760 \text{ torr}}{500 \text{ K}} = \frac{P_2}{100 \text{ K}}$

$\frac{(760 \text{ torr})(100 \text{ K})}{500 \text{ K}} = P_2$

$P_2 = \frac{760 \text{ torr}}{5} = 152 \text{ torr}$

A-25. $\frac{2.5\text{ g}}{\text{cm}^3} = \frac{125\text{ g}}{V}$

$\frac{V(2.5\text{ g})}{\text{cm}^3} = 125\text{ g}$

$V = 125\text{ g}\left(\frac{1\text{ cm}^3}{2.5\text{ g}}\right)$

$V = 50\text{ cm}^3$

A-26. $(760\text{ torr})(22.4\text{ L}) = n\left(\frac{62.3\text{ L torr}}{\text{mol K}}\right)(273\text{ K})$

$(760\text{ torr})(22.4\text{ L})\left(\frac{1\text{ mol}}{62.3\text{ L torr}}\right)\left(\frac{1}{273}\right) = n$

$n = 1.00\text{ mol}$

A-27. $y = 0.02\text{ kL}\left(\frac{1{,}000\text{ L}}{1\text{ kL}}\right)\left(\frac{1{,}000\text{ mL}}{1\text{L}}\right)$

$y = 20{,}000\text{ mL}$ (There are 20,000 milliliters in 0.02 kiloliters.)

A-28. $1\text{ dm} = 0.1\text{ m}$
$(1\text{ dm})^3 = (0.1\text{ m})^3$
$1\text{ dm}^3 = 0.001\text{ m}^3$ (One cubic decimeter is 1/1000 of a cubic meter.)

A-29. $\frac{1.2\text{ g}}{\text{cm}^3} = \frac{M}{(2\text{ cm})(3\text{ cm})(0.5\text{ cm})}$

$\frac{1.2\text{ g}}{\text{cm}^3} = \frac{M}{3\text{ cm}^3}$

$\frac{(1.2\text{ g})(3\text{ cm}^3)}{\text{cm}^3} = M$

$M = 3.6\text{ g}$

A-30. $\frac{0.6\text{ g}}{\text{cm}^3} = \frac{60\text{ g}}{(5\text{ cm})(10\text{ cm})(h)}$

$\frac{0.6\text{ g}}{\text{cm}^3} = \frac{60\text{ g}}{(50\text{ cm}^2)h}$

$\frac{(0.6\text{ g})(50\text{ cm}^2)(h)}{\text{cm}^3} = 60\text{ g}$

$\frac{(0.6\text{ g})(50)(h)}{\text{cm}(60\text{ g})} = 1$

$h = \frac{60\text{ cm}}{(0.6)(50)}$

$h = 2\text{ cm}$

A-31. $(760\text{ torr})(11.2\text{ L}) = 1\text{ mol}\left(\frac{62.3\text{ L torr}}{\text{mol K}}\right)T$

$\frac{(760\text{ torr})(11.2\text{ L})\text{K}}{62.3\text{ L torr}} = T$

$T = 136.6\text{ K}$

A-32. $x = \left(\frac{20\text{ kg}}{0.2\text{ m}^3}\right)\left(\frac{1\text{ m}}{100\text{ cm}}\right)^3\left(\frac{1{,}000\text{ g}}{1\text{ kg}}\right)$

$x = \frac{20}{0.2\text{ m}^3}\left(\frac{1\text{ m}^3}{1{,}000{,}000\text{ cm}^3}\right)\left(\frac{1{,}000\text{ g}}{1}\right)$

$x = \frac{0.1\text{ g}}{\text{cm}^3}$

A-33. $30\text{ cm}^2 = \frac{1}{2}(5\text{ cm})h$

$\frac{(30\text{ cm}^2)(2)}{5\text{ cm}} = h$

$h = 12\text{ cm}$

A-34. $H = \left(\frac{0.8\text{ cal}}{\text{g}}\right)(50\text{ g}) + \left(\frac{80\text{ cal}}{\text{g C}^\circ}\right)(500\text{ g})(0.1\text{ C}^\circ)$

$H = 40\text{ cal} + 4{,}000\text{ cal}$

$H = 4{,}040\text{ cal}$

A-35. $500\text{ g}\left(\frac{0.25\text{ cal}}{\text{g}}\right) = \left(\frac{1\text{ cal}}{\text{g C}^\circ}\right)(50\text{ g})(\Delta t)$

$125\text{ cal} = \left(\frac{50\text{ cal}}{\text{C}^\circ}\right)\Delta t$

$(125\text{ cal})\left(\frac{\text{C}^\circ}{50\text{ cal}}\right) = \Delta t$

$\Delta t = 2.5\text{ C}^\circ$

A-36. $(1\text{ atm})(22.4\text{ L}) = 1\text{ mol}\,(R)(273\text{ K})$

$\frac{(1\text{ atm})(22.4\text{ L})}{(1\text{ mol})(273\text{ K})} = R$

$R = 0.082\ \frac{\text{atm L}}{\text{mol K}}$ (This is one value for the ideal gas constant, which you will use in this course.)

Appendix B

Decimal Numbers, Multiplication by Ten, and Rules of Signed Numbers

DECIMAL NUMBERS

Long ago it was discovered that counting could be simplified by grouping into piles of ten, and our number system is based on this idea. When we write a number like 236, each digit except the 6 represents the number of piles containing ten of the next smaller unit.

Figure B.1

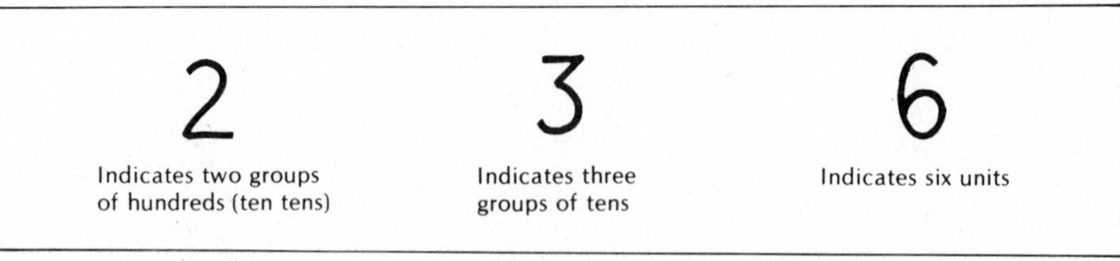

The 6 represents just that, six. The 3 says that there are three piles, each containing ten units. In the same way, the 2 says that there are two piles, each containing ten groups of ten (one hundred). A digit to the left of the 2 would tell us how many piles of ten hundred (one thousand) had been counted. Every digit indicates groups of ten of the previous place value.

Quite a long time after this system of counting was invented, someone decided to extend the system to include smaller numbers. The decimal point was invented to indicate the units (ones) place, and digits were added to the right of the decimal point to represent numbers smaller than 1. Every number still represents a group of ten of the place value shown to its right.

Figure B.2

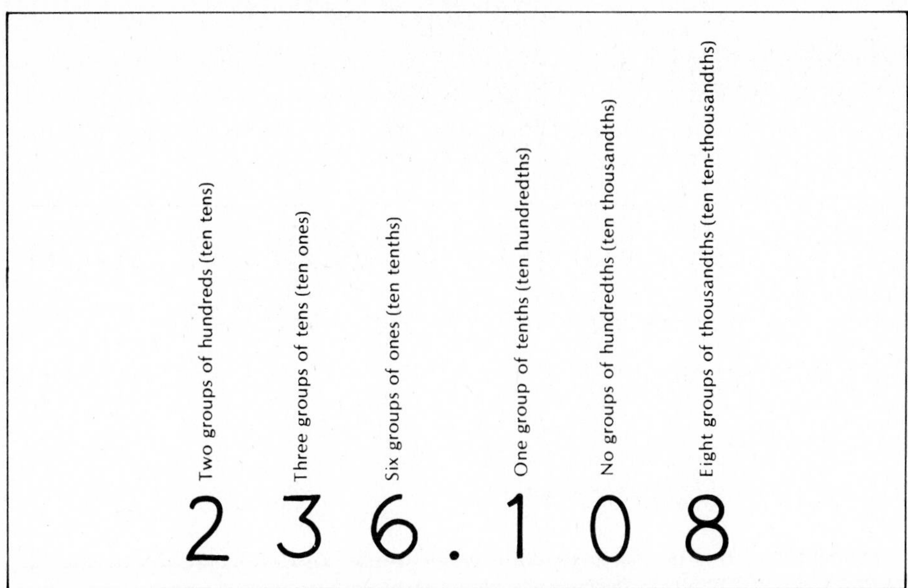

Here the decimal point indicates that the 6 is the digit describing the "ones." Once this is established, we know the value of all other digits. They are shown in Figure B.2.

MULTIPLICATION BY POWERS OF TEN

Because our number system is based on groups of ten, multiplication and division by 10, 100 (10 × 10), 1,000 (10 × 10 × 10), or any other power of ten are very simple. All we need do is move the decimal point to the right or left to give each digit in the number a new place value. For example, when we multiply 236 by 10, we get

$$\begin{array}{r} 236. \\ \times\ 10 \\ \hline 2360. \end{array}$$

Figure B.3 *In multiplying by 10, we move the decimal one place to the right.*

All that we have done is move the decimal one place to the right and insert a 0 to make sure we know the new place value of the original digits. When we multiply 236 by 100, we are multiplying by 10 twice, so the decimal point is moved *two* places to the right:

$$\begin{array}{r} 236. \\ \times\ 100 \\ \hline 23600. \end{array}$$

23600.

Figure B.4
In multiplying *by 100, we move the decimal* two *places to the* right.

The rule for multipying by 10, 100, 1,000, or any power of 10 is that the decimal point is moved to the *right* the same number of places as there are zeros in the multiplier. *Without rewriting,* give the answers to the following multiplication problems.

B-1. 24 × 1,000 =
B-2. 2.45 × 10 =
B-3. 2,408 × 1,000,000 =
B-4. 0.218 × 100 =
B-5. 0.0001 × 10,000 =

If you are not accustomed to multiplying in this way, practice with other numbers until it comes naturally.

DIVISION BY POWERS OF TEN

Division by 10 is just as easy as multiplication. For example, 236 divided by 10 is

$$\begin{array}{r} 23.6 \\ 10\,\overline{)236.} \\ \underline{20} \\ 36 \\ \underline{30} \\ 60 \\ \underline{60} \end{array}$$

Figure B.5
In dividing *by 10, we move the decimal* one *place to the* left.

All of that arithmetic is unnecessary. We could get the answer simply by moving the decimal point one place to the left to give the digits a new place value. Dividing by 100 is the same as dividing by 10 twice, so the decimal point is moved *two* places to the left:

$$\begin{array}{r} 2.36 \\ 100\,\overline{)236.} \\ \underline{200} \\ 360 \\ \underline{300} \\ 600 \\ \underline{600} \end{array}$$

Figure B.6
In dividing by 100, we move the decimal two *places to the* left.

The rule for dividing by 10, 100, 1,000, or any power of ten is that the decimal point is moved to the *left* the same number of places as there are zeros in the divisor. *Without rewriting,* give the answers to the following division problems.

B-6. 24,000 ÷ 1,000 =
B-7. 24.5 ÷ 10 =
B-8. 2,408,000,000 ÷ 1,000,000 =
B-9. 21.8 ÷ 100 =
B-10. 1 ÷ 10,000 =

The answers you just found are the numbers that you were given in the multiplication problems; the numbers given here are the answers to the multiplication problems. Multiplication and division are inverse operations. When you multiply by a number and then divide your answer by the same number, you are back to where you started. This is what we mean by inverse operations; the second operation "undoes" the effect of the first operation. Once again, if you have any trouble with these problems, practice until you gain confidence.

SIGNED NUMBERS

Multiplying and dividing by moving the decimal point is particularly useful when you write numbers in exponential notation. (Exponential notation is discussed in Appendix C.) When you multiply and divide exponential numbers, you need to recall the rules for adding signed numbers.

Keep in mind that the plus sign and the minus sign are used to represent different things. Two uses of these symbols are illustrated by the number line shown in Figure B.7.

−10 −9 −8 −7 −6 −5 −4 −3 −2 −1 0 +1 +2 +3 +4 +5 +6 +7 +8 +9 +10

Figure B.7
A number line.

The scale on the number line shows both magnitude and direction. A plus sign is used to indicate positions to the right of zero, and a minus sign is used to indicate positions to the left of zero. Thus the plus and minus signs are used to represent positions relative to zero. A plus sign indicates that the number is to the right of zero; a minus sign indicates that the number is to the left of zero.

A number line can be used to represent many situations in real life. For example, it could be used to represent financial condition. The zero on the number line could represent having no money and no debts. Numbers to the right (positive numbers) would then represent your financial condition when you have money in your pocket. Numbers to the left (negative numbers) would represent your financial condition when you are in debt.

In addition to representing position relative to zero, the plus and minus signs are used in mathematics to represent operations. For example, a plus sign is used to represent movement to the right on the number line, and a minus sign is used to represent movement to the left. This is illustrated in Figure B.8.

Figure B.8 *Plus means movement to the right on a number line; minus means movement to the left.*

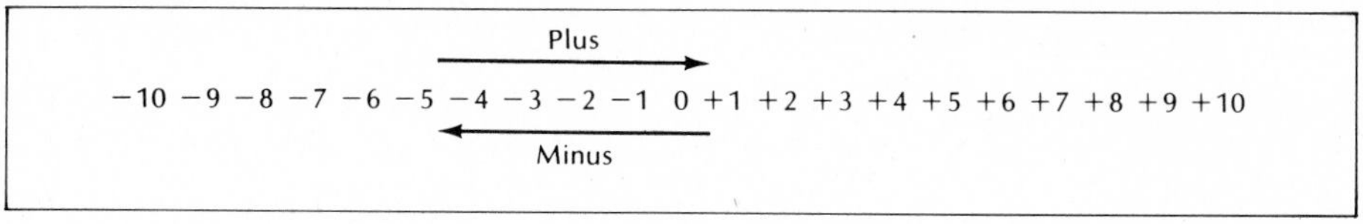

Plus →

$-10\ -9\ -8\ -7\ -6\ -5\ -4\ -3\ -2\ -1\ \ 0\ +1\ +2\ +3\ +4\ +5\ +6\ +7\ +8\ +9\ +10$

← Minus

A number such as -3 can be interpreted in two ways. We can interpret it to mean the -3 position on the number line (using an everyday example, being three dollars in debt). This interpretation is illustrated in Figure B.9. The other interpretation is that -3 indicates the *operation* of moving three spaces to the left (getting rid of three dollars). When we assume that we start at zero, this operation is illustrated by Figure B.10.

Figure B.9

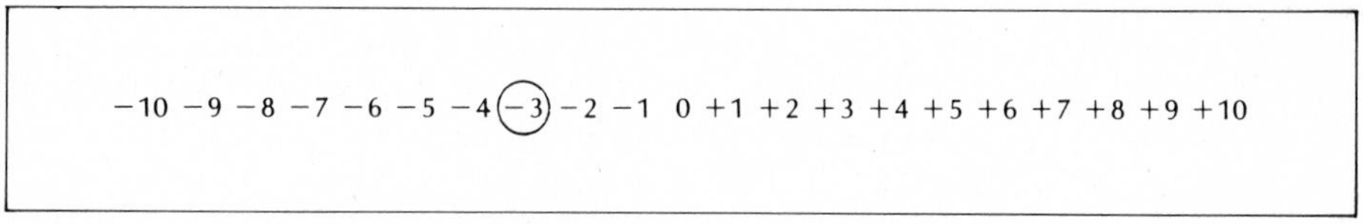

$-10\ -9\ -8\ -7\ -6\ -5\ -4\ \textcircled{-3}\ -2\ -1\ \ 0\ +1\ +2\ +3\ +4\ +5\ +6\ +7\ +8\ +9\ +10$

Figure B.10

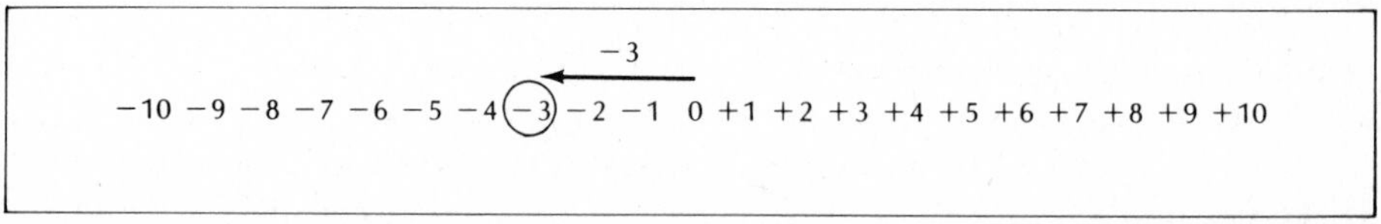

← -3

$-10\ -9\ -8\ -7\ -6\ -5\ -4\ \textcircled{-3}\ -2\ -1\ \ 0\ +1\ +2\ +3\ +4\ +5\ +6\ +7\ +8\ +9\ +10$

Now we will use the number line to illustrate the addition of signed numbers, beginning with the simple addition of 4 and 2 to get 6. This would normally be represented as

$$\textcircled{1} \qquad 4 + 2 = 6$$

When a number is written without a sign, it is understood to be positive. In this section there will be less confusion if the sign of the number is shown. Using this complete notation, we would write the equation as

$$+4 + (+2) = +6$$

This can be interpreted to mean, "Start at zero and move four spaces to the right followed by a move of two spaces to the right, and you end up six spaces to the right of zero." In terms of an everyday example, it could mean getting four dollars followed by getting two dollars. This is illustrated in Figure B.11.

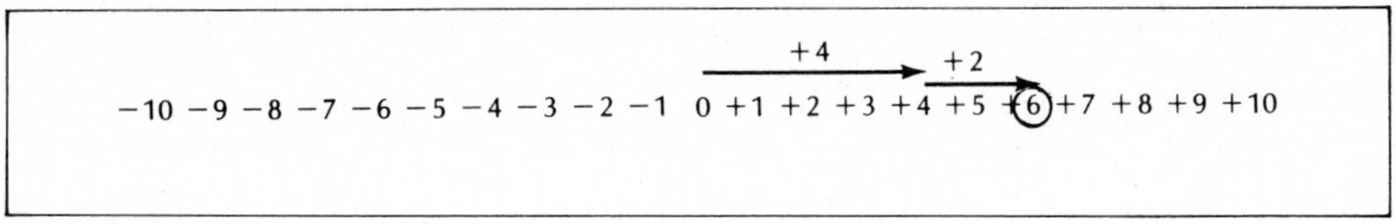

+4 +2
−10 −9 −8 −7 −6 −5 −4 −3 −2 −1 0 +1 +2 +3 +4 +5 +6 +7 +8 +9 +10

Figure B.11

(2) $-4 + (-2) = -6$

This can be interpreted to mean, "Start at zero and move four spaces to the left followed by a move of two spaces to the left, and you end up six spaces to the left of zero." In terms of the everyday example, if you are flat broke, spending four dollars followed by spending two dollars leaves you six dollars in debt. This is illustrated in Figure B.12.

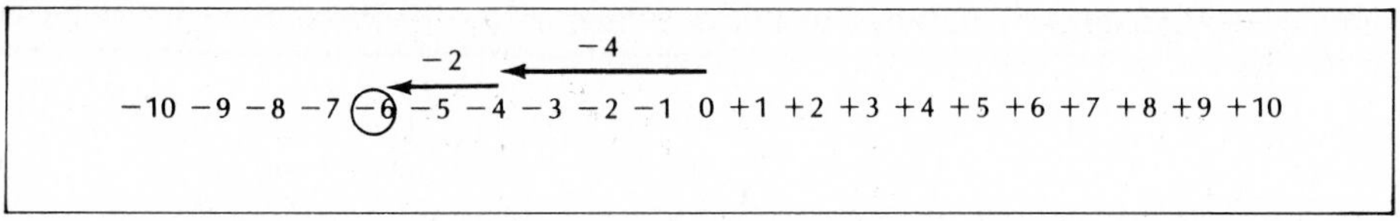

−2 −4
−10 −9 −8 −7 −6 −5 −4 −3 −2 −1 0 +1 +2 +3 +4 +5 +6 +7 +8 +9 +10

Figure B.12

(3) $+4 + (-2) = +2$

This means, "Start at zero and move four spaces to the right followed by a move of two spaces to the left, and you end up two spaces to the right of zero." In terms of our monetary example, earning four dollars followed by spending two dollars leaves you two dollars in your pocket. This is illustrated in Figure B.13.

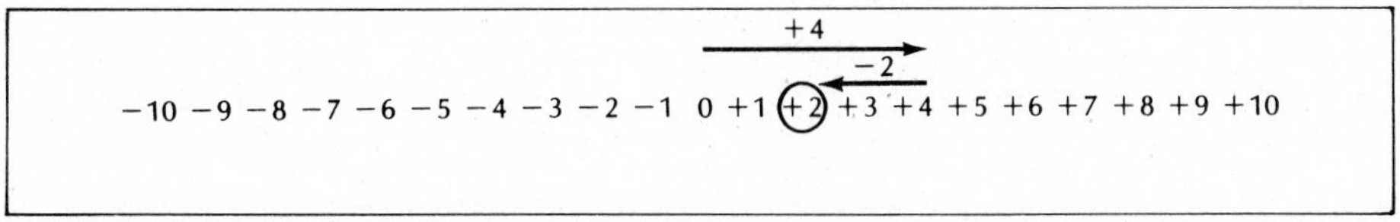

+4 −2
−10 −9 −8 −7 −6 −5 −4 −3 −2 −1 0 +1 +2 +3 +4 +5 +6 +7 +8 +9 +10

Figure B.13

(4) $-4 + (+2) = -2$

This means, "Start at zero and move four spaces to the left followed by a move of two spaces to the right, and you end up two spaces to the left of zero." In terms of money, spending four dollars that you don't have followed by earning two dollars leaves you two dollars in debt. This is illustrated in Figure B.14.

Figure B.14

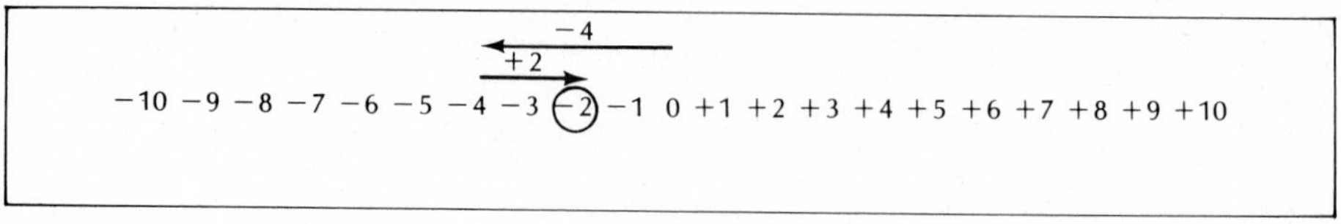

−4
+2
−10 −9 −8 −7 −6 −5 −4 −3 (−2) −1 0 +1 +2 +3 +4 +5 +6 +7 +8 +9 +10

What we have just illustrated in terms of a number line and an everyday example is usually summarized by the following rules for adding signed numbers.

To add numbers with the same sign, find the sum and affix the sign of the numbers. Refer to examples ① and ②.

To add numbers with opposite sign, take the difference and affix the sign of the larger number. Refer to examples ③ and ④.

Addition of signed numbers usually causes little problem, but subtraction is sometimes more difficult. However, if you remember that subtraction is the inverse of addition, the following rule is sensible.

To subtract signed numbers, change the sign of the number to be subtracted and add.

The following examples illustrate subtraction of signed numbers.

⑤ $+4 - (+2) = +4 + (-2) = +2$

This can be interpreted to mean, "Start at zero and move four spaces to the right followed by the *opposite* of moving two spaces to the right. This is the same as moving four spaces to the right followed by moving two spaces to the left, so I end up two spaces to the right of zero." In terms of an everyday example, this could mean earning four dollars and losing two dollars. You end up with two dollars.

Figure B.15

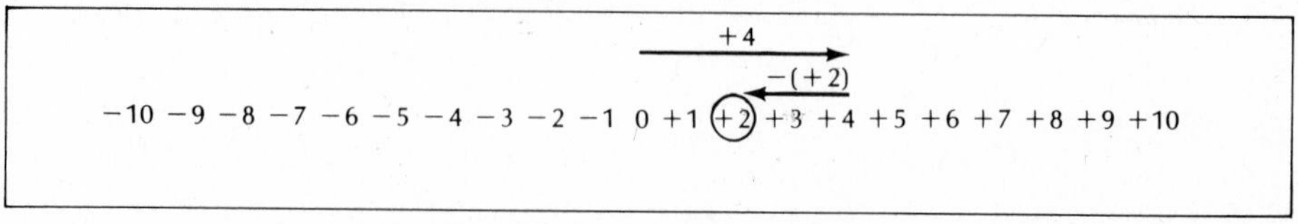

+4
−(+2)
−10 −9 −8 −7 −6 −5 −4 −3 −2 −1 0 +1 (+2) +3 +4 +5 +6 +7 +8 +9 +10

⑥ $-4 - (-2) = -4 + (+2) = -2$

This can be interpreted to mean, "Start at zero and move four spaces to the left followed by the *opposite* of moving two spaces to the left. This is the same as moving four spaces to the left followed by moving two spaces to the right, so I end up two spaces to the left of zero." In terms of the everyday example, if you go in debt four dollars followed by the opposite of going in debt two more dollars, you have gone in debt four dollars and then earned two dollars. You are now only two dollars in debt.

Figure B.16

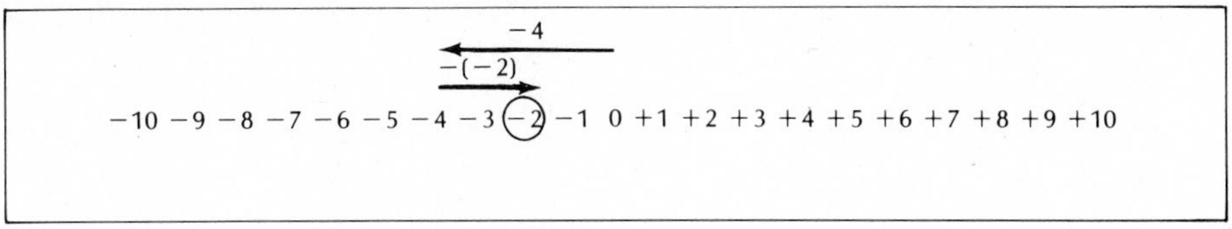

−4
−(−2)
−10 −9 −8 −7 −6 −5 −4 −3 (−2) −1 0 +1 +2 +3 +4 +5 +6 +7 +8 +9 +10

(7) $+4 - (-2) = +4 + (+2) = +6$

This means, "Start at zero and move four spaces to the right followed by the *opposite* of moving two spaces to the left. The opposite of moving two spaces to the left is moving two spaces to the right, so I end up six spaces to the right of zero." In terms of money, you have earned four dollars and then done the opposite of going in debt two dollars more. (In other words, you have earned two more dollars.) You end up six dollars ahead.

Figure B.17

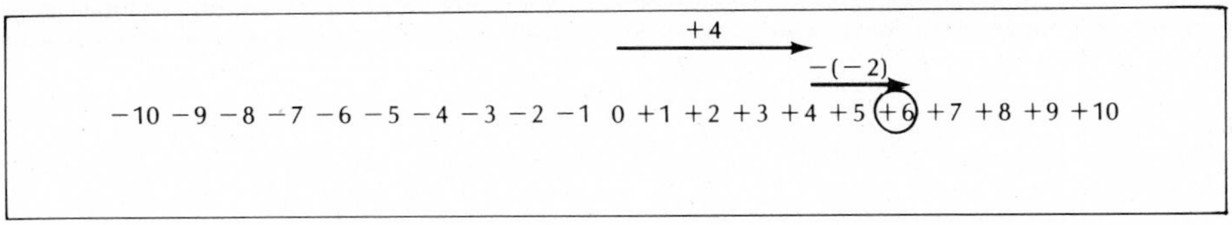

+4
−(−2)
−10 −9 −8 −7 −6 −5 −4 −3 −2 −1 0 +1 +2 +3 +4 +5 (+6) +7 +8 +9 +10

(8) $-4 - (+2) = -4 + (-2) = -6$

This means, "Start at zero and move four spaces to the left followed by the *opposite* of moving two spaces to the right. I end up moving a total of six spaces to the left." The opposite of earning two dollars is going in debt two dollars. If you go in debt four dollars followed by going in debt two dollars, you are six dollars in debt.

Figure B.18

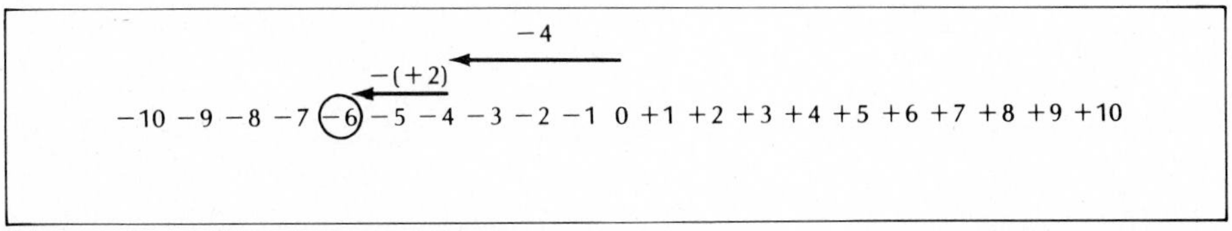

−4
−(+2)
−10 −9 −8 −7 (−6) −5 −4 −3 −2 −1 0 +1 +2 +3 +4 +5 +6 +7 +8 +9 +10

Multiplication and division of signed numbers are more difficult to illustrate, and no attempt will be made to do so here. The rules are as follows:

To multiply or divide numbers that have the same sign, find the product or quotient of the numbers and affix a positive sign.

To multiply or divide numbers that have opposite signs, find the product or quotient of the numbers and affix a negative sign.

The following examples illustrate these rules.

$$(+4) \times (+2) = +8$$
$$(-4) \times (-2) = +8$$
$$(+4) \times (-2) = -8$$
$$(-4) \times (+2) = -8$$

Practice using the rules of signs by solving the following problems.

B-11. $(2.5) + (3.5) =$
B-12. $(-2.3) + (-1.7) =$
B-13. $(5.0) - (3.0) =$
B-14. $(5.0) - (-3.0) =$
B-15. $(4) \times (-20) =$
B-16. $(-16) \div (4) =$
B-17. $(-16) \div (-4) =$
B-18. $(-5) + (-5) =$
B-19. $(-3) \times (-2) =$
B-20. $-18 - 5 =$

Answers to Questions in Appendix B

B-1. 24,000
B-2. 24.5
B-3. 2,408,000,000
B-4. 21.8
B-5. 1
B-6. 24
B-7. 2.45
B-8. 2,408
B-9. 0.218
B-10. 0.0001
B-11. 6.0
B-12. −4.0
B-13. 2.0
B-14. 8.0
B-15. −80
B-16. −4
B-17. +4
B-18. −10
B-19. +6
B-20. −23

Appendix C

Exponential Notation

You are probably familiar with exponents. You have seen numbers like 3^2, and you know that this number means "three squared," or 3×3. The 2 is called the exponent; the 3 is called the base. The exponent indicates the number of times the base would appear as a factor in an indicated product. For example, 6^4 means $6 \times 6 \times 6 \times 6$; the 6 appears four times.

When numbers are written in exponential notation, the base is normally 10, because our number system is in base 10. It is easy to convert from exponential numbers to ordinary numbers by moving a decimal point. Table C.1 gives some examples.

Table C.1. Powers of Ten Written in Decimal and Exponential Form

Number	*Indicated product*	*Exponential form*
1,000,000	$= 10 \times 10 \times 10 \times 10 \times 10 \times 10$	$= 10^6$
100,000	$= 10 \times 10 \times 10 \times 10 \times 10$	$= 10^5$
10,000	$= 10 \times 10 \times 10 \times 10$	$= 10^4$
1,000	$= 10 \times 10 \times 10$	$= 10^3$
100	$= 10 \times 10$	$= 10^2$
10	$= 10$	$= 10^1$
1	$= 1$	$= 10^0$
.1	$= \frac{1}{10}$	$= 10^{-1}$
.01	$= \frac{1}{10} \times \frac{1}{10}$	$= 10^{-2}$
.001	$= \frac{1}{10} \times \frac{1}{10} \times \frac{1}{10}$	$= 10^{-3}$
.0001	$= \frac{1}{10} \times \frac{1}{10} \times \frac{1}{10} \times \frac{1}{10}$	$= 10^{-4}$
.00001	$= \frac{1}{10} \times \frac{1}{10} \times \frac{1}{10} \times \frac{1}{10} \times \frac{1}{10}$	$= 10^{-5}$
.000001	$= \frac{1}{10} \times \frac{1}{10} \times \frac{1}{10} \times \frac{1}{10} \times \frac{1}{10} \times \frac{1}{10}$	$= 10^{-6}$

Any number made up of the digit 1 followed by zeros can be written as a power of 10 (that is, 10 to some exponent), as shown in Table C.1. Any other number can be written as a number between 1 and 10 multiplied by 10 to some power. Here is an example:

$$17{,}600 = 1.76 \times 10 \times 10 \times 10 \times 10 = 1.76 \times 10^4$$

All we are saying is that 17,600 is the number you will get when you multiply 1.76 by 10 four times. Then 1.76×10^4 is just another way of representing 17,600. In Table C.1, we could have written 1,000 as 1×10^3. After all, 1,000 is simply 1 multiplied by 10 three times. We did not write the 1 because 1 times any number is that number. In other words,

$$1{,}000 = 10 \times 10 \times 10 = 10^3 = 1 \times 10^3$$

We have written the *same* number in four different ways. The four expressions are equivalent.

For some reason, students seem to be confused about what is happening when a number is converted to exponential notation. It is not mysterious. Multiplication and division are inverse operations. When you first divide a number by 2 and then multiply the answer by 2, you get the same number you started with, as shown in Figure C.1.

The number we start with → $6 \div 2 = 3$

$3 \times 2 = 6$ ← The same number, after the inverse operations of multiplication and division

Figure C.1
Multiplication and division are inverse operations. Dividing by and then multiplying by the same number result in the number that you started with.

It works with any number. This is what we do to convert to exponential notation:

Figure C.2

The number we start with → $17{,}600 \div 10{,}000 = 1.76$

$1.76 \times 10{,}000 = 1.76 \times 10^4$ ← The same number expressed in exponential form.

We first divided by 10,000 (or divided by 10 four times) and then reversed the operation and multiplied by 10,000 (or multiplied by 10 four times).

Instead of actually carrying out the second operation, we left the number as an indicated product for convenience. In practice, we do the division the simple way—by moving the decimal point to the left. In Figure C.2, note that in dividing 17,600 by 10,000, we move the decimal *four* places to the left. Instead of multiplying by 10,000 when we reverse the process, we indicate this multiplication by writing 10^4. The exponent of 10 is *four,* the same as the number of places the decimal point was moved when the division was done.

A number written in exponential notation can be written as an ordinary number by reversing the process.

$$1.76 \times 10^4 = 17,600$$

Here we move the decimal point in 1.76 *four* places to the right, adding zeros as needed to locate the decimal. Practice converting the following numbers to exponential notation, or vice versa.

	Number	*Exponential form*
C-1.	1,240,000	______________
C-2.	______________	1.32×10^5
C-3.	1,280	______________
C-4.	______________	2.06×10^3
C-5.	1,000,000,000,000,000	______________
C-6.	______________	6.02×10^{23}

So far we have discussed only large numbers, but the same principle applies to small numbers such as 0.0000025. In this case, the number is multiplied by some power of 10 to get a number between 1 and 10 and then divided by the same power of 10, as shown in Figure C.3.

Figure C.3

The original number → $0.00000025 \times 10,000,000 = 2.5$

$2.5 \div 10,000,000 = 2.5 \times 10^{-7}$ ← The same number in exponential form

The only thing new here is the negative exponent. A negative exponent indicates division. The second step shown in Figure C.3 could be broken down like this:

$$2.5 \div 10,000,000 = \frac{2.5}{10,000,000} = \frac{2.5}{10^7} = 2.5 \times 10^{-7}$$

These are four different ways of saying "2.5 divided by 10 million" or "2.5 divided by 10 seven times."

When converting a small number to exponential notation, we do the initial multiplication the easy way—by moving the decimal point to the right. You will notice that, in the example, we moved the decimal point *seven* places to get 2.5. You will also notice that the exponent in the final number is negative *seven.* If you can count, you can convert numbers to exponential notation and back again. Do some counting on the following problems.

	Number	*Exponential form*
C-7.	0.000124	__________
C-8.	__________	1.32×10^{-3}
C-9.	0.000000000210	__________
C-10.	__________	3.45×10^{-6}

MULTIPLICATION AND DIVISION OF EXPONENTIAL NUMBERS

Arithmetic operations with exponential numbers are simple. In multiplication and division, there are two rules to remember.

1. To multiply exponential numbers, add the exponents.
2. To divide exponential numbers, subtract the exponents.

We will illustrate the rules first with numbers that are powers of 10.

$$10^4 \times 10^3 = ?$$

The rule says that we add exponents to multiply exponential numbers. Then,

$$10^4 \times 10^3 = 10^7$$

To see that this is true, recall what an exponential number means.

$$10^4 \times 10^3 = (10 \times 10 \times 10 \times 10) \times (10 \times 10 \times 10) = 10^7$$

Some students have trouble when negative exponents are involved, but the rules are exactly the same.

$$10^{-4} \times 10^3 = 10^{-1}$$

A negative four (−4) added to a positive three (+3) results in a negative one (−1). (If this is not clear, review the section on signed numbers found in Appendix B.) Again, we can see that this is correct by recalling what the exponential numbers represent.

$$10^{-4} \times 10^{3} = (\tfrac{1}{10} \times \tfrac{1}{10} \times \tfrac{1}{10} \times \tfrac{1}{10}) \times (10 \times 10 \times 10) = \tfrac{1}{10} = 10^{-1}$$

In division, the exponents are subtracted:

$$10^{4} \div 10^{3} = 10^{1}$$

Again, it is easy to see that the rule leads to the correct result by recalling the meaning of exponential numbers.

$$10^{4} \div 10^{3} = \frac{10 \times 10 \times 10 \times 10}{10 \times 10 \times 10} = 10$$

The rule is applied in exactly the same way when negative exponents are involved:

$$10^{-4} \div 10^{3} = 10^{-7}$$

Translating the exponential numbers, we have

$$10^{-4} \div 10^{3} = \frac{\tfrac{1}{10} \times \tfrac{1}{10} \times \tfrac{1}{10} \times \tfrac{1}{10}}{10 \times 10 \times 10} = \tfrac{1}{10} \times \tfrac{1}{10} \times \tfrac{1}{10} \times \tfrac{1}{10} \times \tfrac{1}{10} \times \tfrac{1}{10} \times \tfrac{1}{10} = 10^{-7}$$

Practice multiplication and division on the following problems.

C-11. $10^{3} \times 10^{5} =$
C-12. $10^{-5} \times 10^{8} =$
C-13. $10^{-3} \times 10^{-4} =$
C-14. $10^{3} \div 10^{5} =$
C-15. $10^{-5} \div 10^{8} =$
C-16. $10^{-3} \div 10^{-4} =$

So far all of our examples have involved only numbers that are even powers of ten. Multiplying or dividing numbers like 3.14×10^{23} is no different. Remember that the product is the same, regardless of the order in which a series of numbers are multiplied. The following example illustrates this principle.

Example C.1

Find the following product:

$$(2 \times 3)(4 \times 5) =$$

The normal way to find the product would be to first find the products within the parentheses, like this:

$$(2 \times 3)(4 \times 5) =$$
$$6 \times 20 = 120$$

However, a basic rule in mathematics is that the result is the same, regardless of the order in which you do successive multiplications. The problem could be rewritten as

$$(2 \times 3)(4 \times 5) = 2 \times 3 \times 4 \times 5 = (2 \times 4)(3 \times 5) = (2 \times 5)(4 \times 3)$$

and in any other arrangement you can think of. If you carry out the indicated multiplications, you will see that the product is always 120.

This same principle is used to find the product of exponential numbers:

$$(2 \times 10^3)(3 \times 10^2) = 2 \times 10^3 \times 3 \times 10^2$$

Regrouping, we get

$$(2 \times 3) \times (10^3 \times 10^2)$$

Carrying out the indicated operations, we get

$$6 \times 10^5$$

Practice this rearrangement by working the following problems.

C-17. $(2.1 \times 10^3) \times (1.86 \times 10^{-2}) =$
C-18. $(4.5 \times 10^{-4}) \times (6.44 \times 10^{15}) =$
C-19. $(1.11 \times 10^{23}) \times (5.68 \times 10^{-20}) =$
C-20. $(5.5 \times 10^{-8}) \times (3 \times 10^{-10}) =$
C-21. $(1.09 \times 10^{-25}) \times (2.99 \times 10^3) =$

Division is similar, but we must pay attention to what is multiplied and what is divided. An example will show how it is done.

Example C.2

Find the following quotient.

$$(2 \times 3) \div (4 \times 6) =$$

Let us first work this problem as one would normally work it to see what the answer should be.

$$(2 \times 3) \div (4 \times 6) = \frac{2 \times 3}{4 \times 6} = \frac{6}{24} = \frac{1}{4}$$

$$6 \div 24 = \frac{6}{24} = \frac{1}{4}$$

Now notice that we get the same result when we rearrange terms and carry out the two operations of multiplication and division as follows:

$$(2 \times 3) \div (4 \times 6) = \frac{2 \times 3}{4 \times 6} = \frac{1}{2} \times \frac{1}{2} = \frac{1}{4}$$

$$(2 \div 4) \times (3 \div 6) = \frac{2}{4} \times \frac{3}{6}$$

$$\frac{1}{2} \times \frac{1}{2} = \frac{1}{4}$$

In division of exponential numbers, the first terms are divided, next the second terms are divided, and then the product of these quotients is written as the final answer:

$(2 \times 10^3) \div (4 \times 10^2) =$

$(2 \div 4) \times (10^3 \div 10^2) =$

$$(2 \times 10^3) \div (4 \times 10^2) =$$

$$(2 \div 4) \times (10^3 \div 10^2) =$$

$$0.5 \times 10^1 = 0.5 \times 10^1 = 5$$

Practice this rearrangement by working the following problems.

C-22. $(2.1 \times 10^3) \div (1.86 \times 10^{-2}) =$

C-23. $(4.5 \times 10^{-4}) \div (6.44 \times 10^{15}) =$

C-24. $(1.11 \times 10^{23}) \div (5.68 \times 10^{20}) =$

C-25. $(5.5 \times 10^{-8}) \div (3 \times 10^{-10}) =$

C-26. $(1.09 \times 10^{-25}) \div (2.99 \times 10^3) =$

ADDITION AND SUBTRACTION OF EXPONENTIAL NUMBERS

Addition and subtraction of exponential numbers are just as simple as multiplication and division, so long as you remember not to mix apples and pears. The exponents of the numbers added or subtracted must be the same.

$$(\mathbf{2} \times 10^3) + (\mathbf{4} \times 10^3) = \mathbf{6} \times 10^3$$

The arithmetic is the same as in the following problem.

$$(\mathbf{2} \times 8) + (\mathbf{4} \times 8) = \mathbf{6} \times 8$$

But there is no way to do the following addition without multiplying first.

$$(2 \times 8) + (4 \times 6) =$$

There is also no way to add the following exponential numbers directly.

$$(2 \times 10^3) + (4 \times 10^4) =$$

The addition *can* be done if one of the numbers is rewritten so that the exponents are the same. The following solutions are equivalent.

$$(0.2 \times 10^4) + (4 \times 10^4) = 4.2 \times 10^4 \quad (\text{or } 42 \times 10^3)$$

$$(2 \times 10^3) + (40 \times 10^3) = 42 \times 10^3 \quad (\text{or } 4.2 \times 10^4)$$

Subtraction of exponential numbers is done in exactly the same way. As in any subtraction, the answer may be a negative number:

$$(2 \times 10^3) - (4 \times 10^3) = -2 \times 10^3$$

Practice the following addition and subtraction problems. After you have done the problem with the numbers written in exponential form, check your work by first converting the numbers to the decimal equivalent and then doing the arithmetic. If you have worked the problem correctly, your answers in exponential form and in decimal form should agree.

C-27. $(2 \times 10^3) + (4 \times 10^3) =$
C-28. $(3 \times 10^2) - (2 \times 10^2) =$
C-29. $(5 \times 10^{-3}) + (4.0 \times 10^{-2}) =$
C-30. $(6 \times 10^3) - (2.5 \times 10^4) =$
C-31. $(2.2 \times 10^{-3}) + (1.1 \times 10^{-4}) =$
C-32. $(5.6 \times 10^2) - (4.5 \times 10^3) =$

The following problems give additional practice in the arithmetic of exponential numbers.

C-33. $(2.24 \times 10^{-3}) \times (4.56 \times 10^4) =$
C-34. $(8.9 \times 10^5) \div (1.345 \times 10^9) =$
C-35. $(8.9 \times 10^5) \times (1.345 \times 10^9) =$
C-36. $(8.9 \times 10^5) + (1.345 \times 10^6) =$
C-37. $(8.9 \times 10^5) - (1.345 \times 10^6) =$
C-38. $(6.65 \times 10^{-8}) - (6.65 \times 10^{-9}) =$
C-39. $(3.30 \times 10^{-5}) \times (8.74 \times 10^{-2}) =$
C-40. $(1.05 \times 10^{15}) \div (4.2 \times 10^{-3}) =$
C-41. $(5 \times 10^{25}) \times (10^{-23}) =$
C-42. $(3.4 \times 10^{22}) \div (10^{20}) =$
C-43. $(1.0 \times 10^{26}) \div (2.0 \times 10^{24}) =$
C-44. $(4.44 \times 10^{-8}) \times (2 \times 10^{-4}) =$
C-45. $(3.4 \times 10^5) + (5 \times 10^4) =$

Answers to Questions in Appendix C

C-1. 1.24×10^6
C-2. 132,000
C-3. 1.28×10^3
C-4. 2,060
C-5. 10^{15}
C-6. 602,000,000,000,000,000,000,000
C-7. 1.24×10^{-4}
C-8. 0.00132
C-9. 2.10×10^{-10}
C-10. 0.00000345
C-11. 10^8
C-12. 10^3
C-13. 10^{-7}
C-14. 10^{-2}
C-15. 10^{-13}
C-16. 10
C-17. $3.9 \times 10^1 = 39$
C-18. 2.9×10^{12}
C-19. 6.30×10^3
C-20. 2×10^{-17}
C-21. 3.26×10^{-22}
C-22. 1.1×10^5
C-23. 7.0×10^{-20}
C-24. 1.95×10^2
C-25. 2×10^2
C-26. 3.65×10^{-29}
C-27. 6×10^3
C-28. 1×10^2
C-29. 4.5×10^{-2}
C-30. -1.9×10^4
C-31. 2.3×10^{-3}
C-32. -3.9×10^3
C-33. 1.02×10^2
C-34. 6.6×10^{-4}
C-35. 1.2×10^{15}
C-36. 2.24×10^6
C-37. -4.5×10^5
C-38. 5.98×10^{-8}
C-39. 2.88×10^{-6}
C-40. 2.5×10^{17}
C-41. 5×10^2
C-42. 3.4×10^2
C-43. 50
C-44. 9×10^{-12}
C-45. 3.9×10^5

Appendix D

Concentration Terms

Concentration is an intensive property that indicates the amount of solute in some given amount of solution. The most commonly used measure of concentration is molarity, which is discussed in Chapter 10. **Molarity** is defined as follows:

$$\text{molarity} = \frac{\text{moles of solute}}{\text{liter of solution}}$$

Here, amount of solute is measure in moles and amount of solution is measured in liters.

When the amount of solute or the amount of solution is measured in different terms, we use other concentration terms. Some of the concentration terms that you may encounter in other books are defined here.

MOLALITY

Molality is closely related to molarity. There are two differences between the terms. In both cases the amount of solute is measured in moles. In molality, however, the *mass* of the *solvent* is found in the denominator of the ratio, rather than the *volume* of the *solution*. **Molality** is defined as follows:

$$\text{molality} = \frac{\text{moles of solute}}{\text{kilogram of solvent}}$$

The most common use of molality is in expressions dealing with the lowering of the freezing point or the elevation of the boiling point that results from dissolving a solid in a liquid solvent.

PERCENT COMPOSITION

Percent composition is a concentration and may be given in terms of mass or volume.

$$\text{percent (by mass)} = \frac{\text{mass of solute}}{\text{mass of solution}} \times 100$$

$$\text{percent (by volume)} = \frac{\text{volume of solute}}{\text{volume of solution}} \times 100$$

MOLE FRACTION OR MOLE PERCENT

Mole fraction or **mole percent** is similar to the previous terms, except that the ratio is expressed in moles rather than in terms of mass or volume. The expressions are

$$\text{mole fraction} = \frac{\text{moles of solute}}{\text{moles of solution}}$$

$$\text{mole percent} = \frac{\text{moles of solute}}{\text{moles of solution}} \times 100$$

The quantity "moles of solution" is the sum of the moles of each component in the solution.

NORMALITY

Normality is related to molarity. In both cases the amount of solute is compared to the volume of solution expressed in liters. The difference is that, in expressing normality, we express the amount of solute as the number of equivalents of solute rather than the number of moles of solute. The expression that defines **normality** is

$$\text{normality} = \frac{\text{equivalents of solute}}{\text{liter of solution}}$$

Equivalent is a new term, and it is difficult to define. The number of equivalents in a given mass of a compound is always a simple whole-number multiple of the number of moles in the sample. However, what that simple whole number is depends on the reaction under consideration. In general, the number of equivalents of an acid is the number of moles of that acid multiplied by the number of replaceable hydrogen ions in the acid. Thus the number of equivalents of HCl would be the same as the number of moles of HCl in solution, because there is only one replaceable hydrogen. However, the number of equivalents of H_2SO_4 (for most reactions) is twice the number of moles of H_2SO_4 in solution, because there are two replaceable hydrogens in the compound.

In a similar manner, the number of equivalents of a base is generally equal to the number of moles of that base multiplied by the number of basic OH groups in the compound. Thus the number of equivalents of NaOH is the same as the number of moles of NaOH, because there is only one OH group in the molecule. However, the number of equivalents of $Ca(OH)_2$ is twice the number of moles, because there are two OH groups that can act as a base for every molecule of the compound.

What we have just said is a slight oversimplification of equivalents, but a more complete explanation is beyond the scope of this book.

Appendix E

Rules for Good Graphing

You have probably made a graph before, but you may have forgotten some important points. This appendix reviews some important guidelines. It does not teach you to plot points. If you have never made a graph, ask your instructor for help.

1. The purpose of a graph is to communicate information in a concise manner. The graph will tell your readers nothing if they don't know what is represented. Rule number 1 is to *give your graph a title.* The title normally appears at the top of the graph.
2. *Indent the axes from the edge of the graph paper* and draw them with a straightedge. (Some graph paper is printed with an adequate margin, making this step unnecessary.)
3. *Label each axis and indicate the units used.* Numbers along the axis tell very little if you don't indicate what they represent.
4. *Choose an appropriate scale.* What is appropriate? First, the scale must allow you to get all data on the graph. Check your largest and smallest values. For most graphs, there is no need to begin the scale at zero. Just be sure you begin with a number lower than your lowest data point. Before plotting any points, see whether the largest value will fall on the graph when you use the scale you have selected. If not, change the scale so that both the largest and the smallest values will appear. The scale should produce a graph large enough to read conveniently. As a rule of thumb, the scale should be selected so that your graph covers at least half of the sheet of graph paper.
5. *Choose a convenient scale.* Most people find a graph easier to read (and plot) when they let each square represent a value of 1, 2, 5, or a multiple of 10 times these numbers: 10, 20, 50 or 0.1, 0.2, 0.5, etc. Most of us count by ones, twos, and fives fairly well, but we count by threes, fours,

sixes, and sevens rather poorly. This is why giving each square a value of 1, 2, or 5 makes a more convenient scale.

6. *Maintain the same scale for the length of the graph.* Some students solve the problem of running off the edge of the graph paper by changing the scale when they see that the last few points will not go on the paper. This distorts the shape of the graph and gives misleading information. It is absolutely taboo!
7. *Locate points with an* X *or a dot with a small circle around it.* After you finish plotting points and draw a line through them, you may need to locate those points again. If each point is marked by a small dot, the line you drew may have obscured them. An X or a dot with a circle around it can be located.
8. *Draw a smooth curve to represent the general tendency of the data points.* Graphs drawn in math classes represent absolute numbers, and each data point falls on the curve being plotted. When you draw these graphs, you make the curve pass through every point plotted. This is not the case in science, where the data points represent experimental measurements. In science, the data used to plot the graph are uncertain, and we try to reduce the uncertainty when we plot the graph. The following example shows how.

Example E.1

Several pieces of glass were weighed and their volume measured. A graph was then made with the mass of each piece of glass plotted along the y axis of the graph paper and the corresponding volume plotted along the x axis. As with any measurement, the data used to plot the graph are uncertain.

One piece of glass had the following mass and volume, with the indicated uncertainty:

$$\text{mass} = 53.2\ \text{g} \pm 0.2\ \text{g} \qquad \text{volume} = 21.2\ \text{cm}^3 \pm 0.2\ \text{cm}^3$$

A portion of a sheet of graph paper is shown in Figure E.1, and the point for this piece of glass is properly plotted with an X.

We do not know that the X in Figure E.1 is where the point should be located. The problem is not in plotting the graph; it is in making the measurements. The mass that was measured was uncertain. The true mass could be as small as 53.0 g or as large as 53.4 g. Furthermore, the true volume could be as small as 21.0 cm^3 or as large as 21.4 cm^3. This means that, if you were able to make these measurements with *no* uncertainty, this data point could fall anywhere within the shaded area on the graph.

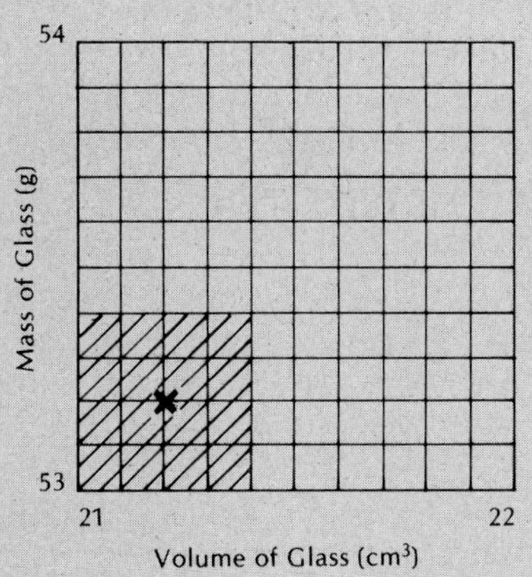

Figure E.1
Because of uncertainty in measurement, the point indicated by the X could be anywhere within the shaded region.

When we draw a *smooth* curve through as many points as possible, letting an equal number of points fall above and below the curve, we are trying to eliminate uncertainty in the measurement. We believe that the points that fall on a line drawn in this manner are closer to true values than any single point obtained by measurement.

Now that you have reviewed rules for preparing a graph, plot the following values for mass and volume of glass on a full sheet of graph paper. Be sure that you follow all of the rules, and then compare your graph with Figure E.7 on page 498.

Volume of glass (cm^3)	*Mass of glass* (g)
0.4	0.73
0.7	1.51
1.0	2.25
1.3	3.03
1.9	3.74
2.0	4.47
2.2	5.21
2.9	5.97
3.2	6.89
3.5	7.72
4.0	8.45
4.2	9.27
4.6	9.98
4.9	10.70
5.1	11.44
5.5	12.24
6.0	12.99
6.2	13.73
6.8	14.45
7.0	15.19

Now that you have plotted a graph, look at some others and tell what is wrong with them. Figures E.2 through E.6 are graphs prepared by students in my class. All graphs are for the *same* experiment, in which students measured the temperature of moth flakes and the temperature of water used to cool the moth flakes and then plotted these temperatures vs. time.

E-1. Name all of the things that you find wrong with the graph shown in Figure E.2. Check your list against the one at the end of Appendix E before answering the next question.
E-2. Name all of the things that you find wrong with the graph shown in Figure E.3.
E-3. Name all of the things that you find wrong with the graph shown in Figure E.4.
E-4. Name all of the things that you find wrong with the graph shown in Figure E.5.
E-5. Name all of the things that you find wrong with the graph shown in Figure E.6.

Figure E.2

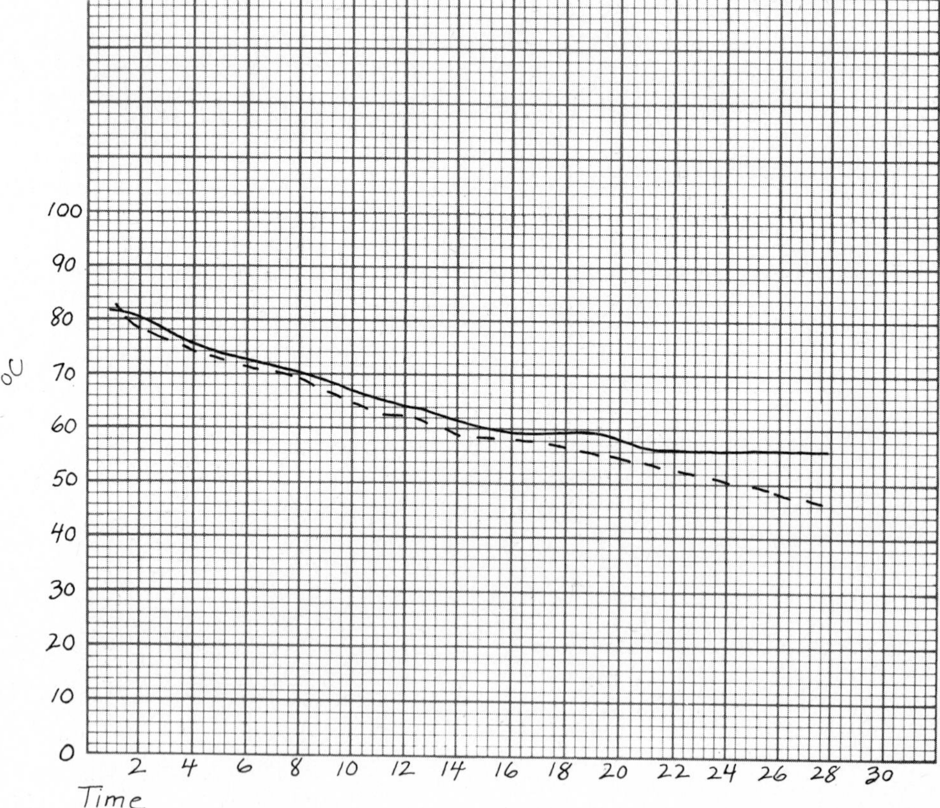

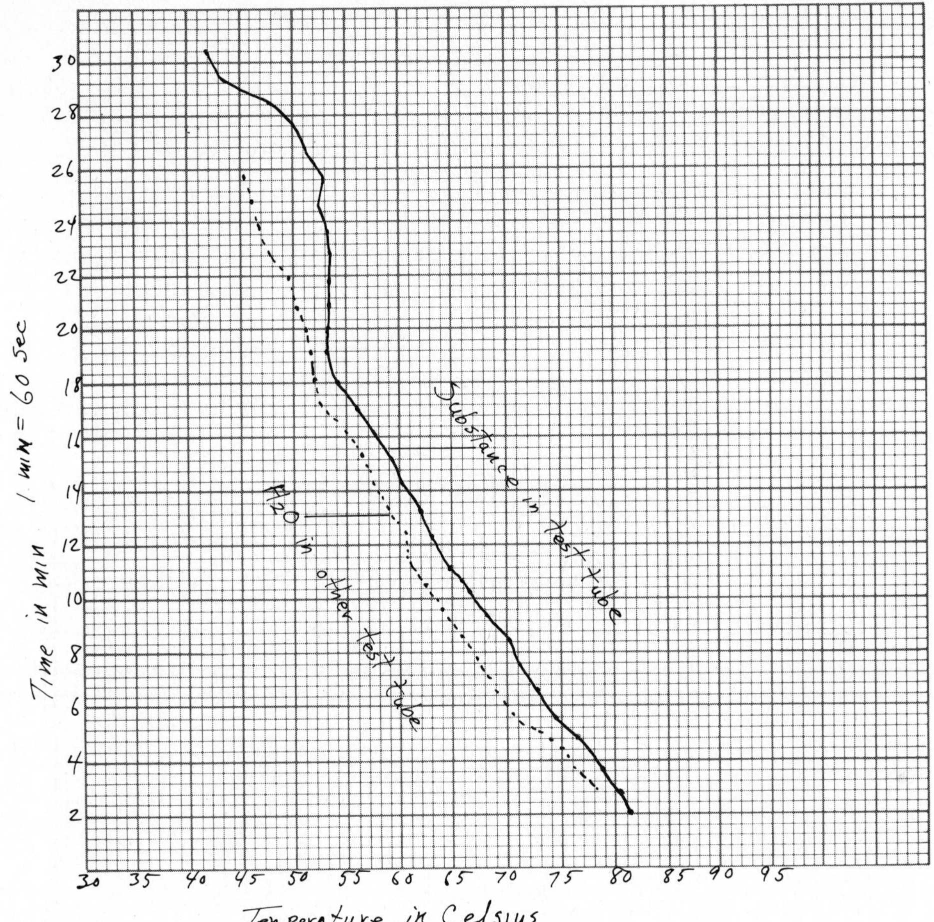
30
28
26
24
22
20
18
16
14
12
10
8
6
4
2
Time in min 1 min = 60 sec
30 35 40 45 50 55 60 65 70 75 80 85 90 95
Temperature in Celsius
Substance in test tube
H2O in other test tube

Figure E.3

Figure E.4

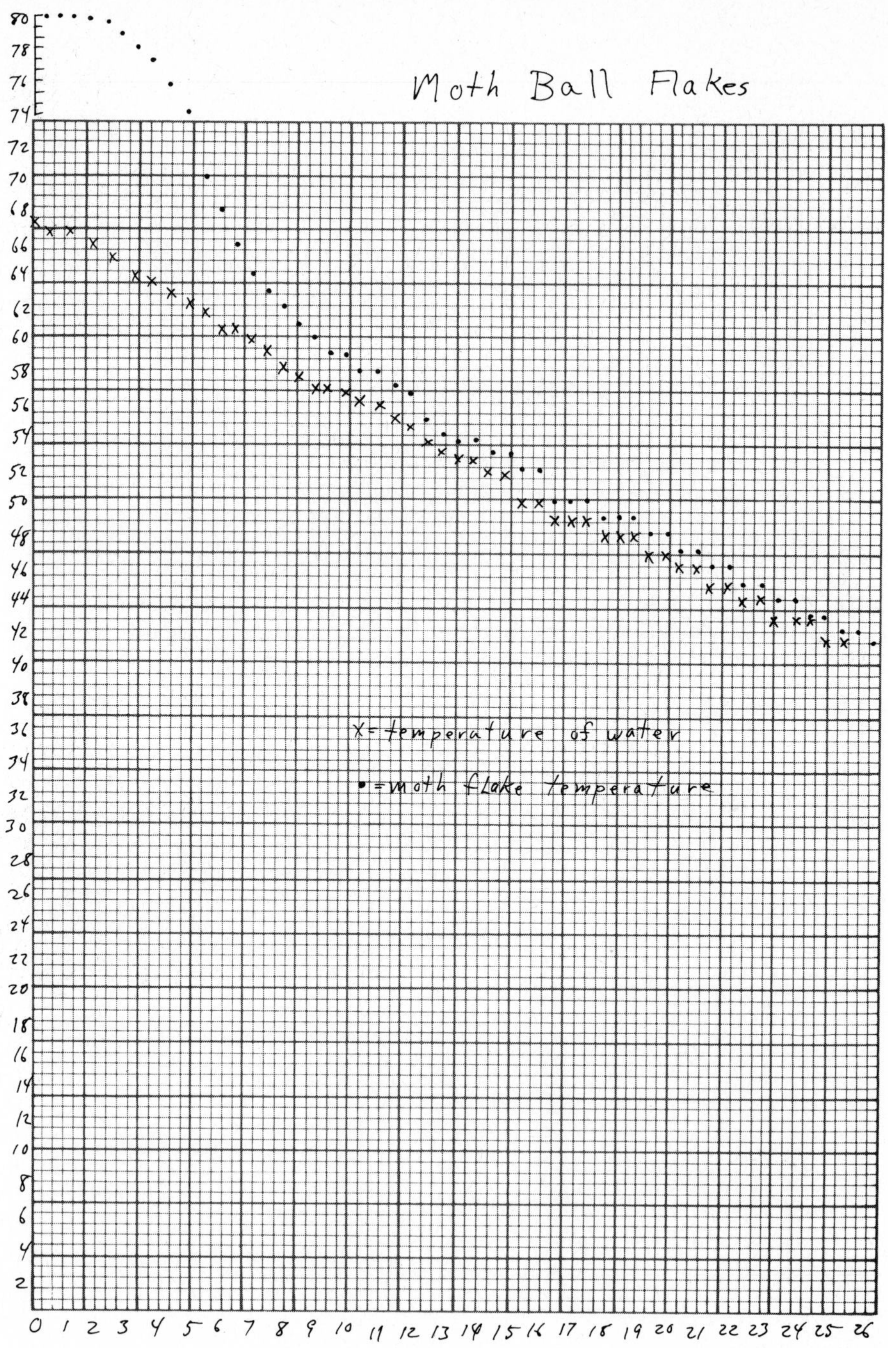
Moth Ball Flakes
X = temperature of water
• = moth flake temperature
80
78
76
74
72
70
68
66
64
62
60
58
56
54
52
50
48
46
44
42
40
38
36
34
32
30
28
26
24
22
20
18
16
14
12
10
8
6
4
2
0 1 2 3 4 5 6 7 8 9 10 11 12 13 14 15 16 17 18 19 20 21 22 23 24 25 26

Figure E.5

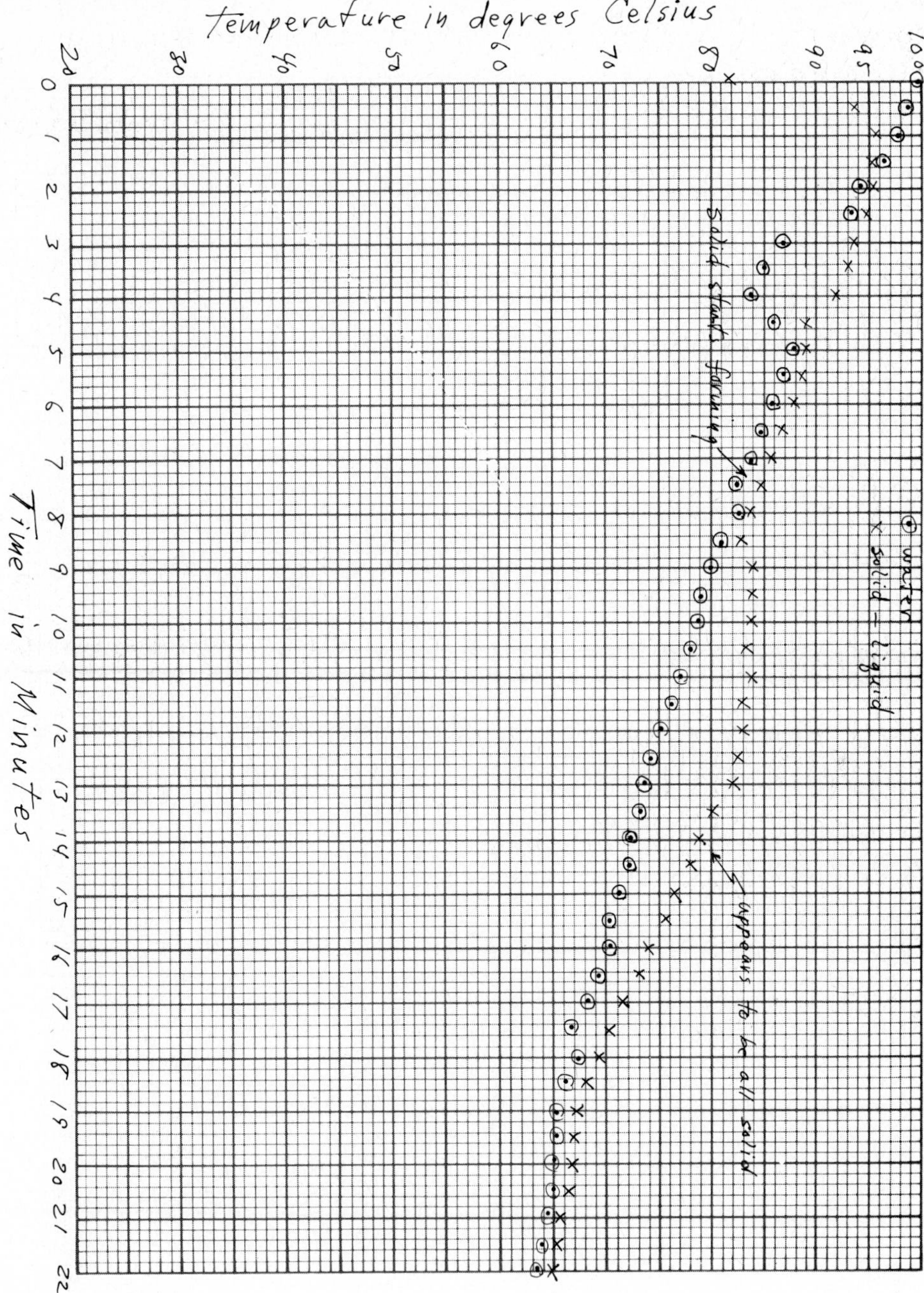
temperature in degrees Celsius
20
30
40
50
60
70
80
90
95
100
Time in Minutes
0
1
2
3
4
5
6
7
8
9
10
11
12
13
14
15
16
17
18
19
20
21
22
⊙ water
× solid – liquid
solid starts forming
appears to be all solid

Figure E.6

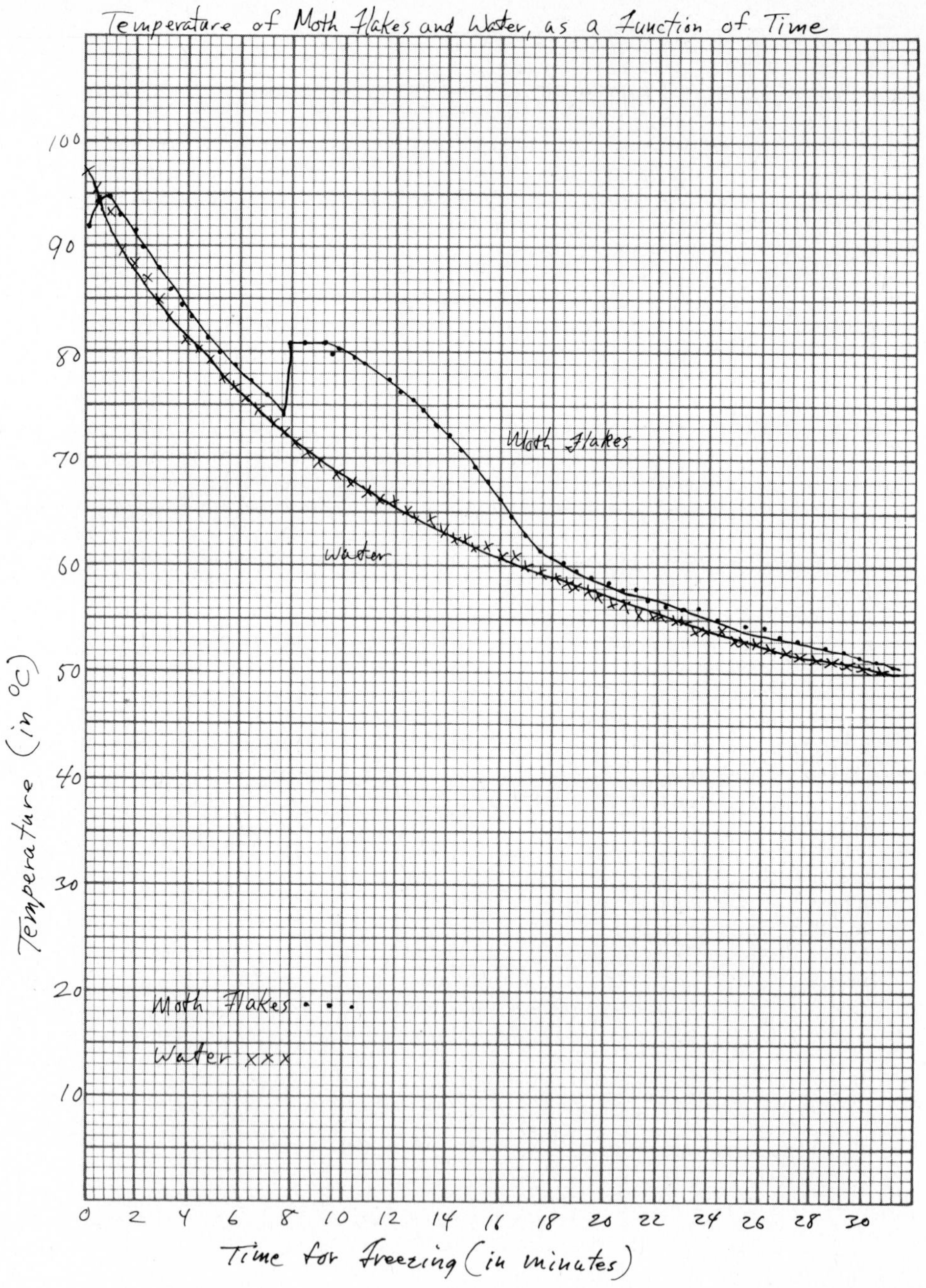
Temperature of Moth Flakes and Water, as a Function of Time
Temperature (in °C)
100
90
80
70
60
50
40
30
20
10
0 2 4 6 8 10 12 14 16 18 20 22 24 26 28 30
Time for Freezing (in minutes)
Moth Flakes
Water
Moth Flakes • • •
Water x x x

Answers to Questions in Appendix E

E-1. (1) The graph has no title. The reader cannot tell what the graph represents.
(2) Labels for the axes are inadequate. In particular, you do not know whether the time is in seconds, minutes, or hours. It is likely that a reader would interpret "°C" as the temperature of the moth flakes and water in Celsius, but it would be better for the graph to say so.
(3) The scale selected for the *y* axis is too small. Letting each division represent 5 degrees rather than 10 would have been much better.
(4) There are no X's or dots with circles to indicate where the points were located.
(5) Obviously two things are plotted, but we aren't told what is represented by the solid line and what is represented by the dotted line.
(6) The curves don't appear to be very smooth, but without being able to see where the original points were located, we can't tell how the curves were drawn.

E-2. This is a definite improvement over the previous graph, but there are still some problems.
(1) There is still no title.
(2) The temperature scale could have been selected to utilize more of the graph paper.
(3) Data points were not circled or marked with X's. They are hard to see.
(4) Obviously the curves were drawn so that they passed through every point. As a result, they are "bumpy." Curves should be smooth, and they should pass through as many points as possible with equal numbers of points falling above and below the smooth curve that emerges.
(5) Note that this student plotted temperature along the *x* axis and time along the *y* axis. There is nothing wrong with this except that the instructions were to plot the graph the other way around. It certainly makes this graph look different from the others!

E-3. (1) There is a title, but it doesn't tell us much.
(2) There are no labels for the axes.
(3) The data points look good, but no curve was drawn.
(4) The scale for the *y* axis is poor. This student wasted the bottom half of the graph paper by starting the scale on the *y* axis at zero, even though the lowest temperature reading is about 40°C. It would have been much better to start the scale at around 35 and use a larger scale.
(5) Under no circumstances should you extend the data beyond the edge of the graph paper, as this student has done at the top. It is impossible to plot points accurately this way.
(6) The scale selected is inconvenient. Each small block is $\frac{1}{3}$ of a unit. This makes the scale difficult to read and makes it hard to plot the data without error.
(7) No margin was left for labeling the axes (not obvious from the figure).

E-4. (1) Still no title. However, the other labels are good.
(2) The points are properly located, but there is no curve.
(3) The temperature scale is poor. The student could easily have begun with 55 and used a larger scale.

E-5. Finally, a fairly good graph! Still, there is room for improvement.
(1) The dots should be smaller with circles around them.
(2) Although acceptable, the temperature scale could have been made larger by starting the scale at 40 rather than zero.

Figure E.7

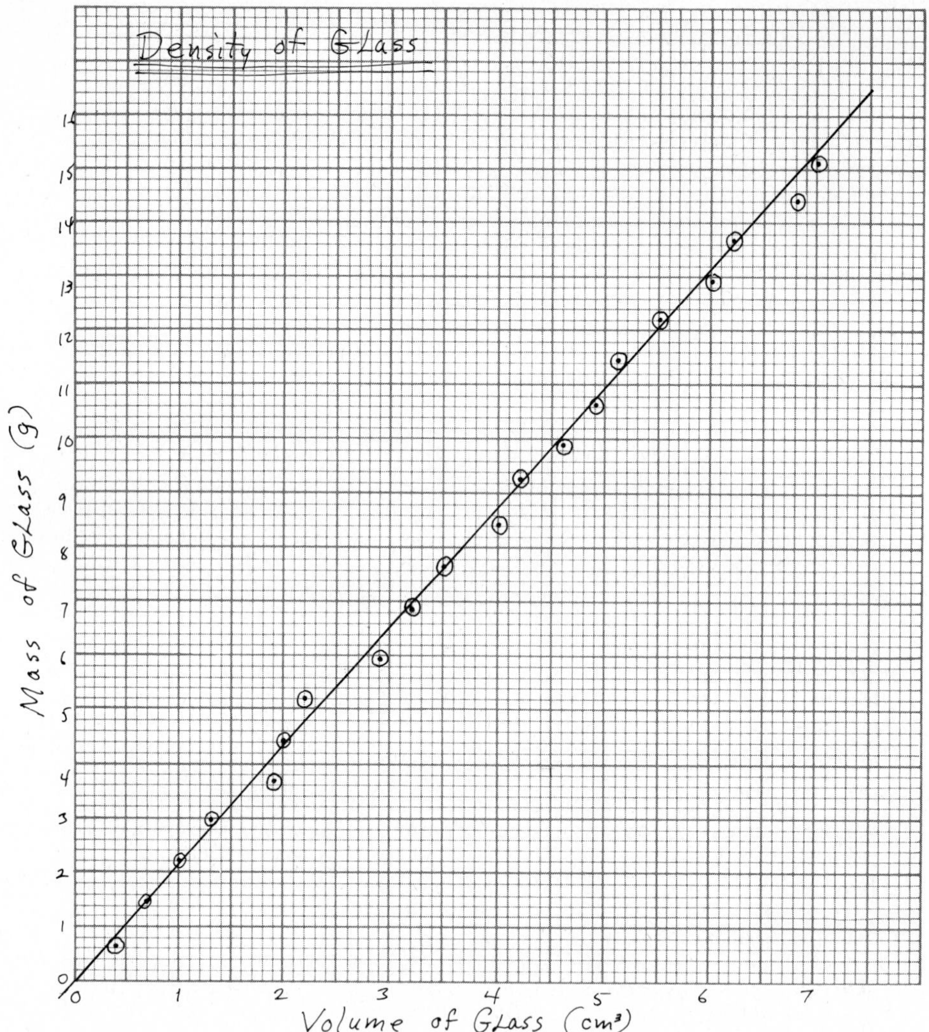
Density of Glass
Mass of Glass (g)
Volume of Glass (cm³)
0
1
2
3
4
5
6
7
8
9
10
11
12
13
14
15

Appendix F

Solving Word Problems

Many students have words for word problems. Difficult, awful, impossible, and nonsense are some of the *nicer* things they say. Perhaps you have felt the same way. If so, it is time to get over these feelings, because you can and must learn to solve word problems. It will take practice and hard work, but you can certainly do it.

Working word problems is basically a matter of translating from English words and sentences to mathematical words and sentences. It requires careful reading, patience, and logical thinking. In this appendix you will learn some of the things that can make you a better problem solver. However, you will not be an expert when you finish. Problem solving is a skill at least as complex as playing tennis. You would not expect to be a tennis expert after one or two lessons. You need practice over a long period of time. The same is true of problem solving.

RESEARCH ON PROBLEM SOLVING

A great deal of research has been done on problem solving, and we can say several things about good problem solvers and poor problem solvers.

Poor problem solvers

1. Poor problem solvers don't believe that they can solve problems. They seem to feel that you either know it or you don't.
2. Poor problem solvers are impatient. If they don't see the answer quickly, they give up.

Good problem solvers

1. Good problem solvers believe that they can solve just about any problem if they work at it long enough.
2. Good problem solvers are persistent. They will work for a long time on a problem before giving up.

3. Poor problem solvers are careless readers. They often misread what is written. They may begin working the problem before they know what the problem says.

3. Good problem solvers read carefully. They often read the problem several times before beginning to work it. They are certain that they know what is said before they begin.

4. Poor problem solvers jump to conclusions and guess. They expect to go immediately from what is given to the answer. If they can't, they guess or give up.

4. Good problem solvers break complex problems into small steps and solve them one step at a time. If they don't see a path to a solution immediately, they look at what is given and try to find new relationships that will lead to a solution.

5. Poor problem solvers organize their work carelessly or not at all. They are unable to retrace their thinking at the end of the problem.

5. Good problem solvers organize their work so that they can come back at any point and follow the steps they have taken. They often draw pictures to describe the problem.

6. Poor problem solvers seldom check their work.

6. Good problem solvers habitually check what they have done, both at the end of the problem and at various steps along the way.

7. Poor problem solvers have only one approach to problems. Usually they try to recall a "formula." If they can't, they don't know what to do and give up.

7. Good problem solvers do many things when they try to solve problems. They sometimes draw pictures or try to visualize a concrete example. They jot down notes that translate what they have read into words or symbols that are easier to understand. They may try out many different approaches as they solve the problem.

If you are not a good problem solver, you probably recognized some of the things that you do in the list on the left. You can become better by doing the things mentioned on the right. However, it is not enough to know the difference between a good problem solver and a poor problem solver. Many people who know the difference between a good tennis player and a poor one still can't play tennis! Once you know you are not a good problem solver, you must do something about it. That will take some effort and possibly some help.

HOW TO WORK ON PROBLEM SOLVING

There are several things that you can do to improve your problem solving. Here are some things that you might try.

1. Get someone to work with you. When you work together, one of you should act as problem solver and the other as listener. When you are solving a problem, read the problem aloud and talk aloud as you work the problem. The listener should read along with you and stop you when you misread something. The listener should *not* tell you the correct reading. You should go back and find it yourself. (The idea is to encourage you to read carefully. *You* must find what you read incorrectly.)
2. When you see your instructor or another student work a problem that you find difficult, pay attention to how they work the problem. *How are they reasoning?* If you have trouble following their work, ask them to explain their thinking. Better, see if you can get them to think aloud as they solve the problem.
3. When you get help with a difficult problem, don't ask the instructor to work the problem for you. Rather, tell the instructor how you would go about working the problem and ask him or her to tell you where your reasoning is wrong. Tennis players get better by playing the game and asking the pro to tell them what they are doing wrong. They do not get better by just watching the pro.
4. Practice the strategies outlined in the following section.

HOW I WORK WORD PROBLEMS

When I work word problems, I ask several questions. I am usually not conscious of asking these questions, however. They are automatic. For someone just learning to work word problems, it is a good idea to write them down. Here are the questions that I usually ask in the order in which I usually ask them.

1. What does the problem ask me to find?
2. What information is given in the problem?
3. Do I seem to have all the information that I need?
4. If not, are there facts I should remember that I can use in the problem?
5. If I can't remember facts that help, are there facts given that I can use to *get* the facts I need?
6. Can I ignore some information that is given?
7. What are the important relationships among the facts that I have?

The following example shows how I might reason through a problem. You work it with me.

Example F.1

Large oranges sell for 60¢ a dozen, while small ones cost 55¢ a dozen. What would you pay for four large oranges?

1. What are we asked to find?

- - -

"What would you pay for four large oranges?"

For such a problem, I would usually translate this into the format that I use for factor-label:

$$? ¢ = 4 \text{ large oranges}$$

2. What facts are given?

- - -

a) 1 doz large oranges cost 60¢.
b) 1 doz small oranges cost 55¢.
c) I am to buy four large oranges.

Before going on to answer other questions, note the translation that I did. I said that I was told *one* dozen oranges cost 60¢. The problem actually says "60¢ *a* dozen." The word *a* has been translated as the number *one.* Note also that I was not asked how many *cents* I would pay for four oranges but, when I translated, I wrote "? ¢ = 4 large oranges." It often helps to translate statements of "how much" into appropriate units when you can. The units can be a useful clue to calculation.

3. Do I have all the information I need?

- - -

Yes, I think so, but I won't be sure until I see how to work the problem.

4. Can we ignore some information that is given?

- - -

Since we are asked about *large* oranges, it would appear that we can ignore the price of the *small* oranges.

5. What are the important relationships among the facts given?

- - -

To decide what relationships are important, I look again at what I want to find:

$$? ¢ = 4 \text{ large oranges}$$

I need some relationship between cost and number of large oranges. I notice that 1 doz large oranges cost 60¢.

It just dawned on me that I have the price of a dozen oranges and that

I am asked the cost of 4 oranges—not the cost of 4 doz oranges. I *do* need more information than is given. What is it?

---

1 doz oranges = 12 oranges

Now I can proceed. First, I'll find out how many dozen oranges I am buying.

$$? \text{¢} = 4 \text{ large oranges} \times \frac{1 \text{ doz oranges}}{12 \text{ oranges}} = \tfrac{1}{3} \text{ doz large oranges}$$

Now, since I know how many dozen I am buying, I will multiply by the cost of 1 dozen.

$$? \text{¢} = \tfrac{1}{3} \text{ doz large oranges} \times \frac{60\text{¢}}{1 \text{ doz large oranges}} = 20\text{¢}$$

Actually, I would work the problem in one step, like this:

$$? \text{¢} = 4 \text{ large oranges} \times \frac{1 \text{ doz oranges}}{12 \text{ oranges}} \times \frac{60\text{¢}}{1 \text{ doz large oranges}} = 20\text{¢}$$

You might have used a different strategy. There are usually several ways to attack a problem. If the solution makes sense to you and is well organized, you probably worked it correctly.

The following three examples are similar to the one we just worked. After each example, the questions that need to be asked are repeated. Answer all of the questions pertaining to each example, and then compare your answers with mine at the end of Appendix F.

Example F.2

A foreign car has a fuel tank that holds 80 liters of gasoline. How much would it cost to fill the tank when gasoline costs 62¢ per gallon?

F-1. What are you asked to find?
F-2. What facts are you told?
F-3. Do you need information that isn't given? If so, what?
F-4. Do you have any information that you can ignore? If so, what?
F-5. What relationships do you see? Use them to write unit-factors and solve the problem.

Before going on to the next example, I want to call your attention to the way words were used in the first two examples. In particular, look at the phrase "60¢ a dozen" in Example F.1 and "62¢ per gallon" in Example F.2. Both phrases tell you the cost of *one* unit of material, and both phrases imply some kind of ratio. The word "per" means "for each." It tells us that we will pay 62¢ *for each* gallon. "Per" is implied in the phrase "60¢ a dozen." We are told that we will pay 60¢ *for each* dozen oranges.

Many people would recognize "62¢ per gallon" as the ratio it represents and immediately write down the unit-factor, 62¢/1 gal as information given. In similar fashion, they would translate "60 miles per hour" as 60 mi/1 hr and "2.5 grams per milliliter" as 2.5 g/1 mL. When you see "per" in a word problem or even when you don't see the word but know that it is implied, you can be confident that some ratio is involved.

Some students get fooled by %. This symbol means *per*cent, and its literal meaning is "for each hundred parts of the whole." In the next example, "70% iron" means that there are 70 parts of iron for each hundred parts of iron ore. The ratio implied is 70 lb iron/100 lb iron ore. With this helpful hint on translation, answer the questions for the following example.

Example F.3

A certain iron ore contains 70% iron. How many pounds of this ore are needed to yield 10,000 lb of iron?

F-6. What are you asked to find?
F-7. What are you told?
F-8. Do you need any information that isn't given? If so, what?
F-9. Do you have any information that you can ignore? If so, what?
F-10. What relationships do you see? Use them to solve the problem.

In this example we encountered another word that usually implies a particular mathematical operation. The word is "of," and when you see it, you can usually expect to do some multiplication in order to solve the problem. "And" is another word that usually implies a particular mathematical operation. "Ten pounds of fish *and* five pounds of shrimp" suggests that the weight of the fish and the shrimp will be added.

Example F.4

How many seconds are there in a leap year?

F-11. What are you asked to find?
F-12. What are you told?
F-13. Do you need any information that isn't given? If so, what?
F-14. Do you have any information that you can ignore? If so, what?
F-15. What relationships do you see? Use them to solve the problem.

We have considered only one type of problem in this appendix. It is a type of problem that you will encounter often in chemistry. Other kinds of problems call for different strategies. If you have difficulty with word problems such as these, you may want to work through *Problem Solving and Comprehension: A Short Course in Analytical Reasoning.** It offers many kinds of practice problems and excellent suggestions for improving your problem-solving skills. Many of the suggestions in this appendix were adapted from material in that book.

Answers to Questions in Appendix F

F-1. ? ¢ = 80 L (I have done some translation here. The actual question is how much it would cost to fill the tank. I have decided to express cost in cents, and I noted the relationship that filling the tank requires 80 liters.)

F-2. a) It is a foreign car.
b) Its fuel tank holds 80 liters.
c) Gasoline costs 62¢ for each gallon.

F-3. Yes. I need a relationship between liters and gallons.

$$1 \text{ gal} = 3.78 \text{ L}$$

F-4. I think we can ignore the fact that the car is foreign.

F-5. Fact c implies that 62¢ = 1 gal, so I get the ratio 62¢/1 gal. The relationship between gallons and liters gives me the unit-factor 1 gal/3.78 L.
The problem is solved like this:

$$? ¢ = 80 \not{L} \times \frac{1 \not{\text{gal}}}{3.78 \not{L}} \times \frac{62¢}{1 \not{\text{gal}}} = 1{,}302¢ \ (\$13.02)$$

F-6. ? lb iron ore = 10,000 lb iron

F-7. a) We start with iron ore.
b) The iron ore is 70% iron.
c) We want to get 10,000 lb of iron.

F-8. No

F-9. No

F-10. The only relationship needed is the one derived from 70%. It says: 70 lb iron/100 lb iron ore. As this is written, however, it cannot be used as a unit-factor in the problem. The units will not cancel. Since the inverse of any unit-factor is also a unit-factor, I will use 100 lb iron ore/70 lb iron.

$$? \text{ lb iron ore} = 10{,}000 \not{\text{lb iron}} \times \frac{100 \text{ lb iron ore}}{70 \not{\text{lb iron}}}$$

$$= 14{,}000 \text{ lb iron ore}$$

F-11. ? sec = 1 leap year

F-12. None

F-13. Yes. 1 leap year = 366 days
1 day = 24 hr
1 hour = 60 min
1 minute = 60 sec
(You may not see that you need these relationships immediately. If you don't, ask, "How can I get from what I have (year) to what I want to find (seconds)?" A little thought suggests that a day is closer to a second than is a year, so write the first relationship. Then repeat the question. "How can I get from days to seconds?" An hour is closer to a second than is a day, so write the next relationship. Keep it up until you get to what you want.)

F-14. No.

F-15.

$$? \text{ sec} = 1 \not{\text{leap year}} \times \frac{366 \not{\text{days}}}{1 \not{\text{leap year}}} \times \frac{24 \not{\text{hours}}}{1 \not{\text{day}}} \times \frac{60 \not{\text{min}}}{1 \not{\text{hour}}} \times \frac{60 \text{ sec}}{1 \not{\text{min}}}$$

$$= 31{,}600{,}000 \text{ sec}$$

* Authored by Lochhead, J., and Whimbey, A., and published by The Franklin Institute Press, Philadelphia, Pa. (1979).

Index

A

Absolute temperature, *see* Kelvin
Accuracy, 34, 35
Acids
  common reactions of, 306
  conjugate, 317
  definition of, 306, 313, 332
  dissociation of, 313
  properties of, 305
  relative strengths of, 317
  strong and weak, 311, 318
Acid-base theories
  of Arrhenius, 332
  of Brønsted-Lowry, 332
  of Lewis, 332
Activity series, 260
Adhesives, 452–458
Alkali metals, 407
Aluminum
  Julia B. Hall and, 268
  corrosion of, 226
Aqueous, 225
Area
  meaning of, 22
Arrhenius, Svante, 332
Atmospheric pressure, 350
Atom
  definition of, 100
  description of, 187–190
  electron arrangement in, 393
  size of, 128–129
  structure of, 391
Atomic mass (atomic weight)
  definition of, 153
  history of, 164
  and mass number, 393
  of oxygen-16, 414
Atomic model, 111
Atomic number
  definition of, 190
  and protons, 391
Atomic orbitals, 417, 423
Atomic radius
  periodic trends in, 128
  table of, 128–129
Atomic structure, 391
Atomic weight, *see* Atomic mass
Aufbau principle, 400
Avogadro, Amadeo, 166
  hypothesis, 166
Avogadro's number, 166
  new Avogadro's number, 182
Azeotropic, 82

B

Baking, 231
Balances (common), 34
Barometer
  description of, 350

Bases
as proton acceptors, 307
conjugate, 317
definition of, 307
properties of, 310
relative strengths of, 317
strong and weak, 311–313, 318
Batteries
car, 234
new types, 206
Binary compounds, 131
Boiling point, 54, 76, 82, 102
of various hydrocarbons, 133
Bond
adhesive, 452
angle, 435
covalent, 438
double, 427
energy, 421
ionic, 441
length, 429
polar-covalent, 440
reasons for, 422
single, 427
triple, 427
Bonding electrons, 426
Boyle, Robert, 354
Boyle's Law, 354, 358
unstated conditions, 356
Brønsted, Nicholas, 332
BTU, 51

C

Calculations from chemical equations, *see* Stoichiometry
calorie, 51
Calorie, 52
Cave formation, 233
Cavendish, Henry, 72
Celsius temperature, 50, 58
Certain digits, 38
Change
chemical, 85, 387
physical, 86, 387
Charge
electrical, 185
positive, 186, 191
negative, 186, 191
Charles' Law, 365
Chemical bond, *see* Bond
Chemical change, 85, 387–389
common, 224–234
involving compounds, 115
involving elements, 114
microscopic nature of, 106
Chemical equation
balancing, 216–218
definition of, 214
and quantities, 246–251
reading of, 218–220
symbols used in, 236
Chemical equilibrium, *see* Equilibrium
Classification
of compounds, 302
of matter, 93, 117
Coefficients
in equations, 216
Colloids, 273
Color, 54, 82, 102
Colorless, 272
Combining number, 138
of common ions, 199
and oxidation number, 443
rules for assigning, 140
using to write formulas, 141
of various groups, 139–140
Combustion, 229
magnesium in water, 230
Compound, 111
binary, 131
classification of matter, 93
definition of, 90
microscopic nature of, 101, 104
vs. mixture, 91
Concentration of solutions, 277, 486
Condenser, 85
Conductivity (ionic), 193
Constant
gas, 368, 370
proportionality, 342–343

Conversion factor, *see* Unit-factor
Converting units, 17
Corrosion
of aluminum, 226
of iron, 224
protection from, 240–243
of silver, 226
Crystal structure, 80

D

Dalton, John, 164
atomic theory, 164
law of partial pressures, 373, 375
Decompose, 108
Decomposition
reactions, 235, 254
of soda, 231
Density, 53, 54, 68
of sulfuric acid, 234
of various hydrocarbons, 133
Destructive distillation of wood, 84
Dielectric constant, 80
Diffraction grating, 395
Diffusion of gases, 374
Dilute, 277
Dimensional analysis, *see* Factor-label
Direct proportionality, 342
Dissociation, 286
of acids, 313
Dissolve, 277
Distillation
definition of, 85
destructive, 84
microscopic nature of, 112–113

E

Electrical conductivity, 80
Electricity
conductors, 130, 193
current, 193
insulators, 130, 193
semiconductors, 130
static, 193
Electrolysis, 197
of salt, 195
of water, 108, 171
Electrolyte, 318
Electron
arrangement in atoms, 393
bonding, 426, 432
configuration, 400, 404
configuration of ions, 411
definition of, 186
nonbonding, 426
valence, 406, 409
Electron-dot structures
for atoms, 413
for molecules, 425–434
Electronegativity, 197, 418
definition of, 441
values for elements, 440
Electropositive, 197
Element, 104
class of matter, 93
definition of, 90
microscopic nature of, 101
names and symbols for, 125
Emission spectrum, 396
Energy
bond, 421
definition of, 50
forms of, 51
levels of electrons, 397
levels and periodic table, 410
quantization of, 397
sublevels of electrons, 399–400
units of, 51–52
Equation, *see* Chemical equation
Equilibrium
definition of, 316
in acids, 317
Equivalence point, 327
Equivalent, 487
Eutectic, 83
Exponent, 22
Exponential notation, 62, 477–485
Extensive property, *see* Properties, extensive

F

Factor-label, 10
  application of, 27–30
  example problems, 13
Fahrenheit, 50, 58
Formula
  of common ions, 199
  of compounds containing polyatomic ions, 220
  definition of, 131
  empirical, 169, 175
  molecular, 175
  rules for writing, 131, 141
Formula unit, 168
Fraction, 7
Franklin, Benjamin, 186
Freezing point, 76, 79
  of mercury, 72

G

Gas, 103
  discharge tube, 396
  effect of pressure on, 354
  effect of temperature on, 361
  microscopic description, 348–349
  solubility of, 283
Gay-Lussac, 165
Geiger, Hans, 188
Gram, 5, 9, 10, 16
Graphing, 489–496
Groups of the periodic table, 139

H

Halide
  definition of, 261
Hall, Charles Martin, 268
Hall, Julia, B., 268
Halogen
  definition of, 261
  and periodicity, 408
Heat, 50
  of combustion, 80
  of formation, 80
  of fusion, 79, 102
  specific, 80
  units of, 51
  of vaporization, 78
  Δ, 231
Heating curves, 83
Heterogeneous, 81
Homogeneous, 81
Hydrocarbon, 132
Hydrogen ion
  concentration in water, 320
  scale for large values, 321
Hydrogen peroxide, 132
Hydronium ion
  definition of, 309
  see also Hydrogen ion

I

Ideal gas, 365
  constant, 368, 370
  equation, 368
Index of refraction, 80
Indicator
  behavior of, 319
  common acid-base, 306
  definition of, 305
  pH range of, 327
  and titration, 327
Intensive property, *see* Properties, intensive
Inverse operations, 471
Inversely proportional, 347
Ion
  common polyatomic, 199
  definition of, 191
  polyatomic, 198
Ionic bonds
  and ionic compounds, 442
Ionic compound
  definition of, 197
  and ionic bonds, 442
Ionic equation, 290
  limitations of net, 293
  net, 291

Isomers, stereo, 135
Isotopes
of carbon, 393
definition of, 392
standard representation of, 414

J

Joule, 51, 52

K

Kelvin
graphic representation, 58
temperature, 50, 363
Kilowatt-hours, 51
Kinetic energy, 366

L

Law
of conservation of mass, 215
of definite composition, 110, 165
of multiple proportions, 132, 165
Length, 20
Lewis, G. N., 332
Lewis structures
atoms, 413
molecules, 425–434
rules for writing, 430
Light year, 4
Limestone, 233
Limiting reagent, 257
Liquid, 102
solubility of, 283
Lowry, Thomas Martin, 332

M

Macroscopic, 100
Mass
atomic, 152
definition of, 77
instruments to measure, 34
mole, 159, 162
molecular, 160
number, 392
relative, 153–157
SI (metric) units of, 16
Math review, 461
Matter
classification of, 93, 117
Measurement, of length, 21
Measuring devices, 47–48
Melting
microscopic nature of, 111
Melting points, 54, 76, 79, 82, 102
of various hydrocarbons, 133
Meniscus, 49
Metabolism, 231
Metals, 130
naming of, 143
Metathesis, defintion of, 287
Meter, 20
Metric
common units, 9
length units, 20
mass units, 16
prefixes, 10
Microscopic, 100
Mixtures, 82, 111, 270, 276
azeotropic, 82
vs. compounds, 91, 109
eutectic, 83
Molality, 486
Molarity, 278, 486
definition of, 280
Mole
definition of, 157
fraction, 487
and molecule, 168
relationship of, 163
of solute, 278
Mole fraction, 487
Mole mass
calculation of, 162
definition of, 159
Molecular mass
calculation of, 161, 203
definition of, 161
of gases, 379

Molecule, 168–169
  definition of, 104
  vs. ionic crystal, 167
  vs. mixture, 109
  and mole, 108
  shapes of, 434
  of various elements, 104
Moth flakes, 78–79

N

Naming, *see* Nomenclature
Neutralization, 310, 327
Neutron
  and atomic structure, 391
  definition of, 189
  mass of, 414
Newton
  definition of, 349
Nobel gas, 407
Nomenclature, 135
  of compounds containing polyatomic ions, 200
  Greek prefixes and, 137
  of metallic compounds, 141–143
  of nonmetalic compounds, 136–138
  old, 144
  subscripts and, 131
Nonbonding electrons, 426
Nonmetals, 130
  naming of, 136–138
Normality, 487
Nucleus, 187
Number(s)
  Avogadro's, 166
  calculations with, 59–62
  decimal, 10, 468
  exact, 43
  relative size, 6
  rounding, 40, 42, 43
  whole, 174

O

Octet rule
  application, 425–429
  and Lewis structure, 424
  statement of, 425
Opaque, 272
Orbit, 423
Orbitals
  atomic, 417, 423
  shapes of *s* and *p*, 422, 423
Oxidation number, 443
  and combining number, 443
  rules for assigning, 446

P

Parallax, 37
Paradichlorobenzene, 78–79
Pascal
  definition of, 350
  relation to other units, 353
Percent
  composition of compounds, 176
  composition of solutions, 487
Periodic table, 127
  and electron arrangement, 407–409
  with emphasis, 148
Periodic trends
  and electron arrangement, 407–409
  in metallic character, 130
  in size, 128
Periodicity
  and electron arrangement, 407–409
pH
  defintion of, 325
  at which indicators change color, 327
  meter, 322
  scale, 321
Phase change, 111
Physical change, 86, 387–389
  microscopic nature of, 106
Polar bonds and polar molecules, 438–440
Positive charge, 186, 191
Precipitate, 288
Precision, 33, 35
Pressure
  atmospheric, 350
  microscopic explanation of, 359

standard atmospheric, 353
units of, 349–350, 353
vapor, 376
of water vapor, 377
Principal energy level, 399
Principal quantum number, 399
Problem solving, 499
Product
in equations, 216
predicting, 235, 288, 289
Properties, 49, 74
defintion of, 76
extensive, 49, 68
intensive, 49, 54, 68, 79, 82, 102
names of, 76, 80
representative, of some pure substances, 89
of simple hydrocarbons, 133
of water, 96
Proportionality, 57
constant, 343–344
direct, 342
inverse, 347
Proportions, in an equation, 223
Proton
acceptor, 307
and atomic number, 391
definition of, 189
donor, 307
mass of, 414
Proust, Louis Joseph, 121
Purity, 87
checking, 82
meaning of, 80

Q

Qualitative, 419
Quantitative, 246
Quantized, 189, 397
Quicklime, 233

R

Ratio, 15
as unit-factor, 14
working with, 54–57
Reactant
in equations, 216
Reaction, chemical
combination, 235, 246
combustion, 246
common, 224–234
decomposition, 235, 254
ionic, 291
magnesium burning, 251
metathesis, 287
replacement, 255, 260, 292
reversible, 314
to form SbI_3, 221–223
Reagent, 200
Refluxing
apparatus for, 221
definition of, 221
Relative mass
of atoms, 153
calculation of, 154
Rusting, 224
Rutherford, Ernest, 188
experiment, 188–189

S

Saturated solution, definition of, 283
Scientific notation, 32, 44, 62
SI
common units, 9
definition, 9
length units, 20
mass units, 16
prefixes, 10
volume units, 46
Significant digits, 38
definition of, 38
infinite, 43
rules for, 40, 42, 43, 44
zeros as, 43
Slaked lime, 234
Solid, 102
solubility of, 283
Solubility
of common compounds, 284

Solubility—*Continued*
of liquids, 283
rules, 289–290
of solids, 283
and temperature, 282
Solute
definition of, 275
Solutions
characteristics of, 275
concentration, 277
definition of, 271–272
types of, 273
Solvent
definition of, 275
Specific heat, 80
Spectrum
definition of, 396
line, 394
Square root, 20, 23
State, change of, 102–103
Stereo-, 135
Stoichiometry
definition of, 253
involving decomposition, 255
involving limiting reagents, 257–260
STP, 373
Sublimation, 133
Supercooling, 79
Surface tension, 80
Suspension
definition of, 271–272
Symbols
chemical, 124
list of atomic, 125

T

Temperature
Celsius, 50, 58
comparison of scales, 58
effect on atoms, 102–103
effect on solubility, 282–283
Fahrenheit, 50, 58
Kelvin, 50, 58, 363
microscopic explanation of, 366
Tetrahedron, 434–435
Thermal conductivity, 80
Thompson, J. J., 189
Titration
apparatus for, 330
description of, 326, 331
Torr
definition of, 353
relation to other units, 353
Toothpaste, chemistry of, 297
Transition metals, 411

U

Uncertain digit, 39
in calculations, 40
Uncertainty, 32
calculations involving, 40
and empirical formulas, 174
reasons for, 35–38
reducing, 65
Unit-factor, 7–8, 10, 27–28
ratios as, 14
Units, 24, 463
common metric, 9
converting, 17
of energy, 51–52
of length, 20
of mass, 16
of volume, 46
why important, 8

V

Valence electrons, 406, 409

Viscosity, 80
Volume
  combining, of gases, 165
  instruments for measuring, 47, 48
  metric (SI) units of, 46
  molar (gas), 373

W

Water, wonder of, 96
Weight
  definition of, 77
  finding molecular, 379
  see also Mass

About the Author

J. Dudley Herron was born in Providence, Kentucky. He received his B.A. from the University of Kentucky, his M.S. from the University of North Carolina, and his Ph.D. from Florida State University. Since 1965, Professor Herron has taught Chemistry at Purdue University. He has been active in the field of science education at all levels, and has been a frequent lecturer on issues of science education throughout the United States. Professor Herron is presently a member of the Executive Committee, Division of Chemical Education, of the American Chemical Society. He is also a member of the Executive Board of the National Association for Research in Science Teaching. Professor Herron has authored numerous articles on science education, which have been published in the United States and abroad.

Names, Symbols, and Masses* of

Name	Symbol	Atomic No.	Atomic Mass
Actinium	**Ac**	89	(227)
Aluminum	**Al**	13	26.98154
Americium	**Am**	95	(243)
Antimony	**Sb**	51	121.75
Argon	**Ar**	18	39.948
Arsenic	**As**	33	74.9216
Astatine	**At**	85	(210)
Barium	**Ba**	56	137.34
Berkelium	**Bk**	97	(247)
Beryllium	**Be**	4	9.01218
Bismuth	**Bi**	83	208.9808
Boron	**B**	5	10.81
Bromine	**Br**	35	79.904
Cadmium	**Cd**	48	112.41
Calcium	**Ca**	20	40.08
Californium	**Cf**	98	(249)
Carbon	**C**	6	12.011
Cerium	**Ce**	58	140.12
Cesium	**Cs**	55	132.9054
Chlorine	**Cl**	17	35.453
Chromium	**Cr**	24	51.996
Cobalt	**Co**	27	58.9332
Copper	**Cu**	29	63.546
Curium	**Cm**	96	(247)
Dysprosium	**Dy**	66	162.50
Einsteinium	**Es**	99	(254)
Erbium	**Er**	68	167.26
Europium	**Eu**	63	151.96
Fermium	**Fm**	100	(253)
Fluorine	**F**	9	18.99840
Francium	**Fr**	87	(223)
Gadolinium	**Gd**	64	157.25
Gallium	**Ga**	31	69.72
Germanium	**Ge**	32	72.59
Gold	**Au**	79	196.967
Hafnium	**Hf**	72	178.49
Helium	**He**	2	4.00260
Holmium	**Ho**	67	164.9304
Hydrogen	**H**	1	1.0079
Indium	**In**	49	114.82
Iodine	**I**	53	126.9045
Iridium	**Ir**	77	192.2
Iron	**Fe**	26	55.847
Krypton	**Kr**	36	83.80
Lanthanum	**La**	57	138.91
Lawrencium	**Lr**	103	(257)
Lead	**Pb**	82	207.2
Lithium	**Li**	3	6.94
Lutetium	**Lu**	71	174.97
Magnesium	**Mg**	12	24.305
Manganese	**Mn**	25	54.9380
Mendelevium	**Md**	101	(256)

*Based on assigned relative atomic mass $^{12}C = 12$. Values are reliable to ± 1 in the last digit. Numbers in parentheses are the mass numbers of the most stable isotopes.